Fifth Edition

Using Information Technology

A Practical Introduction to Computers & Communications

Introductory Version

Stacey C. Sawyer

Brian K. Williams

McGraw-Hill
Irwin

Boston   Burr Ridge, IL   Dubuque, IA   Madison, WI   New York   San Francisco   St. Louis
Bangkok   Bogotá   Caracas   Kuala Lumpur   Lisbon   London   Madrid   Mexico City
Milan   Montreal   New Delhi   Santiago   Seoul   Singapore   Sydney   Taipei   Toronto

McGraw-Hill Higher Education

*A Division of The **McGraw-Hill** Companies*

USING INFORMATION TECHNOLOGY: A PRACTICAL INTRODUCTION
TO COMPUTERS & COMMUNICATIONS

Published by McGraw-Hill/Irwin, a business unit of The McGraw-Hill Companies, Inc., 1221 Avenue of the Americas, New York, NY, 10020. Copyright © 2003, 2001, 1999, 1997, 1995 by The McGraw-Hill Companies, Inc. All rights reserved. No part of this publication may be reproduced or distributed in any form or by any means, or stored in a database or retrieval system, without the prior written consent of The McGraw-Hill Companies, Inc., including, but not limited to, in any network or other electronic storage or transmission, or broadcast for distance learning.

Some ancillaries, including electronic and print components, may not be available to customers outside the United States.

This book is printed on acid-free paper.

international   1 2 3 4 5 6 7 8 9 0   QPD/QPD   0 9 8 7 6 5 4 3 2 1
domestic   2 3 4 5 6 7 8 9 0   QPD/QPD   0 9 8 7 6 5 4 3 2

ISBN 0-07-248464-0

Publisher: *George Werthman*
Sponsoring editor: *Steve Schuetz*
Developmental editors: *Craig S. Leonard/Burrston House, Ltd.*
Manager, Marketing and Sales: *Paul Murphy*
Senior project manager: *Christine A. Vaughan*
Lead production supervisor: *Heather D. Burbridge*
Freelance design coordinator: *Laurie J. Entringer*
Production and Quark makeup: *Stacey C. Sawyer*
Photo research coordinator: *Judy Kausal*
Photo researcher: *Susan Friedman*
Lead supplement producer: *Marc Mattson*
Media producer: *Greg Bates*
Cover design: *Asylum Studios*
Compositor: *GTS Graphics, Inc.*
Typeface: *10/12 Trump Mediaeval*
Printer: *Quebecor World/Dubuque*

Library of Congress Control Number: 2002100742

INTERNATIONAL EDITION ISBN 0-07-115104-4
Copyright © 2003. Exclusive rights by The McGraw-Hill Companies, Inc., for manufacture and export. This book cannot be re-exported from the country to which it is sold by McGraw-Hill. The International Edition is not available in North America.

www.mhhe.com

Photo and other credits are listed in the back of the book.

Brief Contents

1 INTRODUCTION TO INFORMATION TECHNOLOGY: Mind Tools for Your Future   1

2 THE INTERNET & THE WORLD WIDE WEB: Exploring Cyberspace   31

3 SOFTWARE: The Power behind the Power   83

4 HARDWARE—THE CPU & STORAGE: How to Buy a Multimedia Computer System   141

5 HARDWARE—INPUT & OUTPUT: Taking Charge of Computing & Communications   181

6 TELECOMMUNICATIONS—Networks & Communications: The "New Story" in Computing   211

7 FILES, DATABASES, & E-COMMERCE: Digital Engines for the New Economy   245

8 SOCIETY & THE DIGITAL AGE: Challenges & Promises   271

APPENDIX
SYSTEMS & PROGRAMMING: Development, Programming, & Languages   299

About the Authors

Who are **Stacey Sawyer and Brian Williams?** We are a married couple living near Lake Tahoe, Nevada, with an avid interest in seeing students become well educated—especially in information technology.

What best describes what we do? We consider ourselves **watchers and listeners.** We spend our time watching what's happening in business and society and on college campuses and listening to the views expressed by instructors, students, and other participants in the computer revolution. We then try to translate those observations into meaningful language that can be easily understood by students.

Over the past two decades, we have individually or together **authored more than 20 books** (and over 30 revisions), most of them on computers and information technology. Both of us have **a commitment to helping students succeed in college.** Brian, for instance, has co-authored five books in the college success field: *Learning Success, The Commuter Student, The Urban Student, The Successful Distance Learning Student,* and *The Career Student Handbook.* Stacey has an interest in language education and has worked on several college textbooks in English as a Second Language (ESL) and in Spanish, German, French, and Italian. We thus bring to our information-technology books an awareness of the needs of the increasingly diverse student bodies now in our colleges.

Stacey has a B.A. from Ohio Wesleyan and an M.A. from Middlebury College and the University of Mainz, Germany. She has taught at Ohio State University and managed and consulted for a number of for-profit and nonprofit health, educational, and publishing organizations. Brian has a B.A. and M.A. from Stanford University and has held managerial jobs in education, communications, and publishing.

In our spare time, we enjoy travel, music, philosphy, cooking, and exploring the American West.

To the Instructor

Introduction

As authors, we are enormously gratified by the continued endorsement of *USING INFORMATION TECHNOLOGY* as a teaching tool for the introductory college course on computers. Over 500,000 students have been introduced to this dynamic and exciting subject through UIT's four earlier editions, and instructors in over 500 schools have selected it for use in their courses.

What are the reasons for this acceptance? One is that UIT was the first textbook to foresee and define the impact of digital convergence—the fusion of computers and communications—as the new and broader foundation for this course. And we have continued to try to pioneer in coverage of new developments. Thus, we are extremely pleased to hear reviewers label UIT as the most up-to-date text published for this course.

The UIT Difference: Motivating the Unmotivated

But there is another important reason, we think, for UIT's frequent adoption. We often ask instructors what their most significant challenge is in teaching this course. One professor at a state university seems to speak for most when she says: "Making the course interesting and challenging." Others echo her with remarks such as "Keeping students interested in the material enough to study" and "Many students take the course because they must, instead of because the material interests them." Another speaks about the need to address a "variety of skill/knowledge levels while keeping the course challenging and interesting."

Our experience with reviews, surveys, and focus groups, then, suggests that **the number one challenge to instructors is *motivating the unmotivated*.** As authors, we find information technology tremendously exciting, but we have long recognized that many students come to the subject with attitudes ranging from, on the one hand, complete apathy and even abject terror to, on the one hand, a high degree of experience and technical understanding (such as those taking the course for a certificate).

We address the problem of motivating the unmotivated by offering unequaled treatment of the following:

1. **Practicality**
2. **Readability**
3. **Currentness**

We explain these features below.

Feature #1: Emphasis on Practicality

This popular feature received overwhelming acceptance by both students and instructors in past editions. **Practical advice,** of the sort found in computer magazines, newspaper technology sections, and general-interest computer books, is expressed not only in the text but also in the following:

- **Bookmark It! Practical Action Box:** This box consists of optional material on practical matters. *Examples:* "Tips for Managing Your E-Mail." "Choosing an Internet Service Provider.""Web Research, Term Papers, & Plagiarism." "How to Buy a Notebook." "Preventing Your Identity from Getting Stolen."

  New to this edition: "Should You Upgrade to Windows XP or Mac OS X?"

Survival Tip

- *New to this edition!* **Survival Tips:** In the margins throughout we present utilitarian **Survival Tips** to aid students' explorations of the infotech world. *Examples:* "Broadband: Riskier for Security." "Accessing E-Mail While Traveling Abroad." "Handling the Annoyance of Spam." "Information on Web Radio." "Don't Trash Those Icons." "XP Installation." "Ready for Linux?" "Need Info on RAM?" "What to Do When Your Floppy Jams." "Backing Up on Zip." "How Do I Use the Prnt Scrn Key?" "Setting Mouse Properties." "Coping with Cookies." "Guard Your Social Security Number." "Some Records Have to Be Hardcopy." "Music File Sharing." "Some Websites about Privacy." "Keep Antivirus Software Updated." "Deal with Secure Websites." "Financial Portals." "Online Government Help."

- **Early discussion of Internet:** Many instructors have told us they like having **"e-concepts" treated earlier and more extensively** in this text compared with other books. Accordingly, the Internet and World Wide Web are discussed in Chapter 2 instead of in a later chapter, reflecting their importance in students' daily lives.

- **How to understand a computer ad:** In the hardware chapters (Chapters 4 and 5), we explain important concepts by showing students **how to understand the hardware components in a hypothetical PC ad.**

Feature #2: Emphasis on Readability & Reinforcement for Learning

We offer the following features for reinforcing student learning:

- **Interesting writing:** Studies have found that textbooks **written in an imaginative style** significantly improve students' ability to retain information. Both instructors and students have commented on the distinctiveness of the writing in this book. We employ a number of journalistic devices—colorful anecdotes, short biographical sketches, interesting observations in direct quotes—to make the material as interesting as possible. We also use real anecdotes and examples rather than fictionalized ones.

- *New to this edition!* **"Click-along" web connection for student "multitasking" for learning reinforcement:** Today's students often do "multitasking"—many tasks at once, such as studying while talking on the phone, watching TV, or surfing the Web. Educators say because the brain has limits, the distraction in attention often means less learning takes place (although the trend among students may be irreversible). This book addresses this impulse by **harnessing multitasking in the service of student motivation and learning.** Wherever the Click-along icon (shown at left) appears in the book, readers are invited to use their computers to go to our website *(www.mhhe.com/cit/uit5e/intro/clickalong)* and use their mouse to "click along" while reading the text. Principal uses of the Click-along website include updates, elaboration, examples, more practical advice, and access to the Student Online Learning Center.

Ethics

- **Emphasis throughout on ethics:** Many texts discuss ethics in isolation, usually in one of the final chapters. We believe this topic is too important to be treated last or lightly, and users have agreed. Thus, **we cover ethical matters throughout the book,** as indicated by the special logo shown here in the margin. *Example:* We discuss such all-important questions as copying of Internet files, online plagiarism, privacy, computer crime, and netiquette.

- **Key terms AND definitions emphasized:** To help readers avoid any confusion about which terms are important and what they actually mean, we print each key term in ***bold italic underscore*** and its definition in **boldface.** *Example* (from Chapter 1): "***Data*** **consists of raw facts and figures that are processed into information.**"

- **Material in bite-size portions:** Major ideas are presented in **bite-size form,** with generous use of advance organizers, bulleted lists, and new paragraphing when a new idea is introduced. Most **sentences have been kept short,** the majority not exceeding 22–25 words in length.

- **Key Questions—to help students read with purpose:** We have **crafted the learning objectives as Key Questions** to help readers focus on essentials. Each Key Question appears in two places: on the first page of the chapter and beneath the section head. Key Questions are also tied to the end-of-chapter summary, as we will explain.

- **Concept Checks:** Appearing periodically throughout the text, **Concept Checks** spur students to recall facts and concepts they have just read.

- **Summary:** Each chapter ends with a **Summary** of important terms, with an explanation of **what they are and why they are important.** The terms are accompanied, when appropriate, by a picture. Each concept or term is also given a cross-reference page number that refers the reader to the main discussion within the text. In addition, the term or concept is given a Key Question number corresponding to the appropriate Key Question (learning objective).

Feature #3: Currentness

Reviewers have applauded previous editions of UIT for being more up to date than other texts. Among the new topics and terms covered in this edition are *HomeRF, iDEN, intelligent smart cards, Intel P4 chip, Internet help sites, Mac OS X, Office XP, touch-screen voting, video/audio editing software, VRAM, WAP, webcams, WiFi, Windows XP.*

In addition, in this latest edition, **we have taken the feature of currentness to another level through use of the Click-along feature to offer updates** to new material throughout the life of the book.

Feature #4: Three-Level System to Help Students Think Critically about Information Technology

This is a feature first created for the last edition that we have tried to make even more prominent and useful in this one. More and more instructors seem to have become familiar with Benjamin **Bloom's *Taxonomy of Educational Objectives,*** describing a hierarchy of six critical-thinking skills: (a) two lower-order skills—*memorization* and *comprehension;* and (b) four higher-order skills—*application, analysis, synthesis,* and *evaluation.*

Drawing on our experience in writing books to guide students to college success, **we have implemented Bloom's ideas in a three-stage pedagogical approach,** using the following hierarchical approach in the Chapter Review at the end of every chapter:

- **Stage 1 learning—memorization:** "I can recognize and recall information." Using self-test questions, multiple-choice questions, and true/false questions, we enable students to test how well they recall basic terms and concepts.
- **Stage 2 learning—comprehension:** "I can recall information in my own terms and explain them to a friend." Using open-ended short-answer questions, we enable students to re-express terms and concepts in their own words.
- **Stage 3 learning—applying, analyzing, synthesizing, evaluating:** "I can apply what I've learned, relate these ideas to other concepts, build on other knowledge, and use all these thinking skills to form a judgment." In this part of the Chapter Review, we ask students to put the ideas into effect using the activities described. The purpose is to help students take possession of the ideas, make them their own, and apply them realistically to their own ideas.

New to this edition! **Many new and different Internet activities have been created as Stage 3 learning activities** for this edition of the book.

Resources for Instructors

The instructor supplements for this edition HAVE UNDERGONE A MAJOR REVISION, with a focus on enhancing instructors' ability to understand and utilize all the resources provided for the text.

- **Instructor's Manual:** The Instructor's Manual now **incorporates all of the resources available to the instructor for each chapter.** With a lecture outline on the left-hand page and available resources pertaining to the topics in the outline on the right-hand page, instructors now have the ability to harness these assets to create effective lectures. It works like a Web page that is easy to navigate and simple to understand because it provides links to appropriate assets elsewhere on the Web or CD-ROM. Each chapter contains an overview of the changes to this edition, a chapter overview, teaching tips, PowerPoint slides with speaker's notes, group projects, Click-alongs, outside projects, web exercises, text figures, and links to appropriate information and games on the book's website.
- **Testbank:** The Testbank now has **a new format that allows instructors to effectively pinpoint areas of content within each chapter on which to test students.** Each chapter starts off with a "Test Table" that provides a convenient guide for finding questions that pertain to chapter objectives and difficulty level. The Test Table also indicates the type of question so that instructors can create exams using the question types of their choice. The test questions are first organized by chapter objectives and then learning level; they include answers, Key Question numbers, learning levels, page references from the text, and rationales. Following each chapter's test bank questions is a Quick Quiz, designed for use when instructors don't have time to tailor an exam.

    Diploma by Brownstone: Diploma is the most flexible, powerful, and easy-to-use computer-based testing system available for higher education. The Diploma system allows instructors to create an exam as a printed version, as a LAN-based online version, or as an Internet version. Diploma also includes grade book features, which automate the entire testing process.
- **PowerPoint presentation:** The PowerPoint presentation **includes additional material** that expands upon important topics from the text,

allowing instructors to create interesting and engaging classroom presentations. Each chapter of the presentation includes important illustrations, and animations to enable instructors to emphasize important concepts in memorable ways. **Each slide of the presentation is integrated into the Instructor's Manual** so that instructors can quickly and effectively determine which slides they would like to use in their presentations.

- Figures from the book: All of the photos, illustrations, screenshots, and tables are available electronically for use in presentations, transparencies, or handouts.
- Online Learning Center: *(www.mhhe.com/cit/uit5e)* Designed to provide a wide variety of learning opportunities for students, the website for the fifth edition now includes a Web Summary for each chapter, with all of the key terms linked to relevant exercises, games, web links, and self-quizzes. Additional end-of-chapter exercises, web exercises, group projects, outside projects, Instructor's Manual, and PowerPoint presentations are also available online for instructors to download.
- Interactive companion CD-ROM: This free CD-ROM includes a collection of 20 interactive tutorial labs on some of the most popular topics. This CD expands the reach and scope of the text by combining video, interactive exercises, animation, additional content, and actual "lab" tutorials. The CD can be used in class, in the lab, or at home by students and instructors. The labs include the following topics:

Using Information Technology Interactive Companion Labs

Lab	Function	Chapter
Binary Numbers	Explore binary numbers including such topics as binary numbers as switches, how to make a binary number, binary addition, and binary logic.	4
Basic Programming	Learn about the thought processes and tools used to instruct computers to perform our work. This lab includes topics on basic computer tasks (input, processing, and output), variables, constants, assignment, mathematical calculations, and reusing code.	App.
Computer Anatomy	Learn the parts that make up a personal computer, including Input, Output, Storage, and Processing devices.	4 & 5
Disk Fragmentation	Understand how data and programs are stored and accessed. Includes concepts such as disk storage, blocks, fragmentation, defragmentation, media types, and data storage.	4
E-mail Essentials	Learn the tools, techniques, and etiquette needed to communicate by e-mail.	2 & 6
Multimedia Tools	Learn the basics of creating a simple multimedia presentation by understanding the types of media, virtual reality, interactivity, multimedia applications, and the uses of multimedia.	2 & 3
Workplace Issues	Learn how Ethics, Privacy, Security, and Time Wasters affect you, either as an employer or as an employee.	All
Introduction to Databases	This lab will introduce you to the many concepts involved in making, maintaining, and using a database to store large amounts of related data. You will have the opportunity to design a database, create the database, and use the information you entered to generate useful reports.	7
Programming II	Learn some of the essentials of visual programming, then implement them to build a working program.	N/A

Using Information Technology Interactive Companion Labs (continued)

Lab	Function	Chapter
Network Communications	Explore the many types of computer-based communications; how they work and how to use them effectively.	2 & 6
User Interfaces	Learn the basics of user interface elements, key Windows interface features, customizing the Windows interface, key Macintosh interface features, customizing the Macintosh interface, and key Unix (Linux) interface features.	3
Purchasing Decisions	Explore the factors you should consider when deciding what computer to buy, including software, the differences between PCs and Macs, the myths about CPU power, internal upgrades, and external upgrades.	4
File Organization	Learn about the way files are stored on your hard drive and how you can configure this storage to help you work more efficiently.	7
Word Processing and Spreadsheets	Learn the common features of word processing and spreadsheet programs, the basic features of word processing programs, and the basic features of spreadsheet programs.	3
Internet Overview	Explore features of the Internet including communication, browsing, sharing, and how to get connected.	2
Computer Troubleshooting	Learn how to avoid, repair, and troubleshoot computer problems.	All
Presentation Techniques	Learn the tricks for making effective presentations such as focusing your presentation to reach your audience, creating effective graphics, using sound and video in your presentation, and creating auxiliary materials to help augment your presentation or generate discussions.	3
Photo Editing	Learn how to edit digital photos including such topics as capturing digital images, storing devices, resizing and enhancement, and other manipulation techniques.	3
Programming Overview	Learn how code is written and changed into machine language by compilers or interpreters. Learn the different types of languages including object-oriented, procedural, and declarative.	N/A
SQL Queries	Learn what a relational database is (RDBMS), what a structured language is (SQL), how to use SQL to build a database, and how to use SQL to retrieve data from a database.	7

Digital Solutions to Help You Manage Your Course

PageOut: PageOut is our Course Web Site Development Center and offers a syllabus page, URL, McGraw-Hill Online Learning Center content, online exercises and quizzes, gradebook, discussion board, and an area for student Web pages.

Available free with any McGraw-Hill/Irwin product, PageOut requires no prior knowledge of HTML, no long hours of coding, and a way for course coordinators and professors to provide a full-course website. PageOut offers a series of templates—simply fill them with your course information and click on one of 16 designs. The process takes under an hour and leaves you with a professionally designed website. We'll even get you started with sample websites, or enter your syllabus for you! PageOut is so straightforward and intuitive, it's little wonder why over 12,000 college professors are using it. For more information, visit the PageOut website at *www.pageout.net*

The Online Learning Center can be delivered through any of these platforms:

McGraw-Hill Learning Architecture (TopClass)
Blackboard.com
Ecollege.com (formerly Real Education)
WebCT (a product of Universal Learning Technology)

McGraw-Hill has partnerships with WebCT and Blackboard to make it even easier to take your course online. Now you can have McGraw-Hill content delivered through the leading Internet-based learning tool for higher education. At McGraw-Hill, we have the following service agreements with WebCT and Blackboard:

- **Instructor Advantage:** Instructor Advantage is a special level of service McGraw-Hill offers in conjunction with WebCT designed to help you get up and running with your new course. A team of specialists will be immediately available to ensure everything runs smoothly through the life of your adoption.

- **Instructor Advantage Plus:** Qualified McGraw-Hill adopters will be eligible for an even higher level of service. A certified WebCT or Blackboard specialist will provide a full day of on-site training for you and your staff. You will then have unlimited e-mail and phone support through the life of your adoption. Contact your local McGraw-Hill representative for more details.

SimNet XPert: This is the TOTAL solution for learning and assessment in introductory applications and computer courses. The next generation of computer-based training/learning, SimNet XPert offers several ways for students to learn the skills or concepts being covered. With over 900 Learning Tasks throughout, SimNet XPert covers the concepts you need in multiple ways: (1) reading and interacting (*Teach Me* mode), (2) hearing (*Show Me* mode), and (3) practicing (*Let Me Try* mode).

The Assessment Component, which now uses a significantly deeper simulated interface, has two different pools of questions: one for Pre-Tests or Practice Tests, and one for the actual Exams. And SimNet XPert will also include Live-In-The-Application exams! Experience the future of training and assessment: eXPerience SimNet XPert for yourself!

Powerweb: Powerweb for Information Technology is an exciting online product available for use with *Using Information Technology*. A nominally priced token grants students access through our website to a wealth of resources—all corresponding to the text. Features include an interactive glossary; current events with quizzing, assessment, and measurement options; Web survey; links to related text content; and WWW searching capability via Northern Lights, an academic search engine.

Microsoft Applications Manuals

The following list presents McGraw-Hill/Irwin's Microsoft Applications books that are available for use with *Using Information Technology*. For more information about these books, visit the McGraw-Hill/Irwin Computer and Information Technology Supersite at *www.mhhe.com/it* or call your McGraw-Hill campus representative.

Windows Applications

Advantage Series by Hutchinson/Coulthard

Office XP
- Microsoft Office XP Volume I[1]
- Microsoft Office XP Volume II[2]
- Microsoft Word 2002 (Brief, Intro, Complete)
- Microsoft Excel 2002 (Brief, Intro, Complete)
- Microsoft PowerPoint 2002 (Brief, Intro)
- Microsoft Access 2002 (Brief, Intro, Complete)
- Integrating and Extending Microsoft Office XP (Brief)
- Microsoft FrontPage 2002 (Brief, Intro)

Microsoft Office 2000
- Microsoft Office 2000
- Microsoft Word 2000 (Brief, Intro, Complete)
- Microsoft Excel 2000 (Brief, Intro, Complete)
- Microsoft PowerPoint 2000 (Brief, Intro, Complete)
- Microsoft Access 2000 (Brief, Intro, Complete)
- Integrating and Extending Microsoft Office 2000 (Brief)
- Microsoft Outlook 2000 (Brief)
- Microsoft Internet Explorer 5.0 (Brief)

Microsoft Windows
- Microsoft Windows XP (Brief, Intro)
- Microsoft Windows 2000 (Brief, Intro)
- Microsoft Windows 98 (Brief, Intro)

Interactive Computing Series by Laudon

Microsoft Office XP
- Microsoft Office XP Volume I[1]
- Microsoft Office XP Volume II[2]
- Microsoft Word 2002 (Brief, Intro)
- Microsoft Excel 2002 (Brief, Intro)
- Microsoft PowerPoint 2002 (Brief, Intro)
- Microsoft Access 2002 (Brief, Intro)
- Microsoft FrontPage 2002 (Brief, Intro)
- Microsoft Internet Explorer 6.0 (Brief)

Microsoft Office 2000
- Microsoft Office 2000
- Microsoft Office 2000 Advanced
- Microsoft Word 2000 (Brief, Intro)
- Microsoft Excel 2000 (Brief, Intro)
- Microsoft PowerPoint 2000 (Brief, Intro)
- Microsoft Access 2000 (Brief, Intro)
- Integrating and Extending Microsoft Office 2000 (Brief)
- Microsoft Outlook 2000 (Brief)
- Microsoft Internet Explorer 5.0 (Brief)
- Netscape Communicator 6.0 (Brief)

Microsoft Windows
- Microsoft Windows XP (Brief, Intro)
- Microsoft Windows 2000 (Brief, Intro)
- Microsoft Windows 98 (Brief, Intro)

[1] Volume I contains all the Brief-level books.
[2] Volume II contains all the Intro-level books.

Acknowledgments

Two names are on the front of this book, but a great many others are important contributors to its development. First, we wish to thank our publisher, George Werthman, and our editor, Steve Schuetz, for their support and encouragement during this fast-moving revision process. Thanks also go to our marketing champions, Jeff Parr and Paul Murphy, for their enthusiasm and ideas. Craig Leonard deserves our special thanks for his excellent handling of the supplements program. Everyone in production provided support and direction: Christine Vaughan, Laurie Entringer, Heather Burbridge, Judy Kausal, and Marc Mattson.

Outside of McGraw-Hill we were fortunate—indeed, blessed—to once again have the most professional of all development services, those of Burrston House, specifically the help of Glen and Meg Turner. Jonathan Lippe provided creative ideas for the Chapter Review exercises, especially Internet activities. Susie Friedman, photo researcher; Patterson Lamb, copyeditor; Martha Ghent, proofreader and James Minkin, indexer, all gave us valuable assistance. Glen Coulthard did us a favor that we are anxious to repay, and we thank him. Pat and Michael Rogondino also provided terrific support. Thanks also to all the extremely knowledgeable and hardworking professionals at GTS Graphics, who provided so many of the prepress services.

Finally, we are grateful to the following people for their participation in manuscript reviews, and we want to single out their special contributions, which help to make this the most market-driven book possible.

Debra Harper, Montgomery County Community College–North Harris: For pointing out where more detailed information was needed and for acknowledging the importance of an interesting writing style.

Richard Hewer, Ferris State University: For reinforcing our views about discussing the Internet in the second chapter and our emphasis on ethics and critical thinking, as well as our attempt to strive for continual reinforcement to support learning.

Dana Lasher, North Carolina State University: For reinforcing our views about the importance of good examples and for pointing out the need for improvements in application software coverage.

Mary Levesque, University of Nebraska–Omaha: For her support for our level of detail, as well as her remarks for improving software coverage and her fine attention to detail on the book's final draft, which led to many improvements in quality.

Diane Mayne-Stafford, Grossmont College: For her favorable comments about our attempts to tackle the challenges of the 21st century and for her suggestions for improvements to Chapter 7.

Bruce Neubauer, Pittsburgh State University: For reinforcing our opinions about the importance of good illustrations and for pointing out the wide variability in student interest and backgrounds.

William Pritchard, Wayne State University: For his review of the book's early chapters.

Dick Schwartz, Macomb County Community College: For pointing out the challenge of changing student educational backgrounds, as well as suggesting improvements for the chapter on databases.

Susan Sells, Wichita State University: For pointing out the need for more instructor's resources and for acknowledging the importance of our early coverage of the Internet.

James Sidbury, University of Scranton: For pointing out the need to decrease certain technical aspects and for pointing out the importance of social and ethical issues coverage.

Maureen Smith, Saddleback College: For acknowledging our attempts to reach students through readability and writing style.

Angela Tilaro, Butte College: For her many extremely detailed comments on the final draft of the manuscript, which helped us make quality improvements.

Student's Guide

A One-Minute Course on How to Succeed in This Class

Got one minute to read this section? It could mean the difference between getting an A instead of a B, or a B instead of a C. Or even passing instead of failing.

Here Are the Rules

There are only four rules, and they aren't difficult.

Rule 1. You have to attend every class. (But that alone won't get you an A, as some students think.)

Rule 2. You can't put off studying, then cram the night before a test. This may work in high school, but college isn't high school.

Rule 3. You have to read or repeat material more than once. The important thing isn't reading. It's *re*reading.

Rule 4. You have to learn the secrets to using your textbook. It would be nice if all textbooks were organized the same way, but they aren't. Different texts have different features.

Getting the Most Information in the Least Time from This Book

Let's consider how you can best read *Using Information Technology*.

- Check the Key Questions in each section before you read it
- Read the section, trying to answer the Key Question(s)
- Do the Concept Checks
- Go the Extra Mile, if you have time
- Read the Summary at the end of the section
- Answer the questions in the Chapter Review

A look through the next seven pages will show you what the features we discussed look like.

Get an Overview of the Chapter First

Before you set out on a trip to a place you've never been to before, you would probably look at a map so you would get a "big picture" view of the route. Reading is the same way.

Scan the first page of the chapter and look at the **Chapter Outline** and the **Key Questions.**

Chapter 2

The Internet & the World Wide Web

Exploring Cyberspace

Key Questions
You should be able to answer the following questions.

2.1 **Choosing Your Internet Access Device & Physical Connection: The Quest for Broadband** What are the means of connecting to the Internet, and how fast are they?

2.2 **Choosing Your Internet Service Provider (ISP)** What is an Internet service provider, and what kinds of services do ISPs provide?

2.3 **Sending & Receiving E-Mail** What are the options for obtaining e-mail software, what are the components of an e-mail address, and what are netiquette and spam?

2.4 **The World Wide Web** What are websites, web pages, browsers, URLs, and search engines?

2.5 **The Online Gold Mine: More Internet Resources, Your Personal Cyberspace, E-Commerce, & the E-conomy** What are FTP, Telnet, newsgroups, real-time chat, and e-commerce?

Chapter Outline

Each chapter begins with an outline of the section headings in the chapter.

Key Questions

Use these Key Questions to help you read with purpose. Key Questions are repeated throughout the text.

Student's Guide

xvi

Check the Key Question in Each Section before You Read It. Then Read the Section, Trying to Answer the Key Question(s)

Look at the **Key Question** near the section heading. Read this aloud (or beneath your breath) or write it down. Next read the section, trying to answer the Key Question or Key Questions as you go. Make marks in the book if this helps you answer the question. In particular, look at the **key terms and definitions**, which appear in boldface. Look at the **graphics** (artwork and photos), which help to clarify the discussion.

Key Questions

Use these Key Questions presented at the start of each section to help you read with purpose.

Key Terms and Definitions

Throughout the text, key terms and definitions are easily identified by distinctive type.

Graphics

Many concepts and procedures are best explained through the use of artwork and photographs.

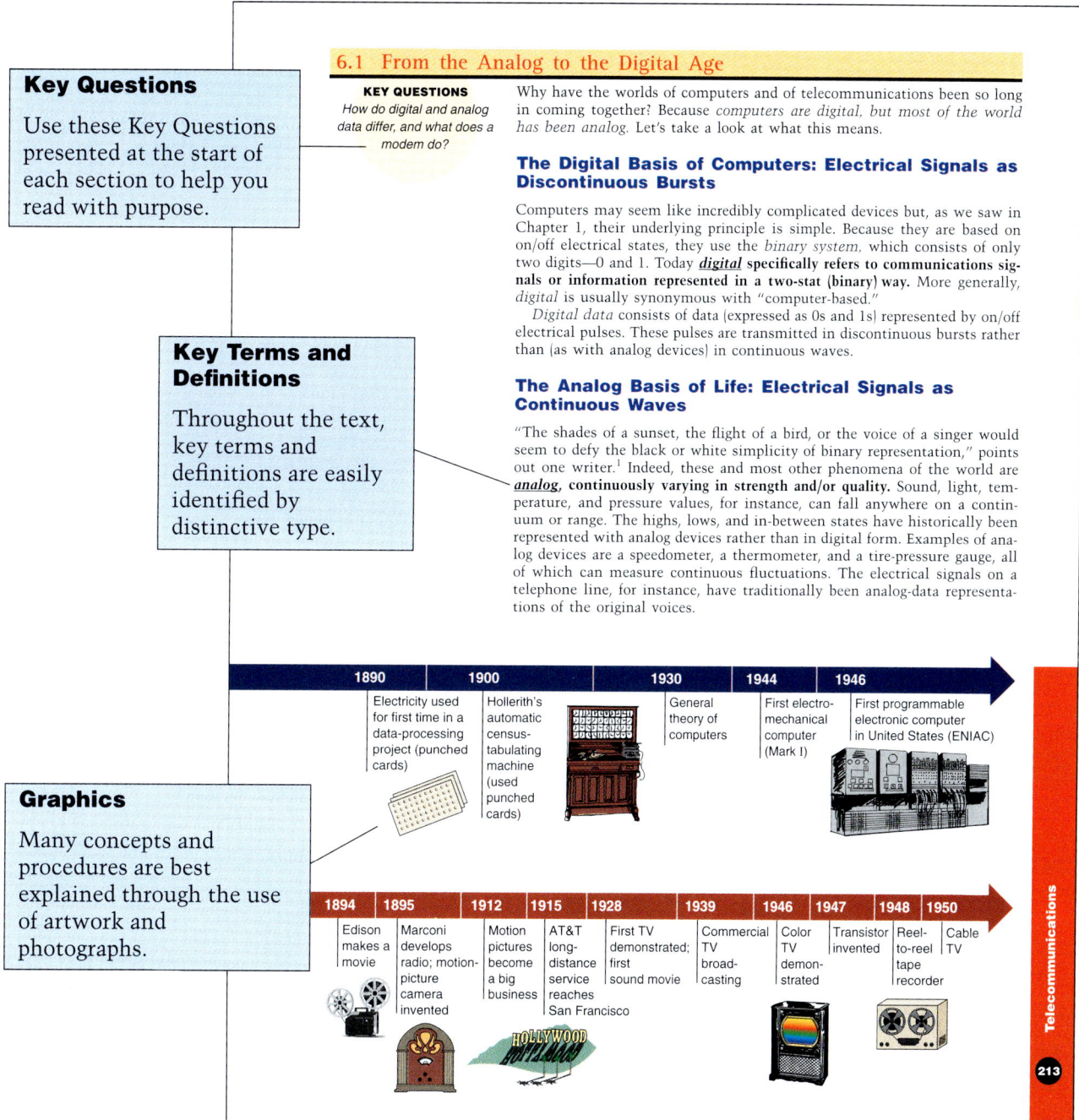

6.1 From the Analog to the Digital Age

KEY QUESTIONS
How do digital and analog data differ, and what does a modem do?

Why have the worlds of computers and of telecommunications been so long in coming together? Because *computers are digital, but most of the world has been analog.* Let's take a look at what this means.

The Digital Basis of Computers: Electrical Signals as Discontinuous Bursts

Computers may seem like incredibly complicated devices but, as we saw in Chapter 1, their underlying principle is simple. Because they are based on on/off electrical states, they use the *binary system,* which consists of only two digits—0 and 1. Today **_digital_** specifically refers to communications signals or information represented in a two-stat (binary) way. More generally, *digital* is usually synonymous with "computer-based."

Digital data consists of data (expressed as 0s and 1s) represented by on/off electrical pulses. These pulses are transmitted in discontinuous bursts rather than (as with analog devices) in continuous waves.

The Analog Basis of Life: Electrical Signals as Continuous Waves

"The shades of a sunset, the flight of a bird, or the voice of a singer would seem to defy the black or white simplicity of binary representation," points out one writer.[1] Indeed, these and most other phenomena of the world are **_analog_**, **continuously varying in strength and/or quality.** Sound, light, temperature, and pressure values, for instance, can fall anywhere on a continuum or range. The highs, lows, and in-between states have historically been represented with analog devices rather than in digital form. Examples of analog devices are a speedometer, a thermometer, and a tire-pressure gauge, all of which can measure continuous fluctuations. The electrical signals on a telephone line, for instance, have traditionally been analog-data representations of the original voices.

xvii

BOOKMARK IT!

PRACTICAL ACTION BOX
Choosing an Internet Service Provider

If you belong to a college or company, you may get an ISP free. Some public libraries also offer freenet connections.

If these options are not available to you, be sure to ask these questions when you're making phone calls to locate an Internet service provider.[a] (Also see *www.thelist.com*, which provides a comprehensive list of 9700 ISPs.)

Costs

- Is there a contract, and for what length of time? That is, are you obligated to stick with the ISP for a while even if you're unhappy with it?
- Is there a setup fee? (Most ISPs no longer charge this, though some "free" ISPs will.)
- How much is unlimited access per month? (Most charge about $20 for unlimited usage. But inquire if there are free or low-cost trial memberships or discounts for long-term commitments.)
- If access is supposedly free, what are the trade-offs besides putting up with heavy advertising? (For instance, if the ISP closely monitors your activity in order to accurately target ads, what guarantees do you have that information about you will be kept private? What charges will you face if you try to scrap the advertising window or drop the service?)

Access

- Is the access number a local phone call? (If not, your monthly long-distance phone tolls could exceed the ISP fee.)
- Is there an alternative dial-up number is out of service?
- Is access available when you're tra provider should offer either a wide access numbers in the cities you te free 800 numbers.

Support

- What kind of help does the ISP give in setting up your connection?
- Is there free, 24-hour technical support? Is it reachable through a toll-free number?
- How difficult is it to reach tech support? (Try calling the number before you sign up for the ISP and see how long it takes to get a response. Many ISPs keep customers on hold for a long time.)

Reliability

- What is the average connection success rate for users trying to connect on the first try? (You can try dialing the number during peak hours, to see if you get a modem screech, which is good, rather than a busy signal, which is bad.)
- Will the ISP keep up with technology? (Are they planning to offer broadband technology such as DSL for speedier access?)
- Will the ISP sell your name to marketers or bombard you with junk messages (spam)?

> **Practical Action Box**
>
> These boxes present material on practical matters that students find useful.

software for setting up your computer and modem to dial into their network of servers. For this you use your *user name* ("user ID") and your *password*, a secret word or string of characters that enables you to **log on, or make a connection to the remote computer.** You will also need to get yourself an e-mail address, as we discuss next.

CONCEPT CHECK

What is an Internet service provider?

Describe the different services Internet services provide. What is a point of presence?

> **Concept Check**
>
> These questions, which appear throughout the text, encourage you to take a moment to see how well you understand the concepts you have read in the preceding material.

Do the Concept Checks

The **Concept Checks** are questions that appear throughout the text that encourage you to see how well you understand the concepts you have read. Take a break from your reading and try to answer as many of these as possible. If you have trouble with the questions, you probably should go back and review the section before proceeding.

Go the Extra Mile: Check out the Boxes, Survival Tips, & Click-Alongs

Successful students don't just do the minimum. They do the kind of further exploration that helps really fix the material in their minds. Read the **Practical Action Boxes** and **Survival Tips**. Use a computer to check out the **Click-Alongs**—that is, click on your mouse and go to the Click-Along website.

Users of the *Introductory* version of this book should go to *www.mhhe.com/cit/uit5e/intro/clickalong*

Users of the *Complete* version of this book should go to *www.mhhe.com/cit/uit5e/complete/clickalong*

Survival Tip

These tips provide helpful advice for avoiding problems and handling difficulties.

Click-Along

To satisfy your curiosity about further information on the text material, go to the appropriate website (see text above).

Wireless Systems: Satellite & Other Through-the-Air Connections

Suppose you live out in the country and you're tired of the molasses-like speed of your cranky local phone system. You might consider taking to the air.

- **Satellite:** With a pizza-size satellite dish on your roof, you can receive data at the rate of 400 Kbps from a **communications satellite**, **a space station that transmits radio waves called microwaves from earth-based stations.** Unfortunately, your outgoing transmission will still be only 56 Kbps, because you'll have to use your phone line for that purpose, although genuine two-way satellite service is under development (for example, by I.S.I.S. AXXESS). Equipment available from InfoDish or PC Connection costs about $190; installation runs $100–$250; and monthly charges (including ISP charges) are $30–$130, depending on how much time you spend online.

- **Other wireless connections:** In urban areas, some businesses are using radio waves transmitted between towers that handle cellular phone calls, which can send data at up to 155 Mbps (but commonly 10 Mbps) and are not only fast and dependable but also always on. The equipment costs from $200 to $2500, and the operating cost is $159–$1400 a month, depending on speed.

Universal Broadband Is Coming

Most PC users employ the physical connections we have described, but there are other possibilities. With WebTV, for instance, you use your television to access the Internet. The Palm VII and the PalmVx, personal digital assistants from Palm Computing, enable you to get on the Net, but the small screen and slow connection accommodate only e-mail and limited Web access (no graphics).

We are living in a time of rapid changes. Already, nearly a third of the online households in the U.S. have DSL, cable, or wireless services. Most of them, admittedly, are in and around major cities, and the companies providing connections have been slow to expand them to the rest of North America. But broadband is coming. When the telcos (telephone companies) finish scrambling to upgrade their phone lines to DSL, the cable companies make their "pipes" better handle two-way data, and the wireless companies refine their through-the-air links, ordinary Internet users will probably have connections *100 times faster* than they are today.[6]

Survival Tip

Broadband: Riskier for Security

Unlike dial-up services, broadband services, because they are always switched on, make your computer vulnerable to over-the-Internet security breaches. Solution: Install firewall software (p. 301).

CLICK-ALONG 2–1
The quest for broadband

CONCEPT CHECK

What are the measures of data transmission speed?

What does "broadband" mean?

What are some of the differences between a dial-up modem, DSL, and a cable modem?

Read the Summary

After you read the whole chapter, go through the **Summary,** which gives the important concepts and terms of the chapter in alphabetical order and tells you why they are important.

Summary

> **Summary**
>
> Each chapter ends with an innovative Summary of important terms and concepts, with an explanation of what they are and why they are important. Graphics provide visual reinforcement.

analytical graphics (p. 109, KQ 3.4) Also called *business graphics;* graphical forms that make numeric data easier to analyze than when it is organized as rows and columns of numbers. The principal examples of analytical graphics are bar charts, line graphs, and pie charts. Why it's important: *Whether viewed on a monitor or printed out, analytical graphics help make sales figures, economic trends, and the like easier to comprehend and analyze.*

Cell:
Formed by intersection of row and column (letter and number)

cell (p. 107, KQ 3.4) Place where a row and a column intersect in a spreadsheet worksheet; its position is called a *cell address.* Why it's important: *The cell is the smallest working unit in a spreadsheet. Data and formulas are entered into cells. Cell addresses provide location references for spreadsheet users.*

computer-aided design (CAD) programs (p. 121, KQ 3.6) Programs intended for the design of products, structures, civil engineering drawings, and maps. Why it's important: *CAD programs, which are available for microcomputers, help architects design buildings and workspaces and help engineers design cars, planes, electronic devices, roadways, bridges, and subdivisions. While similar to drawing programs, CAD programs provide precise dimensioning and positioning of the elements being drawn, so that they can be transferred later to computer-aided manufacturing programs; in addition, they lack special effects for illustrations. One advantage of CAD software is that three-dimensional drawings can be rotated on screen, so the designer can see all sides of the product.*

computer-aided design/computer-aided manufacturing (CAD/CAM) software (p. 122, KQ 3.6) Programs allowing products designed with CAD to be input into an automated manufacturing system that makes the products. Why it's important: *CAM systems have greatly enhanced efficiency in many industries.*

copyright (p. 89, KQ 3.1) Exclusive legal right that prohibits copying of intellectual property without the permission of the copyright holder. Why it's important: *Copyright law aims to prevent people from taking credit for and profiting from other people's work.*

Cursor

cursor (p. 101, KQ 3.3) Movable symbol on the display screen that shows where you may next enter data or commands. The symbol is often a blinking rectangle or an I-beam. You can move the cursor on the screen using the keyboard's directional arrow keys or a mouse. The point where the cursor is located is called the *insertion point.* Why it's important: *All application software packages use cursors to show the current work location on the screen.*

database (p. 110, KQ 3.5) Collection of interrelated files in a computer system. These computer-based files are organized according to their common elements, so that they can be retrieved easily. Why it's important: *Businesses and organizations build databases to help them keep track of and manage their affairs. In addition, online database services put enormous resources at the user's disposal.*

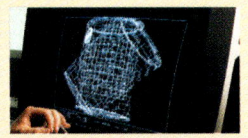

database file (p. 91, KQ 3.1) File created by database management programs; it consists of organized data that can be analyzed and displayed in various useful ways. Why it's important: *Database files make up a database.*

Application Software

125

Student's Guide

xx

Answer the Questions in the Chapter Review

The **Chapter Review** at the end of the chapter offers a three-stage process that helps you truly understand and "take ownership" of the material. Here's how it works:

- **First-Stage Review—Memorization:** Stage 1 questions test how well you recall basic terms and concepts. They include **Self-Test**, **Multiple-Choice**, and **True/False** questions.

Chapter Review

The Chapter Review follows a three-stage process that helps students truly understand and "take ownership" of the chapter material they have just completed reading.

First-Stage Review: Memorization

Stage 1 questions test how well you recall basic terms and concepts. They include **Self-Test, Multiple-Choice,** and **True/False** questions.

Chapter Review

stage 1 LEARNING — MEMORIZATION
"I can recognize and recall information."

Self-Test Questions

1. _____ enables the computer to perform essential operating tasks.
2. A(n) _____ is a firm that leases software over the Internet.
3. _____ is the activity in which a computer works on more than one process at a time.
4. Windows and Mac OS are generally used on _____ computers.
5. A(n) _____ is an inexpensive, stripped-down computer that connects people to networks and runs applications tied to servers.
6. A(n) _____ will find all the scattered files on your hard disk and reorganize them as contiguous files.
7. The _____ is the component of system software that comprises the master system of programs that manage the basic operations of the computer.
8. The _____ is the user-controllable display screen that allows you to communicate, or interact, with your computer.
9. Disk scanner and disk cleanup utilities detect and correct certain types of common problems on hard disks, such as removing unnecessary files called _____ files that are created by Windows only for short tasks and auto-recovery.
10. OSs allow users to control access to their computers via use of a _____ and a _____.

Multiple-Choice Questions

1. Which of the following are functions of the operating system?
    a. file management
    b. CPU management
    c. task management
    d. booting
    e. all of the above

2. Which of the following was the first major microcomputer OS?
    a. Mac OS
    b. Windows
    c. DOS
    d. Unix
    e. Linux

3. Which of the following is a prominent network operating system?
    a. Linux
    b. Unix
    c. Windows NT
    d. DOS
    e. Mac OS

4. Which of the following is the newest Microsoft Windows operating system?
    a. Windows Me
    b. Windows XP
    c. Windows 2000
    d. Windows NT
    e. Windows CE

True/False Questions

T F 1. The supervisor manages the CPU.
T F 2. The first graphical user interface was provided by Microsoft Windows.
T F 3. Multiprocessing is processing done by two or more computers or processors linked together to perform work simultaneously.
T F 4. Formatting a floppy disk will not affect any data already written on it.
T F 5. All operating systems are mutually compatible.

- **Second-Stage Review—Comprehension:** Stage 2 questions test how well you understand concepts and integrate ideas. They include **Short-Answer Questions** and **Concept Mapping** assignments.
- **Third-Stage Review—Application, Analysis, Synthesis, Evaluation:** Stage 3 questions and assignments appear under **Knowledge in Action.** They show your mastery of the material, including your ability to use it to solve problems and make decisions.

> **Second-Stage Review: Comprehension**
>
> Stage 2 questions test how well you understand concepts and integrate ideas. They include **Short-Answer Questions** and **Concept Mapping** assignments.

stage LEARNING — COMPREHENSION

"I can recall information in my own terms and explain them to a friend."

Short-Answer Questions

1. Briefly define *booting*.
2. What is the difference between a command-driven interface and a graphical user interface (GUI)?
3. Why can't you run your computer without system software?
4. Why is multitasking useful?
5. What is a *device driver*?
6. What is a *utility program*?
7. What is a *platform*?
8. What are the three components of system software? What is the basic function of each?
9. What is Rule No. 1?

Concept Mapping

On a separate sheet of paper, draw a concept map, or visual diagram, linking concepts. Show how the following terms are related.

backup	multiprogramming
booting	multitasking
data compression	NetWare
data recovery	operating system
defragmenter	platform
device driver	supervisor
DOS	time-sharing
fragmentation	Unix
Linux	utilities
Mac OS	Windows
multiprocessing	

> **Third-Stage Review: Application, Analysis, Synthesis, Evaluation**
>
> Stage 3 questions and assignments appear under **Knowledge in Action.** They test higher-order critical-thinking skills, including the ability to solve problems and make decisions.

stage LEARNING — APPLYING, ANALYZING, SYNTHESIZING, EVALUATING

"I can apply what I've learned, relate these ideas to other concepts, build on other knowledge, and use all these thinking skills to form a judgment."

Knowledge in Action

1. Here's an exercise in defragmenting your hard disk drive. Defragmenting is a housekeeping procedure that will speed up your system and often free up more hard-disk space.

   Double-click on My Computer on your Windows desktop (opening screen). Now use your right mouse button to click on C drive, then right-click on Properties, then left-click on the Tools tab. You will see the status of your system (error checking, backup, and defragmenting) and the last time the task was performed on the system. To clear out any errors, click the Check Now button; this will run a scan.

   Once the scan is complete, return to the Tools window and click the Defragment Now button. Click on Show Details. This will visually display on the screen the process of your files being reorganized into a contiguous order.

   Many times when your PC isn't preforming right, such as running sluggishly, a combination of running Scandisk and Defragment will solve the problem.

2. Go to the box "Comparison of Task Management," page 142 in this chapter. Create noncomputer analogies for *multitasking, multiprogramming, time-sharing,* and *multiprocessing*. For instance, for time-sharing, you could imagine a waiter taking menu requests, because he or she spends a fixed amount of time with each customer (program) before going on to the next one.

3. What do you think is the future of operating systems? Look up Yale computer scientist David Gelernter's paper "The Second Coming—A Manifesto" on the technology forum *www.edge.org*. Do you agree with him that data and computer processing will be increasingly spread across thousands, if not millions, of interconnected computers so that it is less likely that Microsoft or any other company will dominate the field?

4. What do you think is the future of Linux? Experts currently disagree about whether Linux will become a serious competitor to Windows. Research Linux on the Web. Which companies are creating application software to run on Linux? Which businesses are adopting Linux as an OS? What are the predictions about Linux use?

5. How do you think you will be obtaining software for your computer in the future? Explain your answer.

Web Exercises

1. Did your computer come with a Windows Startup disk and have you misplaced it? If your computer crashes, you'll need this disk to reinstall the operating system. This exercise shows you how to create your own Startup disk. Insert a blank disk in your floppy disk drive. From your Windows desktop, click on Start, Settings, and then Control Panel. Now click on Add/Remove. Click on the tab Startup Disk, then click on the Create Disk button.

   After the disk is created, label it *Startup Disk for Windows* and write the date on the disk. Also note the

Chapter 4 — 166

> **Investigating the World Wide Web**
>
> Activities for exploring the Web appear under **Web Exercises** in the Chapter Review.

Clearly, this method takes longer than simply reading the material once and rapidly underlining it. But because it requires your involvement and understanding, it is a better way to learn.

Contents

Chapter 1
INTRODUCTION TO INFORMATION TECHNOLOGY: MIND TOOLS FOR YOUR FUTURE 1

1.1 Infotech Becomes Commonplace: Cellphones, E-Mail, the Internet, & the E-World 2

The Telephone Grows Up 3
"You've Got Mail!" E-Mail's Mass Impact 3
The Internet, the World Wide Web, & the "Plumbing of Cyberspace" 4

Bookmark It! Practical Action Box:
"Managing Your E-Mail" 5
The E-World & Welcome to It 6

1.2 The "All-Purpose Machine": The Varieties of Computers 6

All Computers, Great & Small: The Categories of Machines 7
Servers 9

1.3 Understanding Your Computer: What If You Custom-Ordered Your Own PC? 10

How Computers Work: Three Key Concepts 10
Pretending to Order a Custom-Built Desktop Computer 11
Input Hardware: Keyboard & Mouse 11
Processing & Memory Hardware: Inside the System Cabinet 13
Storage Hardware: Floppy Drive, Hard Drive, & CD/DVD Drive 14
Output Hardware: Video & Sound Cards, Monitor, Speakers, & Printer 15
Communications Hardware: Modem 17
Software 17
Is Getting a Custom-Built PC Worth the Effort? 18

1.4 Where Is Information Technology Headed? 19

Three Directions of Computer Development: Miniaturization, Speed, & Affordability 19
Three Directions of Communications Development: Connectivity, Interactivity, & Multimedia 20

When Computers & Communications Combine: Convergence, Portability, & Personalization 21
"E" Also Stands for Ethics 22
Onward: Handling Information in the Era of Pervasive Computing 22

Chapter 2
THE INTERNET & THE WORLD WIDE WEB: EXPLORING CYBERSPACE 31

2.1 Choosing Your Internet Access Device & Physical Connection: The Quest for Broadband 32

Telephone (Dial-Up) Modem: Low Speed but Inexpensive & Widely Available 34
High-Speed Phone Lines: More Expensive but Available in Most Cities 36
Cable Modem: Close Competitor to DSL 37
Wireless Systems: Satellite & Other Through-the-Air Connections 38
Universal Broadband Is Coming 38

2.2 Choosing Your Internet Service Provider (ISP) 39

Bookmark It! Practical Action Box:
"Choosing an Internet Service Provider" 40

2.3 Sending & Receiving E-Mail 41

E-Mail Software & Carriers 42
E-Mail Addresses 42
Attachments 45
Instant Messaging 49
Mailing Lists: E-Mail–Based Discussion Groups 47
Netiquette: Appropriate Online Behavior 48
Sorting Your E-Mail 49
Spam: Unwanted Junk E-Mail 49

2.4 The World Wide Web 51`

The Web & How It Works 52
Using Your Browser to Get around the Web 54
Web Portals: Starting Points for Finding Information 60
Four Types of Search Engines: Human-Organized, Computer-Created, Hybrid, & Metasearch 62

xxiii

Tips for Smart Searching 63
Multimedia on the Web 65
Push Technology & Webcasting 68
The Internet Telephone & Videophone 68
Designing Web Pages 69

2.5 The Online Gold Mine: More Internet Resources, Your Personal Cyberspace, E-Commerce, & the E-conomy 69

Other Internet Resources: FTP, Telnet, Newsgroups, & Real-Time Chat 69
Your Personal Cyberspace 71
E-Commerce 72

Bookmark It! Practical Action Box: *"Web Research, Term Papers, & Plagiarism"* 74

Chapter 3
SOFTWARE: THE POWER BEHIND THE POWER 83

3.1 System Software 84

The Operating System: What It Does 85
Device Drivers: Running Peripheral Hardware 87
Utilities: Service Programs 87
The Operating System User Interface: Your Computer's Dashboard 87
Common Desktop & Laptop Operating Systems: DOS, Macintosh, & Windows 95
Network Operating Systems: Netware, Windows NT/2000/XP, Unix, & Linux 96
Operating Systems for Handhelds: Palm OS & Windows CE/Pocket PC 98

Bookmark It! Practical Action Box: *"Should You Uprgrade to Windows XP or Mac OS X?"* 100

3.2 Application Software: Getting Started 101

Tutorials & Documentation 103
Files of Data—& the Usefulness of Importing & Exporting 104
The Types of Software 104

3.3 Word Processing 105

Features of the Keyboard 106
Creating Documents 107
Editing Documents 108
Formatting Documents with the Help of Templates & Wizards 110
Printing, Faxing, or E-Mailing Documents 112
Saving Documents 112

Tracking Changes & Inserting Comments 112
Web Document Creation 112

3.4 Spreadsheets 113

The Basics: How Spreadsheets Work 113
Analytical Graphics: Creating Charts 115

3.5 Database Software 116

The Benefits of Database Software 116
The Basics: How Databases Work 116
Personal Information Managers 118

3.6 Specialty Software 119

Presentation Graphics Software 119
Financial Software 121
Desktop Publishing 123
Drawing & Painting Programs 124
Web Page Design/Authoring Software 125
Video/Auditing Editing Software 125
Project Management Software 125
Computer-Aided Design 125

3.7 Online Software & Application Service Providers: Turning Point for the Software Industry 127

Online Software & the Application Service Provider 127
Network Computers Revisited: "Thin Clients" versus "Fat Clients" 127
From ERP to ASP: The Evolution of "Rentalware" 129

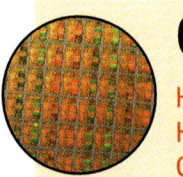

Chapter 4
HARDWARE—THE CPU & STORAGE: HOW TO BUY A MULTIMEDIA COMPUTER SYSTEM 141

4.1 Microchips, Miniaturization, & Mobility 142

From Vacuum Tubes to Transistors to Microchips 142
Miniaturization Miracles: Microchips, Microprocessors, & Micromachines 143
Mobility 143
Buying an Inexpensive Personal Computer: Understanding Computer Ads 145

4.2 The System Unit 146

The Binary System: Using On/Off Electrical States to Represent Data & Instructions 146
The Computer Case: Bays, Buttons, & Boards 148
Power Supply 150

The Motherboard & the Microprocessor Chip   150
Processing Speeds: From Megahertz to Picoseconds   151
How the Processor or CPU Works: Control Unit, ALU, & Registers   152
How Memory Works: RAM, ROM, CMOS, & Flash   154
How Cache Works: Level 1 (Internal) & Level 2 (External)   156
Ports & Cables   156
Expandability: Buses & Cards   158

5.3 Secondary Storage   160

Floppy Disks   161
Hard Disks   163
Optical Disks: CDs & DVDs   165
Magnetic Tape   168
Smart Cards   168
Flash Memory Cards   169
Online Secondary Storage   169

Bookmark It! Practical Action Box: *"How to Buy a Notebook"*   170

Chapter 5
HARDWARE—INPUT & OUTPUT: TAKING CHARGE OF COMPUTING & COMMUNICATIONS   181

5.1 Input & Output   183

5.2 Input Hardware   184

Keyboards   184
Pointing Devices   185
Source Data-Entry Devices   188

5.3 Output Hardware   193

Softcopy Output: Display Screens   193
Hardcopy Output: Printers   196
Other Output: Sound, Voice, Animation, & Video   198

Bookmark It! Practical Action Box: *"Good Habits: Protecting Your Computer System, Your Data, & Your Health"*   200

Chapter 6
TELECOMMUNICATIONS—NETWORKS & COMMUNICATIONS: THE "NEW STORY" IN COMPUTING   211

6.1 From the Analog to the Digital Age   213

The Digital Basis of Computers: Electrical Signals as Discontinuous Bursts   213
The Analog Basis of Life: Electrical Signals as Continuous Waves   213
Purpose of the Modem: Converting Digital Signals to Analog Signals & Back   214
Converting Reality to Digital Form   216

6.2 The Practical Uses of Communications   217

Videoconferencing & Videophones: Video/Voice Communication   217

Bookmark It! Practical Action Box: *"Web-Authoring Tools: How to Create Your Own Simple Website, Easily & for Free"*   217

Workgroup Computing & Groupware   218
Telecommuting & Virtual Offices   219
Home Networks   219
The Information/Internet Appliance   220
Smart Television: DTV, HDTV, & SDTV   220

6.3 Communications Channels: The Conduits of Communications   221

The Radio Spectrum & Bandwidth   221
Wired Communications Channels: Transmitting Data by Physical Means   223
Wireless Communications Channels: Transmitting Data through the Air   223
Types of Long-Distance Wireless Communications   225
Short-Range Wireless Communications at 2.4 Gigahertz: Bluetooth, WiFi, & HomeRF   229
Compression & Decompression: Putting More Data in Less Space   229

6.4 Networks   231

The Benefits of Networks   231
Types of Networks: WANs, MANs, & LANs   232
Types of LANs: Client-Server & Peer-to-Peer   233
Components of a LAN   234
Intranets, Extranets, & Firewalls: Private Internet Networks   234

6.5 Cyberethics: Controversial Material, Censorship, & Privacy Issues   236

Controversial Material & Censorship   236
Privacy   237

Chapter 7
FILES, DATABASES, & E-COMMERCE: DIGITAL ENGINES FOR THE NEW ECONOMY 246

Chapter 8
SOCIETY & THE DIGITAL AGE: CHALLENGES & PROMISES 271

7.1 Managing Files: Basic Concepts 246

How Data Is Organized: The Data Storage Hierarchy 246
The Key Field 247
Types of Files: Program Files, Data Files, & Others 247
Two Types of Data Files: Master File & Transaction File 249
Data Access Methods: Sequential versus Direct Access 249
Offline versus Online Storage 249

7.2 Database Management Systems 250

Four Types of Database Access 250

7.3 Database Models 252

Hierarchical Database 252
Network Database 252
Relational Database 254
Object-Oriented Database 254

7.4 Databases & the New Economy: E-Commerce, Data Mining, & B2B Systems 254

E-Commerce 255
Data Mining 255
Business-to-Business (B2B) Systems 257

7.5 The Ethics of Using Storage & Databases: Concerns about Accuracy & Privacy 258

Manipulation of Sound 259
Manipulation of Photos 259
Manipulation of Video & Television 260
Accuracy & Completeness 260
Matters of Privacy 261

Bookmark It! Practical Action Box: "Preventing Your Identity from Getting Stolen" 263

8.1 The Digital Environment: Is There a Grand Design? 272

The National Information Infrastructure 273
The New Internet: VBNS, Internet2, and NGI 273
The 1996 Telecommunications Act 274
The 1997 White House Plan for Internet Commerce 274
ICANN: The Internet Corporation for Assigned Names & Numbers 274

8.2 Security Issues: Threats to Computers & Communications Systems 275

Errors & Accidents 275
Natural & Other Hazards 276
Crimes Against Computers & Communications 276
Crimes Using Computers & Communications 277
Worms & Viruses 277
Computer Criminals 279

8.3 Security: Safeguarding Computers & Communications 260

Identification & Access 280
Encryption 281
Protection of Software & Data 282
Disaster-Recovery Plans 282

8.4 Quality-of-Life Issues: The Environment, Mental Health, & the Workplace 283

Environmental Problems 283
Mental-Health Problems 284
Workplace Problems: Impediments to Productivity 285

8.5 Economic Issues: Employment & the Haves/Have-Nots 285

Bookmark It! Practical Action Box: "When the Internet Isn't Productive: Online Addiction & Other Time Wasters" 286
Technology, the Job Killer? 287
Gap between Rich & Poor 287

8.6 Artificial Intelligence 288

Robotics 288
Expert Systems 289

Natural Language Processing 289
Artificial Life, the Turing Test, & AI Ethics 289

8.7 The promised Benefits of the Digital age 291

Information & Education 291
Health 291
Commerce & Money 291
Entertainment 293
Government & Electronic Democracy 293

Appendix
SYSTEMS & PROGRAMMING: DEVELOPMENT, PROGRAMMING, & LANGUAGES 299

A.1 Systems Development: The Six Phases of Systems Analysis & Design 300

The Purpose of a System 300
Getting the Project Going: How It Starts, Who's Involved 300
The Six Phases of Systems Analysis & Design 301
The First Phase: Conduct a Preliminary Investigation 302
The Second Phase: Do an Analysis of the System 302
The Third Phase: Design the System 303
The Fourth Phase: Develop the System 303
The Fifth Phase: Implement the System 304
The Sixth Phase: Maintain the System 305

A.2 Programming: A Five-Step Procedure 305

A.3 Programming Languages Used Today 306

A.4 Object-Oriented Programming & Visual Programming 308

Object-Oriented Programming: 308
Visual Programming 309

A.5 Internet Programming: HTML, XML, VRML, Java, & ActiveX, 310

HTML—For Creating 2-D Web Documents & Links 310
XML—For Making the Web Work Better 310
VRML—For Creating 3-D Web Pages 310
Java—For Creating Interactive Web Pages 311
ActiveX—Also for Creating Interactive Web Pages 311

Bookmark It! Practical Action Box: *"More Tips for Creating Your Own Website"* 312

Notes 467
Index 475
Credits 482

Chap

Introduction to Information Technology

Mind Tools for Your Future

Key Questions
You should be able to answer the following questions.

1.1 **Infotech Becomes Commonplace: Cellphones, E-Mail, the Internet, & the E-World** How does information technology facilitate e-mail, networks, and the use of the Internet and the Web; what is the meaning of the term *cyberspace*?

1.2 **The "All-Purpose Machine": The Varieties of Computers** What are the five sizes of computers, and what are clients and servers?

1.3 **Understanding Your Computer: What If You Custom-Ordered Your Own PC?** What four basic operations do all computers follow, and what are some of the devices associated with each operation? How does communications affect these operations?

1.4 **Where Is Information Technology Headed?** What are three directions of computer development and three directions of communications development?

"**S**ay goodbye to the personal computing era," writes technology journalist Kevin Maney. "Just on the horizon is the era that comes next—the personal information era."[1]

You could also call it the era of *pervasive computing.* The world is moving on beyond boxy computers that sit on desks or even on laps. We are entering a time in which handheld computers, two-way wireless pagers, and beefed-up cellphones (not to mention terminals everywhere—libraries, airports, cafes) will let you access information anytime anywhere. And not just general information but *your personal information*—the electronic correspondence, documents, appointments, photos, songs, money matters, and other data important to you.

Central to this concept, as you might expect, is the Internet—the "Net," that sprawling collection of data residing on computers around the world and accessible by high-speed connections. Everything that presently exists on a personal computer or a laptop, experts suggest, will move onto the Net, giving us greater mobility and wrapping the Internet around our lives.[2] This is why the Internet is featured so prominently in this book, starting on this page and presented fully in Chapter 2.

With so much information available everywhere all the time, what will this do to us as human beings? We can already see the outlines of the future. One result is *information overload:* The typical American office worker, for instance, sends or receives 201 messages daily—e-mail, voice mail, faxes, Post-its, and so on—with e-mail alone consuming an average of 49 minutes a day.[3] Another is *less use of our brains for memorizing:* Familiar phone numbers and other facts are being stored on speed-dial cellphones, pocket computers, and electronic databases, increasing our dependence on technology.[4] A third result is *a surge in "multitasking" activity:* Young people in particular have become highly skilled in performing several tasks at once, such as talking on the phone, watching TV, answering e-mail, surfing the Web, and doing homework.[5]

These three trends pose unique challenges to how you learn and manage information. An important purpose of this book is to give you the tools for doing so, as we explain at the end of this chapter.

First, however, we need to give you an immediate though brief overview of what computing is and how it works. We begin by discussing the Internet and some of its features, particularly e-mail. We next describe the varieties of computers that exist. We then explain the three key concepts behind how a computer works and what goes into a personal computer, both hardware and software. Finally, we address how to survive in the new ocean of 24/7/365 information.

1.1 Infotech Becomes Commonplace: Cellphones, E-Mail, the Internet, & the E-World

KEY QUESTIONS
How does information technology facilitate e-mail, networks, and the use of the Internet and the Web; what is the meaning of the term cyberspace?

This book is about computers, of course. But not *just* about computers. It is also about the way computers communicate with one another. When computer and communications technologies are combined, the result is **information technology**—"infotech"—**technology that merges computing with high-speed communications links carrying data, sound, and video.** Examples of information technology include personal computers, of course, but also new forms of telephones, televisions, and various handheld devices.

Note there are two parts to this definition—*computers* and *communications:*

- **Computer technology:** You have certainly seen, and probably used, a computer. Nevertheless, let's define what it is. **A _computer_ is a programmable, multiuse machine that accepts data—raw facts and figures—and processes, or manipulates, it into information** we can use, such as summaries, totals, or reports. Its purpose is to speed up problem solving and increase productivity.
- **Communications technology:** Unquestionably you've been using communications technology for years. **_Communications technology_, also called _telecommunications technology_, consists of electromagnetic devices and systems for communicating over long distances.** The principal examples are telephone, radio, broadcast television, and cable TV. More recently there has been the addition of communications among computers—which is what happens when people "go online" on the Internet. **_Online_ means using a computer or other information device, connected through a network, to access information and services from another computer or information device.**

As an example of a communication device, let's consider something that seems to be everywhere these days—the cellphone.

The Telephone Grows Up

Seventy-something Louis DeMartino hadn't even had a chance to unwrap the package when a ringing sound came from inside it—a call from his granddaughter as a way of making the family's gift presentation of a cellphone all the more dramatic.[6] DeMartino may have just been getting started on using this apparatus, but Amit Sinai, 18, like many other teenagers, cannot imagine life without it. Flopping down on her towel after a quick dip in the ocean at a New York beach, for instance, the first thing she does, even before her face starts to dry, is reach for her cellphone and start dialing. "Hi, it's me," she says. "I'm on the beach. Where are you?"[7]

Cellphone mania has swept the world. All across the globe, people have acquired the portable gift of gab; Finnish wireless giant Nokia estimates that 1 billion people are currently using cellphones.[8] Some make 45 calls or more a day. It has taken more than 100 years for the telephone to get to this point—getting smaller, acquiring push buttons, losing its cord connection. In 1964, the ★ and # were added to the keypad. In 1973, the first cellphone call was processed. In its standard form, the phone is still so simply designed that even a young child can use it. However, it is now becoming more versatile and complex—a way of connecting to the Internet and the World Wide Web.

Why introduce a book that is about computers with a discussion of telephones? Because Internet phones—such as the Kyocera QCP, the Motorola V60$_c$, and the Ericsson R380 World—represent another giant step for information technology. (See ● Panel 1.1.) Now you no longer need a personal computer to get on the Internet. These infotech phones, with their small display screens, provide a direct, wireless connection that enables you not only to make voice calls and check your daily "to-do" list but also to browse the World Wide Web and receive all kinds of information: news, sports scores, stock prices, term-paper research. And you can also send and receive e-mail.

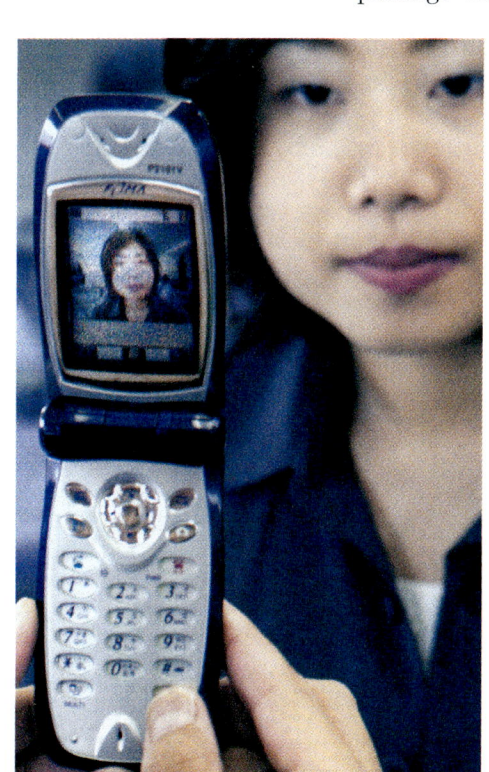

● **PANEL 1.1**
The Internet phone With an Internet-accessible phone, you have e-mail and specific Web services such as news and stock quotes in addition to regular cellphone services—and, in the case of NTT DoCoMo's FOMA, video.

"You've Got Mail!" E-Mail's Mass Impact

It took the telephone 40 years to reach 10 million customers, and fax machines 20 years. Personal computers made it into that many American

E-mailer
A college student sends e-mail.

homes 5 years after they were introduced. E-mail, which appeared in 1981, became popular far more quickly, reaching 10 million users in little more than a year.[9] No technology has ever become so universal so fast. Not surprisingly, then, one of the first things new computer users learn is how to send and receive *e-mail*—"electronic mail," messages transmitted over a computer network, most often the Internet. A *network* is a communications system connecting two or more computers; the Internet is the largest such network.

In 1998, the volume of e-mail in the United States surpassed the volume of hand-delivered mail. In 2002, an estimated average of 8 billion messages a day are zipping back and forth across the United States (which has some 60% of the world's e-mail accounts), up from 3.5 billion three years earlier.[10] Because of this explosion in usage, suggests a *Business Week* report, "E-mail ranks with such pivotal advances as the printing press, the telephone, and television in mass impact."[11]

How is electronic mail different from calling on a telephone or writing a conventional letter? E-mail "occupies a psychological space all its own," says one journalist. "It's almost as immediate as a phone call, but if you need to, you can think about what you're going to say for days and reply when it's convenient."[12] E-mail has blossomed, points out another writer, not because it gives us more immediacy but because it gives us *less*. "The new appeal of e-mail is the old appeal of print," he says. "It isn't instant; it isn't immediate; it isn't in your face." E-mail has succeeded for the same reason that the videophone—which allows callers to see each other while talking—has been so slow to catch on: because "what we actually want from our exchanges is the minimum human contact commensurate with the need to connect with other people."[13]

From this it is easy to conclude, as *New York Times* computer writer Peter Lewis did, that e-mail "is so clearly superior to paper mail for so many purposes that most people who try it cannot imagine going back to working without it."[14] What is interesting, though, is that in these times when images often seem to overwhelm words, e-mail is actually *reactionary*. "The Internet is the first new medium to move decisively backward," points out one writer, because it essentially involves *writing*. "Ten years ago, even the most literate of us wrote maybe a half a dozen letters a year; the rest of our lives took place on the telephone."[15] E-mail has changed all that.

The Internet, the World Wide Web, & the "Plumbing of Cyberspace"

Communications is extending into every nook and cranny of civilization. It has been called the "plumbing of cyberspace." The term *cyberspace* was coined by William Gibson in his novel *Neuromancer*, to describe a futuristic computer network into which users plug their brains. In everyday use, it has a rather different meaning.

Today many people equate cyberspace with the Internet. But it is much more than that, says David Whittle in his book *Cyberspace: The Human Dimension*.[16] Cyberspace includes not only the World Wide Web, chat rooms, online bulletin boards, and member-based services such as America Online—all features we explain in this book—"but also such things as conference calls and automatic teller machines," he says.[17] We may say that *cyberspace* encompasses not only the online world and the Internet in particular but also the whole wired and wireless world of communications in general.

The two most important aspects of cyberspace are the Internet and that part of the Internet known as the World Wide Web.

- **The Internet—"the mother of all networks":** The Internet is at the heart of the "Information Age." Called "the mother of all networks," **the _Internet_ (_Net_) is a worldwide network that connects hundreds of thousands of smaller networks.** These networks link educational, commercial, nonprofit, and military entities.
- **The World Wide Web—the multimedia part of the Net:** The Internet has been around for more than 30 years. But what made it popular, apart from e-mail, was the development of the _World Wide Web_—**an interconnected system of computers all over the world that store information in multimedia form.** The word _multimedia_, from "multiple media," refers to technology that presents information in more than one medium, such as text, still images, moving images, and sound. In other words, the Web provides information in more than one way.

BOOKMARK IT!

PRACTICAL ACTION BOX
Managing Your E-Mail

For many people, e-mail is *the* online environment, more so than the World Wide Web. In one study, respondents reported they received an average of 31 e-mail messages a day.[a] But then there are those such as Jeremy Gross, managing director of technology for Countrywide Home Loans, who gets 300 e-mails a day—and about 200 are junk e-mail (spam), bad jokes, or irrelevant memos (the "cc," or carbon copy).[b] Astronomer Seth Shostak gets 50 electronic messages daily, at least half requiring a reply. "If I spend five minutes considering and composing a response to each correspondence," he complains, "then two hours of my day are busied with e-mail, even when I don't initiate a single one."[c]

It's clear, then, that e-mail will increase productivity only if it is used properly. Overuse or misuse just causes more problems and wastes time. The following are some ideas to keep in mind when using e-mail:[d]

- **Do your part to curb the e-mail deluge:** Put short messages in the subject line so recipients don't have to open the e-mail to read the note. Don't reply to every e-mail message you get. Avoid "cc:ing" (copying to) people unless absolutely necessary. Don't send chain letters or lists of jokes, which just clog mail systems.
- **Be helpful in sending attachments:** Attachments—computer files of long documents or images—attached to an e-mail are supposed to be a convenience, but often they can be an annoyance. Sending a 1-megabyte file to a 500-person mailing list creates 500 copies of that file—and that many megabytes can clog or even cripple the mail system. (A 1-megabyte file is about the size of a 300-page double-spaced term paper.) Ask your recipients beforehand if they want the attachment and tell them how to open it; they may need to know what word-processing program to use, for example.
- **Be careful about opening attachments you don't recognize:** Some dangerous computer viruses—renegade programs that can damage your computer—have been spread by e-mail attachments, which are activated only upon opening.
- **Use discretion about the e-mails you send:** E-mail should not be treated as informally as a phone call. Don't send a message electronically that you don't want some third party to read. E-mail messages are not written with disappearing ink; they remain in a computer system long after they have been sent. Worse, recipients can easily copy and even alter your messages and forward them to others without your knowledge.
- **Don't use e-mail in place of personal contact:** Because e-mail carries no tone or inflection, it's hard to convey personal concern. Avoid criticism and sarcasm in electronic messaging. Nevertheless, you *can* use e-mail to provide quick praise, even though doing it in person will take on greater significance.
- **Be aware that e-mail you receive at work is the property of your employer:** Be careful of what you save, send, and back up.
- **Realize that deleting e-mail messages doesn't totally get rid of them:** "Delete" moves the e-mail from the visible list, but the messages remain on your hard disk and can be retrieved by experts. Special software, such as Spytech Eradicator and Window Washer, will completely erase e-mail from the hard disk.

According to Nua Internet Surveys (NUA.com), a resource for Internet trends and statistics, more than 407.1 million people were using the Internet worldwide in the summer of 2001. Among them:

- 26 million citizens of China
- One-fifth of Irish homes
- All but 1% of students in the United Kingdom
- 10 million British households
- 40% of German citizens
- 16% of Hungarian citizens
- 2 million Argentineans
- 6.7 million Mexicans
- 99% of Canadians ages 9–17
- 3.5 million people in the Arab world
- 3.5 million Australian households
- 7 million, or one-fifth, of Spanish citizens
- 22 million Korean citizens
- 1 million Belgians

There is no doubt the influence of the Net and the Web is tremendous. At present, one-third to one-half of the U.S. population is online; if the Internet industry were a nation, it would be the 18th-largest economy in the world.[18] But just how revolutionary is it? Is it equivalent to the invention of television, as some technologists say? Or is it even more important—equivalent to the invention of the printing press? "Television turned out to be a powerful force that changed a lot about society," says *USA Today* technology reporter Kevin Maney. "But the printing press changed everything—religion, government, science, global distribution of wealth, and much more. If the Internet equals the printing press, no amount of hype could possibly overdo it."[19]

Perhaps in a few years we'll begin to know the answer. No massive study was ever done of the influence of the last great electronic revolution to touch us, namely, television. But the Center for Communication Policy at the University of California, Los Angeles, in conjunction with other international universities, has begun to take a look at the effects of information technology—and at how people's behavior and attitudes toward it will change over a span of years.[20]

The E-World & Welcome to It

One thing we know already is that cyberspace is saturating our lives. More than 52% of American adults are Internet users, according to a recent survey.[21] While the average age of users is rising, there's no doubt that young people love the Net. For instance, an amazingly high percentage of American teenagers—*81%*—use the Internet, according to Teenage Research Unlimited (TRU), a Chicago market research firm.[22] Why is that? E-mail is certainly one important reason—it's all about staying connected.

But it's more than just about e-mail. Teenagers are also big participants in online commerce. "When teens want information on a brand," says TRU's director of research, "they turn to the Internet first." Indeed, teens are voracious consumers of all things electronic.

And not just for teens but for most Americans, the use of the Internet's favorite letter, "e"—as in e-business, e-commerce, e-shopping—is rapidly becoming outmoded. "E" is now part of nearly everything we do. As an executive for a marketing research firm says, "E-business is just business."[23] The electronic world is everywhere. The Net and the Web are everywhere. Cyberspace permeates everything. What editor Frederick Allen has observed is true: Infotech has become . . . ordinary.

CONCEPT CHECK

What are the two parts of information technology?

How does e-mail differ from other means of communication?

What is cyberspace, and what are its two most important aspects?

1.2 The "All-Purpose Machine": The Varieties of Computers

KEY QUESTIONS
What are the five sizes of computers, and what are clients and servers?

When the alarm clock★ blasts you awake, you leap out of bed and head for the kitchen, where you plug in the coffee maker★. After using your electric toothbrush★ and showering and dressing, you stick a bagel in the microwave★, then pick up the TV remote★ and click on the TV★ to catch the weather forecast. Later, after putting dishes in the dishwasher★, you go out and start up the car★ and head toward campus or work. Pausing en route at a traffic light★, you turn on your portable CD player★ to listen to some music.

You haven't yet touched a PC, a personal computer, but you've already dealt with at least 10 computers—as you probably guessed from the ★s. All these familiar appliances rely on tiny "computers on chips" called microprocessors. "These marvels of engineering have been infiltrating our everyday lives for more than a quarter of a century," says technology writer Dan Gillmor, but "in some ways the revolution has only begun."[24]

Maybe, then, the name *computer* is inadequate. As computer pioneer John Von Neumann has said, the device should not be called a computer but rather the "all-purpose machine." It is not, after all, just a machine for doing calculations. The most striking thing about it is that it can be put to *any number of uses*.

What are the various types of computers? Let's take a look.

All Computers, Great & Small: The Categories of Machines

At one time, the idea of having your own computer was almost like having your own personal nuclear reactor. In those days, in the 1950s and '60s, computers were enormous machines affordable only by large institutions. Now they come in a variety of shapes and sizes, which can be classified according to their processing power.

- **Supercomputers: Typically priced from $500,000 to more than $85 million, *supercomputers* are high-capacity machines with hundreds of thousands of processors that can perform over 1 trillion calculations per second.** These are the most expensive but fastest computers available. "Supers," as they are called, have been used for tasks requiring the processing of enormous volumes of data, such as doing the U.S. census count, forecasting weather, designing aircraft, modeling molecules, breaking codes, and simulating explosion of nuclear bombs. More recently they have been employed for business purposes—for instance, sifting demographic marketing information—and for creating film animation. The fastest computer in the world, which cost $85 million and looks like rows of refrigerator-size boxes, is the IBM ASCI White (Accelerated Strategies Computing Initiative White) at the Lawrence Livermore Laboratory in Livermore, California. This supercomputer has 8192 processors, and it can process 12.3 trillion operations per second. It takes up 12,000 square feet—the space required by two NBA basketball courts. It weighs 106 tons. It has 97,000 times the memory of a 64-megabyte microcomputer and 16,000 times the secondary storage capacity of a 10-gigabyte microcomputer hard disk. It uses 83 miles of wiring.

Supercomputer
IBM ASCI White

Mainframe computer

- **Mainframe computers:** The only type of computer available until the late 1960s, _mainframes_ are water- or air-cooled computers that cost $5000–$5 million and vary in size from small, to medium, to large, depending on their use. Small mainframes ($5000–$200,000) are often called _midsize computers;_ they used to be called _minicomputers,_ although today the term is seldom used. Mainframes are used by large organizations—such as banks, airlines, insurance companies, and colleges—for processing millions of transactions. Often users access a mainframe by means of a _terminal_, which has a display screen and a keyboard and can input and output data but cannot by itself process data. Mainframes process billions of instructions per second.

- **Workstations:** Introduced in the early 1980s, _workstations_ are expensive, powerful computers usually used for complex scientific, mathematical, and engineering calculations and for computer-aided design and computer-aided manufacturing. Providing many capabilities comparable to midsize mainframes, workstations are used for such tasks as designing airplane fuselages, prescription drugs, and movie special effects. Workstations have caught the eye of the public mainly for their graphics capabilities, which are used to breathe three-dimensional life into movies such as _Jurassic Park III_ and _Pearl Harbor._ The capabilities of low-end workstations overlap those of high-end desktop microcomputers.

Workstation

- **Microcomputers:** _Microcomputers_, also called _personal computers_, which cost $500–$5000, can fit next to a desk or on a desktop, or can be carried around. They are either stand-alone machines or are connected to a computer network, such as a local area network. A _local area network (LAN)_ connects, usually by special cable, a group of desktop PCs and other devices, such as printers, in an office or a building.

  Microcomputers are of several types: _desktop PCs, tower PCs, laptops_ (or _notebooks_), and _personal digital assistants_—handheld computers or palmtops.

  Desktop PCs are those in which the case or main housing sits on a desk, with keyboard in front and monitor (screen) often on top. _Tower PCs_ are those in which the case sits as a "tower," often on the floor beside a desk, thus freeing up desk surface space.

Three types of microcomputers
Desktop, tower, and notebook

Notebook computers, also called *laptop computers*, are lightweight portable computers with built-in monitor, keyboard, hard-disk drive, battery, and AC adapter that can be plugged into an electrical outlet; they weigh anywhere from 1.8 to 9 pounds.

Personal digital assistants (PDAs), also called *handheld computers* or *palmtops*, combine personal organization tools—schedule planners, address books, to-do lists—with the ability in some cases to send e-mail and faxes. Some PDAs have

Two types of personal digital assistants

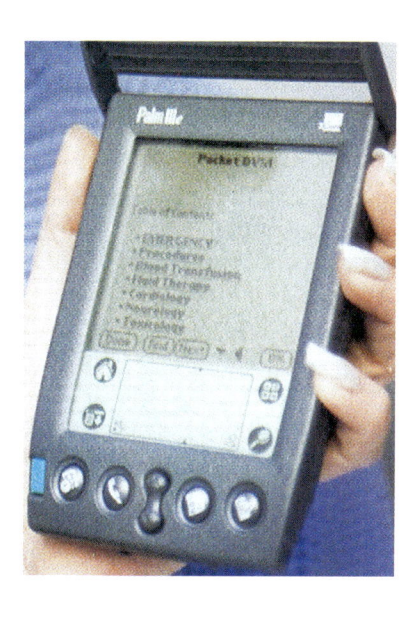

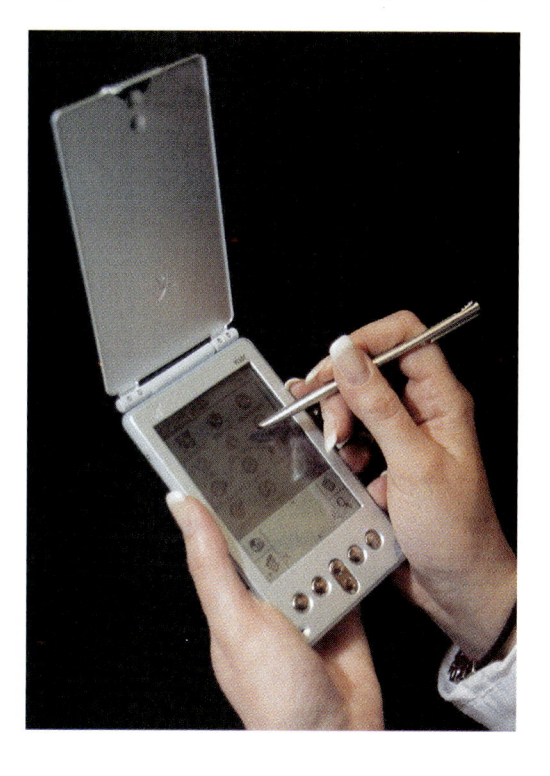

touch-sensitive screens. Some also connect to desktop computers for sending or receiving information. (For now, we are using the word *digital* to mean "computer based.")

- Microcontrollers: *Microcontrollers*, also called embedded computers, are the tiny, specialized microprocessors installed in "smart" appliances and automobiles. These microcontrollers enable microwave ovens, for example, to store data about how long to cook your potatoes and at what temperature. Recently microcontrollers have been used to develop a new universe of experimental electronic appliances—e-pliances—for example, as tiny Web servers embedded in clothing, jewelry, and household appliances such as refrigerators.[25] Microcontrollers are also used in blood pressure monitors, air bag sensors, gas and chemical sensors for water and air, and vibration sensors.

Microcontroller

Servers

The word *server* does not describe a size of computer but rather a particular way in which a computer is used. Nevertheless, because servers have become so important to telecommunications, especially with the rise of the Internet and World Wide Web, they deserve mention here. (Servers are discussed in detail in Chapter 3.)

A *server*, or network server, is a central computer that holds collections of data (databases) and programs for connecting PCs, workstations, and other devices, which are called *clients*. These clients are linked by a wired or wireless network. The entire network is called a *client/server network*. In small organizations, servers can store files and transmit e-mail. In large organizations, servers can house enormous libraries of financial, sales, and product information.

You may never see a supercomputer or mainframe or server or even a tiny microcontroller. But you will most certainly get to know the personal computer, if you haven't already. We consider this machine next.

> **CONCEPT CHECK**
>
> Describe the five sizes of computers.
>
> What are the different types of microcomputers?
>
> Define servers and clients.

1.3 Understanding Your Computer: What If You Custom-Ordered Your Own PC?

KEY QUESTIONS
What four basic operations do all computers follow, and what are some of the devices associated with each operation? How does communications affect these operations?

Could you build your own personal computer? Some people do, putting together bare-bones systems for just a few hundred dollars. "If you have a logical mind, are fairly good with your hands, and possess the patience of Job, there's no reason you can't . . . build a PC," says science writer David Einstein. And, if you do it right, "it will probably take only a couple of hours," because industry-standard connections allow components to go together fairly easily.[26]

Actually, probably only techies would consider building their own PC. But many ordinary users order their own *custom-built* PCs. Let's consider how you might do this.

How Computers Work: Three Key Concepts

We're not actually going to ask you to build or order a PC—just to *pretend* to do so. The purpose of this exercise is to help you understand how a computer works. That information will help you when you go shopping for a new system or, especially, if you order up a custom-built system.

Before you begin, you will need to understand three key concepts.

FIRST: *The purpose of a computer is to process data into information.* **Data consists of the raw facts and figures that are processed into information**—for example, the votes for different candidates being elected to student-government office. **Information is data that has been summarized or otherwise manipulated for use in decision making**—for example, the total votes for each candidate, which are used to decide who won.

SECOND: You should know *the difference between hardware and software.* **Hardware consists of all the machinery and equipment** in a computer system. The hardware includes, among other devices, the keyboard, the screen, the printer, and the "box"—the computer or processing device itself. **Software, or programs, consists of all the instructions that tell the computer how to perform a task.** These instructions come from a software developer in a form (such as a CD, or compact disk) that will be accepted by the computer. Examples you may have heard of are Microsoft Windows or Office XP.

THIRD: Regardless of type and size, *all computers follow the same four basic operations:* (1) *input,* (2) *processing,* (3) *storage,* and (4) *output.* To this we will add (5) *communications.*

Survival Tip

Input is covered in detail in Chapter 5.

1. **Input operation: Input is whatever is put in ("input") to a computer system.** Input can be nearly any kind of data—letters, numbers, symbols, shapes, colors, temperatures, sounds, or whatever raw material needs processing. When you type some words or numbers on a keyboard, those words are considered *input data.*

Survival Tip

Processing is covered in detail in Chapter 4.

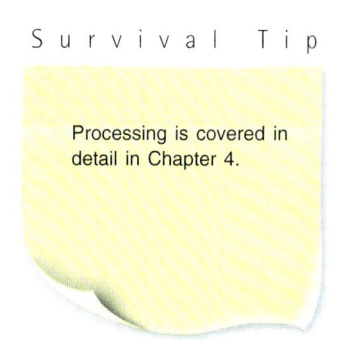

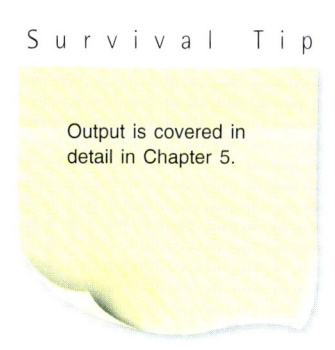

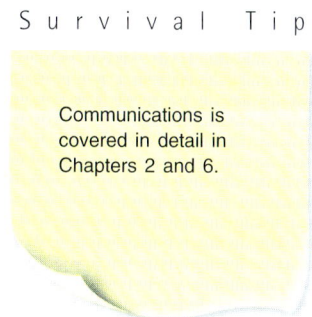

2. **Processing operation:** *Processing is the manipulation a computer does to transform data into information.* When the computer adds 2 + 2 to get 4, that is the act of processing. The processing is done by the *central processing unit*—frequently called just the CPU—a device consisting of electronic circuitry that executes instructions to process data.

3. **Storage operation:** Storage is of two types—temporary storage and permanent storage, or primary storage and secondary storage. ***Primary storage**, or memory, is the computer circuitry that temporarily holds data waiting to be processed.* This circuitry is inside the computer. ***Secondary storage**, simply called storage, is the area in the computer where data or information is held permanently.* A floppy disk or hard disk is an example of this kind of storage. (Storage also holds the *software*—the computer programs.)

Survival Tip

Storage is covered in detail in Chapter 4.

Survival Tip

Output is covered in detail in Chapter 5.

4. **Output operation:** *Output is whatever is output from ("put out of") the computer system, the results of processing,* usually information. Examples of output are numbers or pictures displayed on a screen, words printed out on paper in a printer, or music piped over some loudspeakers.

5. **Communications operation:** These days, most (though not all) computers have *communications* ability, which offers an *extension* capability—in other words, it extends the power of the computer. With wired or wireless communications connections, data may be input from afar, processed in a remote area, stored in several different locations, and output in yet other places. However, you don't need communications ability to write term papers, do calculations, or perform many other computer tasks.

Survival Tip

Communications is covered in detail in Chapters 2 and 6.

These five operations are summarized in the illustration on the next page. (See ● Panel 1.2.)

Pretending to Order a Custom-Built Desktop Computer

Now let's see how you would order a custom-built desktop PC. Remember, the purpose of this is to help you understand the internal workings of a computer so that you'll be better equipped when you go to buy one. (If you were going to build it yourself, you would pretend that someone had acquired the PC components for you from a catalog company and that you're now sitting at a workbench or kitchen table about to begin assembling them. All you would need is a screwdriver, perhaps a small wrench, and a static-electricity-free work area. You would also need the manuals that come with some of the components.) Although prices of components are always subject to change, we have indicated *general ranges* of prices for basic equipment current as of early 2002 so that you can get a sense of the relative importance of the various parts. ("Loaded" components—the most powerful and sophisticated equipment—cost more than the prices given here.)

Input Hardware: Keyboard & Mouse

Input hardware consists of devices that allow people to put data into the computer in a form that the computer can use. At minimum, you will need two things: a *keyboard* and a *mouse*.

● **PANEL 1.2**
The basic operations of a computer

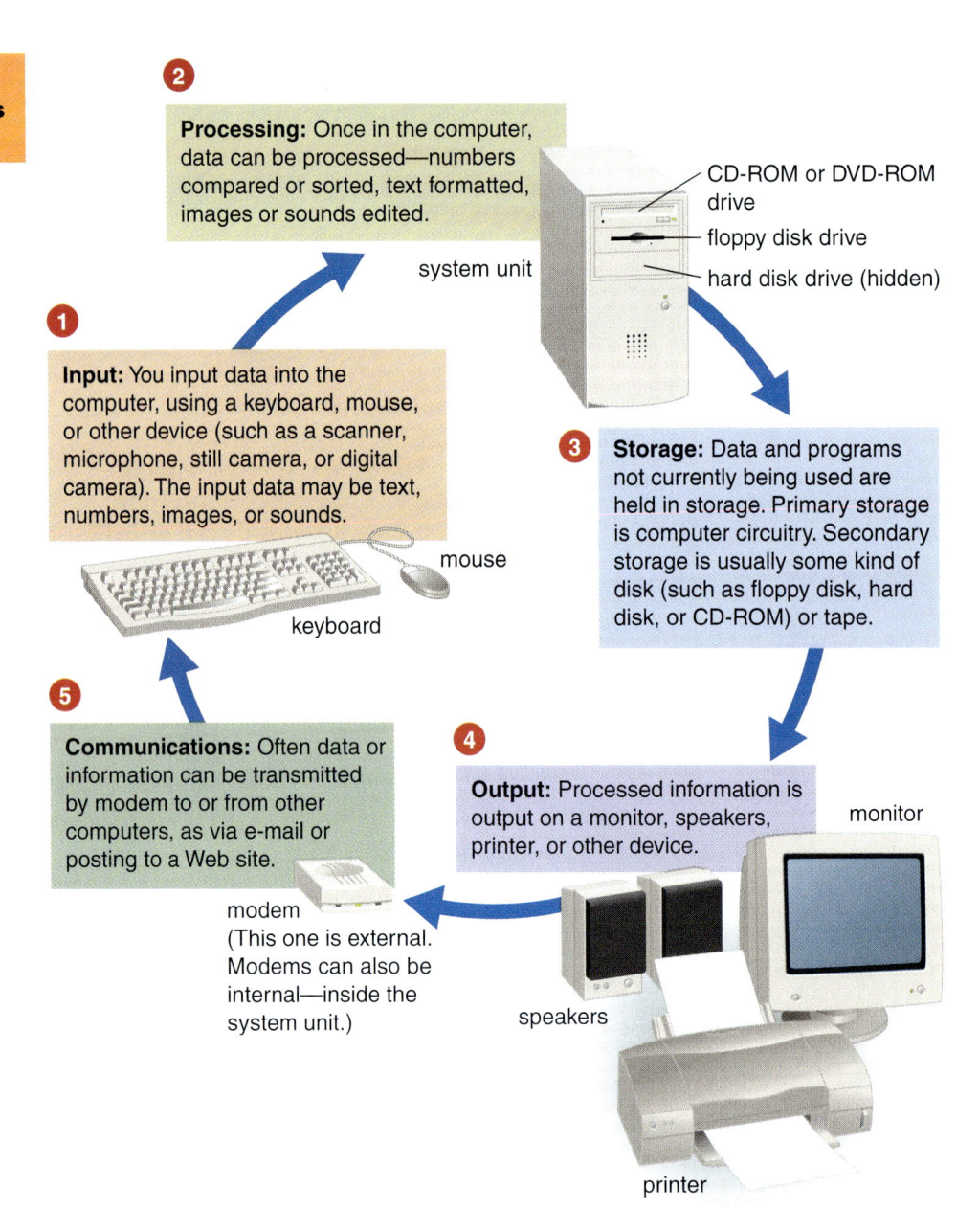

❶ Input: You input data into the computer, using a keyboard, mouse, or other device (such as a scanner, microphone, still camera, or digital camera). The input data may be text, numbers, images, or sounds.

❷ Processing: Once in the computer, data can be processed—numbers compared or sorted, text formatted, images or sounds edited.

system unit — CD-ROM or DVD-ROM drive
— floppy disk drive
— hard disk drive (hidden)

❸ Storage: Data and programs not currently being used are held in storage. Primary storage is computer circuitry. Secondary storage is usually some kind of disk (such as floppy disk, hard disk, or CD-ROM) or tape.

mouse
keyboard

❺ Communications: Often data or information can be transmitted by modem to or from other computers, as via e-mail or posting to a Web site.

modem
(This one is external. Modems can also be internal—inside the system unit.)

❹ Output: Processed information is output on a monitor, speakers, printer, or other device.

monitor
speakers
printer

Keyboard

Mouse

- **Keyboard:** Cost: $35–$70. On a microcomputer, a keyboard is the primary input device. **A _keyboard_ is an input device that converts letters, numbers, and other characters into electrical signals readable by the processor.** A microcomputer keyboard looks like a typewriter keyboard, but besides keys for letters and numbers it has several keys (such as *F* keys and *Ctrl, Alt,* and *Del* keys) intended for computer-specific tasks. After other components are assembled, the keyboard will be plugged into the back of the computer in a socket intended for that purpose (unless you have a cordless keyboard).

- **Mouse:** $30–$120. **A _mouse_ is an input device that is used to manipulate objects viewed on the computer display screen.** The mouse will be plugged into the back of the computer (or on the back of the keyboard) in a socket after the other components are assembled. (Cordless mice are also available.)

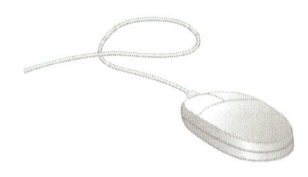

Chapter 1

12

Processing & Memory Hardware: Inside the System Cabinet

This is the part where most of the assembly work will have to be done. The brains of the computer are the *processing* and *memory* devices, which are housed in the case or system cabinet.

Case or system cabinet

- **Case and power supply:** About $60. Also known as the <u>system unit</u>, the *case* or <u>system cabinet</u> is the box that houses the processor chip (CPU), the memory chips, and the motherboard with power supply, as well as some secondary storage devices—floppy-disk drive, hard-disk drive, and CD-ROM or DVD drive, as we will explain. The case comes in desktop or tower models. It includes a power supply unit and a fan to keep the circuitry from overheating.

- **Processor chip:** $75–$750 or more. It may be small and not look like much. But it could be the most expensive hardware component of a build-it-yourself PC—and doubtless the most important. **A processor <u>chip</u> is a tiny piece of silicon that contains millions of miniature electronic circuits.** The speed at which a chip processes information is expressed in *megahertz (MHz)*, millions of processing cycles per second, or *gigahertz (GHz)*, billions of processing cycles per second. For $250, you might get a 600-MHz chip, which is adequate for most student purposes. For top dollar, you might get a 1.7-gigahertz chip, which you would want if you're running software with spectacular graphics and sound, such as those with some video games (Quake).

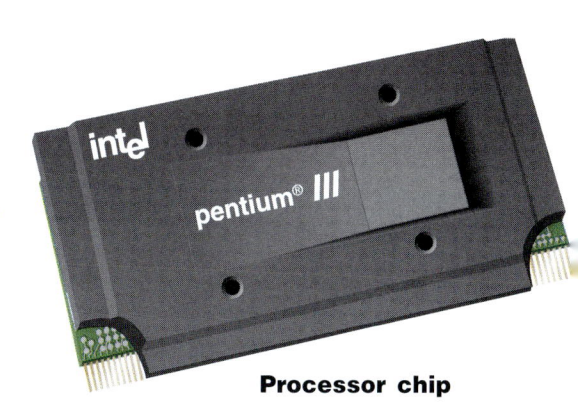

Processor chip

memory chip (RAM chip)

memory chips mounted on module

- **Memory chips:** $100–$150. These are also small. **<u>Memory chips</u>, also known as RAM (random access memory) chips, represent primary storage or temporary storage; they hold data before processing and information after processing,** before it is sent along to an output or storage device. You'll want enough memory chips to hold 128 megabytes, or roughly 128 million characters, of data, which is adequate for most student purposes. If you work with large graphic files, you'll need more memory capacity, 192–256 megabytes. (We explain the numbers used to measure storage capacities in a moment.)

- **Motherboard:** About $100–$150. Also called the *system board*, the <u>motherboard</u> **is the main circuit board in the computer.** This is the big green circuit board to which everything else—such as the keyboard, mouse, and printer—attaches through connections (called *ports*) in the back of the computer. The processor chip and memory chips are also installed on the motherboard.

   Note that the motherboard has <u>expansion slots</u>—**for expanding the PC's capabilities—which give you places to plug in additional circuit boards,** such as those for video, sound, and communications (modem).

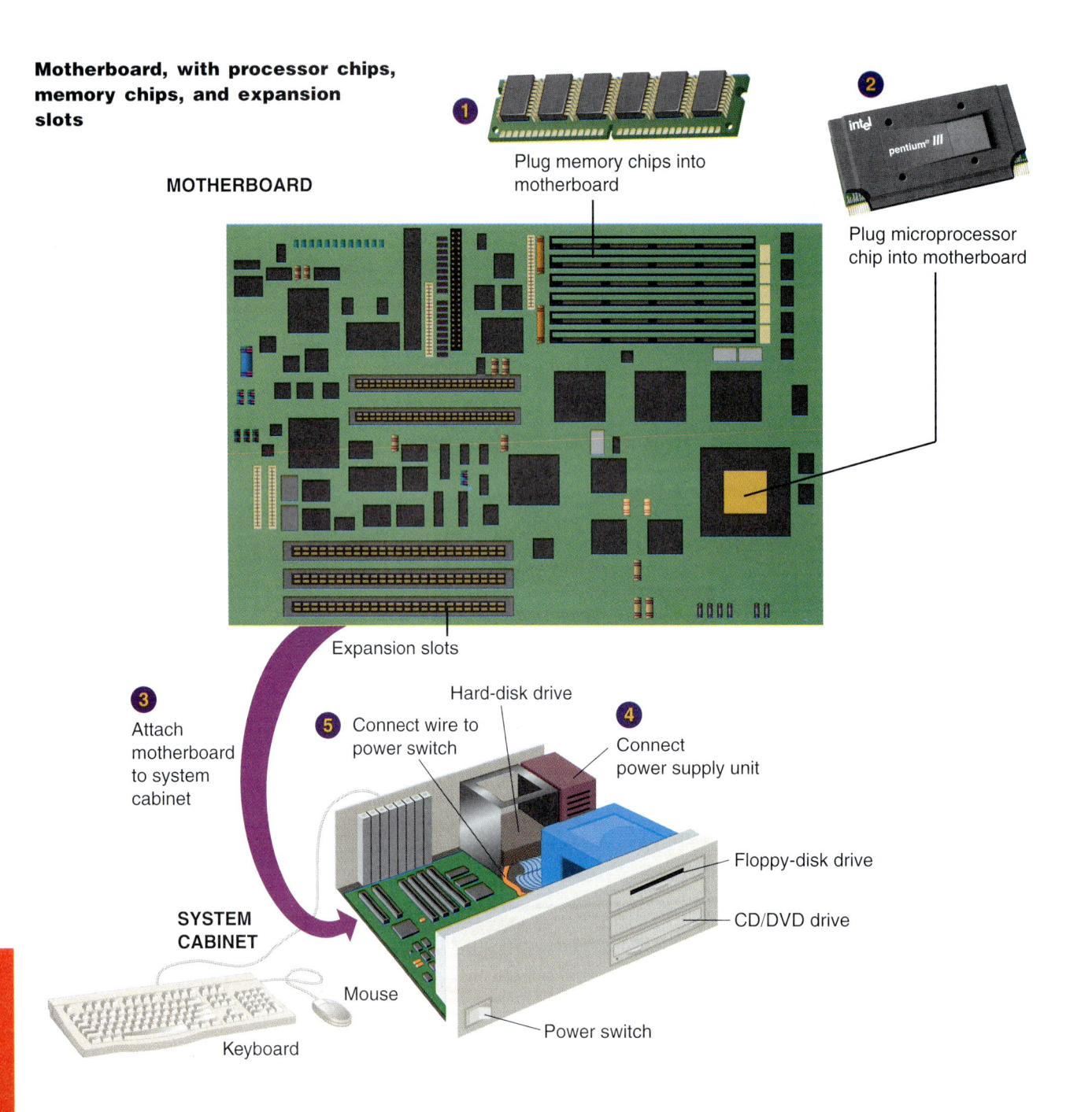

Motherboard, with processor chips, memory chips, and expansion slots

① Plug memory chips into motherboard

② Plug microprocessor chip into motherboard

MOTHERBOARD

Expansion slots

③ Attach motherboard to system cabinet

⑤ Connect wire to power switch

Hard-disk drive

④ Connect power supply unit

Floppy-disk drive

CD/DVD drive

SYSTEM CABINET

Mouse

Keyboard

Power switch

Now the components can be put together. As the illustration above shows, ① the memory chips are plugged into the motherboard. Then ② the processor chip is plugged into the motherboard. Now ③ the motherboard is attached to the system cabinet. Then ④ the power supply unit is connected to the system cabinet. Finally, ⑤ the wire for the power switch, which turns the computer on and off, is connected to the motherboard.

Storage Hardware: Floppy Drive, Hard Drive, & CD/DVD Drive

With the motherboard in the system cabinet, the next step is installation of the storage hardware. Whereas memory chips deal with temporary storage, now we're concerned with *secondary storage* or *permanent storage*, in which you'll be able to store your data for the long haul.

For today's student purposes, you'll need at minimum a floppy-disk drive, a hard-disk drive, and a CD/DVD drive. If you work with large files, you'll also want a Zip-disk drive. These storage devices slide into the system cabinet from the front and are secured with screws. Each drive is attached to the motherboard by a flat cable (called a *ribbon cable*). Also, each drive must be hooked up to a plug extending from the power supply.

A computer system's data/information storage capacity is represented by bytes, kilobytes, megabytes, gigabytes, and terabytes. Roughly speaking, a *byte = 1 character* of data, a *kilobyte = 1000 characters*, a *megabyte = 1 million characters*, a *gigabyte = 1 billion characters*, and a *terabyte = 1 trillion characters*. A character could be alphabetic (A, B, or C) or numeric (1, 2, or 3) or a special character (!, ?, *, $, %).

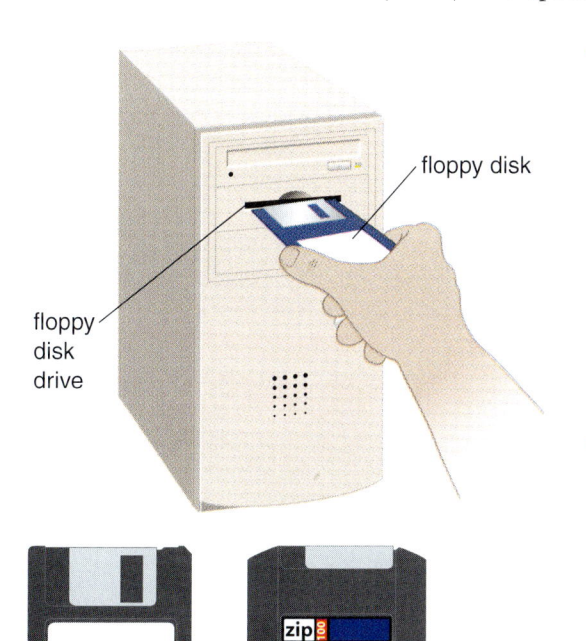

- **Floppy-disk and Zip drives:** $100–$180. A *floppy-disk drive* is a storage device that stores data on removable 3.5-inch-diameter diskettes. These diskettes don't seem to be "floppy," because they are encased in hard plastic, but the mylar disk inside is indeed flexible or floppy. Each can store 1.44 million bytes (characters) or more of data. With the floppy-disk drive installed, you'll later be able to insert a diskette through a slot in the front and remove it by pushing the eject button. A *Zip-disk drive* is a storage device that stores data on floppy-disk cartridges with 70–170 times the capacity of the standard floppy.

- **Hard-disk drive:** $120 for 20 gigabytes of storage (higher-capacity drives are available). A *hard-disk drive* is a storage device that stores billions of characters of data on a nonremovable disk platter. With 20 gigabytes of storage, you should be able to handle most student needs.

- **CD/DVD drive:** $100–$400. A *CD (compact disk) drive*, or its more recent variant, a *DVD (digital video disk) drive*, is a storage device that uses laser technology to read data from optical disks. These days new software is generally supplied on CDs rather than floppy disks. And even if you can get a program on floppies, you'll find it easier to install a new program from one CD rather than repeatedly inserting and removing, say, 10 or 12 floppy disks.

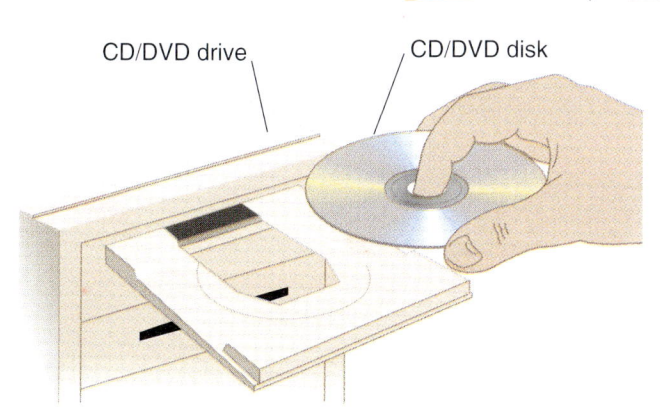

The system cabinet has lights on the front that indicate when these drives are in use. (You must not remove the diskette from the floppy-disk drive until the relevant light goes off, or else you risk damage to both disk and drive.) The wires for these lights need to be attached to the motherboard.

Output Hardware: Video & Sound Cards, Monitor, Speakers, & Printer

Output hardware consists of devices that translate information processed by the computer into a form that humans can understand—print, sound, graphics, or video, for example. Now a video card and a sound card need to be installed in the system cabinet. Next the monitor, speakers, and a printer are plugged in.

Incidentally, this is a good place to introduce the term *peripheral device*. A _peripheral device_ is any component or piece of equipment that **expands a computer's input, storage, and output capabilities**. Examples include printers and disk drives.

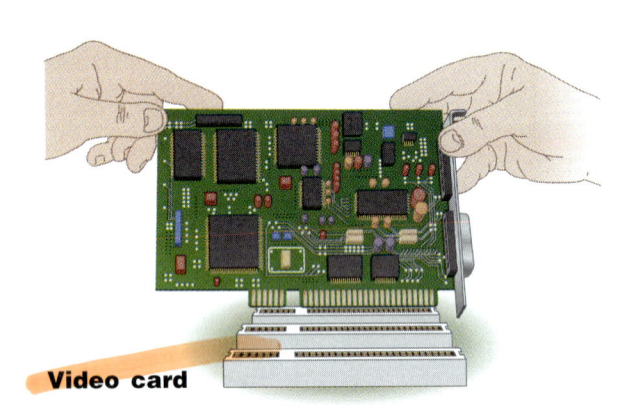

- **Video card:** $50–$650. You doubtless want your monitor to display color (rather than just black-and-white) images. Your system cabinet will therefore need to have a device to make this possible. A _video card_ **converts the processor's output information into a video signal that can be sent through a cable to the monitor.** Remember the expansion slots we mentioned? Your video card is plugged into one of these on the motherboard. (You can also buy a motherboard with built-in video.)

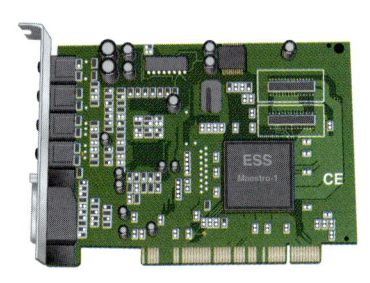

- **Sound card:** $30–$160. You may wish to listen to music on your PC. If so, you'll need a _sound card_, which **enhances the computer's sound-generating capabilities by allowing sound to be output through speakers.** This, too, would be plugged into an expansion slot on the motherboard. (Once again, you can buy a motherboard with built-in sound.) With the CD-ROM drive connected to the card, you can listen to music CDs.

- **Monitor:** $200–$400 for a 17-inch model, $400 or more for a 19-inch model. As with television sets, the "inch" dimension on monitors is measured diagonally corner to corner. **The _monitor_ is the display device that takes the electrical signals from the video card and forms an image using points of colored light on the screen.** Later, after the system cabinet has been closed up, the monitor will be connected by means of a cable to the back of the computer, using the clearly marked connector. The power cord for the monitor will be plugged into a wall plug.

- **Pair of speakers:** $30–$250. _Speakers_—the **devices that play sounds transmitted as electrical signals from the sound card**—may not be very sophisticated, but unless you're into high-fidelity recordings they're probably good enough. The two speakers are connected to a single wire that is plugged into the back of the computer once installation is completed.

- **Printer:** $80–$600. Especially for student work, you certainly need a _printer_, **an output device that produces text and graphics on paper.** There are various types of printers, as we discuss later (Chapter 5). The printer has two connections. One, which relays signals from the computer, goes to the back of the PC, where it connects with the motherboard. The other is a power cord that goes to a wall plug.

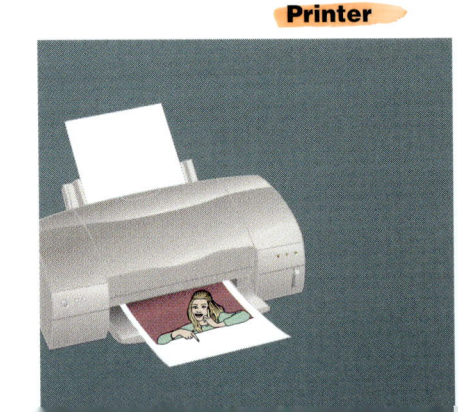

Communications Hardware: Modem

Computers can be stand-alone machines, unconnected to anything else. If all you're doing is word processing to write term papers, that may be fine. As we have seen, however, the *communications* component of the computer system *vastly* extends the range of a PC. Thus, while the system cabinet is still open there is one more piece of hardware to install.

- **Modem:** $40–$100. A <u>modem</u> is a device that sends and receives data over telephone lines to and from computers. The modem is mounted on an expansion card, which is fitted into an expansion slot on the motherboard. Later you can run a telephone line from the telephone wall plug to the back of the PC, where it will connect to the modem.

  Now the system cabinet is closed up. The person building the system will plug in all the input and output devices and turn on the power "on" button. Your microcomputer system will look similar to the one below. *(See ● Panel 1.3.)* Are you now ready to roll? Not quite.

Modem

Software

With all the pieces put together, the person assembling the computer needs to check the motherboard manual for instructions on starting the system. One of the most important tasks is to install software, of which there are two types—system software and application software.

- **System software:** You're really interested in the type of software that allows you to write documents, do spreadsheets, and so on, which is called application software. Before you can get to that, however, system software must be installed. Examples are Windows 98 or Windows XP, which come on CD disks. The assembler will insert these into your CD drive and follow the onscreen directions for installation. (*Installation* refers to the process of copying software programs from secondary storage media—CDs, for example—onto your system's hard disk, so that you can have direct access to the hardware.)

● **PANEL 1.3**
A completely assembled PC hardware system

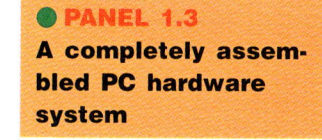

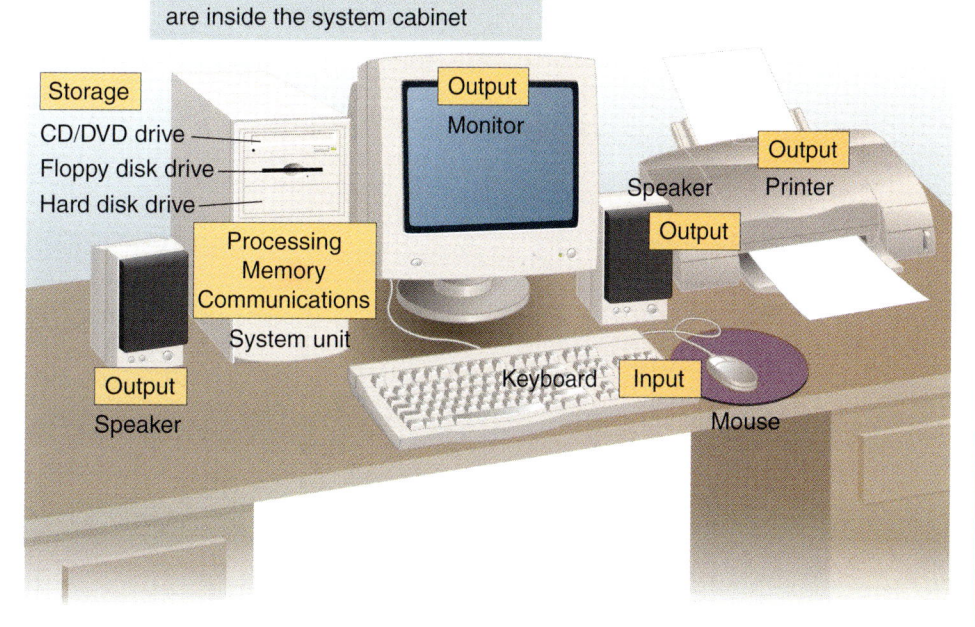

Survival Tip

Hardware Info

Go to *www.bizrate.com/marketplace* for a listing of virtually all types of hardware, their descriptions, ratings, prices, and names of sellers.

Processor, memory, hard-disk drive, video card, sound card, and modem are inside the system cabinet

Storage
- CD/DVD drive
- Floppy disk drive
- Hard disk drive

Processing / Memory / Communications — System unit

Output — Monitor
Output — Speaker
Output — Printer
Output — Speaker
Keyboard
Input — Mouse

Introduction to Information Technology

System software: Windows XP
Microsoft's Bill Gates holds a box of his new XP software.

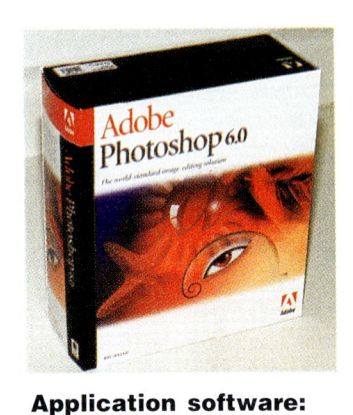

Application software: Adobe Photoshop
The program comes on a CD.

Survival Tip

Recycling Old PCs

Have a new computer? Where to donate your old one? Check with schools, after-school programs, churches. Pep Computer Recycling provides a list of recycling firms: www.microweb.com/pepsite/Recycle/National.html

Share the Technology repairs and upgrades old microcomputers to be donated to others: www.sharetechnology.org

System software helps the computer perform essential operating tasks and enables the application software to run. System software consists of several programs. The most important is the *operating system,* the master control program that runs the computer. Examples of operating system software for the PC are various Microsoft programs (such as Windows 95, 98, Me, XP, and NT/2000), Unix, and Linux. The Apple Macintosh computer is another matter altogether. As we explain in Chapters 3 and 4, it has its own hardware components and software, which aren't directly transferable to the PC, by and large.

After the system software is installed, setup software for the hard drive, the video and sound cards, and the modem must be installed. These setup programs (called *device drivers,* discussed in Chapter 3) will probably come on CDs (or maybe floppy disks). Once again, the installer inserts these into the drive and then follows the instructions that appear on the screen.

- **Application software:** *Now* we're finally getting somewhere! After the application software has been installed, you can start using the PC. *Application software* enables you to perform specific tasks—solve problems, perform work, or entertain yourself. For example, when you prepare a term paper on your computer, you will use a word processing program. (Microsoft Word and Corel WordPerfect are two brands.) Application software is specific to the system software you use. If you want to run Microsoft Word, for instance, you'll need to first have Microsoft Windows on your system, not Unix or Linux.

Application software comes on CDs or floppy disks. You insert them into your computer, and then follow the instructions on the screen for installation. Later on you may obtain entire application programs by getting them off the Internet, using your modem.

Although we have said a lot less about software than about hardware, they are equally important. We discuss software in much more detail in Chapter 3.

Is Getting a Custom-Built PC Worth the Effort?

Does the foregoing description make you want to try putting together a PC yourself? If you add up the costs of all the components (not to mention the value of your time), and then start checking ads for PCs, you might wonder why anyone would bother going to the trouble of building one. And nowadays you would probably be right. "[I]f you think you'd save money by putting together a computer from scratch," says David Einstein, "think again. You'd be lucky to match the price PC-makers are charging these days in their zeal to undercut the competition."[27]

But had you done this for real, it would not have been a futile exercise: By knowing how to build a system yourself, you'd not only be able to impress your friends, but you'd also know how to upgrade any store-bought system to include components that are better than standard. For instance, as Einstein points out, if you're into video games, knowing how to construct your own PC would enable you to make a system that's right for you. You could include the latest three-dimensional graphics video card and a state-of-the-art sound card, for example. More importantly, you'd also know how to *order* a custom-built system (as from Dell, Gateway, or Micron, the mail-order/online computer makers) that's right for you. In Chapters 4 and 5, we'll expand on this discussion so that you can *really know what you're doing* when you go shopping for a microcomputer system.

Before we end this introductory overview of information technology, let's wrap up the chapter with a look at the future.

> **CONCEPT CHECK**
>
> Describe the three key concepts of how computers work.
>
> Name the principal types of input, processing/memory, storage, output, and communications hardware.
>
> Describe the two types of software.

1.4 Where Is Information Technology Headed?

KEY QUESTIONS
What are three directions of computer development and three directions of communications development?

Considering going into hotel work? Then you'll want to know that the hottest new job there is "computer concierge," or *compcierge* (pay: $40,000-plus a year), someone with a knowledge of computer systems who helps the 55% of guests who travel with laptop computers when they have PC or online problems.

Or what about accounting? There the newest specialty is *e-commerce accountant* (pay: up to $54,250), who offers advice on when it makes financial sense to sell goods online, how to find Web developers, and how to manage credit card and sales records.

Today, a knowledge of computers coupled with training in another field offers interesting career paths. Other examples include *bioinformaticist* (biology and computer background, $100,000-plus salary), who studies gene maps, and *virtual set designer* (training in architecture and 3-D computer modeling; pay: up to $150,000), who designs sets for TV shows by combining décors built on computers with physical spaces.[28]

Clearly, information technology is changing old jobs and inventing new ones. To prosper in this environment, you will need to combine a traditional education with training in computers and communications. And you will need to understand what the principal trends of the Information Age are. Let's consider these trends in the development of computers and communications and, most excitingly, the area where they intersect.

Three Directions of Computer Development: Miniaturization, Speed, & Affordability

One of the first computers, the outcome of military-related research, was delivered to the U.S. Army in 1946. ENIAC (short for Electronic Numerical Integrator And Calculator) weighed 30 tons and was 80 feet long and two stories high, but it could multiply a pair of numbers in the then-remarkable time of three-thousandths of a second. This was the first general-purpose, programmable electronic computer, the grandparent of today's lightweight handheld machines.

Grandpa and offspring
ENIAC (left) is the grandfather of today's handheld machines (right).

Since the days of ENIAC, computers have developed in three directions—and are continuing to do so.

- **Miniaturization:** Everything has become smaller. ENIAC's old-fashioned radio-style vacuum tubes gave way to the smaller, faster, more reliable transistor. A *transistor* is a small device used as a gateway to transfer electrical signals along predetermined paths (circuits).

  The next step was the development of tiny integrated circuits. *Integrated circuits* are entire collections of electrical circuits or pathways that are now etched on tiny squares (chips) of silicon half the size of your thumbnail. *Silicon* is a natural element found in sand. In pure form, it is the base material for computer processing devices.

  The miniaturized processor, or microprocessor, in a personal desktop computer today can perform calculations that once required a computer filling an entire room.

- **Speed:** Thanks to miniaturization, computer makers can cram more hardware components into their machines, providing faster processing speeds and more data storage capacity.

- **Affordability:** Processor costs today are only a fraction of what they were 15 years ago. A state-of-the-art processor costing less than $1000 provides the same processing power as a huge 1980s computer costing more than $1 million.

These are the three major trends in computers. What about communications?

Microprocessor with integrated circuits (center of chip)

Three Directions of Communications Development: Connectivity, Interactivity, & Multimedia

Once upon a time, we had the voice telephone system—a one-to-one medium. You could talk to your Uncle Joe and he could talk to you, and with special arrangements (conference calls) more than two people could talk with one another. We also had radio and television systems—one-to-many media (or mass media). News announcers such as Dan Rather, Tom Brokaw, or Peter Jennings could talk to you on a single medium such as television, but you couldn't talk to them.

There have been three recent developments in communications:

- **Connectivity:** *Connectivity* **is the ability to connect computers to one another by communications line, so as to provide online information access.** The connectivity resulting from the expansion of computer networks has made possible e-mail and online shopping, for example.

- **Interactivity:** *Interactivity* **is about two-way communication; a user can respond to information he or she receives and modify the process.** That is, there is an exchange or dialogue between the user and the computer or communications device. The ability to interact means users can be active rather than passive participants in the technological process. On the television networks MSNBC or CNN, for example, you can immediately go on the Internet and respond to news from broadcast anchors. In the future, cars may respond to voice commands or feature computers built into the dashboard.

Auto PC
The Clarion Auto PC is the first personal computer designed for a car.

- **Multimedia:** Radio is a single-dimensional medium (sound), as is most e-mail (mainly text). As we mentioned earlier in this chapter, multimedia refers to technology that presents information in more than one medium—such as text, pictures, video, sound, and animation—in a single integrated communication. The development of the World Wide Web expanded the Internet to include pictures, sound, music, and so on, as well as text.

Exciting as these developments are, truly mind-boggling possibilities emerge as computers and communications cross-pollinate.

When Computers & Communications Combine: Convergence, Portability, & Personalization

Sometime in the 1990s, computers and communications started to fuse together, beginning a new era within the Digital Age. The result was three further developments, which haven't ended yet.

- **Convergence:** <u>Convergence</u> **describes the combining of several industries through various devices that exchange data in the format used by computers. The industries are computers, communications, consumer electronics, entertainment, and mass media.** Convergence has led to electronic products that perform multiple functions, such as TVs with Internet access or phones with screens displaying text and pictures.
- **Portability:** In the 1980s, portability, or mobility, meant trading off computing power and convenience in return for smaller size and weight. Today, however, we are close to the point where we don't have to give up anything. As a result, experts have predicted that small, powerful, wireless personal electronic devices will transform our lives far more than the personal computer has done so far. "[T]he new generation of machines will be truly personal computers, designed for our mobile lives," wrote one journalist back in 1992. "We will read office memos between strokes on the golf course and answer messages from our children in the middle of business meetings."[29] Today such activities are commonplace. The risk they bring is that, unless we're careful, work will invade our leisure time.
- **Personalization:** Personalization is the creation of information tailored to your preferences—for instance, programs that will automatically cull recent news and information from the Internet on just those topics you have designated. Companies involved in e-commerce can send you messages about forthcoming products based on your pattern of purchases, usage, and other criteria. Or they will build products (cars, computers, clothing) customized to your heart's desire.

Portability
A wearable computer

"E" Also Stands for Ethics

Ethics

Every computer user will have to wrestle with ethical issues related to the use of information technology. **_Ethics_ is defined as a set of moral values or principles that govern the conduct of an individual or a group.** Because ethical questions arise so often in connection with information technology, we will note them, wherever they appear in this book, with the symbol shown in the margin. Here, for example, are some important ethical concerns pointed out by Tom Forester and Perry Morrison in their book *Computer Ethics*.[30] These considerations are only a few of many; we'll discuss others in subsequent chapters.

- **Speed and scale:** Great amounts of information can be stored, retrieved, and transmitted at a speed and on a scale not possible before. Despite the benefits, this has serious implications "for data security and personal privacy," as well as employment, they say, because information technology can never be considered totally secure against unauthorized access.
- **Unpredictability:** Computers and communications are pervasive, touching nearly every aspect of our lives. However, at this point, compared to other pervasive technologies—such as electricity, television, and automobiles—information technology seems a lot less predictable and reliable.
- **Complexity:** Computer systems are often incredibly complex—some so complex that they are not always understood even by their creators. "This," say Forester and Morrison, "often makes them completely unmanageable," producing massive foul-ups or spectacularly out-of-control costs.

CONCEPT CHECK

Describe the three directions in which computers have developed.

Describe the three recent directions in which communications have developed.

Discuss three effects of the fusion of computers and communications.

Name three ethical considerations that result from information technology.

Onward: Handling Information in the Era of Pervasive Computing

The mountain of information threatens to overwhelm us. How will we cope? Consider the three challenges we mentioned at the start of this chapter:

- **Learn to deal with information overload:** The volume of available information far exceeds the amount of time needed to absorb it. To avoid being buried in an avalanche of unnecessary data, we must learn to distinguish what we *really* need from what we *think* we need. Throughout this book, we present suggestions for how to deal with too much information. In this first chapter, these are featured in the Practical Action Box ("Managing Your E-Mail," p. 5).
- **Have a strategy for what you memorize and what you don't:** Probably you're already feeling stressed about what you have to memorize. Passing the course for which this book is intended, for example, requires you to absorb and remember a number of facts. Other matters, such as phone numbers, Internet addresses, and birth dates,

can be stored on a cellphone or handheld computer. But don't entrust such information to just one device. The airlines, for instance, recover thousands of cellphones every month that passengers have misplaced (and that most don't claim because they've already bought a new one).[31] Thus, it's best to have your important phone numbers stored elsewhere as well.

- **Learn how to make your personal "multitasking" efficient:**
"Multitasking" describes a computer's ability to run several programs at once (as we describe in Chapter 3). However, it has also become a popular term for people performing several tasks at once, such as studying while eating, listening to music, talking on the phone, and handling e-mail. You may think you're one of those people who has no trouble juggling all this, but the brain has limits and can do only so much at one time. For instance, it has been found that people who do two demanding tasks simultaneously—drive in heavy traffic and talk on a cellphone, for instance—do neither task as well as they do each alone.[32] Indeed, the result of constantly shifting attention is a sacrifice in quality for *any* of the tasks with which one is engaged. This clearly has some consequences for studying.

This book offers a variation on multitasking that is designed to reinforce learning of the book. Whenever you see the mouse icon, such as the one shown at left, you're invited to go to our website and use your mouse to "click along" while reading the text. The material shown on the website won't take the place of the text, but it will add to it and help to reinforce the information—so that you'll better retain it in preparation for a test.

To see how the first "click-along" works, go to *www.mhhe.com/cit/uit5e/intro/clickalong*, and engage the material there. This first instance describes what we mean by multitasking and how it can be used as a positive feature to reinforce learning as you read this book.

CLICK-ALONG 1-1
Understanding your personal multitasking

Go to:
www.mhhe.com/cit/uit5e/intro/clickalong.

Summary

Note to the reader: "KQ" refers to Key Questions on the first page of each chapter. The number ties the summary term to the corresponding section in the book.

application software (p. 18, KQ 1.3) Software that has been developed to solve a particular problem, perform useful work on general-purpose tasks, or provide entertainment. Why it's important: *Application software such as word processing, spreadsheet, database manager, graphics, and communications packages are commonly used tools for increasing people's productivity.*

case (p. 13, KQ 1.3) Also known as the *system unit* or *system cabinet;* the box that houses the processor chip (CPU), the memory chips, and the motherboard with power supply, as well as storage devices—floppy-disk drive, hard-disk drive, and CD or DVD drive. Why it's important: *The case protects many important processing and storage components.*

CD drive (p. 15, KQ 1.3) Storage device that uses laser technology to read data from optical disks. Why it's important: *New software is generally supplied on CD-ROM disks rather than diskettes. And even if you can get a program on floppies, you'll find it easier to install a new program from one CD-ROM disk rather than repeatedly inserting and removing many diskettes. The newest version is called DVD (digital video disk). The DVD format stores even more data than the CD format.*

central processing unit (CPU) (p. 11, KQ 1.3) Device consisting of electronic circuitry that executes instructions to process data. Why it's important: *The CPU is the "brain" of the computer.*

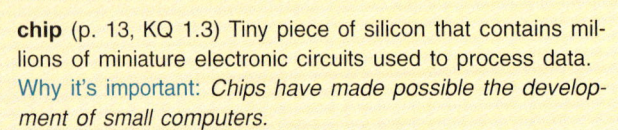

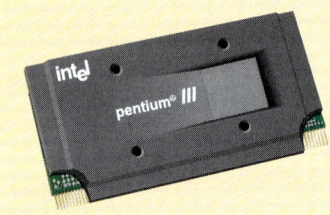

chip (p. 13, KQ 1.3) Tiny piece of silicon that contains millions of miniature electronic circuits used to process data. Why it's important: *Chips have made possible the development of small computers.*

clients (p. 9, KQ 1.2) Computers and other devices connected to a server, a central computer. Why it's important: *Client/server networks are used in many organizations for sharing databases, devices, and programs.*

communications technology (p. 3, KQ 1.1) Also called *telecommunications technology;* consists of electromagnetic devices and systems for communicating over long distances. Why it's important: *Communications systems using electronic connections have helped to expand human communication beyond face-to-face meetings.*

computer (p. 3, KQ 1.1) Programmable, multiuse machine that accepts data—raw facts and figures—and processes (manipulates) it into useful information, such as summaries and totals. Why it's important: *Computers greatly speed up problem solving and other tasks, increasing users' productivity.*

connectivity (p. 20, KQ 1.4) Ability to connect computers to one another by communications lines, so as to provide online information access. Why it's important: *Connectivity is the foundation of the advances in the digital age. It provides online access to countless types of information and services. The connectivity resulting from the expansion of computer networks has made possible e-mail and online shopping, for example.*

convergence (p. 21, KQ 1.4) The combining of several industries through various devices that exchange data in the format used by computers. The industries are computers, communications, consumer electronics, entertainment, and mass media. Why it's important: *Convergence has led to electronic products that perform multiple functions, such as TVs with Internet access or phones with screens displaying text and pictures.*

cyberspace (p. 4, KQ 1.1) Term used to refer to the online world and the Internet in particular but also the whole wired world of communications in general. Why it's important: *More and more human activities take place in cyberspace.*

data (p. 10, KQ 1.3) Raw facts and figures that are processed into information. Why it's important: *Users need data to create useful information.*

desktop PC (p. 8, KQ 1.2) Microcomputer unit that sits on a desk, with the keyboard in front and the monitor often on top. Why it's important: *Desktop PCs and tower PCs are the most commonly used types of microcomputer.*

DVD drive (p. 15, KQ 1.3) *See* CD drive.

e-mail (electronic mail) (p. 4, KQ 1.1) Messages transmitted over a computer network, most often the Internet. Why it's important: *E-mail has become universal; one of the first things new computer users learn is how to send and receive e-mail.*

ethics (p. 22, KQ 1.4) Set of moral values or principles that govern the conduct of an individual or a group. Why it's important: *Ethical questions arise often in connection with information technology.*

Ethics

expansion slots (p. 13, KQ 1.3) Internal "plugs" used to expand the PC's capabilities. Why it's important: *Expansion slots give you places to plug in additional circuit boards, such as those for video, sound, and communications (modem).*

floppy-disk drive (p. 15, KQ 1.3) Storage device that stores data on removable 3.5-inch-diameter flexible diskettes encased in hard plastic. Why it's important: *Floppy-disk drives are included on almost all microcomputers and make many types of files portable.*

floppy disk

floppy disk drive

hard-disk drive (p. 15, KQ 1.3) Storage device that stores billions of characters of data on a nonremovable disk platter inside the computer case. Why it's important: *Hard disks hold much more data than diskettes do. Nearly all microcomputers use hard disks as their principal secondary-storage medium.*

hardware (p. 10, KQ 1.3) All machinery and equipment in a computer system. Why it's important: *Hardware runs under the control of software and is useless without it. However, hardware contains the circuitry that allows processing.*

information (p. 10, KQ 1.3) Data that has been summarized or otherwise manipulated for use in decision making. Why it's important: *The whole purpose of a computer (and communications) system is to produce (and transmit) usable information.*

information technology (p. 2, KQ 1.1) Technology that merges computing with high-speed communications links carrying data, sound, and video. Why it's important: *Information technology is bringing about the fusion of several important industries dealing with computers, telephones, televisions, and various handheld devices.*

input (p. 10, KQ 1.3) Whatever is put in ("input") to a computer system. Input devices include the keyboard and the mouse. Why it's important: *Useful information cannot be produced without input data.*

interactivity (p. 20, KQ 1.4) Two-way communication; a user can respond to information he or she receives and modify the process. Why it's important: *Interactive devices allow the user to actively participate in a technological process instead of just reacting to it.*

Internet (p. 5, KQ 1.1) Worldwide network that connects hundreds of thousands of smaller networks linking computers at academic, scientific, and commercial institutions. Why it's important: *Thanks to the Internet, millions of people around the world can share all types of information and services.*

Introduction to Information Technology

25

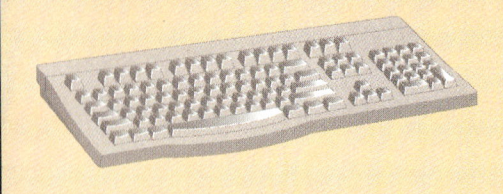

keyboard (p. 12, KQ 1.3) Input device that converts letters, numbers, and other characters into electrical signals readable by the processor. *Why it's important:* Keyboards are the most common kind of input device.

local area network (LAN) (p. 8, KQ 1.2) Network that connects, usually by special cable, a group of desktop PCs and other devices, such as printers, in an office or building. *Why it's important:* LANs have replaced mainframes for many functions and are considerably less expensive.

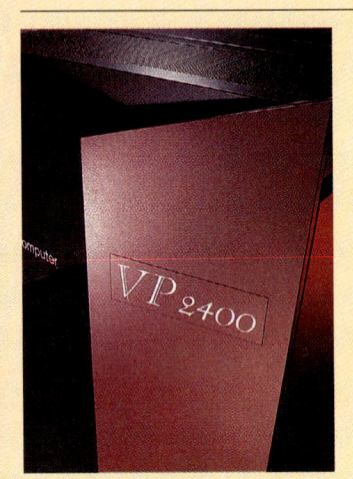

mainframe (p. 8, KQ 1.2) Second-largest computer available, after the supercomputer; capable of great processing speeds and data storage. Costs $5000–$5 million. Small mainframes are often called *midsize computers*. *Why it's important:* Mainframes are used by large organizations (banks, airlines, insurance companies, universities) that need to process millions of transactions.

memory chip (p. 13, KQ 1.3) Also known as *RAM* (for *random access memory*) chip; represents *primary storage* or *temporary storage*. *Why it's important:* Holds data before processing and information after processing, before it is sent along to an output or storage device.

microcomputer (p. 8, KQ 1.2) Also called *personal computer;* small computer that fits on or next to a desktop, or can be carried around. Costs $500–$5000. *Why it's important:* The microcomputer has lessened the reliance on mainframes and has provided more ordinary users with access to computers. It can be used as a stand-alone machine or connected to a network.

microcontroller (p. 9, KQ 1.2) Also called an *embedded computer;* the smallest category of computer. *Why it's important:* Microcontrollers are built into "smart" electronic devices, such as appliances and automobiles.

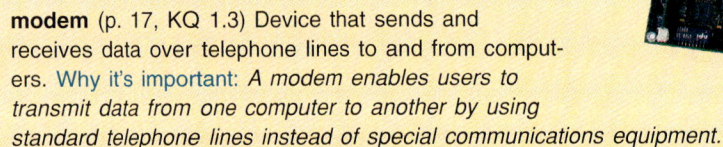

modem (p. 17, KQ 1.3) Device that sends and receives data over telephone lines to and from computers. *Why it's important:* A modem enables users to transmit data from one computer to another by using standard telephone lines instead of special communications equipment.

monitor (p. 16, KQ 1.3) Display device that takes the electrical signals from the video card and forms an image using points of colored light on the screen. *Why it's important:* Monitors enable users to view output without printing it out.

motherboard (p. 13, KQ 1.3) Main circuit board in the computer. *Why it's important:* This is the big green circuit board to which everything else—such as the keyboard, mouse, and printer—is attached. The processor chip and memory chips are also installed on the motherboard.

mouse (p. 12, KQ 1.3) Input device used to manipulate objects viewed on the computer display screen. *Why it's important:* For many purposes, a mouse is easier to use than a keyboard for inputting commands. Also, the mouse is used extensively in many graphics programs.

multimedia (p. 5, KQ 1.1) From "multiple media"; technology that presents information in more than one medium, including text graphics, animation, video, and sound. *Why it's important:* Multimedia is used increasingly in business, the professions, and education to improve the way information is communicated.

network (p. 4, KQ 1.1) Communications system connecting two or more computers. *Why it's important:* Networks allow users to share applications and data and to use e-mail. The Internet is the largest network.

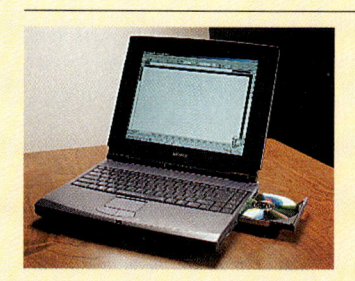

notebook computer (p. 9, KQ 1.2) Also called *laptop computer;* lightweight portable computer with a built-in monitor, keyboard, hard-disk drive, battery, and adapter; weighs 1.8–9 pounds. Why it's important: *Notebook and other small computers have provided users with computing capabilities in the field and on the road.*

online (p. 3, KQ 1.1) Using a computer or other information device, connected through a network, to access information and services from another computer or information device. Why it's important: *Online communication is widely used by businesses, services, individuals, and educational institutions.*

output (p. 11, KQ 1.3) Whatever is output from ("put out of") the computer system; the results of processing. Why it's important: *People use output to help them make decisions. Without output devices, computer users would not be able to view or use the results of processing.*

peripheral device (p. 16, KQ 1.3) Any component or piece of equipment that expands a computer's input, storage, and output capabilities. Examples include printers and disk drives. Why it's important: *Most computer input and output functions are performed by peripheral devices.*

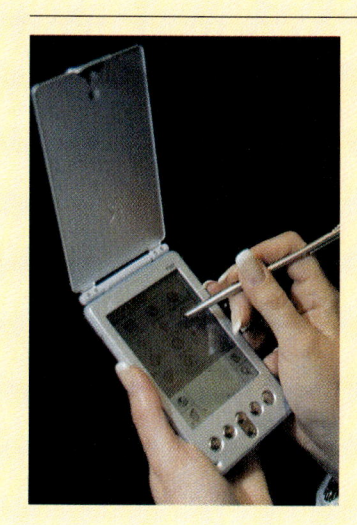

personal digital assistant (PDA) (p. 9, KQ 1.2) Also known as *handheld computer* or *palmtop;* used as a schedule planner and address book and to prepare to-do lists and send e-mail and faxes. Why it's important: *PDAs make it easier for people to do business and communicate while traveling.*

primary storage (p. 11, KQ 1.3) Also called *memory;* computer circuitry that *temporarily* holds data waiting to be processed. Why it's important: *By holding data, primary storage enables the processor to process.*

printer (p. 16, KQ 1.3) Output device that produces text and graphics on paper. Why it's important: *Printers provide one of the principal forms of computer output.*

processing (p. 11, KQ 1.3) The manipulation the computer does to transform data into information. Why it's important: *Processing is the essence of the computer, and the processor is the computer's "brain."*

secondary storage (p. 11, KQ 1.3) Also called *storage;* devices and media that store data and programs *permanently*—such as disks and disk drives, tape and tape drives, CDs and CD drives. Why it's important: *Without secondary storage, users would not be able to save their work. Storage also holds the computer's software.*

server (p. 9, KQ 1.2) Computer in a network that holds collections of data (databases) and programs for connecting PCs, workstations, and other devices, which are called *clients.* Why it's important: *Servers enable many users to share equipment, programs, and data.*

software (p. 10, KQ 1.3) Also called *programs;* step-by-step instructions that tell the computer hardware how to perform a task. Why it's important: *Without software, hardware is useless.*

sound card (p. 16, KQ 1.3) Special circuit board that enhances the computer's sound-generating capabilities by allowing sound to be output through speakers. Why it's important: *Sound is used in multimedia applications. Also, many users like to listen to music CDs on their computers.*

speakers (p. 16, KQ 1.3) Devices that play sounds transmitted as electrical signals from the sound card. Speakers are connected to a single wire plugged into the back of the computer. Why it's important: *See* sound card.

supercomputer (p. 7, KQ 1.2) High-capacity computer with hundreds of thousands of processors that is the fastest calculating device ever invented. Costs $85 million or more. Why it's important: *Supercomputers are used primarily for research purposes, airplane design, oil exploration, weather forecasting, and other activities that cannot be handled by mainframes and other less powerful machines.*

system software (p. 18, KQ 1.3) System software helps the computer perform essential operating tasks. Why it's important: *Application software cannot run without system software. System software consists of several programs. The most important is the operating system, the master control program that runs the computer. Examples of operating system software for the PC are various Microsoft programs (such as Windows 95, 98, NT, Me, and XP), Unix, Linux, and the Macintosh operating system.*

terminal (p. 8, KQ 1.2) Input and output device that uses a keyboard for input and a monitor for output; it cannot process data. Why it's important: *Terminals are generally used to input data to and receive data from a mainframe computer system.*

tower PC (p. 8, KQ 1.2) Microcomputer unit that sits as a "tower" often on the floor, freeing up desk space. Why it's important: *Tower PCs and desktop PCs are the most commonly used types of microcomputer.*

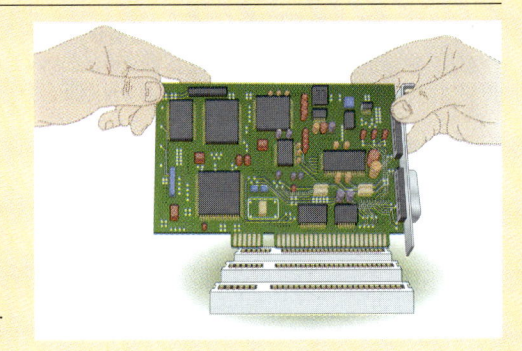

video card (p. 16, KQ 1.3) Circuit board that converts the processor's output information into a video signal for transmission through a cable to the monitor. Why it's important: *Virtually all computer users need to be able to view video output on the monitor.*

workstation (p. 8, KQ 1.2) Smaller than a mainframe; expensive, powerful computer generally used for complex scientific, mathematical, and engineering calculations and for computer-aided design and computer-aided manufacturing. Why it's important: *The power of workstations is needed for specialized applications too large and complex to be handled by PCs.*

World Wide Web (p. 5, KQ 1.1) The part of the Internet that stores information in multimedia form—sounds, photos, and video as well as text. Why it's important: *The Web is the most widely known part of the Internet.*

Zip-disk drive (p. 15, KQ 1.3) Storage device that stores data on floppy-disk cartridges with at least 70 times the capacity of the standard floppy. Why it's important: *Zip drives are used to store large files.*

Chapter Review

More and more educators are favoring an approach to learning (presented by Benjamin Bloom and his colleagues in *Taxonomy of Educational Objectives*) that follows a hierarchy of six critical-thinking skills: (a) two lower-order skills—*memorization* and *comprehension;* and (b) four higher-order skills—*application, analysis, synthesis,* and *evaluation.* While you may be able to get through many introductory college courses by simply memorizing facts and comprehending the basic ideas, to advance further you will probably need to employ the four higher-order thinking skills.

In the Chapter Review following the end of each chapter, we have implemented this hierarchy in a three-stage approach, as follows:

- **Stage 1 learning—memorization:** "I can recognize and recall information." Self-test questions, multiple-choice questions, and true/false questions enable you to test how well you recall basic terms and concepts.
- **Stage 2 learning—comprehension:** "I can recall information in my own terms and explain them to a friend." Using open-ended short-answer questions, we ask you to re-express terms and concepts in your own words.
- **Stage 3 learning—applying, analyzing, synthesizing, evaluating:** "I can apply what I've learned, relate these ideas to other concepts, build on other knowledge, and use all these thinking skills to form a judgment." In this part of the Chapter Review, we ask you to put the ideas into effect using the activities described, some of which include Internet activities. The purpose is to help you take possession of the ideas, make them your own, and apply them realistically to your life.

stage 1 LEARNING MEMORIZATION

"I can recognize and recall information."

Self-Test Questions

1. The _____ refers to the part of the Internet that stores information in multimedia form.
2. _____ and _____ refer to the two types of microcomputer. One sits on the desktop and the other usually is placed on the floor.
3. _____ technology merges computing with high-speed communications lines carrying data, sound, and video.
4. A _____ is a programmable, multiuse machine that accepts data and processes it into information.
5. Messages transmitted over a computer network are called _____.
6. The _____ is a worldwide network that connects hundreds of thousands of smaller networks.
7. _____ refers to technology that presents information in more than one medium.
8. _____ are high-capacity machines with hundreds of thousands of processors.
9. Embedded computers, or _____, are installed in "smart" appliances and automobiles.
10. The kind of software that enables users to perform specific tasks is called _____ software.

Multiple-Choice Questions

1. Which of the following converts computer output into displayed images?
   a. printer
   b. monitor
   c. floppy-disk drive
   d. processor
   e. hard disk drive
2. Which of the following computer types is the smallest?
   a. mainframe
   b. microcomputer
   c. microcontroller
   d. supercomputer
   e. workstation
3. Which of the following is a secondary storage device?
   a. processor
   b. memory chip
   c. floppy-disk drive
   d. printer
   e. monitor

4. Since the days when computers were first made available, computers have developed in three directions. What are they?

   a. increased expense
   b. miniaturization
   c. increased size
   d. affordability
   e. increased speed

5. Which of the following operations constitute the *four basic operations* followed by all computers?

   a. input
   b. storage
   c. programming
   d. output
   e. processing

True/False Questions

T F 1. Mainframe computers process faster than microcomputers.

T F 2. Main memory is a software component.

T F 3. The operating system is part of the system software.

T F 4. Processing is the manipulation by which a computer transforms data into information.

T F 5. Primary storage is the area in the computer where data or information is held permanently.

stage 2 LEARNING — COMPREHENSION

"I can recall information in my own terms and explain them to a friend."

Short-Answer Questions

1. What does *online* mean?
2. What is the difference between system software and application software?
3. Briefly define *cyberspace*.
4. What is the difference between software and hardware?
5. Briefly describe what a local area network is.

stage 3 LEARNING — APPLYING, ANALYZING, SYNTHESIZING, EVALUATING

"I can apply what I've learned, relate these ideas to other concepts, build on other knowledge, and use all these thinking skills to form a judgment."

Knowledge in Action

1. Do you wish there was an invention to make your life easier or better? Describe it. What would it do for you?

2. Determine what types of computers are being used where you work or go to school. In which departments are the different types of computer used? What are they used for? How are they connected to other computers?

3. Imagine a business you could start or run at home. What type of business is it? What type of computer(s) do you think you'll need? Describe the computer system in as much detail as possible, including hardware components in the areas we have discussed so far. Keep your notes and then refine your answers after you have completed the course.

Web Exercises

1. Are computers, cellphones, and other electronic devices bad for our health? You may have heard the term *electromagnetic radiation* and dismissed it as an obscure scientific term not worth understanding. Visit the links below to become educated on a topic that will be discussed more seriously and frequently when our society becomes completely wireless.

   http://gladstone.uoregon.edu/~dgibbens/no_denial/em_radiation.html
   www.nzine.co.nz/articles/Electromagnetic_Radiation/
   www.eos.ncsu.edu/eos/info/computer_ethics/social/workplace/emr/study.html

2. List the pros and cons of a paperless environment. Do you feel this challenge is something to strive for? Run a Web search on what others are doing to implement this idea in their workplaces.

3. Computer pioneer John Von Neumann was one of a group of individuals who conceived the idea of the "stored program." He could also divide two 8-digit numbers in his head. Spend a few hours researching this remarkable man. Add the words Tesla, Einstein, Montauk, Bielek, or Teller to the search, and you will receive bizarre search results. Had you heard of John Von Neumann before reading this textbook?

4. Looking for legally free programs? Some great places to start:

   www.download.com
   www.shareware.com
   www.freeware.com

   (They're all essentially part of *www.cnet.com*.)

Chapter 2

The Internet & the World Wide Web

Exploring Cyberspace

Key Questions

You should be able to answer the following questions.

2.1 **Choosing Your Internet Access Device & Physical Connection: The Quest for Broadband** What are the means of connecting to the Internet, and how fast are they?

2.2 **Choosing Your Internet Service Provider (ISP)** What is an Internet service provider, and what kinds of services do ISPs provide?

2.3 **Sending & Receiving E-Mail** What are the options for obtaining e-mail software, what are the components of an e-mail address, and what are netiquette and spam?

2.4 **The World Wide Web** What are websites, web pages, browsers, URLs, and search engines?

2.5 **The Online Gold Mine: More Internet Resources, Your Personal Cyberspace, E-Commerce, & the E-conomy** What are FTP, Telnet, newsgroups, real-time chat, and e-commerce?

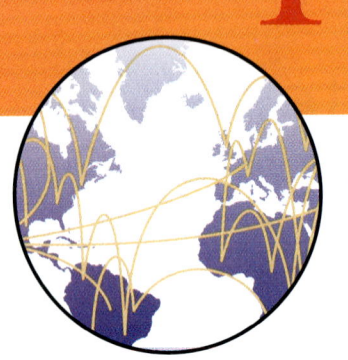

T**he immensity of the changes wrought—and still to come—cannot be underestimated," says futurist Graham Molitor. "This miraculous information channel—the Internet—will touch and alter virtually every facet of humanity, business, and all the rest of civilization's trappings."**[1]

In 1999, there were about 131 million active Internet users worldwide. By 2003, there will be close to 362 million. In 2001, there were about 100 million Internet devices of various kinds. By 2010, there will be an estimated 14 billion.[2]

Today the world of the Internet permits activities hardly imaginable 10 years ago. *(See ● Panel 2.1.)* In 10 more years, pervasive computing will be even more an established fact. By then we will have not only people-to-people connections with cellphones and pagers and wireless personal digital assistants (PDAs). We will also have "everything connected to everything"—for instance, Internet-based remote control devices to regulate your home's climate, lighting, security, and home-entertainment systems, even your refrigerator and oven and other appliances—even, in fact, your car.[3]

Because of its standard interfaces and low rates, the Internet has been the great leveler for communications—just as the personal computer was for computing. Starting in 1969 with four computers linked together by communications lines, the Internet expanded to 62 computers in 1974, 500 computers in 1983, and 28,000 in 1987; but it still remained the domain of researchers and academics. Not until the development of the World Wide Web in the early 1990s, which made multimedia available on the Internet, and the first browser, which opened the Web to commercial uses, did the global network really take off, swelling to 3 million servers or host computers in 1994. *(See ● Panel 2.2 on pages 34–35 for abbreviated history.)* If, as some forecasters believe, in 2005 the Internet reaches 1 billion users, that will be big enough, says Molitor, "to dub the virtual world of the Internet as the eighth continent!"[4] And no one nation, company, or entity really owns it.

How can you become a "citizen of the world," as it were, in this network of networks? To access the Internet, you need three things: (1) an *access device*, such as a personal computer with a modem; (2) a *physical connection*, such as a telephone line; and (3) an *Internet service provider (ISP)*. We cover these subjects in the next two sections.

2.1 Choosing Your Internet Access Device & Physical Connection: The Quest for Broadband

KEY QUESTIONS
What are the means of connecting to the Internet, and how fast are they?

Call it the Bandwidth Gap. In general terms, **_bandwidth_ is an expression of how much data—text, voice, video, and so on—can be sent through a communications channel in a given amount of time.** A college dormitory wired with coaxial or fiber-optic cable for high-speed Internet access will have more bandwidth than will a house out in the country served by conventional copper-wire telephone lines. Access to information will be hundreds of times faster in this kind of dorm. Of course, many students are not privileged to have this kind of **_broadband_—very high speed—connection** (a topic we consider in detail in Chapter 6). But so significant is the difference that some students apparently are making decisions about college housing—even choice of college—based on the availability of high bandwidth. And those who are about to graduate wonder how they will ever survive in a narrow-bandwidth world.

● **PANEL 2.1**
The world of the Internet

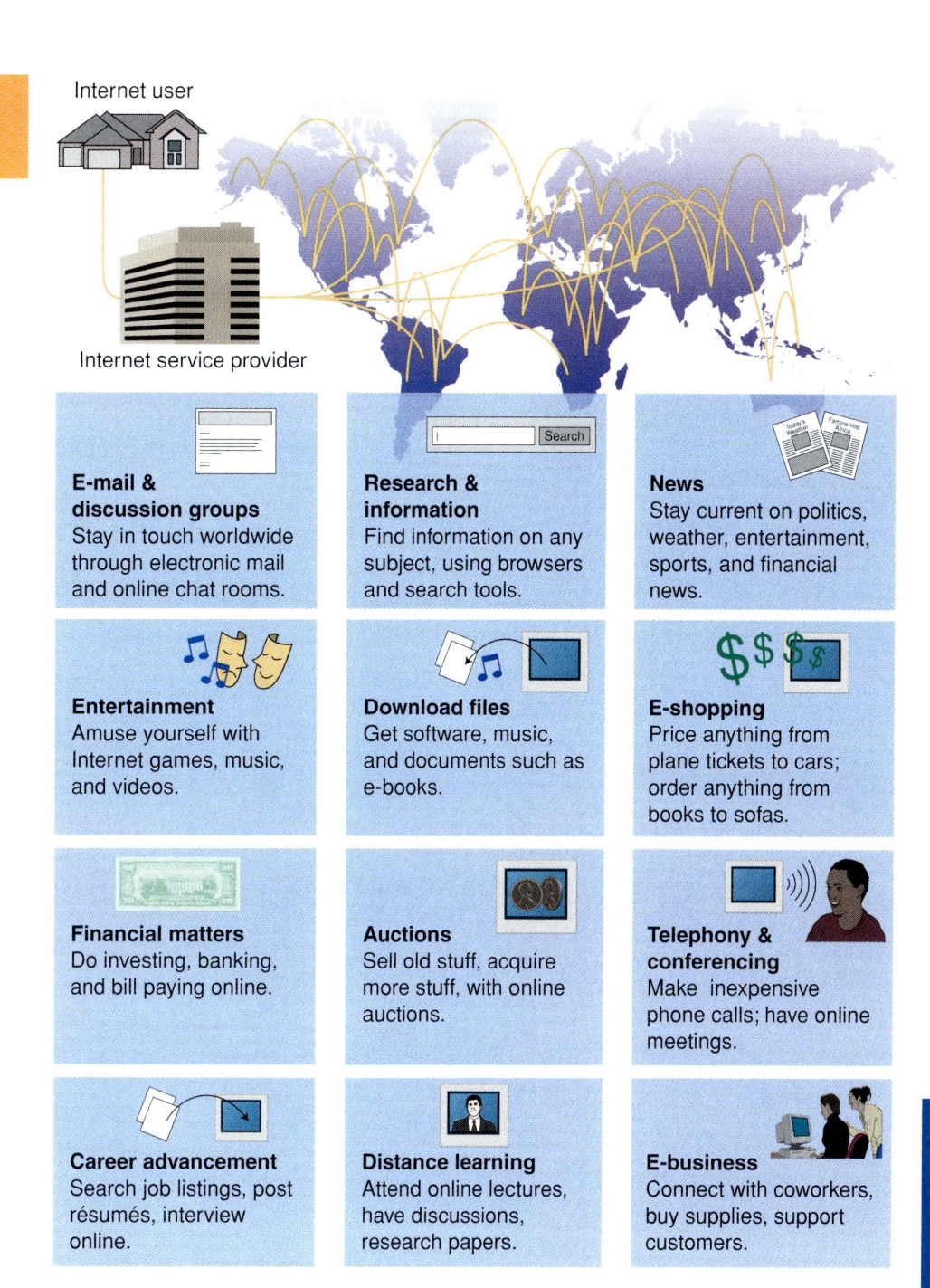

Internet user

Internet service provider

E-mail & discussion groups
Stay in touch worldwide through electronic mail and online chat rooms.

Research & information
Find information on any subject, using browsers and search tools.

News
Stay current on politics, weather, entertainment, sports, and financial news.

Entertainment
Amuse yourself with Internet games, music, and videos.

Download files
Get software, music, and documents such as e-books.

E-shopping
Price anything from plane tickets to cars; order anything from books to sofas.

Financial matters
Do investing, banking, and bill paying online.

Auctions
Sell old stuff, acquire more stuff, with online auctions.

Telephony & conferencing
Make inexpensive phone calls; have online meetings.

Career advancement
Search job listings, post résumés, interview online.

Distance learning
Attend online lectures, have discussions, research papers.

E-business
Connect with coworkers, buy supplies, support customers.

Let's assume that you have no access to a campus network. What are your choices of a *physical connection*—the wired or wireless means of connecting to the Internet? A lot depends on where you live. As you might expect, urban and many suburban areas offer more broadband connections than rural areas do. Among the principal means of connection are (1) telephone (dial-up) modem; (2) several high-speed phone lines—ISDN, DSL, and T1; (3) cable modem; and (4) wireless—satellite and other through-the-air links.

Data is transmitted in characters or collections of bits. A *bit*, as we will discuss later, is the smallest unit of information used by computers. Today's data transmission speeds are measured in bits, kilobits, megabits, and gigabits per second.

- **bps:** A computer with an older modem might have a speed of 28,800 bps, which is considered the minimum speed for visiting websites with graphics. The ***bps*** stands for ***bits per second***. (Eight bits equals one character.)

Satellite

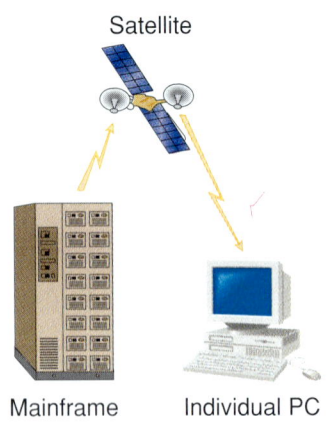

Mainframe   Individual PC

Download
(reverse the direction of data transmission to **upload**)

- **Kbps:** This is the most frequently used measure; *kilobits per second*, or *Kbps*, **are 1000 bits per second.** The speed of a modem that is 28,800 bps might be expressed as 28.8 Kbps.
- **Mbps:** Faster means of connection are measured in *megabits per second*, or *Mbps*—1 million bits per second.
- **Gbps:** At the extreme are *gigabits per second, Gbps*—1 billion bits per second.

Why is it important to know these terms? Because the number of bits affects how fast you can upload and download information from a remote computer. **Download is the transmission of data from a remote computer to a local computer,** as from your college's mainframe to your own PC. **Upload is the transmission of data from a local computer to a remote computer.**

The table opposite shows the transmission rates for various connections, as well as the approximate costs (always subject to change, of course) and their pros and cons.[5] *(See ● Panel 2.3.)* Let's consider each option in turn.

Telephone (Dial-Up) Modem: Low Speed but Inexpensive & Widely Available

The telephone line that you use for voice calls is still the cheapest means of online connections and is available everywhere. As we discussed in Chapter 1, a *modem* is a device that sends and receives data over telephone lines to and from computers. This is known as a *dial-up* connection. These days, the modem is generally installed inside your computer, but there are also external modems. (We explain modems further in Chapter 6.) The modem is attached to the telephone wall outlet. *(See ● Panel 2.4 on page 36.)*

Most modems today have a maximum speed of 56 Kbps. That doesn't mean that you'll be sending and receiving data at that rate. The modem in your computer must negotiate with the modems used by your Internet service provider (ISP), the organization that actually connects you to the Internet, which may have modems operating at slower speeds, such as 28.8 Kbps. In addition, lower-quality phone lines or heavy traffic during peak hours—such as 5 p.m. to 11 p.m. in residential areas—can slow down your rate of transmission.

One disadvantage of a telephone modem is that while you're online you can't use that phone line to make voice calls. In addition, people who try to call you while you're using the modem will get a busy signal. (Call waiting will interrupt an online connection, so you need to talk to your phone

● **PANEL 2.2**
An abbreviated history of the Internet

Origins of the Internet

Early 1960s	Early 1970s	1975	1980
ARPA (Advanced Research Projects Agency): U.S. Defense Department's research organization studies advanced technology that could be used to defend the United States; develops many large databases	ARPANET: ARPA developed a networked communications system that couldn't be knocked out by eliminating computers or links in the system. They also developed the rules by which data was transmitted. By the early 1970s, ARPANET had grown from 4 networked research locations to 20 military sites and universities.	ARPANET was transferred to the U.S. Defense Communications Agency, thus restricting network access to only a few groups.	The National Science Foundation (NSF) started CSnet to provide a network opportunity for computer science researchers at all U.S. universities. By 1986, almost all the country's computer science departments, as well as some private companies, were connected to CSnet.

● **PANEL 2.3**
Methods of going online compared

Service	Cost per Month (Plus Installation, Equipment)	Maximum Speed (Download Only)	Pluses	Minuses
Telephone (dial-up) modem	$0–$30	56 Kbps	Inexpensive, available everywhere	Slow, connection supports only a single user
ISDN	$40–$110 (+ $350–$700 installation cost)	128 Kbps	Faster than dial-up, uses conventional phone lines	More expensive than dial-up, connection supports only a single user, no longer extensively supported by telephone companies
DSL	$40–$300 (+ $100–$200 installation cost and $100–$300 for a special modem)	1.5–8.4 Mbps	Fast download, always on, higher security	Need to be close to phone company switching station, limited choice of service providers, poor customer support, supports only a single user
T1 line	$1500 (+ $1000 installation cost)	1.5 Mbps	Can support many users: 24 separate circuits of 64 Kbps each, reliable high speed both ways	Expensive, best for businesses
Cable modem	$100–$400 (+ installation equipment $300–$500)	10 Mbps	Fast, always on, most popular broadband type of connection, can support many users	Slower service during high-traffic times, vulnerability to hackers, limited choice of service providers, not always available to businesses
Satellite	$30–$130	400 Kbps	Wireless, fast, reliable	Slow (56 Kbps) uploads over phone lines

Late 1980s
After 5 supercomputer centers were built across the United States, the NSF built a very fast connection—called a backbone—among them. Regional companies, schools, and other organizations built their own regional networks and connected them to the backbone.

By 1989
ARPANET had become too expensive and had outlived its usefulness; it was closed down, and many of its sites were connected to the NSF backbone. This vast inter-network became known as the Internet.

By 1995
In its early stages, the Internet was used mainly for research and scientific purposes. Soon, however, it was recognized as a revolutionary information resource, and in 1995 it became known as the Information Superhighway. In 1992, multimedia information became available via the World Wide Web.

1999
Internet access had become virtually universal.

● **PANEL 2.4**

The modem connection

You connect the modem inside your computer from a port (socket) in the back of your computer to a line that is then connected to a wall jack. Your telephone is also connected to your computer so that you can make voice calls.

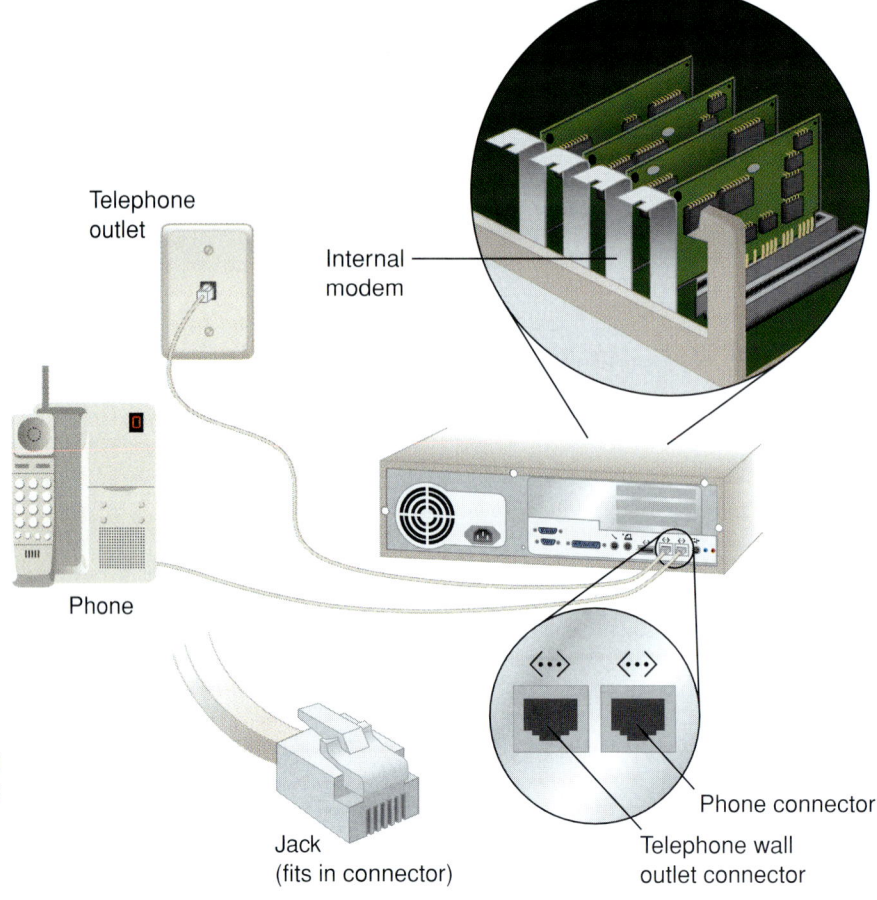

Check online connections

If you're using Windows on your computer, you can check your online connection speed by going to the taskbar and double-clicking the connection icon (bottom right of screen):

The result:

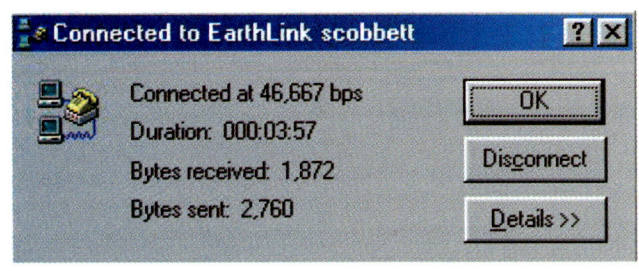

company about disabling it. The Windows operating system also has a feature for disabling call waiting.) As we discuss in a few pages, you probably won't need to pay long-distance phone rates, since most ISPs offer local access numbers. The cost of a dial-up modem connection to the ISP is $10–$30 per month, plus a possible setup charge of $10–$25.

High-Speed Phone Lines: More Expensive but Available in Most Cities

Waiting while your computer's modem takes 25 minutes to transmit a 1-minute low-quality video from a website may have you pummeling the desk in frustration. To get some relief, you could enhance your POTS (for *plain old telephone system*) connection with a high-speed adaptation or get a new, dedicated line. Among the choices are ISDN, DSL, and T1, available in most major cities, though not in rural and many suburban areas.

- **ISDN line:** <u>**ISDN (Integrated Services Digital Network)**</u> **consists of hardware and software that allows voice, video, and data to be communicated over traditional copper-wire telephone lines.** Capable of transmitting up to 128 Kbps, ISDN is able to send signals over POTS lines. If you were trying to download an approximately 6-minute-long music video from the World Wide Web, it would take you about 4 hours and 45 minutes with a 28.8-Kbps modem. An ISDN connection would reduce this to an hour.

ISDN costs $10–$40 a month to the phone company, plus perhaps another $30–$70 per month to the Internet service provider. In addition, you may need to pay your phone company $350–$700 or so to hook up an ISDN connector box, possibly run in a new phone line, and install the necessary software in your PC. We need to point out that many phone companies are no longer strongly supporting ISDN services, because many users prefer DSL and cable modem.

- **DSL line:** **_DSL (digital subscriber line)_ also uses regular phone lines to transmit data in megabits per second.** Incoming data is significantly faster than outgoing data. That is, your computer can *receive* data at the rate of 1.5–8.4 Mbps, but it can *send* data at only 16–640 Kbps. This arrangement may be fine, however, if you're principally interested in obtaining very large amounts of data (video, music) rather than in sending them to others. With DSL, you can download that 6-minute music video in only 11 minutes (compared to an hour with ISDN). A big advantage of DSL is that it is always on and, unlike cable (discussed shortly), its transmission rate is consistent. One-time installation cost is $100–$200 plus $100–$300 for a modem supplied by the phone company, and the monthly cost is $40–$300.

    There is one big drawback to DSL: You have to live within 3.3 miles of a phone company central switching office, because the access speed and reliability degrade with distance. However, phone companies are building thousands of remote switching facilities to enhance service throughout their regions. Another drawback is that you have to choose from a list of Internet service providers that are under contract to the phone company you use, although other DSL providers exist.

- **T1 line:** How important is high speed to you? Is it worth $1500 a month? Then consider getting a **_T1 line_, essentially a traditional trunk line that carries 24 normal telephone circuits and has a transmission rate of 1.5 Mbps.** Generally, T1 lines are used by corporate, government, and academic sites. Another high-speed line, the T3 line, which transmits at 44 Mbps, costs $10,000 or more a month. Telephone companies and other types of companies are making even faster connections available: An STS-1 connection runs at 51 Mbps, and an STS-48 connection speeds data along at 2.5 Gbps (2.5 billion bits per second.)

Cable Modem: Close Competitor to DSL

If DSL's 11 minutes to move a 6-minute video sounds good, 2 minutes sounds even better. That's the rate of transmission for cable modems, which can transmit outgoing data at 500 Kbps and incoming data at 10 Mbps (and eventually, it's predicted, at 30 Mbps). **A _cable modem_ connects a personal computer to a cable-TV system that offers an Internet connection.** Like a DSL connection, it is always on; unlike DSL, you don't need to live near a switching station. Costing $100–$400 a month plus installation and equipment (usually nominal cost), cable is available in most major cities.

A disadvantage, however, is that you and your neighbors are sharing the system and consequently, during peak-load times, your service may be slowed to the speed of a regular dial-up modem. (You're also more vulnerable to attacks from hackers, although there are defensive or "firewall" programs that reduce the risk.) Finally, cable companies may force you to use their own Internet service providers.

Wireless Systems: Satellite & Other Through-the-Air Connections

Suppose you live out in the country and you're tired of the molasses-like speed of your cranky local phone system. You might consider taking to the air.

- **Satellite:** With a pizza-size satellite dish on your roof, you can receive data at the rate of 400 Kbps from a **_communications satellite_, a space station that transmits radio waves called microwaves from earth-based stations.** Unfortunately, your outgoing transmission will still be only 56 Kbps, because you'll have to use your phone line for that purpose, although genuine two-way satellite service is under development (for example, by I.S.I.S. AXXESS). Equipment available from InfoDish or PC Connection costs about $190; installation runs $100–$250; and monthly charges (including ISP charges) are $30–$130, depending on how much time you spend online.
- **Other wireless connections:** In urban areas, some businesses are using radio waves transmitted between towers that handle cellular phone calls, which can send data at up to 155 Mbps (but commonly 10 Mbps) and are not only fast and dependable but also always on. The equipment costs from $200 to $2500, and the operating cost is $159–$1400 a month, depending on speed.

Universal Broadband Is Coming

Most PC users employ the physical connections we have described, but there are other possibilities. With WebTV, for instance, you use your television to access the Internet. The Palm VII and the PalmVx, personal digital assistants from Palm Computing, enable you to get on the Net, but the small screen and slow connection accommodate only e-mail and limited Web access (no graphics).

We are living in a time of rapid changes. Already, nearly a third of the online households in the U.S. have DSL, cable, or wireless services. Most of them, admittedly, are in and around major cities, and the companies providing connections have been slow to expand them to the rest of North America. But broadband is coming. When the telcos (telephone companies) finish scrambling to upgrade their phone lines to DSL, the cable companies make their "pipes" better handle two-way data, and the wireless companies refine their through-the-air links, ordinary Internet users will probably have connections *100 times faster* than they are today.[6]

Survival Tip

Broadband: Riskier for Security

Unlike dial-up services, broadband services, because they are always switched on, make your computer vulnerable to over-the-Internet security breaches. Solution: Install firewall software (covered in Chapter 8).

CLICK-ALONG 2-1
The quest for broadband

CONCEPT CHECK

What are the measures of data transmission speed?

What does "broadband" mean?

What are some of the differences between a dial-up modem, DSL, and a cable modem?

2.2 Choosing Your Internet Service Provider (ISP)

KEY QUESTIONS
What is an Internet service provider, and what kinds of services do ISPs provide?

Suppose you have an access device such as a modem and you've signed up for a wired or wireless connection. Next, unless you're already on a college campus network, you'll need to arrange for an **Internet service provider (ISP), a company that connects you through your communications line to its servers, or central (host) computer, which connect you to the Internet via another company's network access points.** (See ● Panel 2.5.) Some well-known national ISPs, also called online services, are America Online (AOL), EarthLink, Microsoft Network (MSN), AT&T WorldNet, and Prodigy. There are also many local ISPs. The ISP will assign you a user name and a password, as well as an e-mailbox. The ISP's local access number for your area is called its *point of presence (POP)*.

If you've decided simply to use the regular 56-Kbps dial-up modem in your PC connected to a plain old telephone line, you'll quickly notice the fierce competition between companies vying to become your ISP. For instance, perhaps your new PC comes with a keyboard button labeled "Internet," which, when pressed, begins the steps toward connecting you with a service provider—the provider that has come to a financial arrangement with the computer's manufacturer. Beside dealing with the blizzard of ads in magazines and on television, you may also receive promotional ISP start-up disks in the mail. In addition, your phone company probably offers an Internet service. (To do some comparison shopping, go online to *www.thelist.com*, which lists ISPs from all over the world and will guide you through the process of finding one that's best for you.)

Once you have contacted an ISP and paid the required fee (charged to your credit card), the ISP will provide you with information about phone numbers for a local connection. The ISP will also provide you with communications

● **PANEL 2.5**
Your connection to the Internet via an ISP

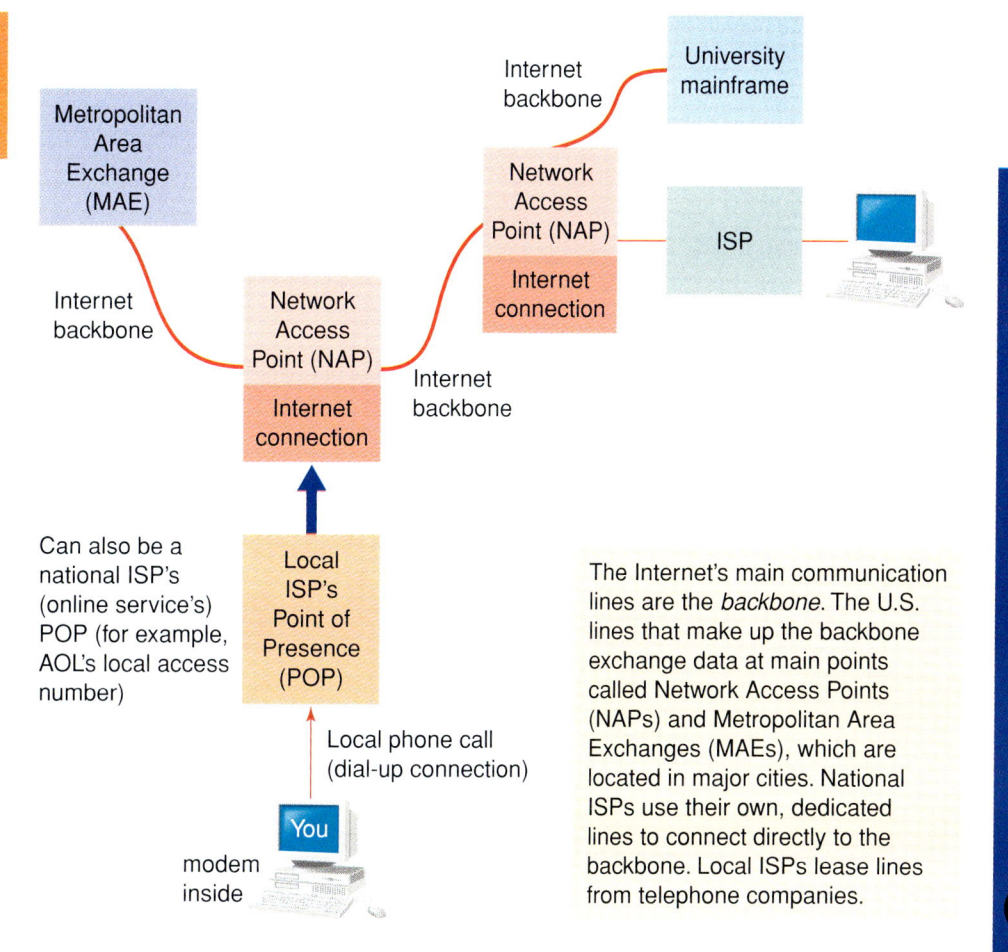

The Internet's main communication lines are the *backbone*. The U.S. lines that make up the backbone exchange data at main points called Network Access Points (NAPs) and Metropolitan Area Exchanges (MAEs), which are located in major cities. National ISPs use their own, dedicated lines to connect directly to the backbone. Local ISPs lease lines from telephone companies.

BOOKMARK IT!

PRACTICAL ACTION BOX
Choosing an Internet Service Provider

If you belong to a college or company, you may get an ISP free. Some public libraries also offer freenet connections.

If these options are not available to you, be sure to ask these questions when you're making phone calls to locate an Internet service provider.[a] (Also see *www.thelist.com,* which provides a comprehensive list of 9700 ISPs.)

Costs

- Is there a contract, and for what length of time? That is, are you obligated to stick with the ISP for a while even if you're unhappy with it?
- Is there a setup fee? (Most ISPs no longer charge this, though some "free" ISPs will.)
- How much is unlimited access per month? (Most charge about $20 for unlimited usage. But inquire if there are free or low-cost trial memberships or discounts for long-term commitments.)
- If access is supposedly free, what are the trade-offs besides putting up with heavy advertising? (For instance, if the ISP closely monitors your activity in order to accurately target ads, what guarantees do you have that information about you will be kept private? What charges will you face if you try to scrap the advertising window or drop the service?)

Access

- Is the access number a local phone call? (If not, your monthly long-distance phone tolls could exceed the ISP fee.)
- Is there an alternative dial-up number if the main number is out of service?
- Is access available when you're traveling? Your provider should offer either a wide range of local access numbers in the cities you tend to visit or toll-free 800 numbers.

Support

- What kind of help does the ISP give in setting up your connection?
- Is there free, 24-hour technical support? Is it reachable through a toll-free number?
- How difficult is it to reach tech support? (Try calling the number before you sign up for the ISP and see how long it takes to get a response. Many ISPs keep customers on hold for a long time.)

Reliability

- What is the average connection success rate for users trying to connect on the first try? (You can try dialing the number during peak hours, to see if you get a modem screech, which is good, rather than a busy signal, which is bad.)
- Will the ISP keep up with technology? (Are they planning to offer broadband technology such as DSL for speedier access?)
- Will the ISP sell your name to marketers or bombard you with junk messages (spam)?

---

software for setting up your computer and modem to dial into their network of servers. For this you use your *user name* ("user ID") and your *password,* a secret word or string of characters that enables you to **log on, or make a connection to the remote computer.** You will also need to get yourself an e-mail address, as we discuss next.

CONCEPT CHECK

What is an Internet service provider?

Describe the different services Internet services provide. What is a point of presence?

Major ISPs compared

Internet Service Provider	Monthly Cost for Unlimited Hours of Use	Number of E-Mail Users per Account	Number of Megabytes for a Website
America Online (AOL) 800-827-6364 www.aol.com	$19.95–$23.90	1 master name, 4 additional names	2 per screen name
AT&T WorldNet 800-967-5363 www.att.net	$21.95	6	5 per user
Earthlink 800-395-8425 www.earthlink.net	$21.95	5	6
Microsoft Internet Access 800-373-3676 www.msn.com	$21.95, first month free	1 primary, up to 5 subaccounts	12
Prodigy 800-776-3439 www.prodigy.com	$19.95	5	6

2.3 Sending & Receiving E-Mail

KEY QUESTIONS
What are the options for obtaining e-mail software, what are the components of an e-mail address, and what are netiquette and spam?

Once connected with an ISP, one of the first things most people want to do is join the millions of users who send and receive electronic mail. E-mail is stored in your mailbox on the ISP's computer, usually a server called a *mail server*. When you use your e-mail software to retrieve your messages, the e-mail is sent from the server to your computer.

It's not necessary, incidentally, to have a PC for e-mail. BeVocal offers a voice dialer, e-mail-over-the-phone service. AOL's Quack is a voice-activated service that lets users access some of their favorite AOL features over any telephone. MailStation, a single-purpose e-mail device about the size of a hardcover book, has a laptop-style keyboard and keys about 85% of normal size plus an adjustable-angle screen. MailStation plugs into a phone jack and allows you to send and receive messages of up to 1000 words. Two-way wireless e-mail pagers, such as RIM's popular Blackberry, use cellphone airwaves to provide e-mail service. Other devices, such as one available from Sharp, enable you to send and receive e-mail from just about any phone—something travelers toting laptop computers may envy.

MailStation

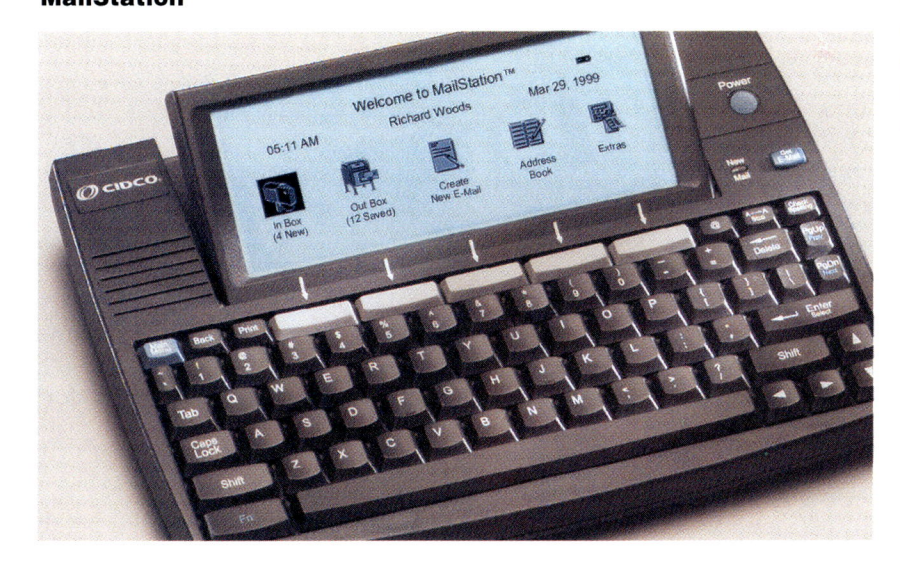

E-Mail Software & Carriers

If you aren't on a campus network, there are four ways to go about getting and sending e-mail:

- **Buy e-mail software:** Popular e-mail software programs are Eudora, Outlook Express, and Lotus Notes. However, there is probably no need for you to spend money on these programs because of the following alternatives.

- **Get e-mail program as part of other computer software:** When you buy a new computer, the system will probably include e-mail software, perhaps as part of the software (called *browsers*) used to search the World Wide Web, such as Internet Explorer or Netscape Navigator. An example is Microsoft's Outlook Express, which is part of its Explorer.

- **Get e-mail software as part of your ISP package:** Internet Service Providers—AOL, Prodigy, EarthLink, AT&T WorldNet—provide e-mail software for their subscribers.

- **Get free e-mail services:** These are available from a variety of sources, ranging from so-called portals or Internet gateways such as Yahoo!, Excite, or Lycos to cable-TV channel CNN's website.

Free e-mail from Yahoo!

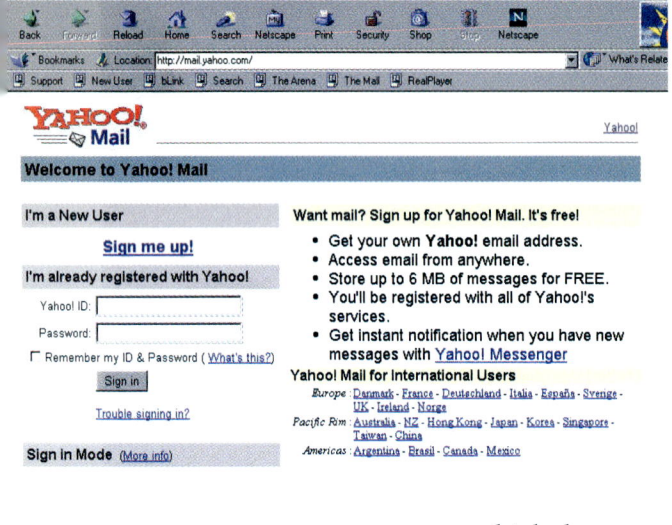

E-Mail Addresses

You'll need an e-mail address, of course, a sort of electronic mailbox used to send and receive messages. All such addresses follow the same approach: *user@domain*. (E-mail addresses are different from Web site addresses, which do not use the symbol @.) A *domain* is simply a location on the Internet. Consider the following address:

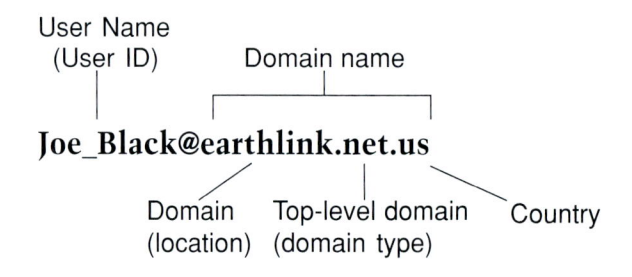

User Name (User ID) | Domain name

Joe_Black@earthlink.net.us

Domain (location) | Top-level domain (domain type) | Country

Let's look at the elements of this address.

Joe_Black The first section, the *user name*, or *user ID*, identifies who is at the address—in this case, *Joe_Black* (note the underscore). (There are many other ways that Joe Black's user name might be designated, with and without capital letters: *Joe_Black, joe_black, joe.black, joeblack, jblack, joeb*, and so on.)

@earthlink The second section, the *domain name*, which is located after the @ (called "at") symbol, tells the location and type of address. Domain name components are separated by periods (called "dots"). The *domain* portion of the address (such as *Earthlink*, an Internet service provider) provides specific information about the *location*—where the message should be delivered.

.net The *top-level domain* is a three-letter extension that describes the *domain type:* .net, .com, .gov, .edu, .org, .mil, .int—network, commercial,

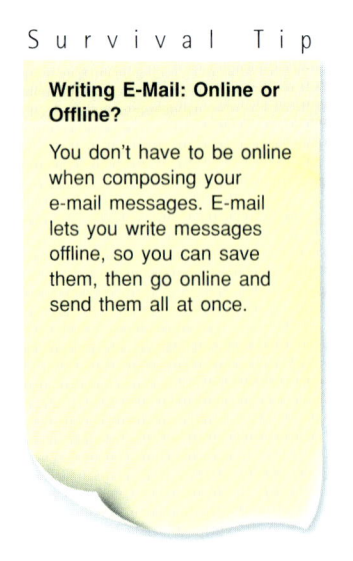

Survival Tip

Writing E-Mail: Online or Offline?

You don't have to be online when composing your e-mail messages. E-mail lets you write messages offline, so you can save them, then go online and send them all at once.

government, educational, nonprofit, military, or international organization. (See ● Panel 2.6.)

.us   Some domain names also include a two-letter extension for the country—for example, *.us* for United States, *.ca* for Canada, *.uk* for United Kingdom, *.jp* for Japan, *.tr* for Turkey.

Sometimes you'll see an address in which people have their own domains—for example, *Joe@Black.com*. However, you can't simply make up a domain name; it has to be registered. (You can check on whether an address is available, as well, and register it by going to *www.register.com* or *www.internic.net/regist/html*).

Incidentally, many people who are unhappy with their ISPs don't change because they don't want to have to notify their friends of a new e-mail address. However, you can switch ISPs and use an e-mail forwarding service. That way, you can keep one e-mail address no matter how many times you change providers.

Some tips about using e-mail addresses:

- **Type addresses carefully:** You need to type the address *exactly* as it appears, including capitalization and all underscores and periods. If you type an e-mail address incorrectly (putting in spaces, for example), your message will be returned to you labeled "undeliverable."
- **Use the "reply" command:** When responding to an e-message someone has sent you, the easiest way to avoid making address mistakes is to use the "Reply" command, which will automatically fill in the correct address in the "To" line (see the illustration on the next page).

Coming soon:
.aero air-transport industry
.name individuals
.coop cooperatives
.museum museums
.pro professions

PANEL 2.6
The meaning of Internet top-level domain abbreviations

Domain	Description	Example	Operated by*
.biz .com	Commercial businesses	Editor@mcgraw-hill.com	NeuLevel, Inc.
.edu	Educational and research institutions	Professor@stanford.edu	Network Solutions
.gov	U.S. government agencies and bureaus	President@whitehouse.gov	U.S. General Services Administration
.info	Research institutions	Contact@research.info	Afilias Limited
.int	International organizations	Secretary_general@unitednations.int	IANA
.mil	U.S. military organizations	Chief_of_staff@pentagon.mil	U.S. Dept. of Defense
.net	Internet network resources	Contact@earthlink.net	VeriSign Global Registry Services
.org	Nonprofit and professional organizations	Director@redcross.org	(same as above)

*For a list of currently operating registrars, where you can register a top-level domain, go to www.internic.com. InterNIC is a registered service mark of the U.S. Department of Commerce. It is licensed to ICANN (Internet Corporation for Assigned Names and Numbers), which operates its website.

Sending e-mail

Send: Command for sending messages

cc: For copying ("carbon copy") message to others

bcc: For copying others ("blind carbon copy") without the primary recipient knowing it

Message area

You can conclude every message with a custom "signature"

Address Book: Lists e-mail addresses you use most; can be attached automatically to messages

Subject line: Preview incoming e-mail by reviewing the subject lines to see if you really need to read the messages

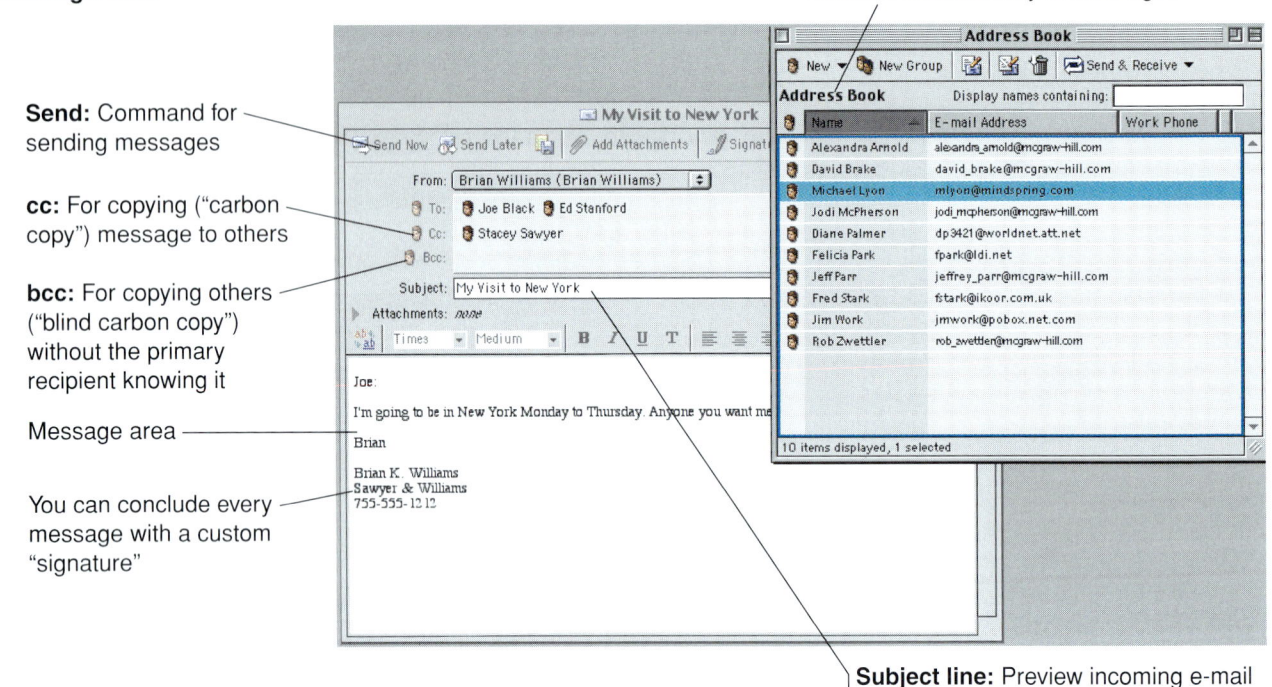

Receiving e-mail

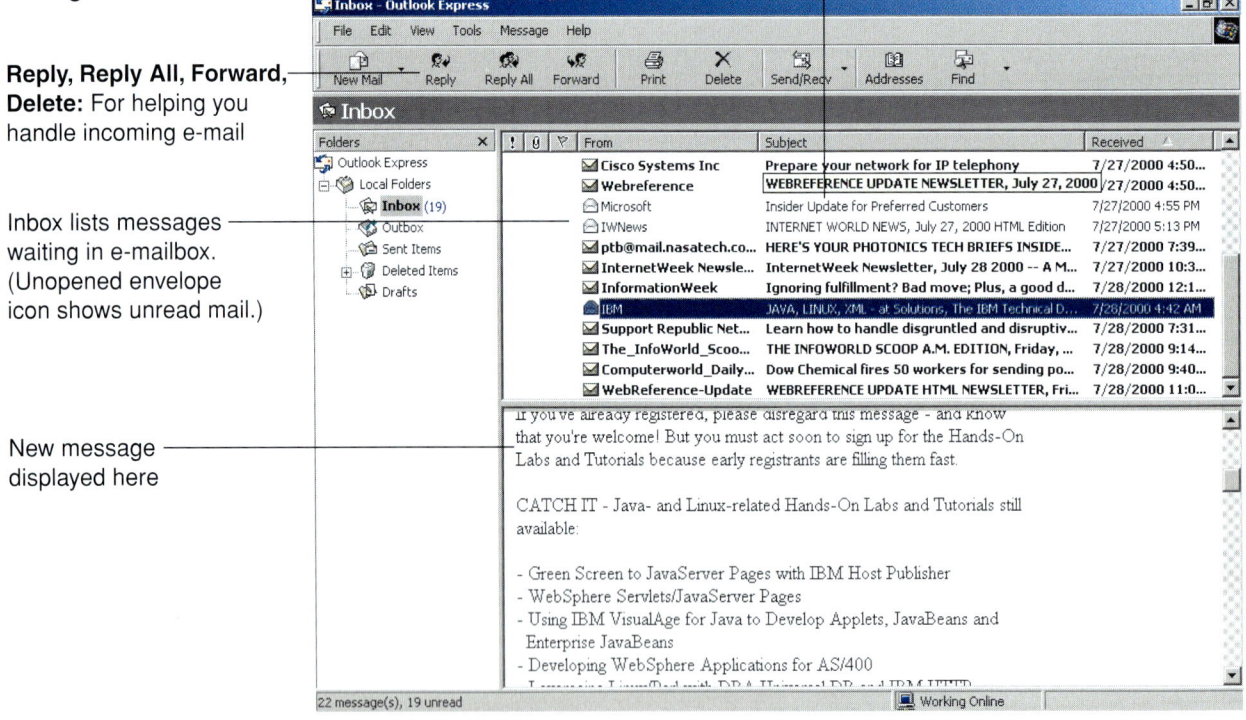

Reply, Reply All, Forward, Delete: For helping you handle incoming e-mail

Inbox lists messages waiting in e-mailbox. (Unopened envelope icon shows unread mail.)

New message displayed here

Replying to e-mail

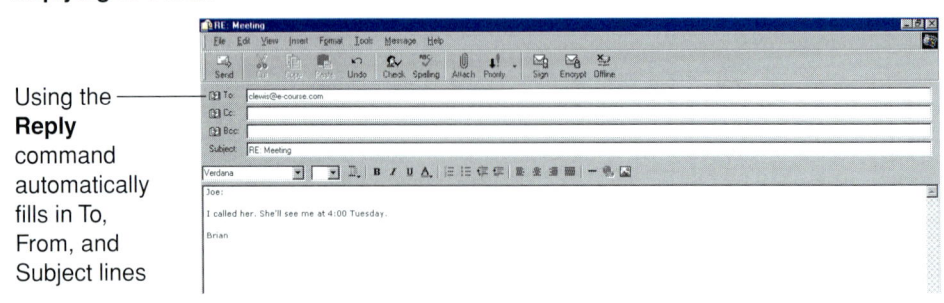

Using the **Reply** command automatically fills in To, From, and Subject lines

Survival Tip

Accessing E-Mail While Traveling Abroad

To access your e-mail using a local call while traveling outside North America, get a free e-mail account with Yahoo! (*http://mail.yahoo*). Hotmail (*www.hotmail.com*), or Mail.com (*www.mail.com*).

- **Use the "address book" feature:** You can store the e-mail addresses of people sending you messages in your program's "address book." This feature also allows you to organize your e-mail addresses according to a nickname or the person's real name so that, for instance, you can look up your friend Joe Black under his real name, instead of under his user name, *bugsme2*, which you might not remember. The address book also allows you to organize addresses into various groups—such as your friends, your relatives, club members—so you can easily send all members of a group the same message with a single command.
- **Deal with each e-mail only once:** When a message comes in, delete it, respond to it, of file it away in a folder. Don't use your inbox for storage.

Attachments

You have written a great research paper and you immediately want to show it off to someone. If you were sending it via the Postal Service, you would write a cover note—"Folks, look at this great paper I wrote about term-paper cheating! See attached"—then attach it to the paper, and stick it in an envelope. E-mail has its own version of this. If the file of your paper exists in the computer from which you are sending e-mail, you can write your e-mail message (your cover note) and then use the Attach File command to attach the document (see illustration on the next page). (Note: It's important that the person receiving the e-mail attachment have the exact same software that created the attached file, such as Microsoft Word 2000, or have software that can read and convert the attached file.)

While you could also copy your document into the main message and send it that way, some e-mail software loses formatting options such as **bold** or *italic* text or special symbols. And if you're sending song lyrics or poetry, the lines of text may break differently on someone else's display screen than they do on yours. Thus, the benefit of the attachment feature is that it preserves all such formatting, provided the recipient is using the same word processing software that you used. You can also attach pictures, sounds, videos, and other files to your e-mail message.

Note: Many *viruses*—those rogue programs that can seriously damage your PC or programs—ride along with e-mail as attached files. Thus, you should never open an attached file from an unknown source. This was what made the so-called May 2000 Love Bug (ILOVEYOU virus), to take one example, such a disaster.

Instant Messaging

Instant messages are like a cross between e-mail and phone, allowing for communication that is far speedier than conventional e-mail. With <u>**instant messaging (IM)**</u>, **any user on a given e-mail system can send a message and have it pop up instantly on the screen of anyone else logged onto that system.** Then, if all parties agree, they can initiate online typed conversations in real time. The messages appear on the display screen in a small <u>**window**</u>—**a rectangular area containing a document or activity**—so that users can exchange messages almost instantaneously while operating other programs.

Examples of present instant-message systems are AOL Instant Messenger, MSN Messenger, ICQ ("I Seek You," also from AOL), Prodigy Instant Messaging, Tribal Voice PowWow, and Yahoo Messenger. Some of these, such as Yahoo!'s, allow voice chats among users, if their PCs have microphones and speakers.

To get instant messaging, which is available free, you download software and register with the service, providing it with a user name and password. You can then create a list of "buddies" with whom you want to communicate regularly. When your computer is connected to the Internet, the

Sending an e-mail attachment

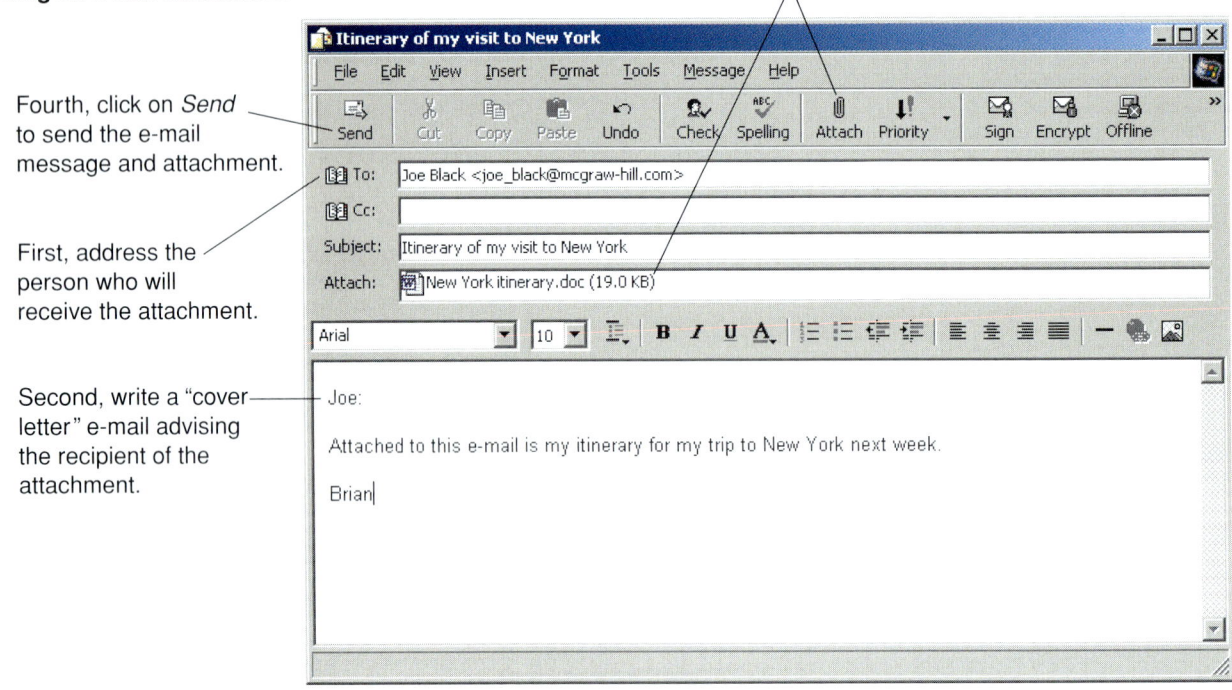

Third, use your e-mail software's toolbar buttons or menus to attach the file that contains the attachment.

Fourth, click on *Send* to send the e-mail message and attachment.

First, address the person who will receive the attachment.

Second, write a "cover letter" e-mail advising the recipient of the attachment.

Receiving an e-mail attachment

When you receive a file containing an attachment, you'll see an icon indicating the message contains more than just text. You can click on the icon to see the attachment. If you have the software the attached file was created in, you can open the attachment immediately to read or print, or you can save the attachment in a location of your choice (on your computer). You can also forward the attachment to another person.

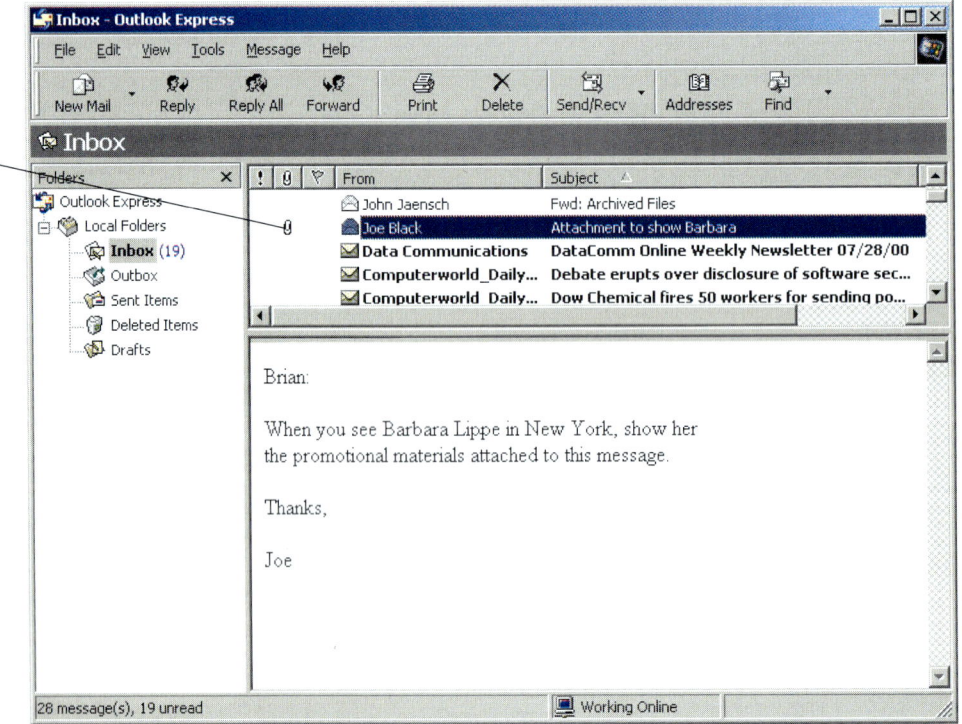

Instant Message Services

Instant Messenger	URL	Users[1]
AOL Buddy List (part of regular AOL service)	NA	40 million
AOL Instant Messenger	www.aim.com	60 million
ICQ	www.icq.com	70 million
MSN Messenger Service	messenger.msn.com	18 million
Tribal Voice PowWow	www.powwow.com	8 million
Yahoo Messenger	messenger.yahoo.com	"millions"[2]
Odigo	www.odigo.com	1.3 million[3]
iCast	www.icast.com	500,000
Prodigy Instant Messaging	pim.prodigy.net	500,000
Imici	www.imici.com	5,000[3]

[1] Number of people who downloaded software; not all users run program on a regular basis.

[2] Yahoo! doesn't release user numbers for specific services.

[3] Allows intermittent access to AOL and ICQ.

software checks in with a central server, which verifies your identity and looks to see if any of your "buddies" are also online. You can then start a conversation by sending a message to any buddy currently online.

IM has become a hit with many users; indeed, users of AOL's product alone exchange more than 500 million messages a day. "The potential of instant messaging is profound," says technology writer Alec Klein. "IMs ricochet across the globe, bringing people that much closer together."[7] Instant messaging is especially useful in the workplace as a way of reducing long-distance telephone bills when you have to communicate with colleagues who are geographically remote but with whom you must work closely.

However, you need to be aware of a couple of drawbacks:

- **Lack of common standards:** As of this writing, most of the existing IM products don't communicate with one another. If you're using AOL's IM, not only can you not communicate with a buddy on Yahoo!—you can't even communicate with a buddy on AOL's ICQ. Perhaps this will have changed by the time you read this.

- **Time wasters when you have to get work done:** An instant message "is the equivalent of a ringing phone because it pops up on the recipient's screen right away," says one writer.[8] Some analysts suggest that, because of its speed, intrusiveness, and ability to show who else is online, IM can destroy workers' concentration in some offices. You can put off acknowledging e-mail, voice mail, or faxes. But instant messaging is "the cyber-equivalent of someone walking into your office and starting up a conversation as if you had nothing better to do," says one critic. "It violates the basic courtesy of not shoving yourself into other people's faces."[9]

You can turn off your instant messages, but that is like turning off the ringer on your phone; after a while people will wonder why you're never available. Buddy lists or other contact lists can also become very in-groupish. When that happens, people are distracted from their work as they worry about staying current with their circle (or being shut out of one). Some companies have reportedly put an end to instant messaging, sending everyone back to the use of conventional e-mail.

Mailing Lists: E-Mail–Based Discussion Groups

Want to receive e-mail from people all over the world who share your interests? You can try finding a mailing list and then "subscribing"—signing up, just as you would for a free newsletter or magazine. **List-serves are e-mail mailing lists of people who regularly participate in discussion topics.** To subscribe, you send an e-mail to the list-serve moderator and ask to become a member, after which you will automatically receive e-mail messages from anyone who responds to the server. A directory of mailing lists is available at Publicly Accessible Mailing Lists *(http://paml.net)* or Yahoo!'s OneList *(www.onelist.com).*

Ethics

Netiquette: Appropriate Online Behavior

You may think etiquette is about knowing which fork to use at a formal dinner. Basically, though, etiquette has to do with politeness and civility—with rules for getting along so that people don't get upset or suffer hurt feelings.

New Internet users, known as *newbies*, may accidentally offend other people in a discussion group or in an e-mail simply because they are unaware of **netiquette**, or "network etiquette"—**guides to appropriate online behavior.** In general, netiquette has two basic rules: (a) don't waste people's time, and (b) don't say anything to a person online that you wouldn't say to his or her face.

Some more specific rules of netiquette are as follows:

- **Consult FAQs:** Most online groups post **FAQs (Frequently Asked Questions) that explain expected norms of online behavior for a particular group.** Always read these first—before someone in the group tells you you've made a mistake.

- **Avoid flaming:** A form of speech unique to online communication, *flaming* **is writing an online message that uses derogatory, obscene, or inappropriate language.** Flaming is a form of public humiliation inflicted on people who have failed to read FAQs or otherwise not observed netiquette (although it can happen just because the sender has poor impulse control and needs a course in anger management). Something that smoothes communication online is the use of **emoticons**, **keyboard-produced pictorial representations of expressions.** (See ● Panel 2.7.)

- **Don't SHOUT:** Use of all-capital letters is considered the equivalent of SHOUTING. Avoid, except when they are required for emphasis of a word or two (as when you can't use italics in your e-messages).

- **Avoid sloppiness, but avoid criticizing others' sloppiness:** Avoid spelling and grammatical errors. But don't criticize those same errors in others' messages. (After all, they may not speak English as a native language.) Most e-mail software comes with spell-checking capability, which is easy to use.

- **Don't send huge file attachments, unless requested:** Your cousin living in the country may find it takes minutes rather than seconds for his or her computer to download a massive file (as of a video that you want to share). This may tie up the system at a time when your relative badly needs to use it. Better to query in advance before sending large files as attachments. Also, whenever you send an attachment, be sure the recipient has the appropriate software to open your attachment (you both are using Microsoft Word 2000, for example).

- **When replying, quote only the relevant portion:** If you're replying to just a couple of matters in a long e-mail posting, don't send back the entire message. This forces your recipient to wade through lots of text to find the reference. Instead, edit his or her original text down to the relevant paragraph and then put in your response immediately following.

● **PANEL 2.7**
Some emoticons

:-)	Happy face	<g>	Grin
:-(	Sorrow or frown	BTW	By the way
:-O	Shock	IMHO	In my humble opinion
:-/	Sarcasm	FYI	For your information
;-)	Wink		

E-mail software lets you create folders for storing mail.

- Mail Folder
  - Business mail
  - Family mail
  - Inbox
  - Mail from friends
  - Sent mail
  - Trash

Sorting Your E-Mail

On an average day, according to one study, 3.4 billion business e-mail messages and 2.7 billion personal e-mails are sent in North America.[10] If, as so many people do, you receive 50–150 e-mails per *day*, you'll have to keep them organized, so you don't lose control.

One way to stay organized is with instant organizers, also called *filters*, which use the name of the person or the mailing list to put each particular mail into one folder. Then you can read e-mails sent to this folder later when you have time, freeing up your inbox for mail that needs your more immediate attention. Instructions on how to set up such organizers are in your e-mail program's Help section.

Spam: Unwanted Junk E-Mail

Several years ago, Monty Python, the British comedy group, did a sketch in which restaurant customers were unable to converse because people in the background (a group of Vikings, actually) kept chanting "Spam, spam, eggs and spam . . ." The term *spam* was picked up by the computer world to describe another kind of "noise" that interferes with communication. Now **spam** **refers to unsolicited e-mail in the form of advertising or chain letters.** Usually you won't recognize the sender on your list of incoming mail, and often the subject line will give no hint, stating something such as "The status of your application" or "It's up to you now." The solicitations can range from money-making schemes to online pornography.

Some ways to deal with this nuisance are as follows:[11]

- **Delete without opening the message:** Opening the spam message can actually send a signal to the spammer that someone has looked at the onscreen message and therefore that the e-mail address is valid—which means you'll probably get more spams in the future. If you don't recognize the name on your inbox directory or the topic on the inbox subject line, you can simply delete the message without reading it. Or you can use a preview feature in your e-mail program to look at the message without actually opening it, then delete it. (Hint: Be sure to get rid of all the deleted messages from time to time; otherwise, they will build up in your "trash" area.)

- **Never reply to a spam message!** The following advice needs to be taken seriously: *Never reply in any way to a spam message!* Replying confirms to the spammer that yours is an active e-mail address. Some spam senders will tell you that if you want to be removed from their mailing list, you should type the word REMOVE or UNSUBSCRIBE in the subject line and use the reply command to send it back. Invariably, however, all this does is confirm to the spammer that your address is valid, setting you up to receive more unsolicited messages.

  Michael Ashley Lopez, an archaeology graduate student at the University of California at Berkeley, found he had been included on an e-mail list for fans of teen idol Britney Spears. He opened the first e-mail message, wasted 13 seconds reading it, then clicked on the link to unsubscribe from it. The result was he couldn't get off. Months later, despite a determined effort to shake this nuisance, he was still receiving invitations to check out Britney's latest single or preview her latest video.[12]

- **Enlist the help of your ISP or use spam filters:** Your ISP may offer a free spam filter (for example, Earthlink's Spaminator) to stop the stuff before you even see it. If it doesn't, you can sign up for a filtering service, such as ImagiNet *(www.imagin.net)* for a small monthly

Survival Tip

Handling the Annoyance of Spam

To better manage spam, some users get *two* e-mail boxes. One is used for on-line shopping, business, research, etc.—that will continue to attract spam. The other is used (like an unlisted phone number) only for personal friends and family—and will probably not receive much spam.

charge. Or there are do-it-yourself spam-stopping programs. Examples: Brightmail *(www.brightmail.com)*, Novasoft SpamKiller *(www.spamkiller.com)*, High Mountain Software SpamEater Pro *(www.hms.com)*.

Be warned, however: Even so-called spam killers don't always work. Certainly it didn't for Michael Lopez, victim of the repeated Britney Spears e-mail, even though he had signed up for an e-mail blocking service. "Nothing will work 100%, short of changing your e-mail address," says the operator of an online service called SpamCop. "No matter how well you try to filter a spammer, they're always working to defeat the filter."[13]

- **Fight back:** If you want to get back at spammers—and other Internet abusers—check with abuse.net *(www.abuse.net)* or Ed Falk's Spam Tracking Page *(www.rahul.net/falk)*. These will tell you where to report spammers, the appropriate people to complain to, and other spam-fighting tips.

What About Keeping E-Mail Private?

Ethics

The single best piece of advice that can be given about sending e-mail is this: *Pretend every electronic message is a postcard that can be read by anyone.* Because the chances are high that it could be. (And this includes e-mail on college campus systems as well.)

Think the boss can't snoop on your e-mail at work? The law allows employers to "intercept" employee communications if one of the parties involved agrees to the "interception." The party "involved" is the employer. And in the workplace, e-mail is typically saved on a server, at least for a while. Indeed, federal laws require employers to keep some e-mail messages for years.

Think you can keep your e-mail address a secret among your friends? You have no control over whether they might send your e-messages on to someone else—who might in turn forward it again. (One thing you can do for them, however, is delete their names and addresses before sending one of their messages on to someone.)

Think your ISP will protect your privacy? Often service providers post your address publicly or even sell their customer lists.

Think spammers can't find you? They will if you post an e-mail to an Internet message or bulletin board, making yourself a target for pieces of software (known as "harvester bots") that scour such boards for active e-mail addresses.

And we have not even mentioned your e-mail being intercepted by those knowledgeable individuals known as hackers or crackers, which we discuss elsewhere.

If you're really concerned about preserving your privacy, you can try certain technical solutions—for instance, installing software that encodes and decodes messages (discussed in Chapter 8). But the simplest solution is the easiest: Don't put any sensitive or embarrassing information in your e-mail. Even deleted e-mail removed from trash can still be traced on your hard disk. Software—for example, Spytech Eradicator and Webroot's Window Washer—is available to completely eliminate deleted files.

CLICK-ALONG 2-2
More on e-mail

CONCEPT CHECK

What are the options for getting and sending e-mail?

What are some features available with e-mail?

What are attachments, and why are they useful?

2.4 The World Wide Web

KEY QUESTIONS
What are websites, web pages, browsers, URLs, and search engines?

"I found my old first-grade teacher by surfing the Internet."

When people talk about the Internet in this way, they really mean the World Wide Web. After e-mail, visiting sites ("surfing") on the Web is the most popular use of the Internet. Among the forces driving its popularity are entertainment and e-commerce. *Entertainment* offerings range from listening to music to creating your own, from playing online games by yourself to playing with others, from checking out local restaurants to researching overseas travel. *E-commerce* offers online auctions, retail stores, and discount travel services as well as all kinds of "B2B," or business-to-business, connections, as when General Motors buys online from its steel suppliers.

What makes the World Wide Web so graphically inviting and easily navigable is that this international collection of servers (1) contains information in multimedia form and (2) is connected by hypertext links, or hyperlinks.

1. **Multimedia form—what makes the Web graphically inviting:**
   Whereas e-mail messages are generally text, the Web provides information in *multimedia* form—graphics, video, and audio as well as text. You can see color pictures, animation, and full-motion video. You can download music. You can listen to radio broadcasts. You can have telephone conversations with others.

2. **Use of hypertext—what makes the Web easily navigable:** Whereas with e-mail you can connect only with specific addresses you know about, with the Web you have hypertext. **Hypertext is a system in which documents scattered across many Internet sites are directly linked—with *hyperlinks*—so that a word or phrase in one document becomes a connection to a document in a different place.**

The format, or language, used on the Web is called hypertext markup language. (It is not, however, a programming language.) **Hypertext markup language (HTML) is the set of special instructions (called "tags" or**

How hypertext-markup language (HTML) works
The coding in the HTML files tells your web browser, first, how to find the files of text, graphics, and multimedia files on the server and, second, how to display them on the web page. The browser also interprets HTML tags, or instructions, as links to other websites or to other web resources, such as files to download.

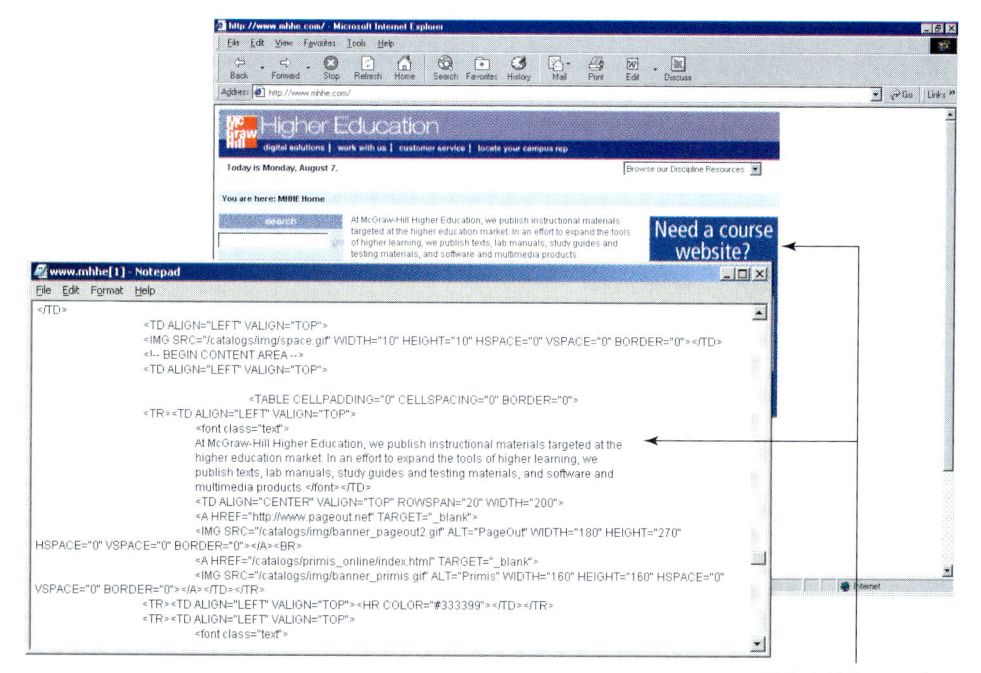

Meaning of tags: Every HTML tag is surrounded by a less-than and greater-than sign—for example, <TR>. Tags often appear as beginning and ending tags, which are identical except for a slash in the end tag—for example, <TR> Paragraph of text. </TR>.

This hidden coding (left) results in the web page above, which you see on your screen "markups") that are used to specify document structure, formatting, and links to other multimedia documents.

For example, if you were reading this book onscreen, you could use your mouse to click on the word *multimedia*—which would be highlighted—in the paragraph above, and that would lead you to another location, where perhaps "multimedia" is defined. Then you could click on a word in that definition, and that would lead you to some related words—or even some pictures.

The result is that one term or phrase will lead to another, and so you can access all kinds of databases and libraries all over the world. Among the droplets in what amounts to a Niagara Falls of information available: *Weather maps and forecasts. Guitar chords. Recipe archives. Sports schedules. Daily newspapers in all kinds of languages. Nielsen television ratings. A ZIP code guide. Works of literature. The Alcoholism Research Data Base. U.S. Government phone numbers. The Central Intelligence Agency world map. The daily White House press releases.* And on and on.

The Web & How It Works

If a Rip Van Winkle fell asleep in 1989 (the year computer scientist Tim Berners-Lee developed the web software) and awoke today, he would be completely baffled by the new vocabulary that we now encounter on an almost daily basis: *website, home page, www.* Let's see how we would explain to him what these and similar web terms mean.

- **Website—the domain on the computer:** You'll recall we described top-level domains, such as .com, .edu, .org, and .net, in our discussion of e-mail addresses. **A computer with a domain name is called a *site*.** When you decide to buy books at the online site of bookseller Barnes & Noble, you would visit its website *www.barnesandnoble.com*; the **website is the location of a web domain name in a computer somewhere on the Internet.** That computer might be located in Barnes & Noble offices, but it might be located somewhere else entirely. (The website for New Mexico's Carlsbad Caverns is not located underground in the caverns, but the website for your college is probably on the campus.)

- **Web pages—the documents on a website:** A website is composed of a web page or collection of related web pages. **A *web page* is a document on the World Wide Web that can include text, pictures, sound, and video.** The first page you see at a website is like the title page of a book. This is the ***home page*, or welcome page, which identifies the website and contains links to other pages at the site.** If you have your own personal website, it might consist of just one page—the home page. Large websites have scores or even hundreds of pages. (Note: The contents of home pages often change. Or they may disappear, so that the connecting links to them in other web pages become links to nowhere.)

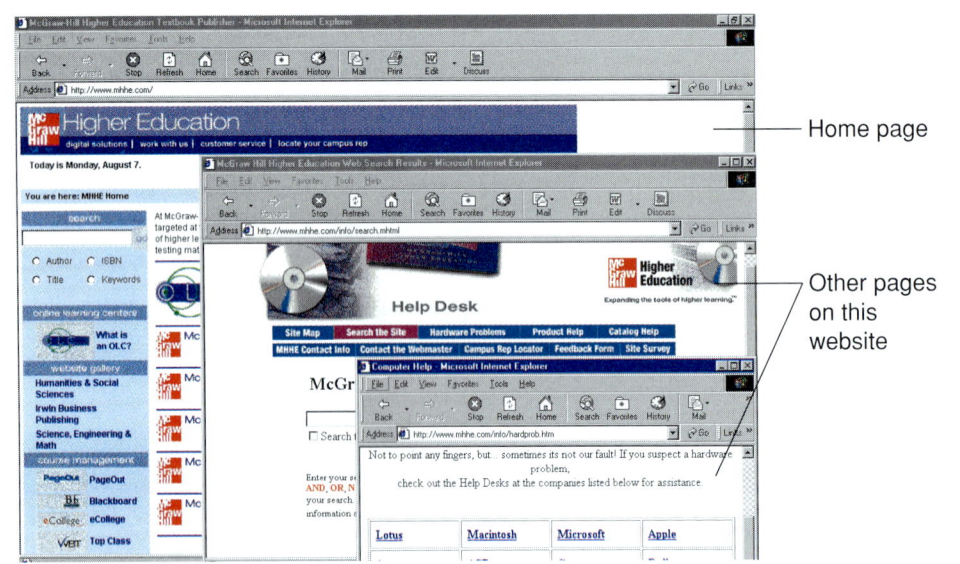

Website home page

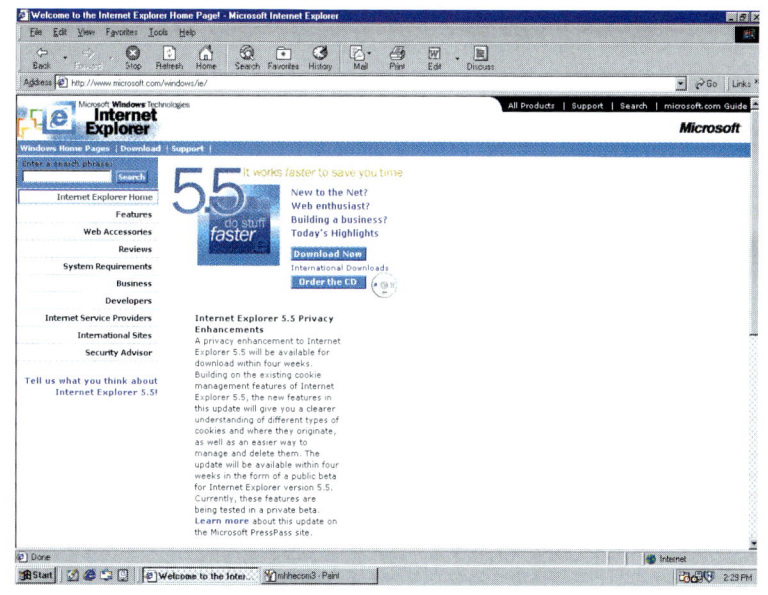

Microsoft Internet Explorer

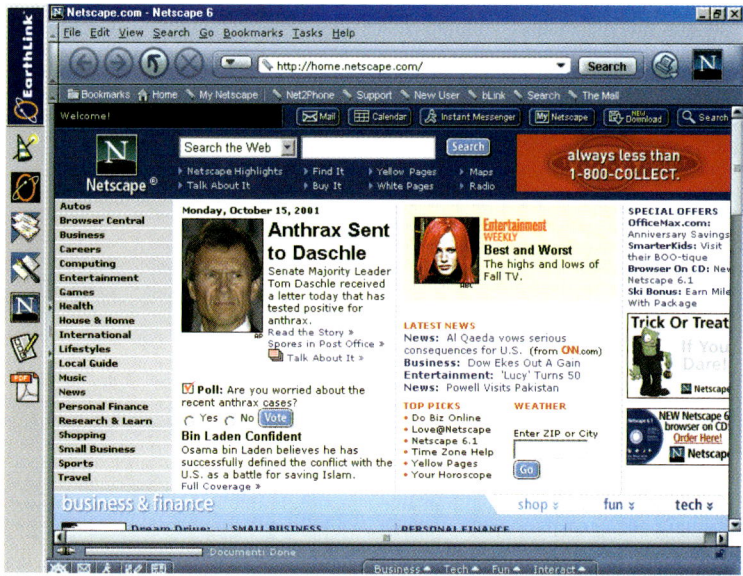

Netscape Navigator

- **Browsers—software for connecting with websites:** A _web browser_, or simply _browser_, is software that enables users to view web pages and to jump from one page to another. The two best-known browsers are Microsoft's Internet Explorer, which most users prefer, and Netscape Navigator, once the leader but now used by only about 13% of consumers.[14] When you connect to a particular website with your browser, the first thing you will see is the home page. Then, using your mouse, you can move from one page to another by clicking on hypertext links.

- **URLs—addresses for web pages:** Before your browser can connect with a website, it needs to know the site's address, the URL. The _URL (Uniform Resource Locator)_ **is a string of characters that points to a specific piece of information anywhere on the Web.** In other words, the URL is the website's unique address. A URL consists of (1) the web _protocol_, (2) the name of the web _server_, (3) the _directory_ (or folder) on that server, and (4) the _file_ within that directory (perhaps with an _extension_ such as _html_ or _htm_). Usually you need to type a URL _exactly_ the way it appears—not type a capital letter, for instance, if a lowercase letter is indicated.

  Consider the following example of a URL for a website offered by the National Park Service for Yosemite National Park:

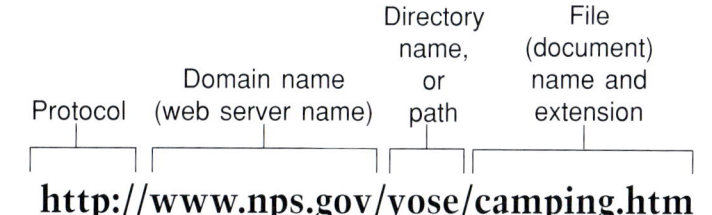

```
        Protocol    Domain name       Directory    File
                   (web server name)  name,       (document)
                                      or path     name and
                                                  extension
```

http://www.nps.gov/yose/camping.htm

Let's look at these elements.

http:// A _protocol_ **is a set of communication rules for exchanging information.** It allows web browsers to communicate with web servers. When you see the _http://_ at the beginning of some web addresses (as in _http://www.mcgraw-hill.com_), that stands for _HTTP (HyperText Transfer Protocol)_, **the communications rules that allow browsers to connect with web servers.** Note: Most browsers assume that all web addresses begin with _http://_ and so you don't need to type this part; just start with whatever follows, such as _www_.

www.nps.gov/ The _web server_ is the particular computer on which this website is located. The _www_ stands for "World Wide

Web," of course; the *.nps* stands for "National Park Service," and *.gov* is the top-level domain name indicating that this is a government website. The server might be physically located in Yosemite National Park in California; in the Park Service's headquarters in Washington, D.C.; or somewhere else entirely.

yose/ The *directory* name is the name on the server for the directory, or folder, from which your browser needs to pull the file. Here it is *yose* for "Yosemite." For Yellowstone National Park, it is *yell.*

camping.htm The *file* is the particular page or document that you are seeking. Here it is *camping.htm*, because you have gone to a web page about Yosemite's camping facilities. The *.htm* is an extension to the file name, and this extension informs the browser that the file is an HTML file.

A URL, you may have observed, is *not* the same thing as an e-mail address. Some people might type in *president@whitehouse.gov.us* and expect to get a website, but it won't happen. The website for the White House (which includes presidential information, history, a tour, and guide to federal services) is *http://www.whitehouse.gov*

Be careful about information found on the Web:

- *Know the source.* To find out who is running a site, go to *www.internic.net* and use the Registry Whois link to search the database of registered domain names.
- *If you use a tilde* (~) in a web address, that's usually a sign of a personal site—run by an individual, not a company, an organization, or an institution.
- Beware of sites with a lot of *grammatical, spelling, and vocabulary errors.*

Using Your Browser to Get Around the Web

As stated, the World Wide Web now consists of an estimated 1 billion web pages. Moreover, the Web is constantly changing; more sites are created and old ones are retired. Without a browser and various kinds of search tools, there would be no way any of us could begin to make any kind of sense of this enormous amount of data.

As we mentioned, a web page may include *hyperlinks*—words and phrases that appear as underlined or color text—that are references to other web pages. On a home page, for instance, the hyperlinks serve to connect the top page with other pages throughout the website. Other hyperlinks will connect to other pages on other websites, whether located on a computer next door or one on the other side of the world.

If you buy a new computer, it will come with a browser already installed. Most browsers have a similar look and feel. On the page opposite we show one of the popular browsers in use—Microsoft Internet Explorer.

Note that the web browser screen has five basic elements: *menu bar, toolbar, address bar, workspace,* and *status bar.* To execute menu bar and toolbar commands, you use the mouse to move the pointer over the word, known as a *menu selection,* and click the left button of the mouse. This will result in a *pull-down menu* of other commands for other options. (See ● Panel 2.8.)

After you've been using a mouse for a while, you may find moving the pointer around somewhat time-consuming. As a shortcut, if you click on the right mouse button, you can reach many of the commands on the toolbar (*Back, Forward,* and so on) via a pop-up menu.

● **PANEL 2.8**
The commands on a browser screen

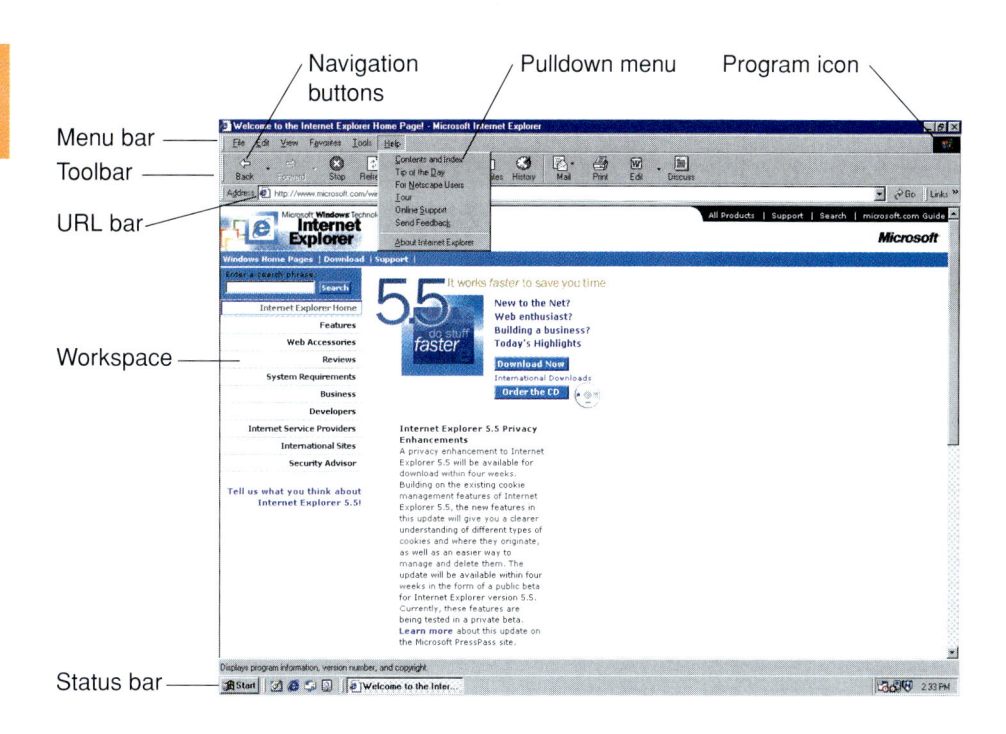

Menu bar —
Toolbar —
URL bar —
Workspace —
Status bar —

Navigation buttons · Pulldown menu · Program icon

- **Starting out from home:** The first page you see when you start up your browser is the *home page* or *start page*. (You can also start up from just a blank page, if you don't want to wait for the time it takes to connect with a home page.) You can choose any page on the Web you want as your start page, but a good start page offers links to sites you want to visit frequently. Often you may find that the ISP with which you arrange your Internet connection will provide its own start page. However, you'll no doubt be able to customize it to make it your own personal home page.

- **Personalizing your home page:** Want to see the weather forecast for your college and/or hometown areas when you first log on? Or your horoscope, "message of the day," or the day's news (general, sports, financial, health, and so on)? Or the websites you visit most frequently? Or a reminder page (as for deadlines or people's birthdays)? You can probably personalize your home page following the directions provided with the first start page you encounter. Or if you have an older Microsoft or Netscape browser you can get a customizing system from either company. A customized start page is also provided by Yahoo!, Excite, AltaVista, and similar services.

- **Getting around—Back, Forward, Home, and Search features:** Driving in a foreign city (or even Boston or San Francisco) can be an interesting experience in which street names change, turns lead into unknown neighborhoods, and signs aren't always evident, so that soon you have no idea where you are. That's what the Internet is like, although on a far more massive scale. Fortunately, unlike being lost in Rome, here your browser toolbar provides navigational aids. *Back* takes you back to the previous page. *Forward* lets you look again at a page you returned from. If you really get lost, you can start over by clicking on *Home,* which returns you to your home page. *Search* lists various other search tools, as we will describe. Other navigational aides are history lists and bookmarks.

The Internet & the World Wide Web

55

Suppose you live in North America and are planning a trip to Europe. While you're in England, you want to visit London, and you're interested in finding an inexpensive place to stay there. The World Wide Web and its hyperlinks can help you achieve this.

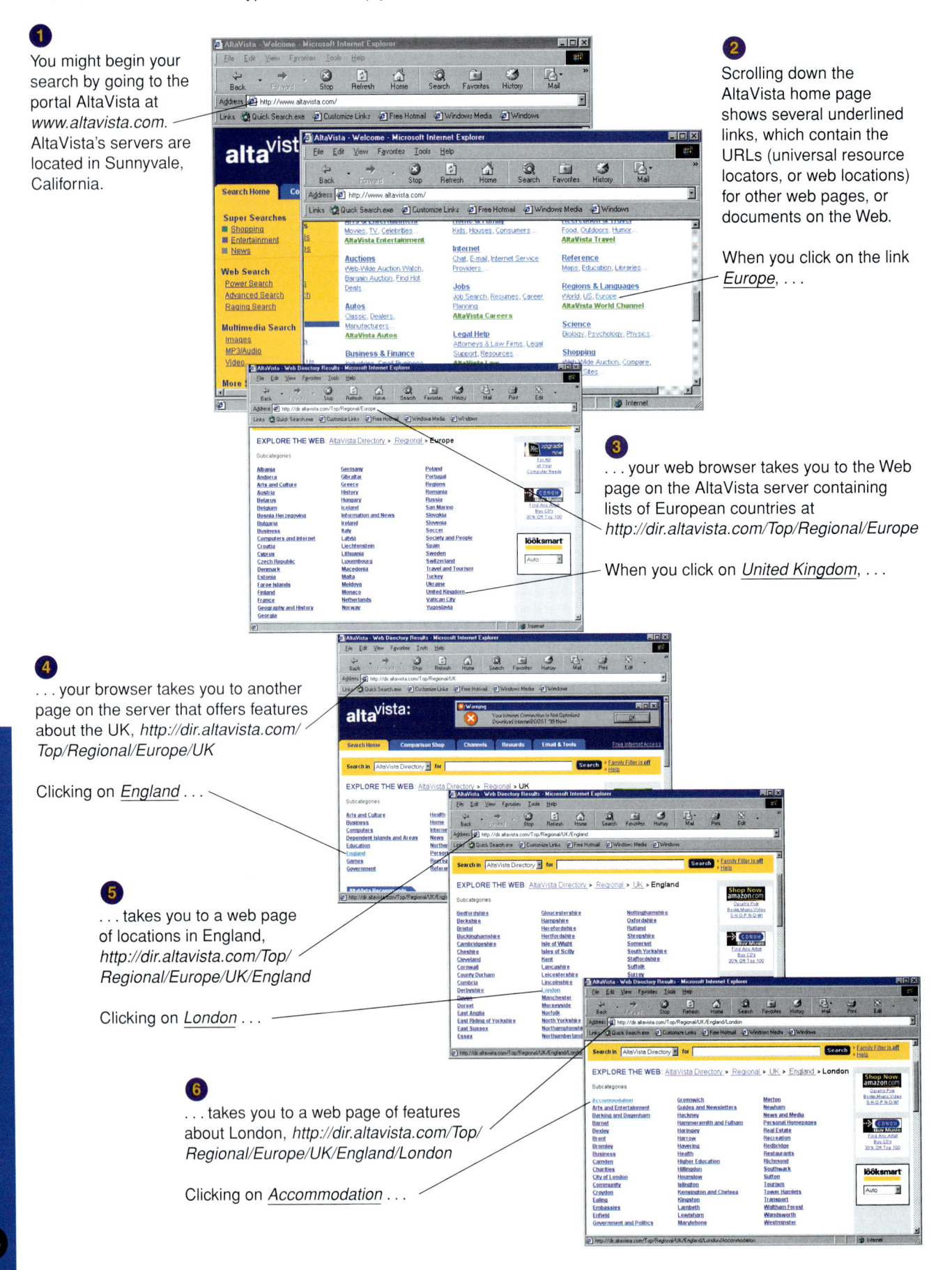

1 You might begin your search by going to the portal AltaVista at *www.altavista.com*. AltaVista's servers are located in Sunnyvale, California.

2 Scrolling down the AltaVista home page shows several underlined links, which contain the URLs (universal resource locators, or web locations) for other web pages, or documents on the Web.

When you click on the link *Europe*, . . .

3 . . . your web browser takes you to the Web page on the AltaVista server containing lists of European countries at *http://dir.altavista.com/Top/Regional/Europe*

When you click on *United Kingdom*, . . .

4 . . . your browser takes you to another page on the server that offers features about the UK, *http://dir.altavista.com/Top/Regional/Europe/UK*

Clicking on *England* . . .

5 . . . takes you to a web page of locations in England, *http://dir.altavista.com/Top/Regional/Europe/UK/England*

Clicking on *London* . . .

6 . . . takes you to a web page of features about London, *http://dir.altavista.com/Top/Regional/Europe/UK/England/London*

Clicking on *Accommodation* . . .

Chapter 2

56

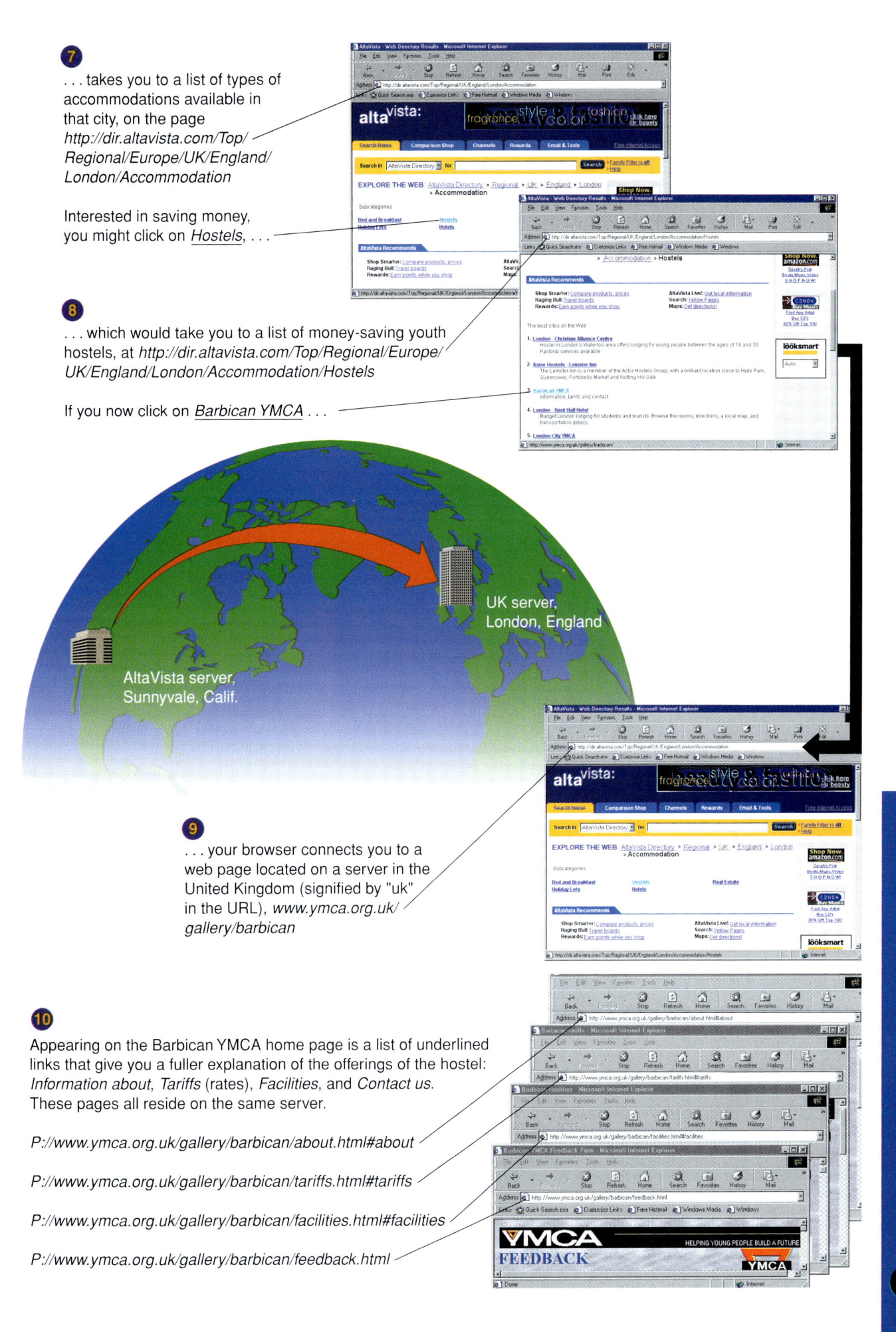

7
... takes you to a list of types of accommodations available in that city, on the page *http://dir.altavista.com/Top/Regional/Europe/UK/England/London/Accommodation*

Interested in saving money, you might click on *Hostels*, ...

8
... which would take you to a list of money-saving youth hostels, at *http://dir.altavista.com/Top/Regional/Europe/UK/England/London/Accommodation/Hostels*

If you now click on *Barbican YMCA* ...

UK server, London, England

AltaVista server, Sunnyvale, Calif.

9
... your browser connects you to a web page located on a server in the United Kingdom (signified by "uk" in the URL), *www.ymca.org.uk/gallery/barbican*

10
Appearing on the Barbican YMCA home page is a list of underlined links that give you a fuller explanation of the offerings of the hostel: *Information about*, *Tariffs* (rates), *Facilities*, and *Contact us*. These pages all reside on the same server.

P://www.ymca.org.uk/gallery/barbican/about.html#about

P://www.ymca.org.uk/gallery/barbican/tariffs.html#tariffs

P://www.ymca.org.uk/gallery/barbican/facilities.html#facilities

P://www.ymca.org.uk/gallery/barbican/feedback.html

The Internet & the World Wide Web

57

Menu bar

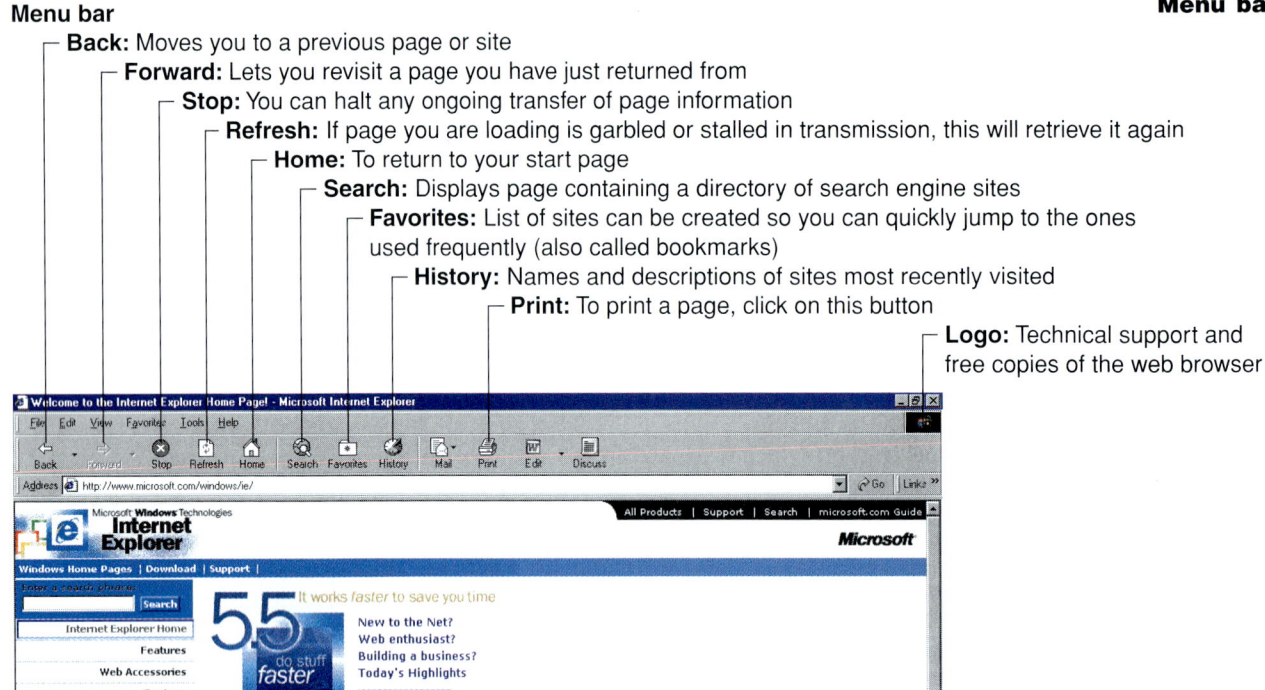

- **Back:** Moves you to a previous page or site
- **Forward:** Lets you revisit a page you have just returned from
- **Stop:** You can halt any ongoing transfer of page information
- **Refresh:** If page you are loading is garbled or stalled in transmission, this will retrieve it again
- **Home:** To return to your start page
- **Search:** Displays page containing a directory of search engine sites
- **Favorites:** List of sites can be created so you can quickly jump to the ones used frequently (also called bookmarks)
- **History:** Names and descriptions of sites most recently visited
- **Print:** To print a page, click on this button
- **Logo:** Technical support and free copies of the web browser

Menu bar

- **History lists:** If you are browsing through many web pages, it can be difficult to keep track of the locations of the pages you've already visited. The *history list* allows you to quickly return to the pages you have recently visited.

- **Bookmarks or favorites:** One great helper for finding your way is the *bookmark* or *favorites* system, which lets you store the URLs of web pages you frequently visit so that you don't have to remember and retype your favorite addresses. Say you're visiting a site that you really like and that you know you'd like to come back to. You click on your *Bookmark* or *Favorites* feature, which displays the URL on your screen, then click on *Add*, which automatically stores the address. Later you can locate the site name on your bookmark menu, click on it, and the site will reappear. (When you want to delete it, you can use the right mouse button and select the delete command.)

The Internet and the Web are now an integral part of student learning.

History List

If you want to return to a previously viewed site and are using Netscape, you click on *Communicator*, then choose *History* from the menu. If you're using Internet Explorer, click on *History*.

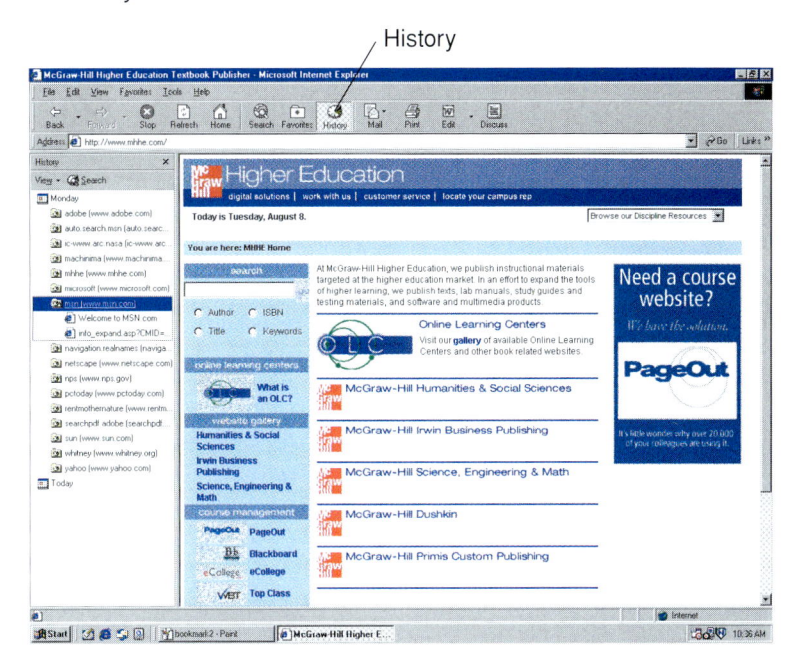

Adding Bookmarks (Favorites)

If you are at a website you may want to visit again, you click on your *Bookmarks* (in Netscape) or *Favorites* (in Internet Explorer) button and choose *Add Bookmark* or *Add to Favorites*. Later, to revisit the site, you can go to the bookmark menu, and the site will reappear.

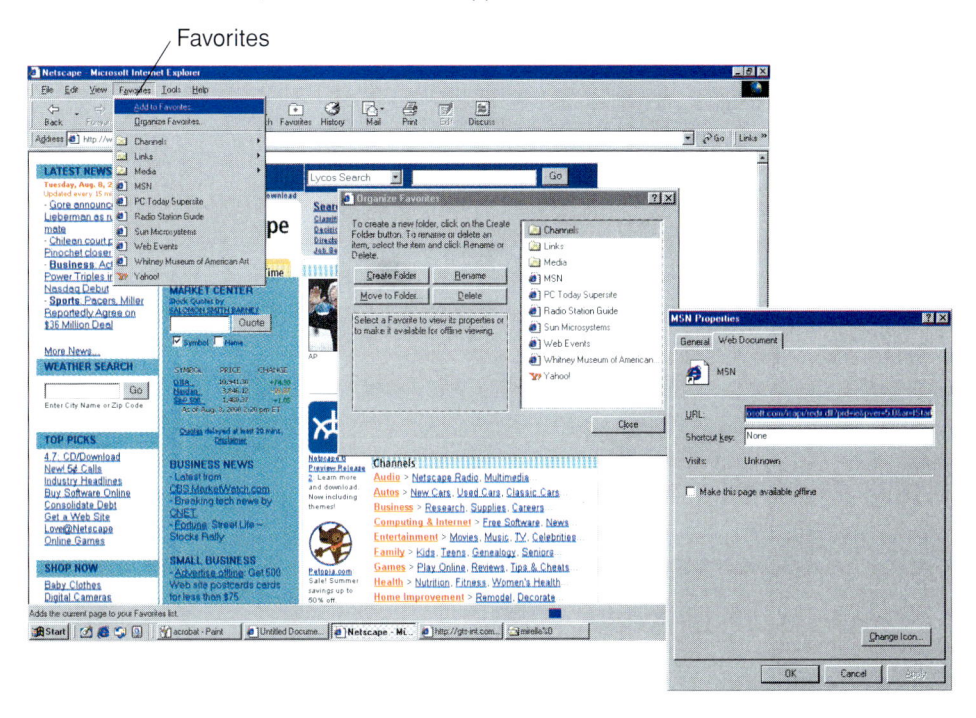

Hyperlinks
Clicking on underlined or color term transfers you to another web page.

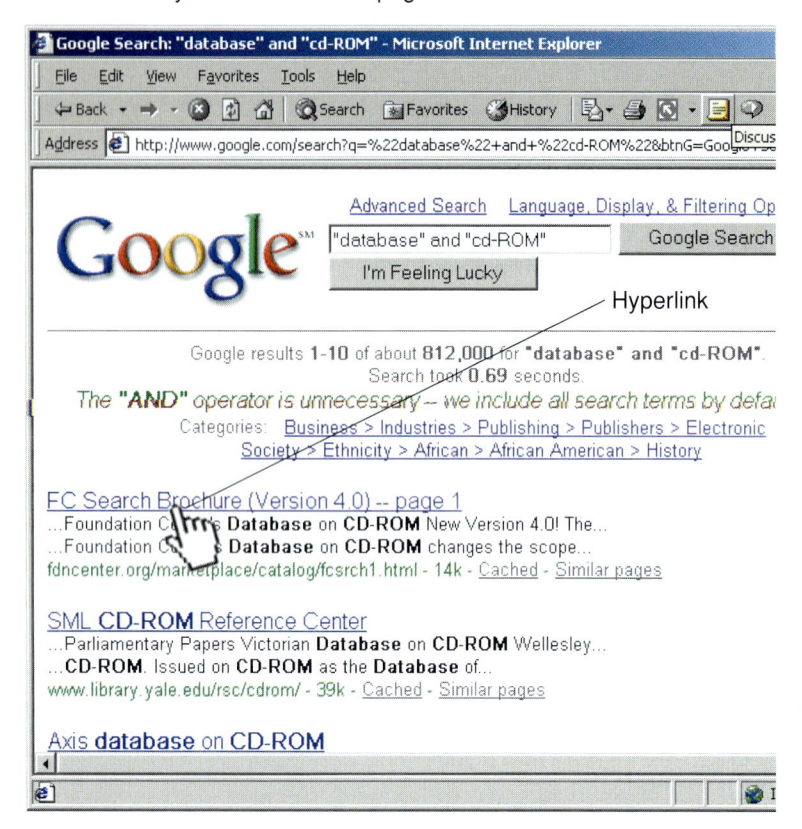

[Screenshot of Google search results page for "database" and "cd-ROM", with a hyperlink indicated by an arrow pointing to "FC Search Brochure (Version 4.0) -- page 1"]

Radio buttons
Act like station selector buttons on a car radio

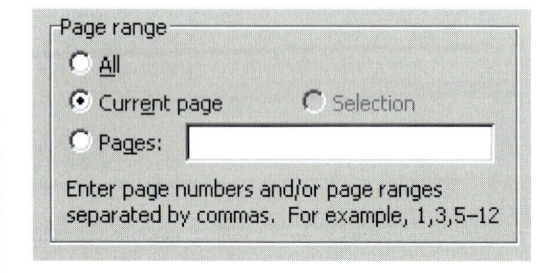

[Screenshot of a Page range dialog with radio buttons: All, Current page, Selection, Pages:]

Text boxes
Require you to type in information

[Screenshot of a Pages text box containing "17-33"]

- **Interactivity—hyperlinks, radio buttons, and fill-in text boxes:** For any given web page that you happen to find yourself on, there may be one of three possible ways to interact with it—or sometimes even all three on the same page.
  (1) By using your mouse to click on the hyperlinks, which will transfer you to another web page.
  (2) By using your mouse to click on a *radio button* and then clicking on a *Submit* command or pressing the Enter key. **Radio buttons are little circles located in front of various options; selecting an option with the mouse places a dot in the corresponding circle.**
  (3) By typing in text in a fill-in text box, then hitting the Enter key or clicking on a *Go* or *Continue* command, which will transfer you to another web page.

- **Scrolling and frames:** To the bottom and side of your screen display, you will note *scroll arrows*, small up/down and left/right arrows. Clicking on scroll arrows with your mouse pointer moves the screen so that you can see the rest of the web page, a movement known as *scrolling*. You can also use the arrow keys on your keyboard for scrolling.

  Some web pages are divided into different rectangles known as frames, each with its own scroll arrows. **A *frame* is an independently controllable section of a web page.** A web page designer can divide a page into separate frames, each with different features or options.

- **Looking at two pages simultaneously:** If you want to look at more than one web page at the same time, you can position them side by side on your display screen. Select *New* from your File menu to open more than one browser window.

Web Portals: Starting Points for Finding Information

Using a browser is sort of like exploring an enormous cave with flashlight and string. You point your flashlight at something, go there, and at that location you can see another cave chamber to go to; meanwhile, you're unrolling the ball of string behind you, so that you can find your way back.

Web portals

▲	America Online (AOL)	www.aol.com
YAHOO!	Yahoo!	www.yahoo.com
msn	Microsoft Network (MSN)	www.msn.com
N	Netscape	www.netscape.com
LYCOS	Lycos	www.lycos.com
Go	Go Network	www.go.com
excite	Excite Network	www.excite.com
▲	AltaVista	www.altavista.com
WEBCRAWLER	WebCrawler	www.webcrawler.com

But what if you want to visit only the most spectacular rock formations in the cave and skip the rest? For that you need a guidebook. There are many such "guidebooks" for finding information on the Web, sort of Internet superstations known as **_web portals_—websites that group together in one convenient location popular features such as search tools, e-mail, electronic commerce, and discussion groups.** Portals can be customized or personalized to fit your interests. The most popular portals are America Online, Yahoo!, Microsoft Network, Netscape, Lycos, Go Network, Infoseek, Snap, Excite Network, AltaVista, and WebCrawler.[15]

When you log on to a portal, you can do three things: (1) check the home page for general information, (2) use the directories to find a topic you want, and (3) use a keyword to search for a topic. (See ● Panel 2.9.)

- **Check the home page for general information:** You can treat a portal's home or start page as you would one of the mass media—something you tune in to in order to get news headlines, weather forecasts,

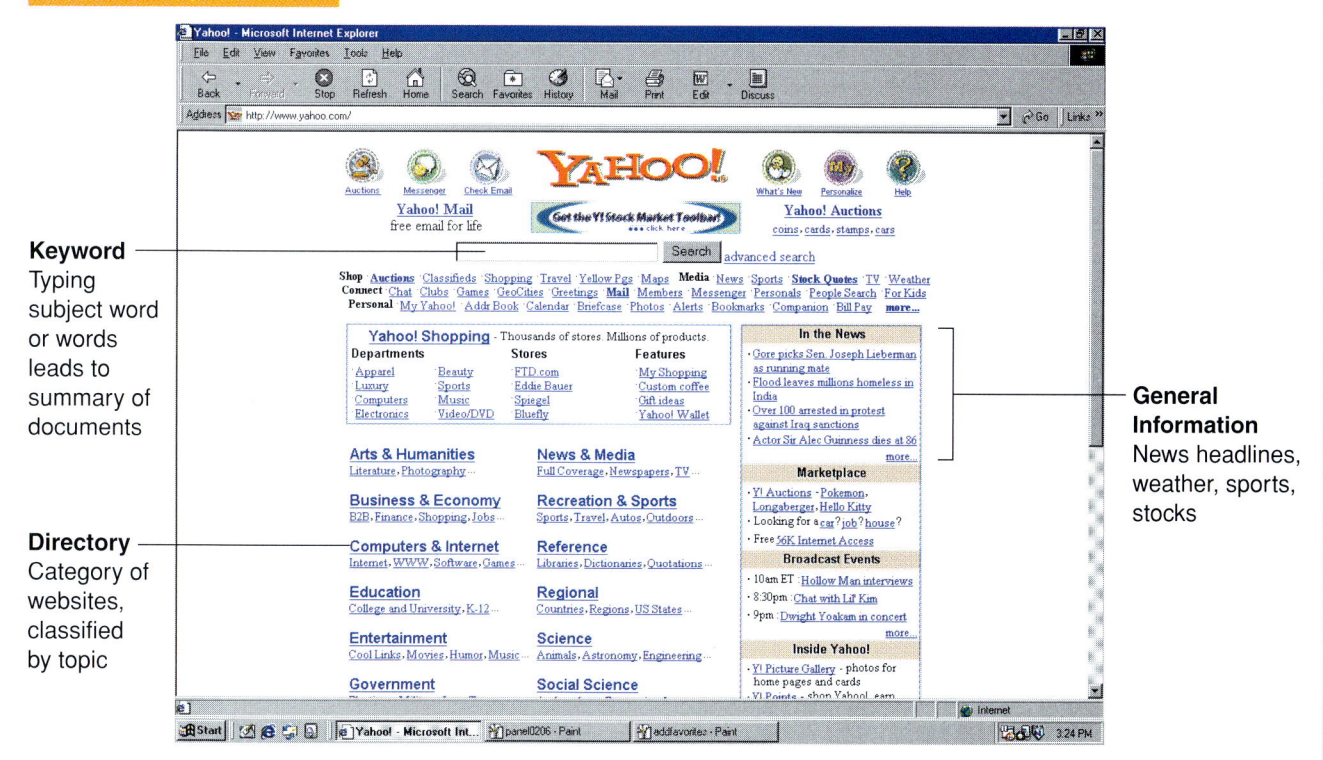

● **PANEL 2.9**
A portal home page

Keyword
Typing subject word or words leads to summary of documents

Directory
Category of websites, classified by topic

General Information
News headlines, weather, sports, stocks

The Internet & the World Wide Web

61

- **Use the directories to find a topic:** Before they acquired their other features, many of these portals began as a type of search tool known as a ***directory***, **providing lists of several categories of websites classified by topic**, such as *Business & Finance* or *Health & Fitness*. Such a category is also called a *hypertext index*, and its purpose is to allow you to access information in specific categories by clicking on a hypertext link.

  The initial general categories in Yahoo!, for instance, are *Arts & Humanities, Business & Economy, Computers & Internet, Education, Government, Health*, and so on. Using your mouse to click on one general category (such as *Recreation & Sports*) will lead you to another category (such as *Sports*), which in turn will lead you to another category (such as *College & University*), and on to another category (such as *Conferences*), and so on, down through the hierarchy. If you do this long enough, you will "drill down" through enough categories that you will find a document (website) on the topic you want.

  Unfortunately, not everything can be so easily classified in hierarchical form. A faster way may be a *keyword search*.

- **Use keyword to search for a topic:** At the top of each portal's home page is a blank space into which you can type a ***keyword*, the subject word or words of the topic you wish to find.** If you want a biography on former San Francisco football quarterback Joe Montana, then *Joe Montana* is the keyword. This way you don't have to plow through menu after menu of subject categories. The results of your keyword search will be displayed in a short summary of documents containing the keyword you typed.

Many users are increasingly bypassing the better known Web portals and going directly to specialty sites or small portals, such as those featuring education, finance, and sports.[16] Examples are Webstart Communications' computer and communications site *(www.cmpcmm.com/cc)*, Travel.com's travel site *(www.travel.com/sitemap.htm)*, and the *New York Times* home page used by the paper's own newsroom staff to find journalism-related sites *(www.nytimes.com/library/tech/reference/cynavi.html)*. Some colleges are also installing portals for their students.

Four Types of Search Engines: Human-Organized, Computer-Created, Hybrid, & Metasearch

When you use a keyword to search for a topic, you are using a piece of software known as a *search engine*. Whereas *directories* are lists of websites classified by topic (as offered by portals), ***search engines*** **allow you to find specific documents through keyword searches and menu choices.** The type of search engine you use depends on what you're looking for.

There are four types of such search tools: (1) human-organized, (2) computer-created, (3) hybrid, and (4) metasearch.[17]

- **Human-organized search sites:** If you're looking for a biography of Apple Computer founder Steve Jobs, a search engine based on human judgment is probably your best bet. Why? Because, unlike a computer-created search site, the search tool won't throw everything remotely associated with his name at you. More and more, the top search sites on the Web are going in the direction of human indexing. Whereas computers can't discriminate in organizing data, humans can judge data for relevance and categorize them in ways that are useful to you. Many of these sites hire people who are subject-area experts (with the idea that, for example, someone interested in gardening would be best

Survival Tip

Don't Be Traced

The graphics and sound files for many websites are stored on *your* computer, on your hard disk in a folder called *cache* that's associated with your browser. Don't want other users of your computer to know where you've been on the Web? Then put all the cached items in the trash and empty the trash bin. (The Cache folder is inside the browser folder—for example, the Netscape folder—in the Program Files folder.)

able to organize gardening sites). Examples of human-organized search sites are Yahoo!, Open Directory, About.com, and LookSmart.

- **Computer-created search sites:** If you want to see what things show up next to Steve Jobs's name or every instance in which it appears, a computer-created search site may be best. These are assembled by software "spiders" that crawl all over the Web and send back reports to be collected and organized with little human intervention. The downside is that computer-created indexes deliver you more information than you want. Examples of this type are Northern Light, Excite, WebCrawler, and Fast Search.

- **Hybrid search sites:** Hybrid sites generally use humans supplemented by computer indexes. The idea is to see that nothing falls through the cracks. All the principal sites are now hybrid: AOL Search, AltaVista, Lycos, MSN Search, and Netscape Search. Others are Ask Jeeves, Direct Hit, Go, GoTo.com, Google, HotBot, and Snap. Ask Jeeves pioneered the use of natural-language queries (you ask a question as you would to a person: "Where can I find a biography of Steve Jobs?"). Google ranks listings by popularity as well as by how well they match the request. GoTo ranks by who paid the most money for top billing.

- **Metasearch sites:** Metasearch sites send your query to several other different search tools and compile the results so as to present the broadest view. Examples are Go2Net/MetaCrawler, SuperCrawler, SavvySearch, Dogpile, Inference Find, ProFusion, Mamma, The Big Hub, and C4 TotalSearch.

More information about these search tools is given in the box on the next page. *(See* ● *Panel 2.10.)*

Tips for Smart Searching

The phrase "trying to find a needle in a haystack" will come vividly to mind the first time you type a word into a search engine and back comes a response on the order of "63,173 listings found." Clearly, it becomes mandatory that you have a strategy for narrowing your search. The following are some tips.[18]

- **Start with general search tools:** Begin with general search tools such as those offered by AltaVista, Excite, GoTo.com, HotBot, Lycos, and Yahoo! (Later, if you haven't been able to narrow your search, you can go to specific search tools, as we'll describe.)

- **Choose your search terms well and watch your spelling:** Use the most precise words possible. If you're looking for information about novelist Thomas Wolfe (author of *Look Homeward Angel*, published 1929) rather than novelist/journalist Tom Wolfe (*A Man in Full*, 1998), details are important: *Thomas*, not *Tom*; *Wolfe*, not *Wolf*. Use *poodle* rather than *dog*, *Maui* rather than *Hawaii*, *Martin guitar* rather than *guitar*, or you'll get thousands of responses that have little or nothing to do with what you're looking for. You may need to use several similar words to explore the topic you're investigating: *car racing, auto racing, drag racing, drag-racing, dragracing,* and so on.

- **Use phrases with quotation marks rather than separate words:** If you type *ski resort,* you could get results of (1) everything to do with skis on the one hand and (2) everything to do with resorts—winter, summer, mountain, seaside—on the other. Better to put your phrase in quotation marks—*"ski resort"*—to narrow your search.

- **Put unique words first in a phrase:** Better to have *"Tom Wolfe novels"* rather than *"Novels Tom Wolfe."* Or if you're looking for the Hoagy Carmichael song rather than the southern state, indicate *"Georgia on My Mind."*

● **PANEL 2.10**
Guide to search sites
Years refer to date launched.

Human-Organized Search Sites

- **Yahoo!** (*www.yahoo.com*). 1994. The most popular search site of all. Has the largest human-compiled web directory. Users should narrow search results category before they begin searching.
- **LookSmart** (*www.looksmart.com*). 1996. One of the easiest directories to use. Supplemented by AltaVista. Used by Excite and MSN Search.
- **About** (*www.about.com*). 1997. Began as The Mining Company. Trained human "guides" cover 50,000 subjects.
- **Open Directory** (*www.dmoz.org*). 1998. Uses 21,500 volunteer indexers. Owned by Netscape. Used by AltaVista, AOL Search, HotBot, Lycos, and Netscape.

Computer-Created Search Sites

- **Webcrawler** (*www.webcrawler.com*). 1994. Began at University of Washington; now owned by Excite.
- **Excite** (*www.excite.com*). 1995. One of the most popular search services. Owns Magellan and WebCrawler.
- **Northern Light** (*www.northernlight.com*). 1997. One of the largest indexes. Also enables users for a fee to get additional documents from sources not easily accessible, such as magazines, journals, and news wires.
- **FAST Search** (*www.alltheweb.com*). 1999. Also powers the Lycos MP3 search service. Has announced plans to index the entire Web.

Hybrid Search Sites

- **Lycos** (*www.lycos.com*). 1995. Began as search engine, shifted to human directory in 1999. Main listings come from Open Directory, secondary results from Direct Hit or Lycos' own index. Good for finding graphics, music files, and other specialized content.
- **AltaVista** (*www.altavista.com*). 1995. One of the largest search engines. Additional listings provided by Ask Jeeves and Open Directory. Multilingual searches, queries entered as simple questions.
- **Ask Jeeves** (*www.ask.com*). 1996. Allows users to ask questions in natural language rather than keywords. Includes Urban Cool (*www.urbancool.com*), which allows users to ask questions posed in the popular language of the street.
- **HotBot** (*www.hotbot.com*). 1996. One of the most well-rounded search engines. Owned by Lycos. First page of results from Direct Hit. Directory information from Open Directory.
- **GoTo** (*www.goto.com*). 1997. Companies pay to be placed higher in search results.
- **Snap** (*www.snap.com*). 1997. Human directory of websites. Owned by CNet and NBC.
- **Direct Hit** (*www.directhit.com*). 1998. Owned by Ask Jeeves. Refines results based on popularity. Highest-ranking results are those most frequently chosen. Used on Ask Jeeves, Lycos, and HotBot and an option on LookSmart and MSN Search.
- **Google** (*www.google.com*). 1998. Links popularity to ranking. The more sites that link to a web page, the higher that page will rank in searches.
- **Go** (*go.com*). 1999. Owned by Disney; offers search service of former Infoseek (1995). Includes human directory. Focuses on entertainment and leisure.
- **AOL Search** (*search.aol.com*). Covers both the Web and America Online's content. Directory listings mainly from the Open Directory.
- **MSN Search** (*search.msn.com*). Results from LookSmart directory, secondary results from AltaVista and Direct Hit.
- **Netscape Search** (*search.netscape.com*). Results from Open Directory and Netscape's own database, secondary results from Google.

Metasearch Sites

- **Go2Net/MetaCrawler** (*www.go2net.com*). 1995. Started at University of Washington.
- **SavvySearch** (*www.savvysearch.com*). 1995. Started at Colorado State University, Fort Collins.
- **Dogpile** (*www.dogpile.com*). Searches a customizable list of search engines.
- **Inference Find** (*www.infind.com*). Lists results grouped by subject, rather than by search engine or in one long list.
- **ProFusion** (*www.profusion.com*).
- **The Big Hub** (*www.thebighub.com*).
- **C4 TotalSearch Technology** (*www.c4.com*).

- **Use operators—AND, OR, NOT, and + and – signs:** Most search sites use symbols called *Boolean operators* to make searching more precise. To illustrate how they are used, suppose you're looking for the song "Strawberry Fields Forever."[19]

  AND connects two or more search words and means that all of them must appear in the search results. Example: *Strawberry AND Fields AND Forever.*

  OR connects two or more search words and indicates that any of the two may appear in the results. Example: *Strawberry Fields OR Strawberry fields.*

  NOT, when inserted before a word, excludes that word from the results. Example: *Strawberry Fields NOT Sally NOT W.C.* (to distinguish from the actress Sally Field and comedian W.C. Fields).

  + (plus sign), like *AND,* precedes a word that must appear: Example: *+ Strawberry + Fields.*

  – (minus sign), like *NOT,* excludes the word that follows it. Example: *Strawberry Field – Sally.*

- **Read the Help or Search Tips section:** All search sites provide a Help section and tips. This could save you time later.

- **Try an alternate general search site or a specific search site:** If you're looking for very specific information, a general type of search site such as Yahoo! or AltaVista may not be the best way to go. Instead you should turn to a specific search site.[20] Examples: To explore public companies, try Company Sleuth *(www.companysleuth.com),* Hoover's Online *(www.hoovers.com),* or KnowX *(www.knowx.com).* For news stories, try Yahoo! News *(http://dailynews.yahoo.com)* or TotalNews *(www.totalnews.com).* For pay-per-view information, try Dialog Web *(www.dialogweb.com),* Lexis-Nexis *(www.lexis-nexis.com),* and Dow Jones Interactive *(www.djnr.com).*

Multimedia on the Web

Many websites (especially those trying to sell you something) are multimedia, using a combination of text, images, sound, video, and animation. While most web browsers can handle basic multimedia elements on a web page, eventually you'll probably want more dramatic capabilities.

- **Plug-ins:** In the 1990s, as the Web was evolving from text to multimedia, browsers were unable to handle many kinds of graphic, sound, and video files. To do so, external application files called plug-ins had to be loaded into the system. **A *plug-in*—also called a *player* or a *viewer*—is a program that adds a specific feature to a browser, allowing it to play or view certain files.** For example, to view certain documents, you may need to download Adobe Acrobat Reader; to listen to CD-quality video you may need to download Liquid MusicPlayer. Plug-ins are required by many websites if you want to fully experience their content.

Common plug-ins

Acrobat Reader	www.adobe.com	View portable document format (PDF) files (chapter 3)
Flash Player	www.macromedia.com	View fancy graphics and animation; hear sound and music
Media Player	www.microsoft.com	See video; hear sound and music
RealPlayer Plus	www.real.com	See video; hear sound and music
QuickTime	www.apple.com	View animation and video; hear sound and music
Shockwave	www.macromedia.com	View animation and video; hear sound and music
Real Jukebox	www.real.com	Create music CDs and play MP3 music files
Liquid Player	www.liquidaudio.com	Create music CDs and play MP3 music files

Recent versions of Microsoft Internet Explorer and Netscape Communicator can handle a lot of multimedia. Now if you come across a file for which you need a plug-in, the browser will ask whether you want it, then tell you how to go about downloading it, usually at no charge.

- **Developing multimedia—applets, Java, JavaScript, and ActiveX:** How do website developers get all those nifty special multimedia effects? Often web pages contain links to <u>**applets**</u>, **small programs that can be quickly downloaded and run by most browsers.** Applets are written in <u>**Java**</u>, **a complex programming language that enables programmers to create animated and interactive web pages.** Java applets enhance web pages by playing music, displaying graphics and animation, and providing interactive games.

    If you are creating your own web multimedia, you may want to learn techniques such as JavaScript and ActiveX, which may be used to create web-page interest and activity, such as scrolling banners, pop-up menus, and the like.

Web page combining text and images
This example shows the National Park Service's opening screen about Mt. Rushmore in South Dakota.

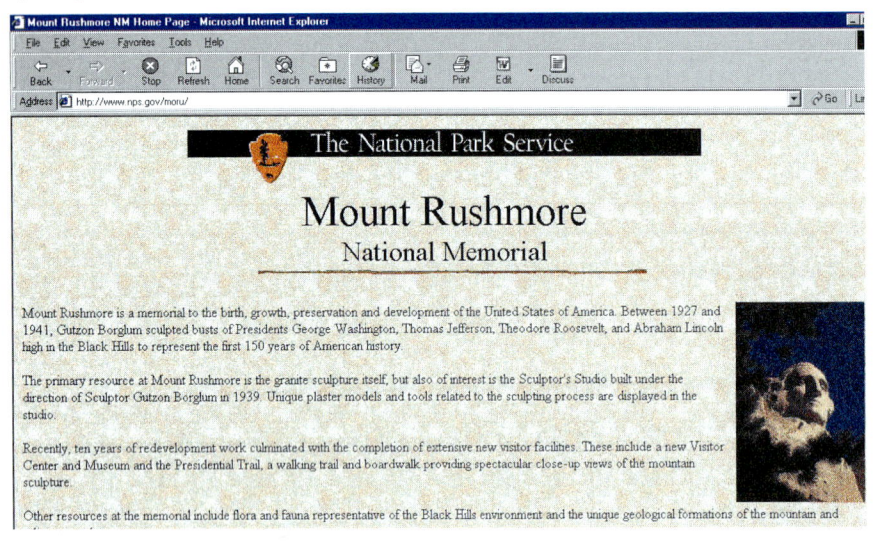

Animation
This web page shows an example of animation in a virtual approach to Mars.

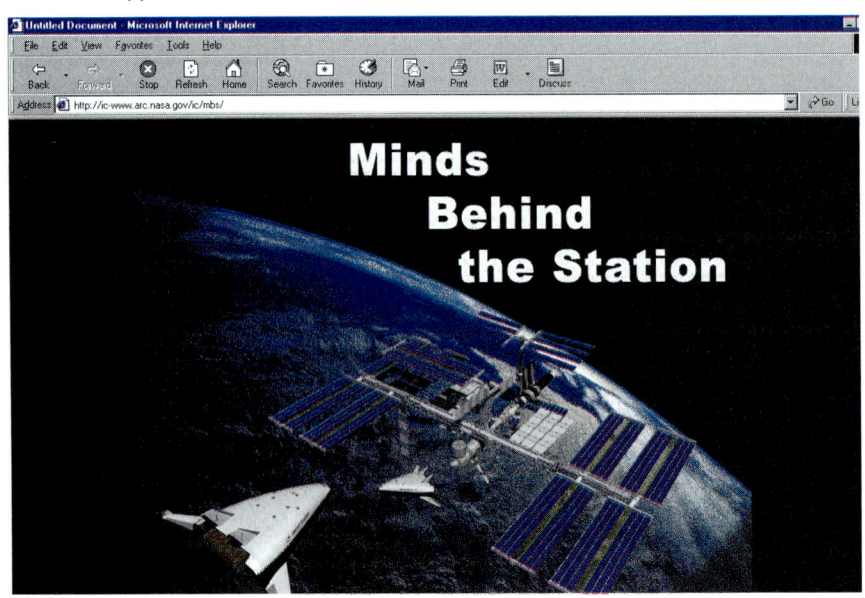

- **Text and images:** You can call up all kinds of text documents on the Web, such as newspapers, magazines, famous speeches, and works of literature. You can also view images, such as scenery, famous paintings, and photographs. Most web pages combine both text and images.

- **Animation:** <u>**Animation**</u> is **the rapid sequencing of still images to create the appearance of motion,** as in a Road Runner cartoon. Animation is used in online video games as well as in moving banners displaying sports scores or stock prices.

- **Video:** Video can be transmitted in two ways. (1) A file, such as a movie or video clip, may have to be completely downloaded before you can view it. This may take several minutes in some cases. (2) A file may be displayed as *streaming video* and viewed while it is still being downloaded to your computer. <u>**Streaming video**</u> **is the process of transferring data in a continuous flow so that you can begin viewing a file even before the end of the file is sent.** For instance, RealPlayer offers live, television-style broadcasts over the Internet as streaming

Survival Tip

Information on Web Radio

For more information, check:
www.diskjockey.com
www.grooveradio.com
www.netradio.com
www.shoutcast.com

video for viewing on your PC screen. You download RealPlayer's software, install it, then point your browser to a site featuring RealVideo. That will produce a streaming-video television image in a window a few inches wide.

- **Audio:** Audio, such as sound or music files, may also be transmitted in two ways: (1) downloaded completely before they can be played or (2) downloaded as *streaming audio*, **allowing you to listen to the file while the data is still being downloaded to your computer.** A popular standard for transmitting audio is RealAudio. Supported by most web browsers, it compresses sound so that it can be played in real time, even though sent over telephone lines. You can, for instance, listen to 24-hour-a-day net radio, which features "vintage rock," or English-language services of 19 shortwave outlets from World Radio Network in London. Many large radio stations outside the United States have net radio, allowing people around the world to listen in.

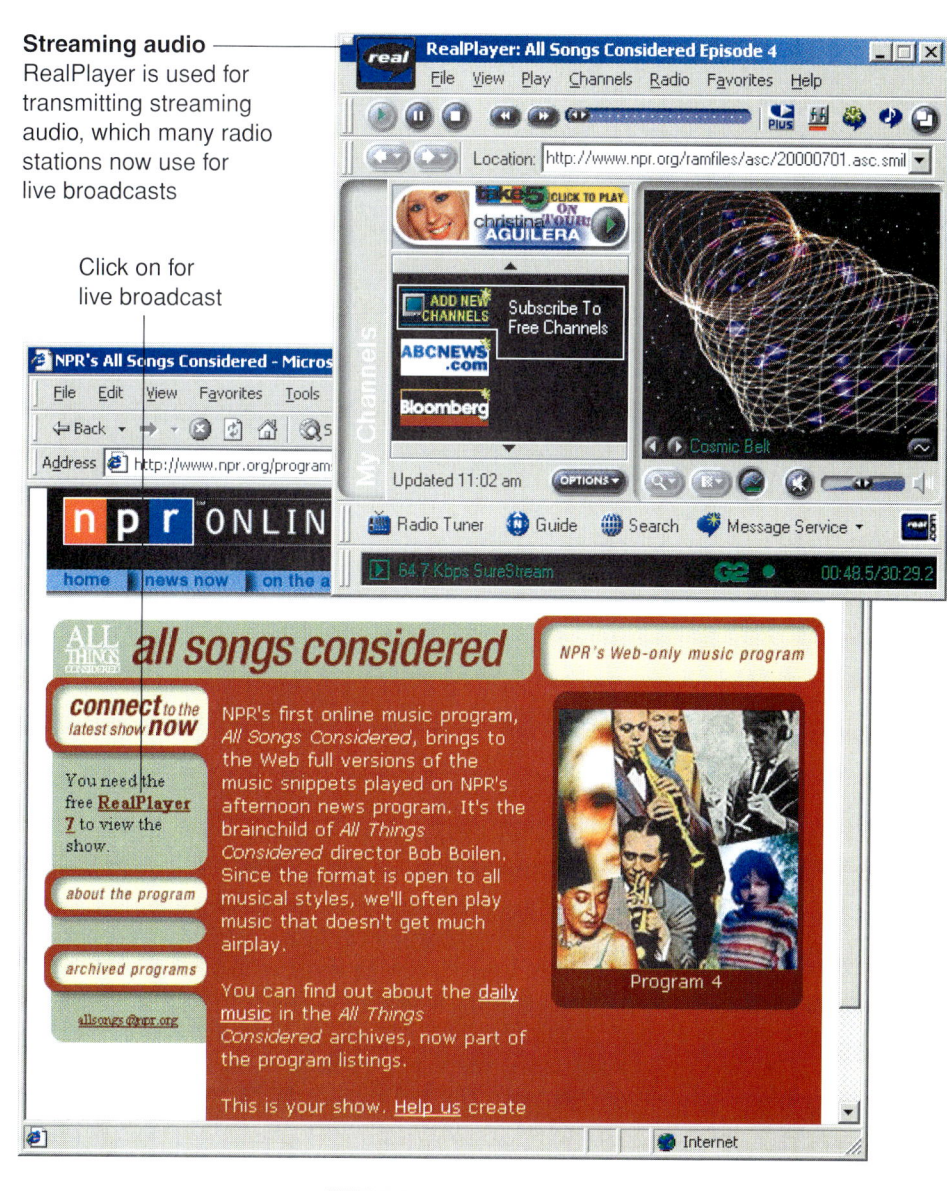

Streaming audio
RealPlayer is used for transmitting streaming audio, which many radio stations now use for live broadcasts

Click on for live broadcast

Streaming video
RealPlayer Plus is used to transmit streaming video (with streaming audio)

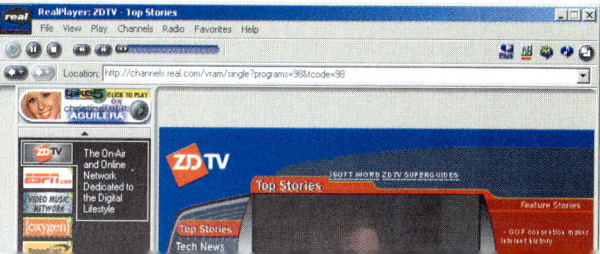

Push Technology & Webcasting

It used to be that you had to do the searching on the World Wide Web. Now, if you wish, the Web will come searching for you. The driving force behind this is **_push technology_, software that automatically downloads information to your computer** (as opposed to "pull" technology, in which you go to a website and pull down the information you want).

One result of push technology is **_webcasting_, in which customized text, video, and audio are sent to you automatically on a regular basis.** You can view or listen to the information immediately or access it later. The idea here is that you choose the categories, or (in Microsoft Internet Explorer) the *channels*, of websites that will automatically send you updated information. Thus, it saves you time because you don't have to go out searching for the information. Several services offer personalized news and information, based on a profile that you define when you register with them. Entrypoint.com, for example, will send news on a particular topic, such as all financial news about sugar beets, the weather in Omaha, or the game results for the Tennessee Titans.

Push technology
The push technology from Entrypoint periodically delivers news that you preset according to your specifications

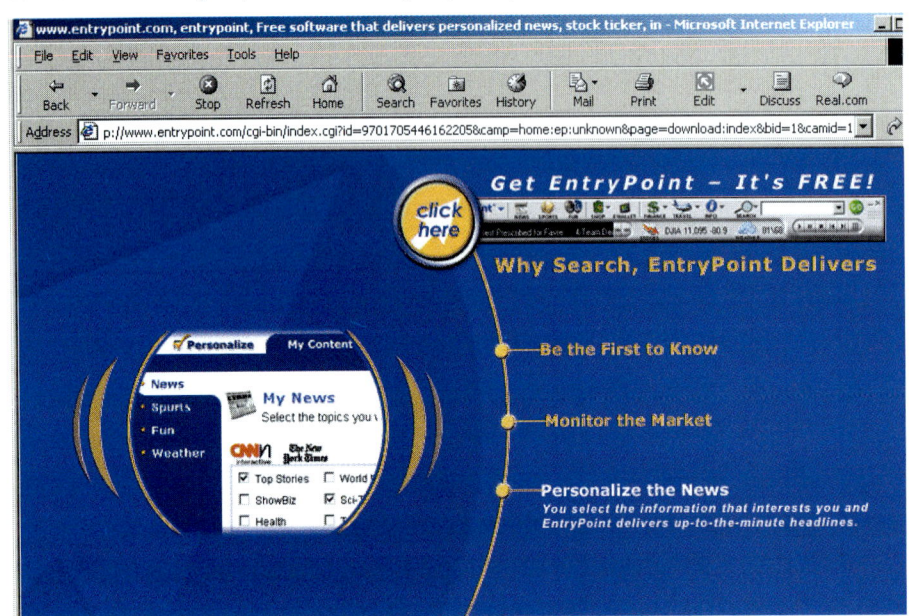

The Internet Telephone & Videophone

The key element is that the Internet breaks up conversations (as it does any other transmitted data) into "information packets," which can be sent over separate lines and then regrouped at the destination, whereas conventional voice phone lines carry a conversation over a single path. Thus, the Internet can move a lot more traffic over a network than the traditional telephone link can. (We describe how telecommunications technology works in Chapter 6.)

With **_Internet telephony_—using the Net to make phone calls, either one-to-one or for audioconferencing**—you can make long-distance phone calls that are surprisingly inexpensive. Indeed, it's theoretically possible to do this without owning a computer, simply by picking up your standard telephone and dialing a number that will "packetize" your conversation. However, it's more common practice to use a PC with a sound card and a microphone, a modem linked to a standard Internet service provider, and Internet telephone software such as Netscape Conference (part of Netscape Communicator) or Microsoft NetMeeting (part of Microsoft Internet Explorer).

Besides carrying voice signals, Internet telephone software also allows videoconferencing, in which participants are linked by a videophone that will transmit their pictures, thanks to a video camera attached to their PCs. It can also allow people to make sketches on "whiteboards" as they talk, as when three people meet online to discuss the floor plan for a new house.

Designing Web Pages

If you want to advertise a business online or just have your own personal website, you will need to design a web page, determine any hyperlinks, and hire 24-hour-a-day space on a web server or buy one of your own. Professional web page designers can produce a website for you, or you can do it yourself using a menu-driven program included with your web browser or a web-page design software package such as Microsoft FrontPage or Adobe PageMill. After you have designed your web page, you can put it on your ISP's server. (We describe web authoring software in Chapter 3.)

CLICK-ALONG 2-3
Keeping up with the changing Web

> **CONCEPT CHECK**
>
> Describe how the World Wide Web works.
>
> What is a browser, and how do you use it to get around the Web?
>
> How do you use a web portal to find information?
>
> Describe the types of search engines and some tips for searching.
>
> What are plug-ins used for?
>
> What is webcasting?

2.5 The Online Gold Mine: More Internet Resources, Your Personal Cyberspace, E-Commerce, & the E-conomy

KEY QUESTIONS
What are FTP, Telnet, newsgroups, real-time chat, and e-commerce?

Deborah Thebes of San Francisco drove from store to store over a period of several months, testing out one couch after another. "I had pretty specific wants and just never saw what I wanted in a store," she said. Then someone told her about Furniture.com on the Internet, and she ended up ordering a custom-built couch from the site. It wasn't quite the size and color she had in mind, but there was no sales tax—a considerable savings on an $800 piece of furniture. "It's very comfortable, it's beautiful, and I'm not sorry I bought it," says Thebes. And at less than $800, it seemed a bargain.[21]

Is this a glimpse of our cyberfuture? Will people be ordering pianos and stoves this way? Certainly the opportunities offered by the Net seem inexhaustible. Let's consider some of them. We'll examine four Internet resources other than e-mail and the Web, we'll look at personal aspects of cyberspace, and we'll explore the ever-expanding realm of e-commerce.

Other Internet Resources: FTP, Telnet, Newsgroups, & Real-Time Chat

E-mail and the World Wide Web seem to attract all the attention. But other cyber resources are also widely used: FTP, Telnet, newsgroups, and real-time chat.

- **FTP—for copying all the free files you want:** Many Net users enjoy "FTPing"—cruising the system and checking into some of the tens of thousands of FTP sites, which predate the Web and offer interesting free files to copy (download). ***FTP (File Transfer Protocol) is a method whereby you can connect to a remote computer called an FTP site and transfer files to your own microcomputer's hard disk.*** Free files

FTP
Downloading free files

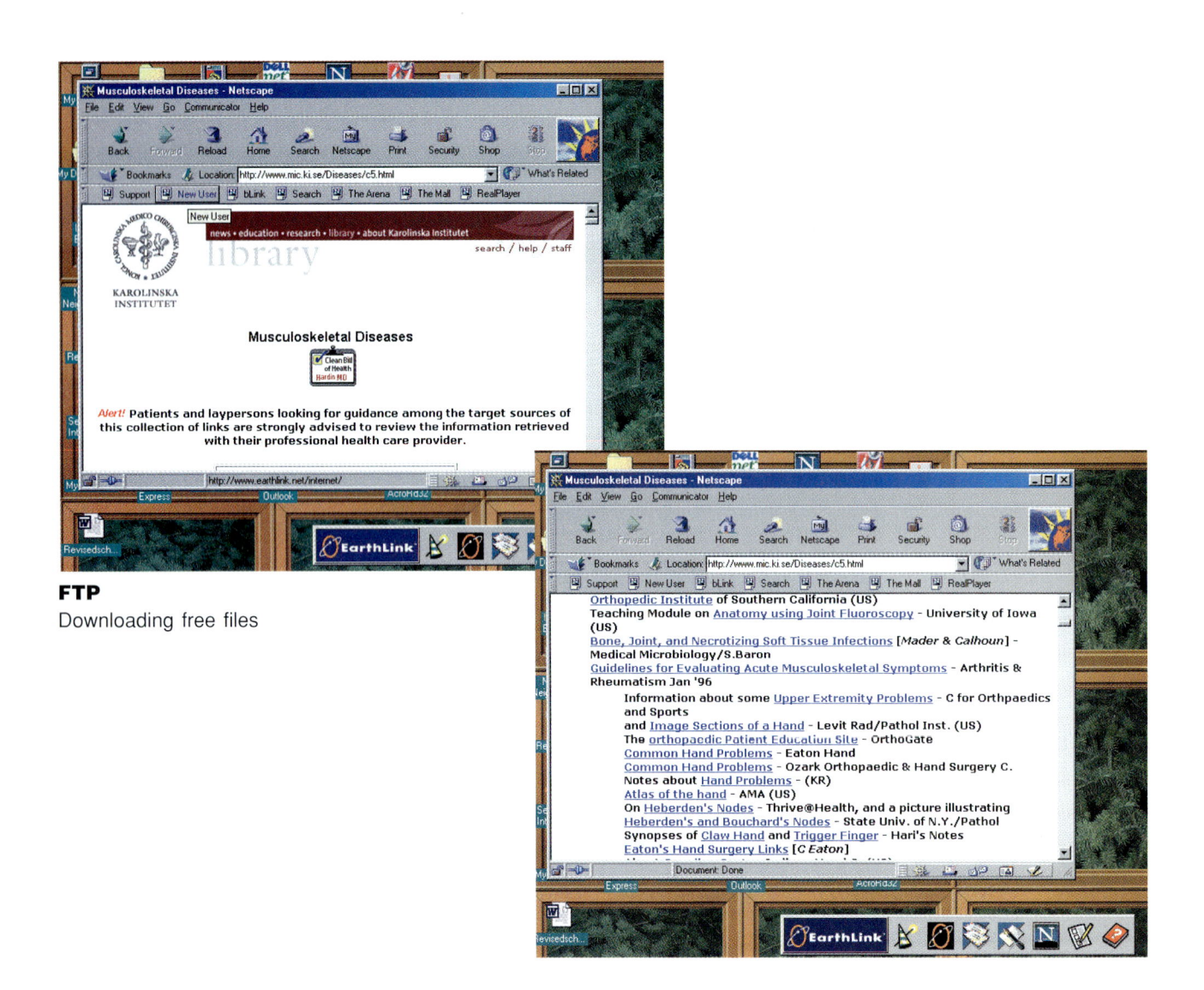

offered cover nearly anything that can be stored on a computer: software, games, photos, maps, art, music, books, statistics.

Some FTP files (called *anonymous FTP sites*) are open to the public, some are not. For instance, a university might maintain an FTP site with private files (such as lecture transcripts) available only to professors and students with assigned user names and passwords. It might also have public FTP files open to anyone with an e-mail address. You can download FTP files using either your web browser or special software (called an *FTP client program*), such as Fetch or Cute.

- **Telnet—to connect to remote computers: Telnet is a program or command that allows you to connect to remote computers on the Internet using a user name and a password.** This feature, which allows microcomputers to communicate successfully with mainframes, enables you to tap into Internet computers and access public files as though you were connected directly instead of, for example, through your ISP site.

    The Telnet feature is especially useful for perusing large databases at universities, government agencies, or libraries. As an electronic version of a library card catalog, Telnet can be used to search most major public and university library catalogs. (See, for example, Internet Public Library, *www.ipl.org*, and Library Spot, *www.library spot.com*)

- **Newsgroups—for online typed discussions on specific topics:** A **newsgroup is a giant electronic bulletin board on which users conduct written discussions about a specific subject.** There are more than

30,000 newsgroup forums—which charge no fee—and they cover an amazing array of topics. Some examples are *www.tlpoe.com,* for oyster lovers, and *www.wdxcyber.com,* for discussions and information on women's health care. (For a small fee, services such as Meganet.news.com and Binaries.net will get you access to 50,000–90,000 newsgroups all over the world.) Newsgroups take place on a special network of computers called **USEnet**, **a worldwide network of servers that can be accessed through the Internet.** To participate, you need a **newsreader**, **a program included with most browsers that allows you to access a newsgroup and read or type messages.** (Messages, incidentally, are known as *articles.*)

One way to find a newsgroup of interest to you is to use a portal such as Yahoo!, Excite, or Lycos to search USEnet for specific topics. Or you can use the search engine Google's Deja USEnet Archive *(http://groups.google.com/googlegroups/deja),* which will present the newsgroups matching the topic you specify. About a dozen major topics, identified by abbreviations ranging from *alt* (alternative topics) to *talk* (opinion and discussion), are divided into hierarchies of subtopics.

- **Real-time chat—typed discussions among online participants:** With newsgroups (and mailing lists, which we described under e-mail), participants may contribute to a discussion, then go away and return hours or days later to catch up on others' typed contributions. With **real-time chat (RTC)**, **participants have a typed discussion ("chat") while online at the same time,** just like a telephone conversation except that messages are typed rather than spoken. Otherwise the format is much like a newsgroup, with a message board to which participants may send ("post") their contributions. To start a chat, you use a service available on your browser such as IRC (Internet Relay Chat) that will connect you to a chat server.

  Unlike instant messaging (discussed under e-mail), which tends to involve one-on-one conversation, real-time chat usually involves several participants. As a result, RTC "is often like being at a crowded party," says one writer. "There are any number of people present and many threads of conversation occurring all at once."[22]

Your Personal Cyberspace

As we mentioned in Chapter 1, information technology has become more personal as it has evolved. Unlike the generally impersonal mass media, the Internet allows you to pursue your personal interests in the areas of relationships, education, health, and entertainment, for example.

- **Relationships—online matchmaking:** It's like walking into "a football stadium full of single people of the gender of your choice," says Trish McDermott, an expert for Match.com, a San Francisco online-dating service. People who connect online before meeting in the real world, she points out, have the chance to base their relationship on personality, intelligence, and sense of humor rather than purely physical attributes. "Online dating allows people to take some risks in an anonymous capacity," she adds. "When older people look back at their lives, it's the risks that they didn't take that they most regret."[23] (Still, there *are* some risks in trying to establish intimacy through online means because people may pretend to be quite different from who they really are.)

  People can also use search sites such as Infospace.com and Switchboard.com to try to track down old friends and relatives.[24] Others find common bonds by joining online communities such as The WELL *(www.well.com)*, the women's site Ivillage.com *(www.ivillage.com)*, the older people's site Third Age *(www.thirdage.com)*, and the gardening site GardenWeb *(www.gardenweb.com)*. Finally, the Net is no longer dominated by English; an estimated 50% of the users are non-English speakers. After English, the most common languages among Internet users are (in order) Japanese, Spanish, and German, with French and Chinese tied for fifth.[25]

- **Education—the rise of distance learning:** Sally Wells of Oregon has four children, a full-time job, and 14 cows to milk. She'd like to get a master's degree, but with no time to drive an hour to campus she takes four marketing courses online.[26] Adult learners—defined by educators as those over age 24—aren't the only ones involved in **distance learning, the name given to online education programs.** Younger college students also like it because they don't have to spend time commuting, the scheduling is flexible, and they often have a greater selection of course offerings. Although for instructors an online class is more labor-intensive than a regular chalk-and-talk class, they often find there is better interaction with students.[27]

- **Health—patient self-education:** Health is one of the most popular subject areas of research on the Web, although it can be difficult to get accurate information. (Many sites are trying to sell you something.) Among the sites offering reputable advice are Intelihealth *(www.intelihealth.com)*, Mayo Clinic Health Oasis *(www.mayohealth.org)*, Cyber Diet *(www.cyberdiet.com)*, the American College of Physicians *(www.acponline.org)*, and Medline *(www.nlm.nih.gov/medlineplus)*.

- **Entertainment—amusing yourself:** Some groups estimate that two-thirds of all Internet users in the United States seek out entertainment on the Web. No wonder so many major media companies have created websites to try to help promote or sell movies, music, TV shows, and the like. Of course, there are many other types of entertainment sites, devoted to games, hobbies, jokes, and so on.

E-Commerce

"What's your opinion on the Internet vs. the real world?" a *USA Today* reader asked. "[T]he Internet *is* the real world," replied reporter Lorrie Grant, who had been assigned to spend a month using the Web for shopping, working, banking, and other activities. "I had everything at my fingertips: office supplies, groceries, stocks, banking and bill payment, apparel, flowers, music, gifts, greeting cards, and more. Just point and click, *voilà*."[28]

The explosion in **electronic commerce (e-commerce)—conducting business activities online—**is not only widening consumers' choice of products and services but also creating new businesses and compelling established businesses to develop Internet strategies. Let's look at some of the developments.

- **E-tailing—retail commerce online:** Is the Internet spawning an entirely new way of doing business? Certainly many so-called *brick-and-mortar* retailers—those operating out of physical buildings—have been surprised at the success of such online companies as Amazon.com, seller of books, CDs, and other products. As a result, traditional retailers from giant Wal-Mart to funky little Buch Spieler Music in Montpelier, Vermont, have rushed to put their products online—and it has helped to revive some small-town main streets that had suffered from plant closings and competition from mega-malls.[29]

Retail goods can be classified into two categories—hard and soft. *Hard goods* are those that can be viewed and priced online, such as computers, clothes, and furniture, but are then sent to buyers by mail or truck. *Soft goods* are those that can be downloaded directly from the retailer's site, such as music, software, and greeting cards.

- **Auctions—linking individual buyers and sellers:** Today millions of buyers and sellers are linking up at online auctions, where everything is available from comic books to wines. The Internet is also changing the tradition-bound art and antiques business (dominated by such venerable names as Sotheby's, Christie's, and Butterfield & Butterfield). There are generally two types of auction sites: (1) person-to-person auctions, such as eBay *(www.ebay.com)*, that connect buyers and sellers for a listing fee and a commission on sold items, and (2) vendor-based auctions, such as OnSale *(www.onsale.com)*, that buy merchandise and sell them at discount. Some auctions are specialized, such as Priceline *(www.priceline.com)*, an auction site for airline tickets and other items.

- **Online finance—trading, banking, and e-money:** The Internet has changed the nature of stock trading. For the first time, says technology observer Denise Caruso, "anyone with a computer, a connection to the global network, and the requisite ironclad stomach for risk has the information, tools, and access to transaction systems required to play the stock market, a game that was once the purview of an elite few."[30] Companies such as E*Trade and Ameritrade are building one-stop financial supermarkets offering a variety of money-related services, including home mortgage loans and insurance. More than 1000 banks have websites, offering services that include account access, funds transfer, bill payment, loan and credit card applications, and investments. You can, for instance, apply for a Visa card called NextCard and get approved (or turned down) in about two minutes.

- **Online job hunting:** There are more than 2000 websites that promise to match job hunters with an employer. Some are specialty "boutique" sites looking for, say, scientists or executives. Some are general sites, the leaders being Monster.com, CareerPath.com, Headhunter.net, CareerMosaic.com, www.usajobs.opm.gov, and CareerBuilder.com.[31] Job sites can help you keep track of job openings and applications by downloading them to your own computer.

- **B2B commerce:** Of course, every kind of commerce has taken to the Web, ranging from travel bookings to real estate. One of the most important variations is **B2B (business-to-business) commerce, the electronic sales or exchange of goods and services directly between companies, cutting out traditional intermediaries.** Expected to grow even more rapidly than other forms of e-commerce, B2B commerce covers an extremely broad range of activities, such as supplier-to-buyer display of inventories, provision of wholesale price lists, and sales of closed-out items and used materials—usually without agents, brokers, or other third parties. Companies say e-purchasing, for example, slashes up to 20% off what they buy from each other.[32]

CLICK-ALONG 2-4
Other online resources

CONCEPT CHECK

Describe FTP, Telnet, newsgroups, and real-time chat.

What are some ways the Internet can be of personal use and of e-commerce use?

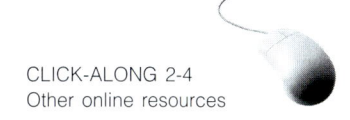

BOOKMARK IT!

PRACTICAL ACTION BOX
Web Research, Term Papers, & Plagiarism

No matter how much students may be able to rationalize cheating in college—for example, trying to pass off someone else's term paper as their own (plagiarism)—ignorance of the consequences is not an excuse. Most instructors announce the penalties for cheating at the beginning of their course—usually a failing grade in the course and possible suspension or expulsion from school.

Even so, probably every student becomes aware before long that the World Wide Web contains sites that offer term papers, either for free or for a price. Some dishonest students may download papers and just change the author's name to their own. Others are more likely just to use the papers for ideas. Perhaps, suggests one article, "the fear of getting caught makes the online papers more a diversion than an invitation to wide-scale plagiarism."[33]

How the Web Can Lead to Plagiarism

Two types of term-paper websites are as follows:

- **Sites offering papers for free:** Such a site requires users to fill out a membership form, then provides at least one free student term paper. (Quality is a crapshoot, since free paper mills often subsist on the submissions of poor students, whose contributions may be subliterate.)

- **Sites offering papers for sale:** Commercial sites may charge $6–$10 a page, which users may charge to their credit card. (Expense is no guarantee of quality. Moreover, the term-paper factory may turn around and make your $350 custom paper available to others—even fellow classmates working on the same assignment—for half the price.)

How Instructors Catch Cheaters

How do instructors detect and defend against student plagiarism? Professors are unlikely to be fooled if they tailor term-paper assignments to work done in class, monitor students' progress—from outline to completion—and are alert to papers that seem radically different from a student's past work.[34]

Eugene Dwyer, a professor of art history at Kenyon College, requires that papers in his classes be submitted electronically, along with a list of World Wide Web site references. "This way I can click along as I read the paper. This format is more efficient than running around the college library, checking each footnote."[35]

Just as the Internet is the source of cheating, it is also a tool for detecting cheaters. Search programs make it possible for instructors to locate texts containing identified strings of words from the millions of pages found on the Web. Thus, a professor can input passages from a student's paper into a search program that scans the Web for identical blocks of text. Indeed, some websites favored by instructors build a database of papers over time so that students can't recycle work previously handed in by others. One system can lock on to a stolen phrase as short as eight words. It can also identify copied material even if it has been changed slightly from the original. (More than 1000 educational institutions have turned to Oakland, California-based Turnitin.com—*www.turnitin.com*—a service that searches documents for unoriginality.)[36]

How the Web Can Lead to Low-Quality Papers

William Rukeyser, coordinator for Learning in the Real World, a nonprofit information clearinghouse, points out another problem: The Web enables students "to cut and paste together reports or presentations that appear to have taken hours or days to write but have really been assembled in minutes with no actual mastery or understanding by the student."[37]

Philosophy professor David Rothenberg, of New Jersey Institute of Technology, reports that as a result of students doing more of their research on the Web he has seen "a disturbing decline in both the quality of the writing and the originality of the thoughts expressed."[38] How does an instructor spot a term paper based primarily on Web research? Rothenberg offers four clues:

- **No books cited:** The student's bibliography cites no books, just articles or references to websites. Sadly, says Rothenberg, "one finds few references to careful, in-depth commentaries on the subject of the paper, the kind of analysis that requires a book, rather than an article, for its full development."

- **Outdated material:** A lot of the material in the bibliography is strangely out of date, says Rothenberg. "A lot of stuff on the Web that is advertised as timely is actually at least a few years old."

- **Unrelated pictures and graphs:** Students may intersperse the text with a lot of impressive-looking pictures and graphs that actually bear little relation to the precise subject of the paper. "Cut and pasted from the vast realm of what's out there for the taking, they masquerade as original work."

- **Superficial references:** "Too much of what passes for information [online] these days is simply *advertising* for information," points out Rothenberg. "Screen after screen shows you where you can find out more, how you can connect to this place or that." Other kinds of information are detailed but often superficial: "pages and pages of federal documents, corporate propaganda, snippets of commentary by people whose credibility is difficult to assess.

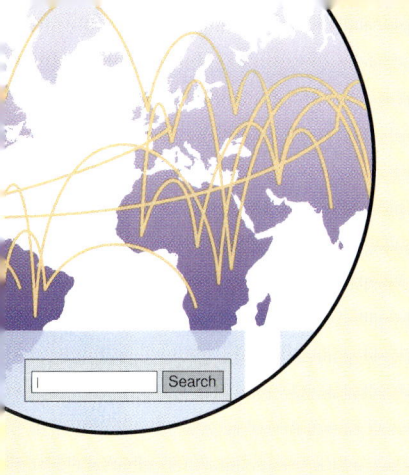

Summary

animation (p. 66, KQ 2.4) The rapid sequencing of still images to create the appearance of motion, as in a cartoon. *Why it's important: Animation is a component of multimedia; it is used in online video games as well as in moving banners displaying sports scores or stock prices.*

applets (p. 66, KQ 2.4) Small programs that can be quickly downloaded and run by most browsers. *Why it's important: Web pages contain links to applets, which add multimedia capabilities.*

B2B (business-to-business) commerce (p. 73, KQ 2.5) Electronic sales or exchange of goods and services directly between companies, cutting out traditional intermediaries. *Why it's important: Expected to grow even more rapidly than other forms of e-commerce, B2B commerce covers an extremely broad range of activities, such as supplier-to-buyer display of inventories, provision of wholesale price lists, and sales of closed-out items and used materials—usually without agents, brokers, or other third parties.*

bandwidth (p. 32, KQ 2.1) Expression of how much data—text, voice, video, and so on—can be sent through a communications channel in a given amount of time. *Why it's important: Different communications systems use different bandwidths for different purposes. The wider the bandwidth, the faster data can be transmitted.*

bps (p. 33, KQ 2.1) Bits per second. *Why it's important: Data transfer speeds are measured in bits per second.*

broadband (p. 32, KQ 2.1) Very high speed connection. *Why it's important: Access to information is much faster than with traditional phone lines.*

cable modem (p. 37, KQ 2.1) Device connecting a personal computer to a cable-TV system that offers an Internet connection. *Why it's important: Cable modems transmit data faster than standard modems.*

communications satellite (p. 38, KQ 2.1) Space station that transmits radio waves called *microwaves* from earth-based stations. *Why it's important: An orbiting satellite contains many communications channels and receives signals from ground microwave stations anywhere on earth.*

directory (p. 62, KQ 2.4) Search tool that provides lists of several categories of websites classified by topic, such as Business & Finance or Health & Fitness. Such a category is also called a *hypertext index,* and its purpose is to allow you to access information in specific categories by clicking on a hypertext link. *Why it's important: Directories are useful for browsing—looking at web pages in a general category and finding items of interest. Search engines may be more useful for hunting specific information.*

distance learning (p. 72, KQ 2.5) Online education programs. *Why it's important: Distance learning provides educational opportunities for people who are not able to get to a campus; also distance-learning students don't have to spend time commuting, the scheduling is flexible, and they often have a greater selection of course offerings.*

domain (p. 42, KQ 2.3) A location on the Internet. *Why it's important: A domain name is necessary for sending and receiving e-mail and for many other Internet activities.*

download (p. 34, KQ 2.1) To transmit data from a remote computer to a local computer. *Why it's important: Downloading enables users to save files on their own computers for later use, which reduces the time spent online and the corresponding charges.*

DSL (digital subscriber line) (p. 37, KQ 2.1) A hardware and software technology that uses regular phone lines to transmit data in megabits per second. *Why it's important: DSL connections are much faster than regular modem connections.*

e-commerce (electronic commerce) (p. 72, KQ 2.5) Conducting business activities online. Why it's important: *E-commerce is not only widening consumers' choice of products and services but is also creating new businesses and compelling established businesses to develop Internet strategies.*

:-)	Happy face
:-(	Sorrow or frown
:-O	Shock
:-/	Sarcasm
;-)	Wink

emoticons (p. 48, KQ 2.3) Keyboard-produced pictorial representations of expressions. Why it's important: *Emoticons can smooth online communication.*

FAQs (Frequently Asked Questions) (p. 48, KQ 2.3) Guides that explain expected norms of online behavior for a particular group. Why it's important: *Users should read a group's/site's FAQs to know how to proceed properly.*

flaming (p. 48, KQ 2.3) Writing an online message that uses derogatory, obscene, or inappropriate language. Why it's important: *Flaming should be avoided. It is a form of public humiliation inflicted on people who have failed to read FAQs or otherwise not observed netiquette (although it can happen just because the sender has poor impulse control and needs a course in anger management).*

frame (p. 60, KQ 2.4) An independently controllable section of a web page. Why it's important: *A web page designer can divide a page into separate frames, each with different features or options.*

FTP (File Transfer Protocol) (p. 69, KQ 2.5) Method whereby you can connect to a remote computer called an *FTP site* and transfer publicly available files to your own microcomputer's hard disk. Why it's important: *The free files offered cover nearly anything that can be stored on a computer: software, games, photos, maps, art, music, books, statistics.*

gigabits per second (Gbps) (p. 34, KQ 2.1) 1 billion bits per second. Why it's important: *Gbps is a common measure of data transmission speed.*

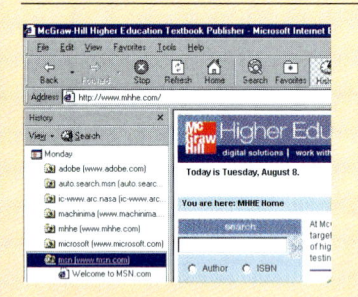

home page (p. 52, KQ 2.4) Also called *welcome page;* web page that, somewhat like the title page of a book, identifies the website and contains links to other pages at the site. Why it's important: *The first page you see at a website is the home page.*

hypertext (p. 51, KQ 2.4) System in which documents scattered across many Internet sites are directly linked, so that a word or phrase in one document becomes a connection to a document in a different place. Why it's important: *Hypertext links many documents by topics, allowing users to find information on topics they are interested in.*

hypertext markup language (HTML) (p. 51, KQ 2.4) Set of special instructions (called "tags" or "markups") used to specify web document structure, formatting, and links to other documents. Why it's important: *HTML enables the creation of web pages.*

HyperText Transfer Protocol (HTTP) (p. 53, KQ 2.4) Communications rules that allow browsers to connect with web servers. Why it's important: *Without HTTP, files could not be transferred over the Web.*

instant messaging (IM) (p. 45, KQ 2.3) Any user on a given e-mail system can send a message and have it pop up instantly on the screen of anyone else logged onto that system. Why it's important: *If both parties agree, they can initiate online typed conversations in real time. As they are typed, the messages appear on the display screen in a small window.*

Internet service provider (ISP) (p. 39, KQ 2.2) Company that connects you through your communications line to its server, or central computer, which connects you to the Internet via another company's network access points. Why it's important: *Unless they subscribe to an online information service (such as AOL) or have a direct network connection (such as a T1 line), microcomputer users need an ISP to connect to the Internet.*

Internet service provider

Telephony & conferencing
Make inexpensive phone calls; have online meetings.

Internet telephony (p. 68, KQ 2.4) Using the Net to make phone calls, either one-to-one or for audioconferencing. Why it's important: *Long-distance phone calls by this means are surprisingly inexpensive.*

ISDN (Integrated Services Digital Network) (p. 36, KQ 2.1) Hardware and software that allows voice, video, and data to be communicated over traditional copper-wire telephone lines (POTS). Why it's important: *ISDN provides faster data transfer speeds than do regular modem connections.*

Java (p. 66, KQ 2.4) Complex programming language that enables programmers to create animated and interactive web pages using applets. Why it's important: *Java applets enhance Web pages by playing music, displaying graphics and animation, and providing interactive games.*

keyword (p. 62, KQ 2.4) The subject word or words of the topic you wish to find in a web search. Why it's important: *The results of your keyword search will be displayed in a short summary of documents containing the keyword you typed.*

kilobits per second (Kbps) (p. 34, KQ 2.1) 1000 bits per second. Why it's important: *Kbps is a common measure of data transfer speed. The speed of a modem that is 28,800 bps might be expressed as 28.8 Kbps.*

list-serves (p. 47, KQ 2.3) E-mail mailing lists of people who regularly participate in discussion topics. To subscribe, the user sends an e-mail to the list-serve moderator and asks to become a member, after which he or she automatically receives e-mail messages from anyone who responds to the server. Why it's important: *Anyone connected to the Internet can subscribe to list-serve services. Subscribers receive information on particular subjects and can post e-mail to other subscribers.*

log on (p. 40, KQ 2.2) To make a connection to a remote computer. Why it's important: *Users must be familiar with log-on procedures to go online.*

megabits per second (Mbps) (p. 34, KQ 2.1) 1 million bits per second. Why it's important: *Mbps is a common measure of data transmission speed.*

netiquette (p. 48, KQ 2.3) "Network etiquette"—guides to appropriate online behavior. Why it's important: *In general, netiquette has two basic rules: (a) Don't waste people's time, and (b) don't say anything to a person online that you wouldn't say to his or her face.*

newsgroup (p. 70, KQ 2.5) Giant electronic bulletin board on which users conduct written discussions about a specific subject. Why it's important: *There are more than 30,000 newsgroup forums—which charge no fee—and they cover an amazing array of topics.*

newsreader (p. 71, KQ 2.5) Program included with most browsers that allows users to access a newsgroup and read or type messages. Why it's important: *Users need a newsreader to participate in a newsgroup.*

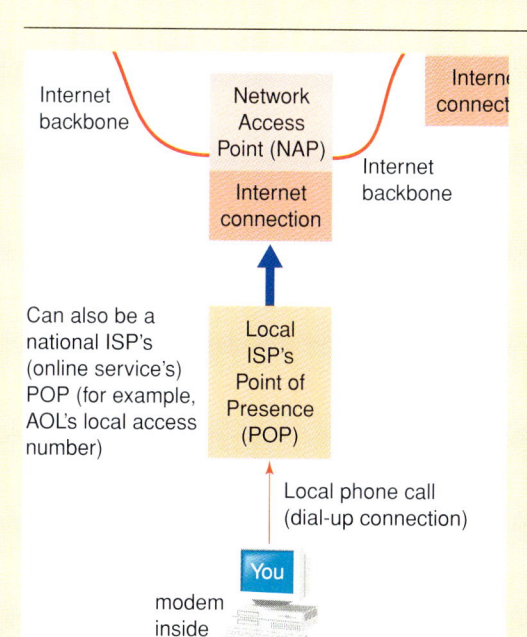

physical connection (p. 33, KQ 2.1) The wired or wireless means of connecting to the Internet. Why it's important: *Without physical connections, the Internet would be impossible, as would telephone and other types of communication connections.*

plug-in (p. 65, KQ 2.4) Also called a *player* or a *viewer*; program that adds a specific feature to a browser, allowing it to play or view certain files. Why it's important: *To fully experience the contents of many web pages, you need to use plug-ins.*

protocol (p. 53, KQ 2.4) Set of communication rules for exchanging information. Why it's important: *HyperText Transfer Protocol (HTTP) provides the communications rules that allow browsers to connect with web servers.*

push technology (p. 68, KQ 2.4) Software that automatically downloads information to your computer, as opposed to "pull" technology, in which you go to a website and pull down the information you want. Why it's important: *With little effort, users can obtain information that is important to them.*

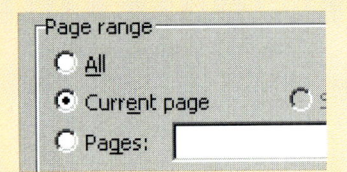

radio buttons (p. 60, KQ 2.4). An interactive tool displayed as little circles in front of options; selecting an option with the mouse places a dot in the corresponding circle. *Why it's important: Radio buttons are one way of interacting with a web page.*

real-time chat (RTC) (p. 71, KQ 2.5) Typed discussion ("chat") among participants who are online at the same time; it is just like a telephone conversation, except that messages are typed rather than spoken. *Why it's important: RTC provides a means of immediate electronic communication.*

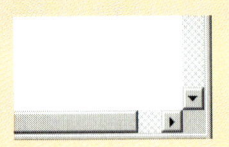

scroll arrows (p. 60, KQ 2.4) Small up/down and left/right arrows located to the bottom and side of your screen display. *Why it's important: Clicking on scroll arrows with your mouse pointer moves the screen so that you can see the rest of the web page, or the content displayed on the screen.*

scrolling (p. 60, KQ 2.4) Moving quickly upward or downward through text or other screen display, using the mouse and scroll arrows (or the arrow keys on the keyboard). *Why it's important: Normally a computer screen displays only part of, for example, a Web page. Scrolling enables users to view an entire document, no matter how long.*

search engine (p. 62, KQ 2.4) Search tool that allows you to find specific documents through keyword searches and menu choices, in contrast to directories, which are lists of websites classified by topic. *Why it's important: Search engines enable users to find websites of specific interest or use to them.*

site (p. 52, KQ 2.4) Computer with a domain name. *Why it's important: Sites provide Internet and web content.*

spam (p. 49, KQ 2.3) Unsolicited e-mail in the form of advertising or chain letters. *Why it's important: Spam filters are available that can spare users the annoyance of receiving junk mail, ads, and other unwanted e-mail.*

streaming audio (p. 67, KQ 2.4) Process of downloading audio in which you can listen to the file while the data is still being downloaded to your computer. *Why it's important: Users don't have to wait until the entire audio is downloaded to hard disk before listening to it.*

streaming video (p. 66, KQ 2.4) Process of downloading video in which the data is transferred in a continuous flow so that you can begin viewing a file even before the end of the file is sent. *Why it's important: Users don't have to wait until the entire video is downloaded to hard disk before watching it.*

T1 line (p. 37, KQ 2.1) Traditional trunk line that carries 24 normal telephone circuits and has a transmission rate of 1.5 Mbps. *Why it's important: High-capacity T1 lines are used at many corporate, government, and academic sites; these lines provide greater data transmission speeds than do regular modem connections.*

Telnet (p. 70, KQ 2.5) Program or command that allows you to connect to remote computers on the Internet. *Why it's important: This feature, which allows microcomputers to communicate successfully with mainframes, enables users to tap into Internet computers and access public files as though they were connected directly instead of, for example, through an ISP site.*

upload (p. 34, KQ 2.1) To transmit data from a local computer to a remote computer. *Why it's important: Uploading allows users to easily exchange files over networks.*

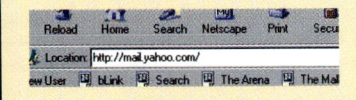

URL (Uniform Resource Locator) (p. 53, KQ 2.4) String of characters that points to a specific piece of information anywhere on the Web. A URL consists of (1) the web protocol, (2) the name of the web server, (3) the directory (or folder) on that server, and (4) the file within that directory (perhaps with an extension such as html or htm). *Why it's important: URLs are necessary to distinguish among websites.*

USEnet (p. 71, KQ 2.5) Worldwide network of servers that can be accessed through the Internet. *Why it's important: Newsgroups take place on USEnet.*

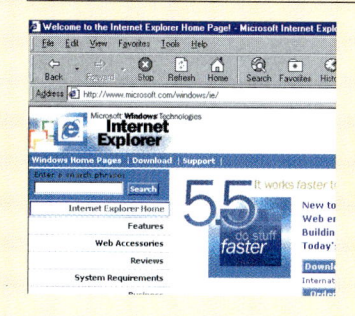

web browser (browser) (p. 53, KQ 2.4) Software that enables users to view web pages and to jump from one page to another. Why it's important: *Users can't surf the Web without a browser. The two most well-known browsers are Microsoft's Internet Explorer and Netscape Communicator.*

webcasting (p. 68, KQ 2.4) Service, based on push technology, in which customized text, video, and audio are sent to the user automatically on a regular basis. Why it's important: *Users choose the categories, or the channels, of websites that will automatically send updated information. Thus, it saves time because users don't have to go out searching for the information.*

web page (p. 52, KQ 2.4) Document on the World Wide Web that can include text, pictures, sound, and video. Why it's important: *A website's content is provided on web pages. The starting page is the home page.*

web portal (p. 61, KQ 2.4) Website that groups together popular features such as search tools, e-mail, electronic commerce, and discussion groups. The most popular portals are America Online, Yahoo!, Microsoft Network, Netscape, Lycos, Go Network, Infoseek, Snap, Excite Network, AltaVista, and WebCrawler. Why it's important: *Web portals provide an easy way to access the Web.*

website (p. 52, KQ 2.4) Location of a web domain name in a computer somewhere on the Internet. Why it's important: *Websites provide multimedia content to users.*

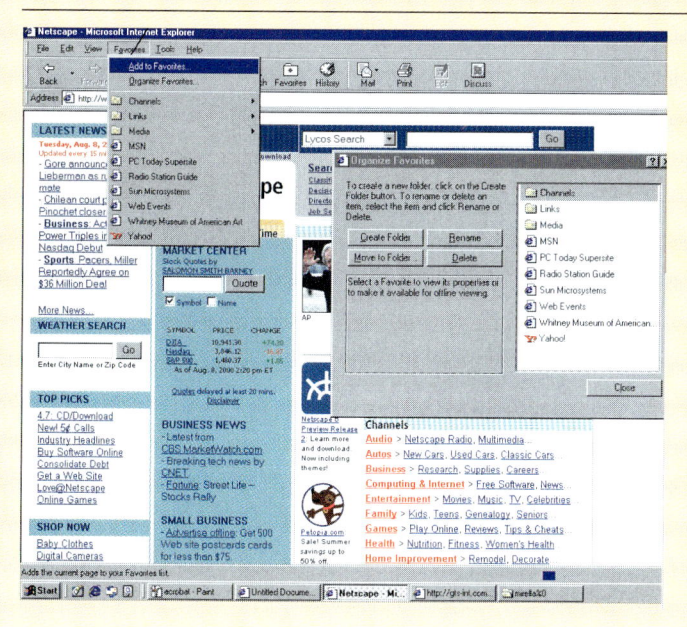

window (p. 45, KQ 2.3) Rectangular area containing a document or activity. Why it's important: *This feature enables different outputs to be displayed at the same time on the screen. For example, users can exchange messages almost instantaneously while operating other programs.*

The Internet & the World Wide Web

79

Chapter Review

stage 1 LEARNING — MEMORIZATION

"I can recognize and recall information."

Self-Test Questions

1. Today's data-transmission speeds are measured in _____, _____, _____, and _____ per second.
2. A(n) _____ connects a personal computer to a cable-TV system that offers an Internet connection.
3. A space station that transmits data as microwaves is a _____.
4. A company that connects you through your communications line to its server, which connects you to the Internet is a(n) _____.
5. A small rectangular area on the computer screen that contains a document or displays an activity is called a(n) _____.
6. _____ is writing an online message that uses derogatory, obscene, or inappropriate language.
7. A(n) _____ is software that enables users to view web pages and to jump from one page to another.
8. A computer with a domain name is called a(n) _____.
9. _____ comprises the communications rules that allow browsers to connect with web servers.
10. A(n) _____ is a program that adds a specific feature to a browser, allowing it to play or view certain files.

Multiple-Choice Questions

1. Kbps means _____ bits per second.
   a. 1 billion
   b. 1 thousand
   c. 1 million
   d. 1 hundred
   e. 1 trillion
2. A location on the Internet is called a(n) _____.
   a. network
   b. user ID
   c. domain
   d. browser
   e. web
3. In the e-mail address *Kim_Lee@earthlink.net.us*, *Kim_Lee* is the _____.
   a. domain
   b. URL
   c. site
   d. user ID
   e. location
4. Which of the following is not one of the four components of a URL?
   a. web protocol
   b. name of the web server
   c. name of the browser
   d. name of the directory on the web server
   e. name of the file within the directory
5. Which of the following is the fastest method of data transmission?
   a. ISDN
   b. DSL
   c. modem
   d. T1 line
   e. cable modem
6. Which of the following is not a netiquette rule?
   a. consult FAQs
   b. flame only when necessary
   c. don't shout
   d. avoid huge file attachments
   e. avoid sloppiness and errors

stage 2 LEARNING: COMPREHENSION

"I can recall information in my own terms and explain them to a friend."

Short-Answer Questions

1. Briefly define *bandwidth*.
2. Name three methods of data transmission that are faster than a regular modem connection.
3. What does *log on* mean?
4. What is netiquette, and why is it important?
5. Many web documents are "linked." What does that mean?

Concept Mapping

On a separate sheet of paper, draw a concept map or visual diagram linking concepts. Show how the following terms are related.

attachment	List-serves
bps	newsgroup
cable modem	newsreader
communications satellite	physical connection
directory	protocol
domain	real-time chat (RTC)
download	site
DSL	search engine
e-mail	Telnet
Explorer	T1 line
FTP	upload
home page	URL
hypertext	USEnet
instant messaging	web browser
Internet service provider	web page
ISDN	website

stage 3 LEARNING: APPLYING, ANALYZING, SYNTHESIZING, EVALUATING

"I can apply what I've learned, relate these ideas to other concepts, build on other knowledge, and use all these thinking skills to form a judgment."

Knowledge in Action

1. Distance learning uses electronic links to extend college campuses to people who otherwise would not be able to take college courses. Is your school or someone you know involved in distance learning? If so, research the system's components and uses. What hardware and software do students need in order to communicate with the instructor and classmates? What courses are offered? Prepare a short report on this topic.

2. It's difficult to conceive how much information is available on the Internet and the Web. One method you can use to find information among the millions of documents is to use a search engine, which helps you find web pages based on typed keywords or phrases. Use your browser to visit the following search sites: *www.yahoo.com* and *www.goto.com*. Click in the Search box and then type the phrase "personal computers," then hit "Go," "Find it!" or the enter key. When you locate a topic that interests you, print it out by choosing File, Print from the menu bar or by clicking the Print button on the toolbar. Report on your findings.

3. As more and more homes get high-speed broadband Internet connections, the flow of data will become exponentially faster and will open up many new possibilities for sharing large files such as video. What types of interactive services can you think of existing once everyone has lightning-speed connections? Will more and more people begin using their PCs as phones? If you were videoconferencing with a family member in another state using a 42-inch flat-screen monitor, would it feel as if the individual was actually there with you in person? How far away (in years) do you think this reality is?

Web Exercises

1. Some websites go overboard with multimedia effects, while others don't include enough. Locate a website that you think makes effective use of multimedia. What is the purpose of the site? Why is the site's use of multimedia effective? Take notes, and repeat the exercise for a site with too much multimedia and one with too little.

2. Visit the following job-hunting websites:

   www.monster.com

   www.occ.com

   www.careerpath.com

   www.cweb.com

   Investigate job offerings in a field you are interested in. Which is the easiest site to use? Why? What recommendations would you make for improving the site?

3. If you have never done a search before and want a specific one to experiment with, try this. Find out how much car dealers pay for cars, how much is a fair price for a car, and what techniques you should use when buying a new car. A company called Edmund publishes a magazine with all that information, but you can get the same information on their website for free. The goal is to use a search engine to get to Edmund's homepage.

Click on the Yahoo! search engine. Search for *automobile buyer's guides.* The search engine may treat this as a request to find references that match the word *automobile* or *buyer's* or *guides.* If this is not what you want, see if you can find instructions on the search engine web pages that have the search engine "find" matches that contain all three words. In Yahoo!, there is a field you can click on called "options." it takes you to a page that allows you to select matches that have all the words in your requested phrase.

Read some of the entries that the search entry returns. If any of them aim you at the Edmund's website, then go there. If not, then experiment with changing your request. If you get tired of playing around, try searching on *Edmunds automobile buyer's guide.* If that fails, then just go to *www.edmunds.com.*

Explore the Edmunds site and see if you get any useful information on car prices.

4. Ever wanted your own dot-com in your name? Visit these sites to see if your name is still available:

   www.register.com

   www.namezero.com

   www.domainname.com

5. If you make PC-to-phone calls through your Internet connection, you use your current connection. There are other connections. Visit these sites and test out their free services:

   www.dialpad.com

   www.hottelephone.com

6. Inspect the accuracy of navigation software. Visit *www.mapquest.com* and get directions from where you're located to a location that you already know how to get to. Compare the computer's directions to your current strategy. Are the computer's directions more efficient, less efficient? If the directions are different, how do you suppose the program decided to choose which routes to take?

7. HTTP (HyperText Transfer Protocol) on the World Wide Web isn't the only method of browsing and transferring data. FTP is the original method and is still useful and an essential Internet practice. To use FTP, you'll need an FTP client software program, just as you would need a web browser to surf the World Wide Web. Download one of these shareware clients and visit their default sites, which come preloaded.

   CuteFtp   *www.cuteftp.com*
   WS_FTP   *www.ipswitch.com*
   FTP Voyager   *www.ftpvoyager.com*

   You will need an FTP client program to upload files to a server if you ever decide to build a website. Some online website builders have browser uploaders, but the conventional method has always been FTP. After you download an FTP client, visit *www.oth.net* to search for FTP servers that house files you would like to download.

8. The videophone isn't an invention of the future. It exists now and is extremely affordable. You can get a PC Camera for about $30–$80 and free software to contact friends. If you and the other party have a high-speed connection, you can have a close to full-motion videoconference for free, even if it's long distance. Here are some sites that have videoconferencing software available for download:

   http://netconference.miningco.com/cs/videoconferencing/index.htm

   www.webattack.com/shareware/comm/swvoice.shtml

   Next time you are going to attach a file to an e-mail and send it to multiple users, try posting the file to a web server and give the URL out instead. This method will ease tension on bandwidth and storage of data and is a more efficient method of sharing data. Your friends will also be happier because they won't have a large file stored in their account.

Websites Especially for Students	
www.student.com *www.studentadvantage.com*	General student information sites covering everything from roommates to romantic relationships to studying (plus free e-mail).
www.collegeclub.com	Connect with students at other campuses.
www.smarterliving.com	No-gimmick site that helps students save money and time. "Students Crossing Cultures" feature gives advice for foreign travel.
www.mountainzone.com	Great coverage of mountain biking, snowboarding, and skiing.
http:/sportsillustrated.cnn.com	Comprehensive sports website.
www.asimba.com	Interactive tools help you design a proper exercise and diet plan.
http://spinner.com	Download the free player software and listen to Europop, new Japanese artists, classic R&B, all types of music from everywhere.
www.live365.com	Tune in to your choice of 18,000 user-created Internet radio stations, or start one of your own.
www.britannica.com	Research almost any topic via the Britannica encyclopedia.
www.ipl.org	Do the same via the Internet Public Library.
www.iaugara.bc.ca/mathstats/resource/onWeb	Internet Resources for Math Students: quick tutorials and online classes.
www.finaid.com	Easy instructions about how to get a loan or a scholarship.

Chapter 3

Software

The Power behind the Power

Key Questions

You should be able to answer the following questions.

- **3.1 System Software** What are three components of system software; what does the operating system (OS) do; what is an OS interface; and what are some common desktop, network, and portable OSs?

- **3.2 Application Software: Getting Started** What are five ways of obtaining application software, tools available to help you learn to use software, three common types of files, and the types of software?

- **3.3 Word Processing** What can you do with word processing software that you can't do with pencil and paper?

- **3.4 Spreadsheets** What can you do with an electronic spreadsheet that you can't do with pencil and paper and a standard calculator?

- **3.5 Database Software** What is database software, and what is personal information management software?

- **3.6 Specialty Software** What are the principal uses of specialty software such as presentation graphics, financial, desktop publishing, drawing and painting, project management, computer-aided design, and video/audio editing software?

- **3.7 Online Software & Application Software Providers: Turning Point for the Software Industry?** What are some recent trends in online software?

"What we need is a science called *practology*, a way of thinking about machines that focuses on how things will actually be used."

So says Alan Robbins, a professor of visual communications, on the subject of *machine interfaces*—the parts of a machine that people actually manipulate.[1] An interface is a machine's "control panel," ranging from the volume and tuner knobs on an old radio to all the switches and dials on the flight-deck of a jetliner. You may have found, as Robbins thinks, that on too many of today's machines—digital watches, VCRs, even stoves—the interface is often designed to accommodate the machine or some engineering ideas rather than the people actually using them. Good interfaces are intuitive—that is, based on prior knowledge and experience—like the twin knobs on a 1950s radio, immediately usable by both novices and sophisticates. Bad interfaces, such as a software program with a bewildering array of menus and icons, force us to relearn the required behaviors every time. Of course, you can prevail over a bad interface if you repeat the procedures often enough.

How well are computer software makers doing at giving us useful, helpful interfaces? The answer is: getting better all the time. In time, as interfaces are refined, it's possible computers will become no more difficult to use than a car. Until then, however, for smoother computing you need to know something about how system software works. Today people communicate one way, computers another. People speak words and phrases; computers process bits and bytes. For us to communicate with these machines, we need to know how to interact with them, through software.

3.1 System Software

KEY QUESTION
What are three components of system software; what does the operating system (OS) do; what is an OS interface; and what are some common desktop, network, and portable OSs?

Software, or *programs*, consist of the instructions that tell the computer how to perform a task. As we've said in Chapter 1, software is of two types. *Application software* is software that can perform useful work on general-purpose tasks, such as word processing or spreadsheets, or that is used for entertainment. Hundreds of application software packages are available for personal computers. *System software,* which you will find already installed if you buy a new computer, enables the application software to interact with the computer and helps the computer manage its internal and external resources. There are only a handful of systems software packages for personal computers.

There are three basic components of system software that you need to know about. (See ● *Panel 3.1.*)

- **Operating systems:** An *operating system* is the principal component of system software in any computing system.
- **Device drivers:** *Device drivers* help the computer control peripheral devices.
- **Utility programs:** *Utility programs* are generally used to support, enhance, or expand existing programs in a computer system.

A fourth type of system software, *language translators*, is described in the appendix.

● **PANEL 3.1**

Three components of system software
System software is the interface between the user and the application software and the computer hardware.

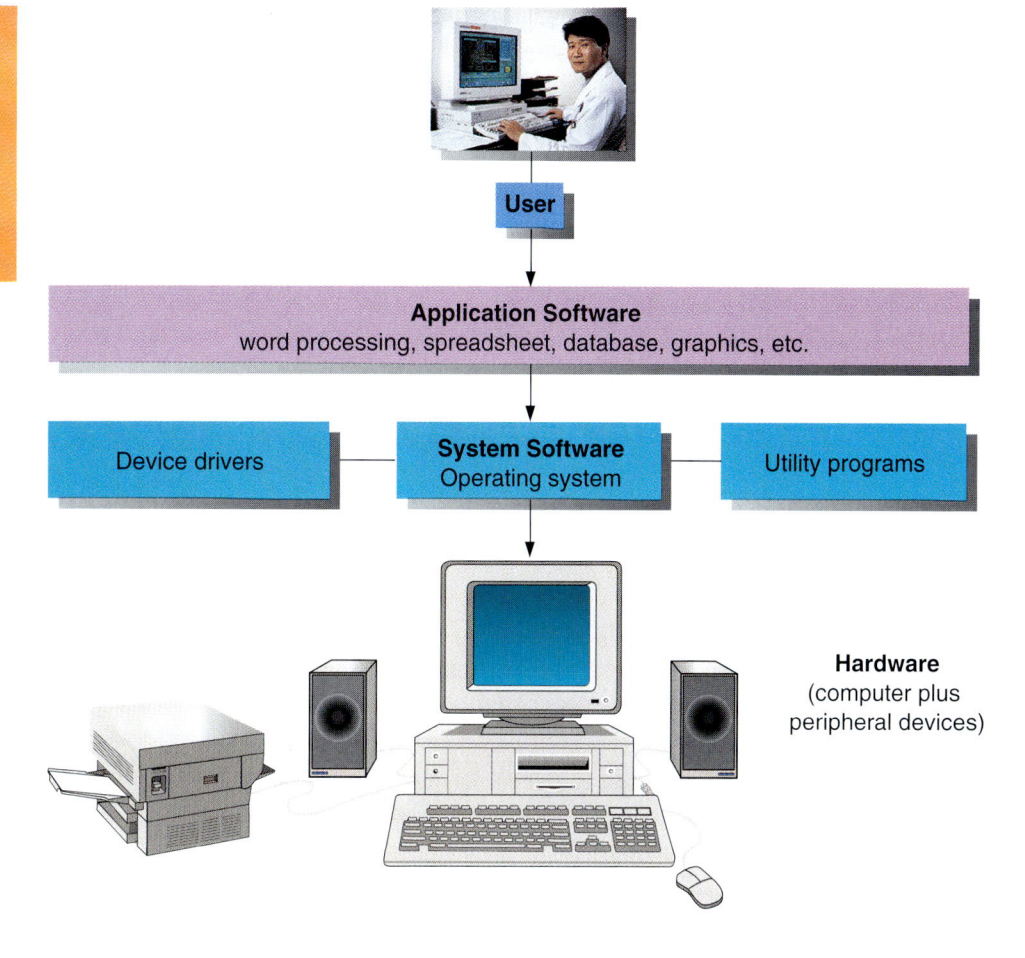

Application Software
word processing, spreadsheet, database, graphics, etc.

Device drivers — **System Software** Operating system — Utility programs

Hardware
(computer plus peripheral devices)

The Operating System: What It Does

The *operating system (OS)*, also called the *software platform*, consists of the **master system of programs that manage the basic operations of the computer.** These programs provide resource management services of many kinds. In particular, they handle the control and use of hardware resources, including disk space, memory, CPU time allocation, and peripheral devices. The operating system allows you to concentrate on your own tasks or applications rather than on the complexities of managing the computer.

Some important functions of the operating system are as follows:

- **Booting:** The work of the operating system begins as soon as you turn on, or "boot," the computer. ***Booting* is the process of loading an operating system into a computer's main memory.** This loading is accomplished by programs stored permanently in the computer's electronic circuitry. When you turn on the machine, programs called *diagnostic routines* test the main memory, the central processing unit, and other parts of the system to make sure they are running properly. Next, BIOS (for basic input/output system) programs are copied to main memory and help the computer interpret keyboard characters or transmit characters to the display screen or to a diskette. Then the boot program obtains the operating system, usually from hard disk, and loads it into the computer's main memory, where it remains until you turn the computer off.

- **CPU management:** The central component of the operating system is the supervisor. Like a police officer directing traffic, the ***supervisor*, or kernel, manages the CPU. It remains in main memory while the**

CLICK-ALONG 3-1
More about the boot process

computer is running and directs other "nonresident" programs (programs that are not in main memory) **to perform tasks that support application programs.** The operating system also manages memory—it keeps track of the locations within main memory where the programs and data are stored.

- **File management:** A *file* is a named collection of related information. A file can be a program, such as a word processing program. Or it can be a data file, such as a word processing document, a spreadsheet, images, songs, and the like. We discuss files in more detail later in the chapter.

  Files containing programs and data are located in many places on your hard disk and other secondary-storage devices. The operating system records the storage location of all files. If you move, rename, or delete a file, the operating system manages such changes and helps you locate and gain access to it. For example, you can *copy,* or duplicate, files and programs from one disk to another. You can *back up,* or make a duplicate copy of, the contents of a disk. You can *erase,* or remove, from a disk any files or programs that are no longer useful. You can *rename,* or give new file names to, the files on a disk.

- **Task management:** In word processing, for example, the operating system accepts input data, stores the data on a disk, and prints out a document—seemingly simultaneously. Some operating systems can also handle more than one program at the same time—word processing, spreadsheet, database searcher. Each program is displayed in a separate window on the screen. Others can accommodate the needs of several different users at the same time. All these examples illustrate *task management.* A "task" is an operation such as storing, printing, or calculating.

  <u>**Multitasking**</u> **is the execution of two or more programs by one user concurrently on the same computer with one central processor.** You may be writing a report on your computer with one program while another program plays a music CD. How does the computer handle both programs at once?

- **Formatting:** <u>*Formatting*</u>, or *initializing*, **a disk is the process of preparing that disk so that it can store data or programs.** Today it is easier to buy preformatted diskettes, which bear the label "Formatted IBM" (for floppies designed to run on PCs) or "Formatted Macintosh." However, it's useful to know how to format a blank floppy disk or reformat a diskette that wasn't intended for your machine.

How to Format a Floppy Disk on the PC

1. Insert unformatted (blank) disk in the floppy disk drive.
2. In My Computer, click once on the icon for the disk you want to format (the floppy disk).
3. On the File menu, click Format.

Notes:

- Be aware that when you format a disk, it *removes all information already on the disk,* such as previous files stored there.
- You can't format a disk if files are already open on that disk.

CONCEPT CHECK

Describe the three components of system software.

Explain the important features of the OS.

Mouse driver

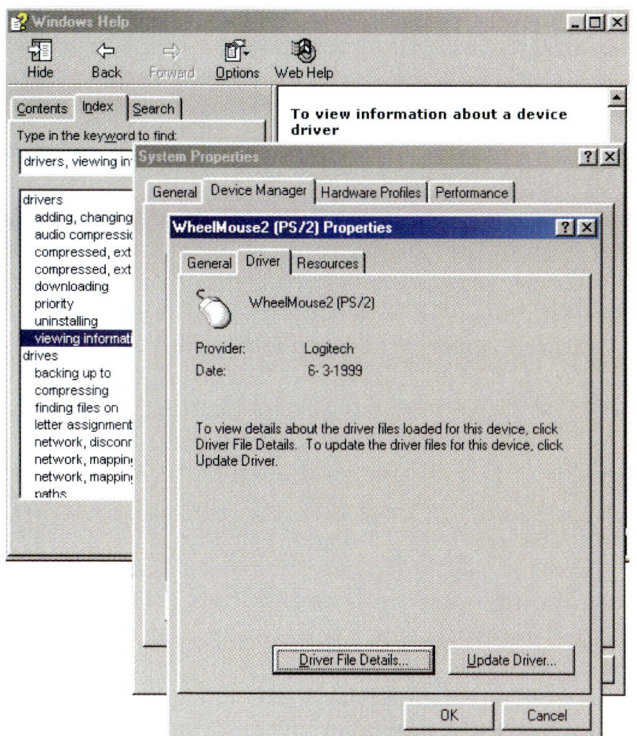

Device Drivers: Running Peripheral Hardware

Device drivers **are specialized software programs that allow input and output devices to communicate with the rest of the computer system.** Many basic device drivers come with system software when you buy a computer, and the system software will guide you through choosing and installing the necessary drivers. If, however, you buy a new peripheral device, such as a mouse, scanner, or printer, the package will include a device driver (probably on a CD-ROM). You'll need to install the driver on your computer's hard-disk drive (by following the manufacturer's instructions) before the device will operate.

Utilities: Service Programs

Utility programs, also known as *service programs*, **perform tasks related to the control and allocation of computer resources. They enhance existing functions or provide services not supplied by other system software programs.** Most computers come with built-in utilities as part of the system software. (Windows 95/98/Me/2000/XP offers several of them.) However, they may also be bought separately as external utility programs (such as Norton Desktop and McAfee utilities).

Among the tasks performed by utilities are backing up data, recovery of lost data, and identification of hardwar problems.

CONCEPT CHECK

What are device drivers?

What are utility programs?

Now that we've described the three components of system software—the operating system, device drivers, and utility programs—let's take a more detailed look at the operating system, which is the most important of the three.

The Operating System's User Interface: Your Computer's Dashboard

When you power up your computer and it goes through the boot process, it displays a starting screen. From that screen you choose the application programs you want to run or the files of data you want to open. Ths part of the operating system is called the *user interface*—**the user-controllable display screen that allows you to communicate, or interact, with the computer.** Like the dashboard on a car, the user interface has gauges that show you what's going on and switches and buttons for controlling what you want to do.

You can interact with this display screen using the keys on your keyboard, but you will also frequently use your mouse. The mouse allows you to direct an on-screen pointer to perform any number of activities. The *pointer* **usually appears as an arrow, although it changes shape depending on the application.** The mouse is used to move the pointer to a particular place on the display screen or to point to little symbols, or icons. You can

87

activate the function corresponding to the symbol by pressing ("clicking") buttons on the mouse. Using the mouse, you can pick up and slide ("drag") an image from one side of the screen to the other or change its size. (See ● Panel 3.2.)

In the beginning, personal computers had *command-driven* interfaces, which required you to type in strange-looking instructions (such as "copy a:\filename c:\" to copy a file from a floppy disk to a hard disk). In the next version, they had *menu-driven interfaces*, in which you could use the arrow keys on your keyboard (or a mouse) to choose a command from a menu, or list of activities. Today the computer's "dashboard" is usually a **graphical user interface (GUI)** (pronounced "gooey"), **which allows you to use a mouse or keystrokes to select icons (little symbols) and commands from menus (lists of activities).** The GUIs on the PC and on the Apple Macintosh (which was the first easy-to-use personal computer available on a wide scale) are somewhat similar. Once you learn one version, it's fairly easy to learn the other. However, the best-known GUI is that of Microsoft Windows. (See ● Panel 3.3, p. 90.)

Three features of a GUI are the *desktop, icons,* and *menus.*

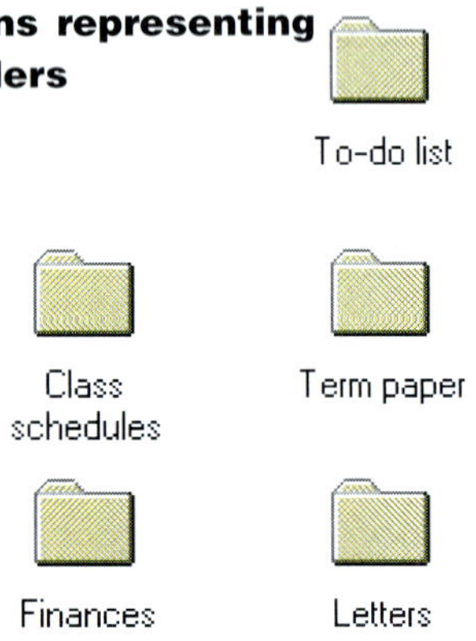

- **Desktop:** After you turn on the computer, the first screen you will encounter is the *desktop,* a term that embodies the idea of folders of work (memos, schedules, to-do lists) on a businessperson's desk. **The *desktop*, which is the system's main interface screen, displays pictures (icons) that provide quick access to programs and information.**

- **Icons and rollovers:** *Icons* **are small pictorial figures that represent programs, data files, or procedures.** For example, a trash can represents a place to dispose of a file you no longer want. If you click your mouse pointer on a little picture of a printer, you can print out a document. One of the most important icons is the *folder,* a representation of a manila folder; folders are the collections of files in which you store your documents and other data.

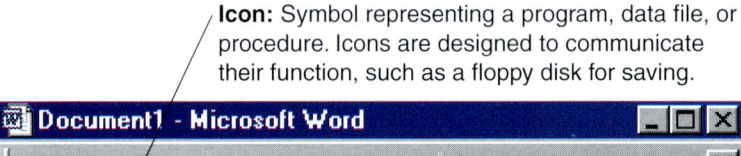

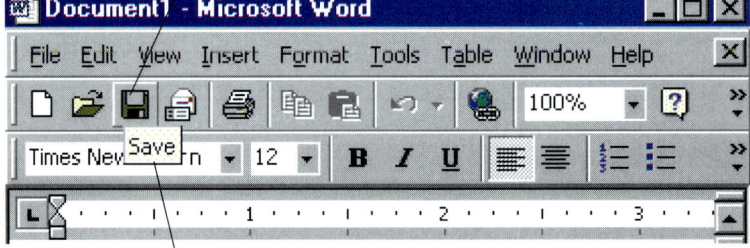

Icons representing folders

To-do list
Class schedules
Term paper
Finances
Letters

Icon: Symbol representing a program, data file, or procedure. Icons are designed to communicate their function, such as a floppy disk for saving.

Rollover: When you roll your mouse pointer over an icon or graphic, a small box with text appears that briefly explains its function.

Survival Tip

Don't Trash Those Icons

Don't delete unwanted software using the mouse to drag their icons to the recycle bin, which might leave behind system files to cause problems.

Best to use an "uninstall" utility. (Go to Start, Settings, Control Panel; double-click Add/Remove Programs. Find the program you want to delete, and click the Add/Remove button.)

Of course, you can't always be expected to know what an icon or graphic means. **A *rollover* feature, a small text box explaining the icon's function, appears when you roll the mouse pointer over the icon.** A rollover may also produce an animated graphic.

PANEL 3.2
Mouse language

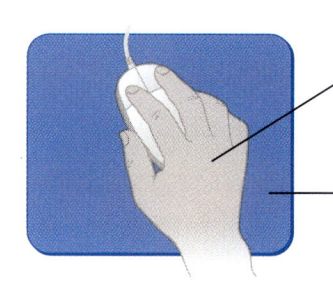

Resting your hand on the mouse, use your thumb and outside two fingers to move the mouse on your desk or mouse pad. Use your first two fingers to press the mouse buttons.

Mouse pad provides smooth surface.

Left button. Click once on an item on screen to select it. Click twice to perform an action.

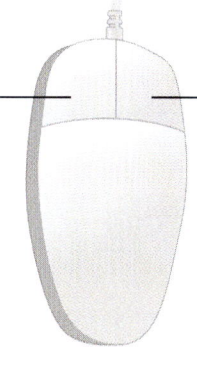

Right button. Click on an object to display a shortcut list of options.

Term	Action	Purpose
Point	Move mouse across desk to guide pointer to desired spot on screen. The pointer assumes different shapes, such as arrow, hand, or I-beam, depending on the task you're performing.	To execute commands, move objects, insert data, or similar actions on screen
Click	Press and quickly release left mouse button.	To select an item on the screen
Double-click	Quickly press and release left mouse button twice.	To open a document or start a program
Drag and drop	Position pointer over item on screen, press and hold down left mouse button while moving pointer to location in which you want to place item, then release.	To move an item on the screen
Right-click	Press and release right mouse button.	To display a shortcut list of commands, such as a pop-up menu of options

Outlook Express: Part of Microsoft's browser, Internet Explorer, that enables you to use e-mail.

Microsoft Network: Click here to connect to Microsoft Network (MSN), the company's online service.

Norton Protected: Click here to activate anti-virus software.

Network Neighborhood: If your PC is linked to a network, click here to get a glimpse of everything on the network.

My Documents: Where your documents are stored unless you specify otherwise.

My Computer: Gives you a quick overview of all the files and programs on your PC.

Documents: Multitasking capabilities allow users to smoothly run more than one program at once.

Start menu: After clicking on the start button, a menu appears, giving you a quick way to handle common tasks. You can launch programs, call up documents, change system settings, get help, and shut down your PC.

Start button: Click for an easy way to start using the computer.

Taskbar: Gives you a log of all programs you have opened. To switch programs, click on the icon buttons on the taskbar.

Multimedia: Windows 98 features sharp graphics and video capabilities.

● **PANEL 3.3**

A graphical user interface

This is for Windows 98. (Icons may differ on your PC.)

- **Menus:** Like a restaurant menu, a _menu_ offers you a list of options to choose from—in this case, a list of commands for manipulating data, such as Print or Edit. Menus are of several types. Resembling a pull-down window shade, a _pull-down menu_, also called a _drop-down menu_, is a list of options that pulls down from the top of the screen. For example, if you use the mouse to "click on" (activate) a command (for example, File) on the menu bar, you will see a pull-down menu offering further commands. Choosing one of these options may produce further menus called _fly-out menus_, menus that seem to explode out to the right. (See ● Panel 3.4.)

A _pull-up menu_ is a list of options that pulls up from the bottom of the screen. In Windows 98, a pull-up menu appears in the lower left-hand corner when you click on the Start button.

● **PANEL 3.4**
Types of menus

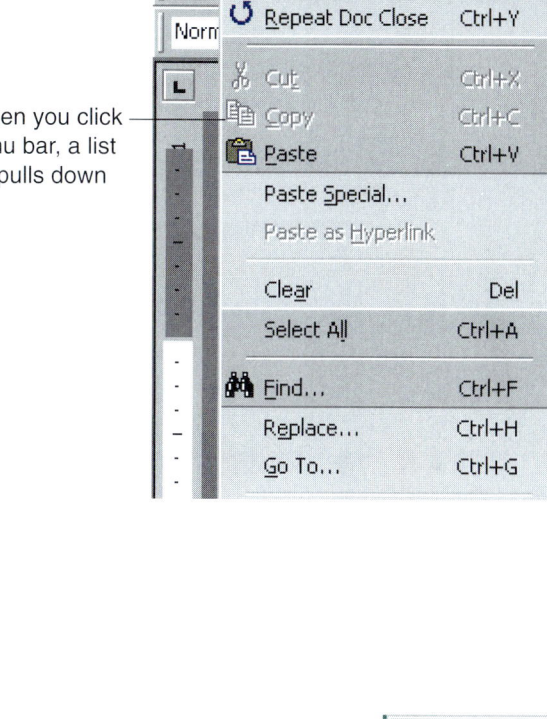

Pull-down menu: When you click the mouse on the menu bar, a list of options appears or pulls down like a shade.

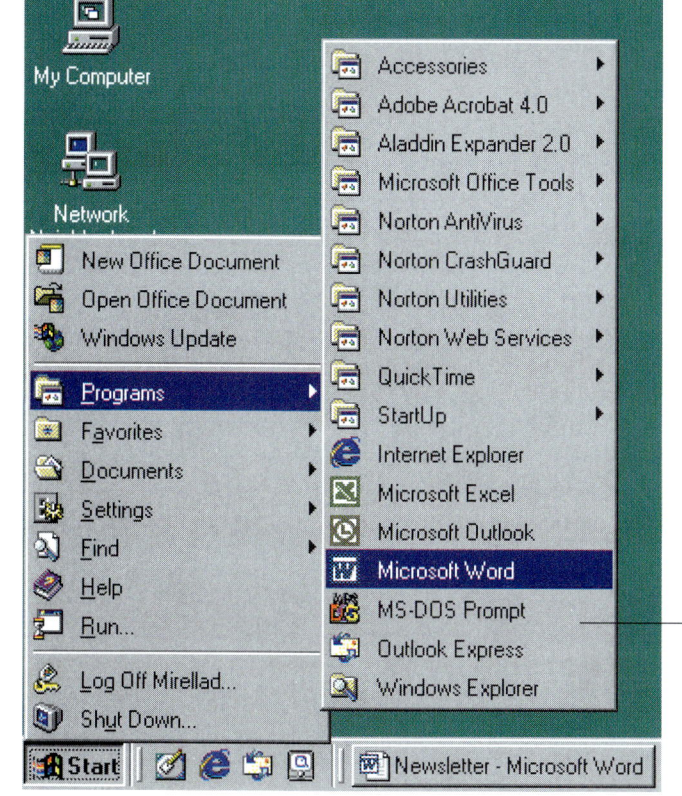

Fly-out menu: Moving the mouse pointer to an option on the pull-up menu produces a flyout menu with more options.

Pull-up menu: When you click the mouse pointer on the *Start* button, it produces a pull-up menu offering access to programs and documents.

A *pop-up menu* is a list of command options that can "pop up" anywhere on the screen when you click the right mouse button. In contrast to pull-down or pull-up menus, pop-up menus are not connected to a toolbar (explained later).

Suppose you want to go to a document—say, a term paper you've been working on. There are two ways to begin working from the Windows 98 desktop: (1) You can click on the Start button at lower left and then make a selection from the pull-up menu that appears. Or (2) you can click on one of the icons on the desktop, probably the most important of which is the *My Computer* icon, and pursue the choices offered there. Either way, the result is the same: The document will be displayed on the window. (See • Panel 3.5.)

Once you're past the desktop, the GUI's opening screen, clicking on the *My Computer* icon reveals title bars, menu bars, toolbars, and windows.

● **PANEL 3.5**
Two ways to go to a document in Windows 98

From Start menu

Click on *Start* button to produce Start menu, then go to *Documents* option, then to *My Documents*.

From My Computer icon

Click on *My Computer* icon, which opens a window that provides access to information on your computer.

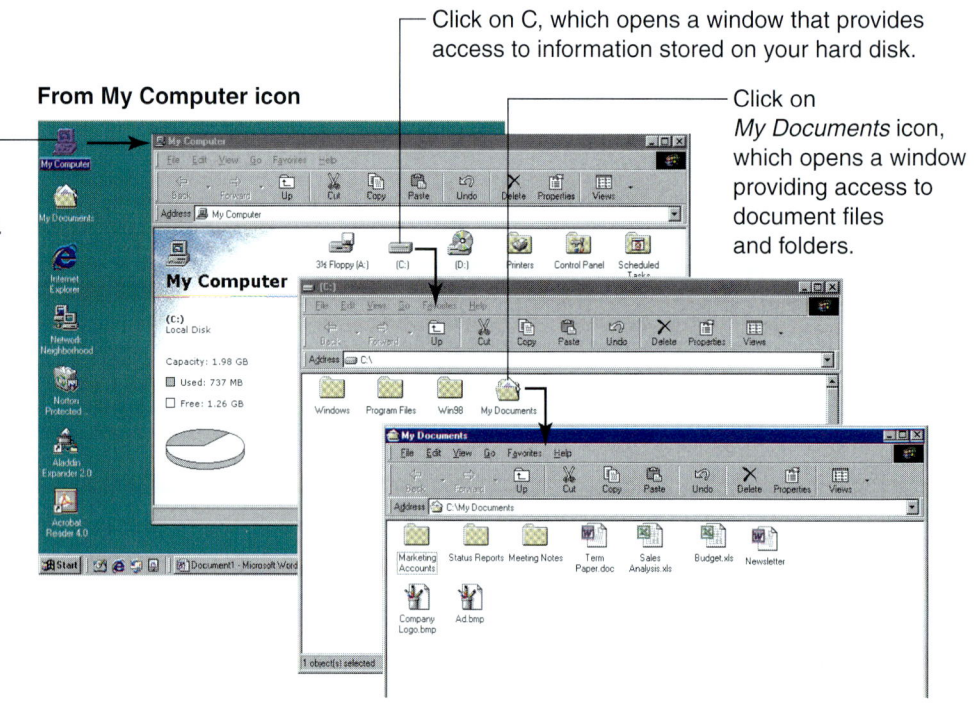

Click on C, which opens a window that provides access to information stored on your hard disk.

Click on *My Documents* icon, which opens a window providing access to document files and folders.

- **Title bar, menu bar, toolbar:** A _title bar_ **contains the name of the window—often the application and the name of the file you're working in. A _menu bar_ contains the names of pull-down menus. A _toolbar_ displays menus and icons representing frequently used options or commands.** Examples of menus are File, Edit, View, Favorites, and Help. An example of an icon is the picture of two overlapping pages (see illustration left), which issues a Copy text command.

In Windows, the toolbar graphic at the bottom of the screen, which shows the applications that are running, is called a _taskbar_.

- **Windows:** When spelled with a capital "W," Windows is the name of Microsoft's system software (Windows 95, 98, Me, XP, and so on). When spelled with a lowercase "w," a _window_ **is a rectangular frame on the computer display screen. Through this frame you can view a file of data—such as a document, spreadsheet, or database—or an application program.**

In the right-hand corner of the Windows 98 toolbar are three icons that represent *Minimize, Maximize,* and *Close.* (See ● Panel 3.6.) By

● **PANEL 3.6**
Some windows

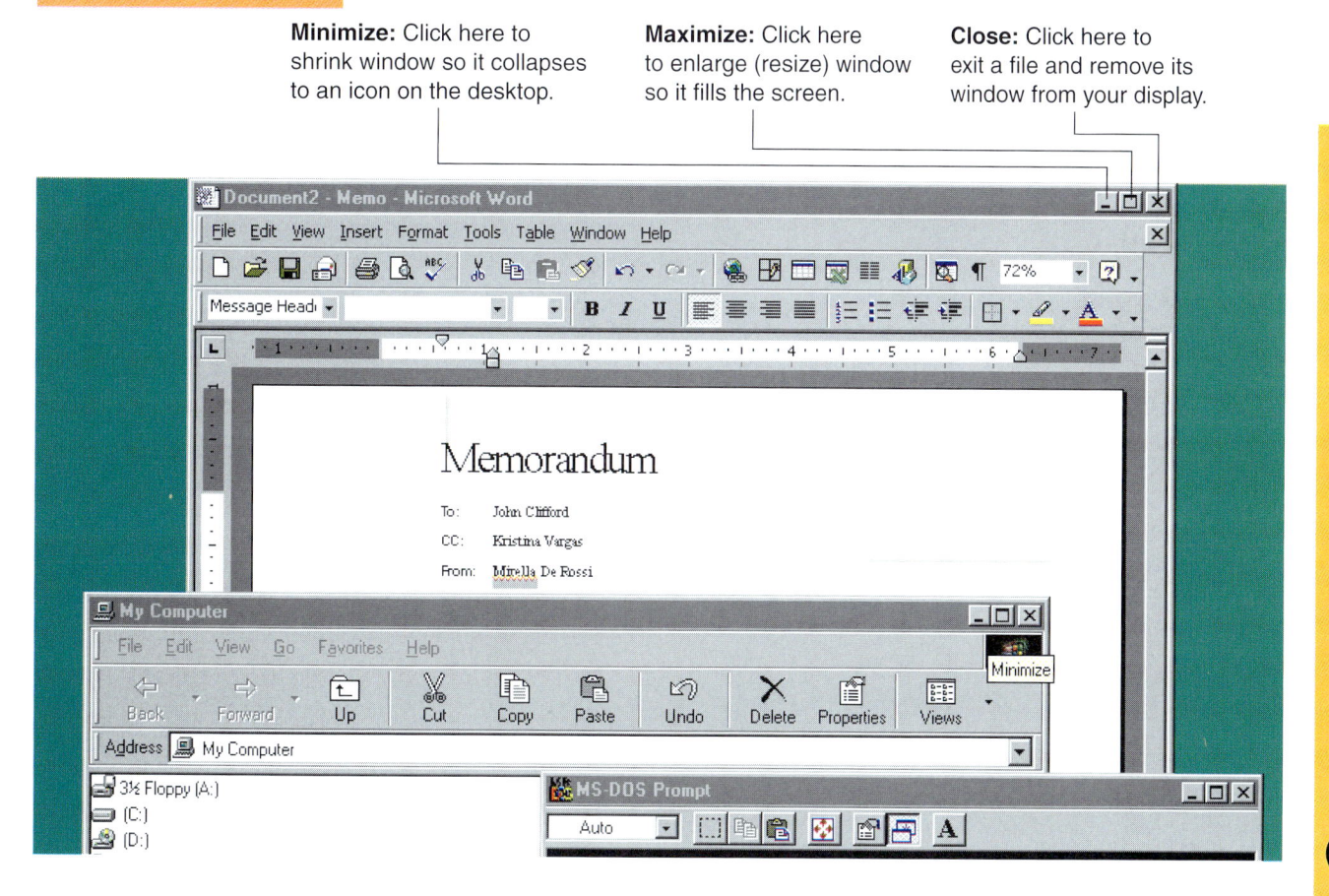

Minimize: Click here to shrink window so it collapses to an icon on the desktop.

Maximize: Click here to enlarge (resize) window so it fills the screen.

Close: Click here to exit a file and remove its window from your display.

Survival Tip

Getting Help

On Windows computers, you can find the Help area by pressing the F1 key. Or use the mouse to click on Start in the lower left screen; then click on Help. On the Macintosh, Help is located under the main menu bar.

---

clicking on these icons, you can *minimize* the window (shrink it down to an icon at the bottom of the screen), *maximize* it (enlarge it), or *close* it (exit the file and make the window disappear).

You can also *move* the window around the desktop, using the mouse.

Finally, you can create *multiple windows* to show operations going on concurrently. For example, one window might show the text of a paper you're working on, another might show the reference section for the paper, and a third might show something you're downloading from the Internet. (Refer back to Panel 3.3, p. 90.)

- **The Help command:** Don't understand how to do something? Forgotten a command? Accidentally pressed some keys that messed up your screen layout and you want to undo it? Most toolbars contain a *Help command*—**a command generating a table of contents, an index, and a search feature that can help you locate answers.** In addition, many applications have *context-sensitive help*, which leads you to information about the task you're performing. (See ● Panel 3.7.)

CONCEPT CHECK

Describe the features of the GUI: desktop, icons, and the various kinds of menus.

What do toolbars and windows enable you to do?

What is the Help command?

● **PANEL 3.7**

Help features

The Help command yields a pull-down menu.

The *Help* menu provides a list of help options.

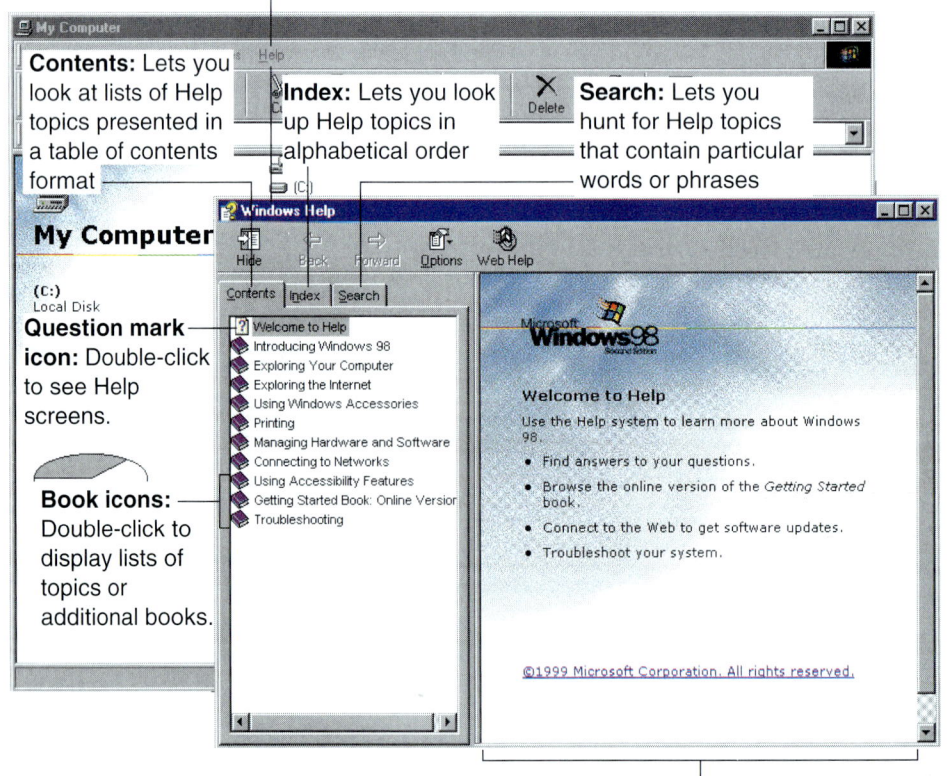

Contents: Lets you look at lists of Help topics presented in a table of contents format

Index: Lets you look up Help topics in alphabetical order

Search: Lets you hunt for Help topics that contain particular words or phrases

Question mark icon: Double-click to see Help screens.

Book icons: Double-click to display lists of topics or additional books.

This window displays the selected Help topic.

Common Desktop & Laptop Operating Systems: DOS, Macintosh, & Windows

The *platform* **is the particular processor model and operating system on which a computer system is based.** For example, there are "Mac platforms" (Apple Macintosh) and "Windows platforms" or "PC platforms" (for personal computers such as Dell, Compaq, Gateway, Hewlett-Packard, or IBM that run Microsoft Windows). Sometimes the latter are called "*Wintel* platforms," for "Windows + Intel," because they often combine the Windows operating system with the Intel processor chip. (We discuss processors in Chapter 5.)

Despite the dominance of the Windows platform, many so-called *legacy systems* are still in use. A legacy system is an older, outdated, yet still functional technology, such as the DOS operating system. You may find yourself having to use DOS at some point.

Let us quickly describe the principal platforms used on desktop computers: DOS, the Macintosh OS, and the Windows series—3.x, 95, 98, NT, 2000, Millennium (Me), and XP. Desktop operating systems are used mainly on single-user computers (both desktops and laptops) rather than on mainframes or servers. We discuss operating systems for servers and for portable information appliances shortly.

- **DOS—the old-timer:** __DOS__ (rhymes with "boss")—for __Disk Operating System__—**was the original operating system produced by Microsoft and had a hard-to-use command-driven user interface.** Its initial 1982 version was designed to run on the IBM PC as PC-DOS. Later Microsoft licensed the same system to other computer makers as MS-DOS. With the growing popularity of cheaper PCs produced by these companies, MS-DOS came to dominate the industry. Two years before the advent of Windows 95, which eventually grew out of DOS, there were reportedly more than 100 million users of DOS, which at that time made it the most popular software ever adopted—of any sort.

- **The Macintosh Operating System—for the love of Mac:** The __Macintosh operating system (Mac OS)__, **which runs only on Apple Macintosh computers, set the standard for icon-oriented, easy-to-use graphical user interfaces.** The software generated a strong legion of fans shortly after its launch in 1984 and inspired rival Microsoft to upgrade DOS to the more user-friendly Windows operating systems. Much later, in 1998, Apple introduced its iMac computer (the "i" stands for Internet), which added capabilities such as small-scale networking.

    The newest version of the operating system, Mac OS X (called "ten"), broke with 15 years of Mac software to offer a dramatic new look and feel. *(See • Panel 3.8, on the next page.)* The user interface known as Aqua has Hollywood-like tricks (thanks to Apple chairman Steve Jobs' experience running Pixar Animation Studios, makers of *Toy Story* and other animated films). "Jelly-colored onscreen buttons pulse as if alive," says one description. "Menu borders are translucent, allowing you to see the documents under them. Sliders glow luminously."[2] In addition, Apple claims that OS X won't allow software conflicts, a frequent headache with Microsoft's Windows operating systems. For example, you might install a game and find that it interferes with the device driver for a sound card. Then, when you uninstall the game, the problem persists. With Mac OS X, when you try to install an application program that conflicts with any other program, the Mac simply won't allow you to run it.

    Mac OS X also offers free universal e-mail services, improved graphics and printing, CD burning capability, DVD player, easier ways to find files, and support for building and storing web pages. (We discuss OS X further in the Practical Action Box, page 100.)

Survival Tip

What to Set Up First?

When you are setting up a new microcomputer system, install your peripherals first—your printer, then your Zip drive, scanner, etc.—and test each one before restarting the computer and installing the next peripheral. Then start installing your applications. Again, test each one to make sure it works properly; then restart your computer and install the next application, and so on.

PANEL 3.8
The Mac OS, version X

The icons and easy-to-use GUI resemble those used in Microsoft Windows 98, although Macintosh popularized these features first. What Apple calls a "dock"—the band of animated icons and miniaturized windows along the bottom—resembles a more graphic version of the taskbar in Windows.

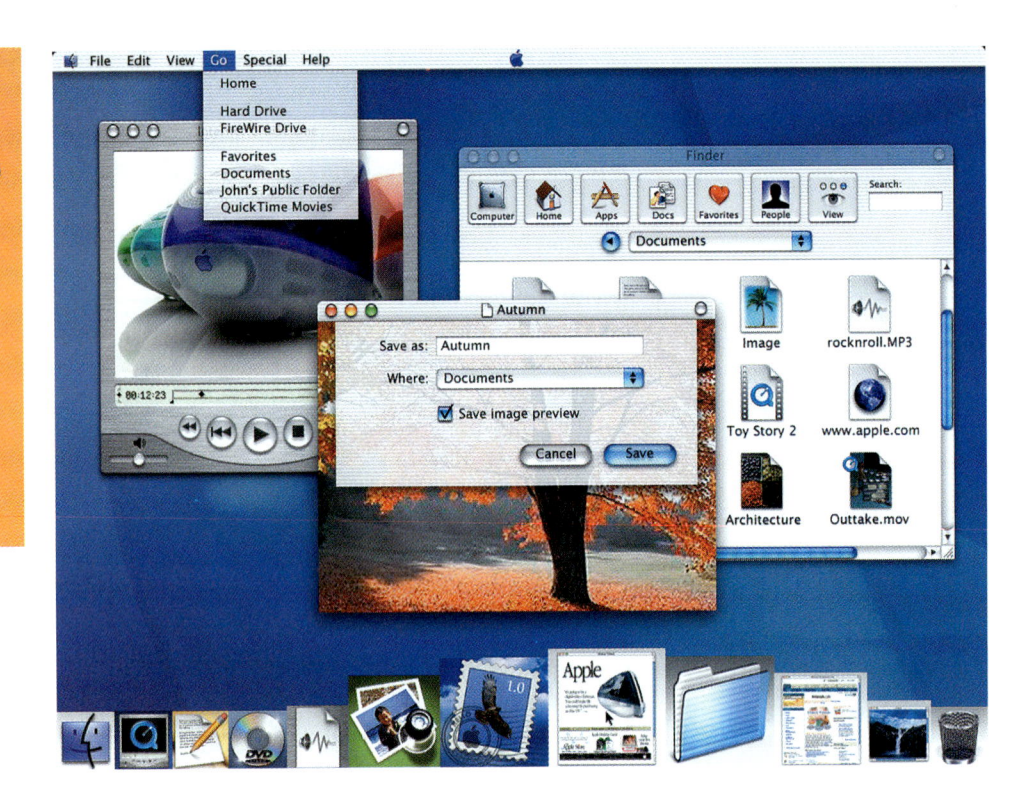

- **Microsoft Windows 3.x, 95, 98, & Me:** In the 1980s, taking its cue from the popularity of Mac's easy-to-use GUI, Microsoft began working on Windows—to make DOS more user-friendly. Early attempts (Windows 1.0, 2.0, 3.0) did not catch on. However, in 1992, Windows 3.x emerged as the preferred system among PC users. (Technically, Windows 3.x wasn't a full operating system; it was simply a layer or shell over DOS.) Later, this version evolved into the Windows 95 operating system, which was succeeded by Windows 98.

  <u>Microsoft Windows 95/98</u> **is still a popular operating system for desktop and portable microcomputers, supporting the most hardware and the most application software.** <u>Microsoft Windows Millennium</u>, or <u>WinMe</u> (the Me stands for Millennium Edition), **is the successor for home users to Windows 95 and 98, designed to support desktop and portable computers.**

Network Operating Systems: NetWare, Windows NT/2000/XP, Unix, & Linux

The operating systems described so far were principally designed for use with stand-alone desktop machines. Now let's consider the important operating systems designed to work with networks: NetWare, Windows NT/2000/XP, Unix/Solaris, and Linux.

- **Novell's NetWare—PC networking software:** <u>NetWare</u> **has long been a popular network operating system for coordinating microcomputer-based local area networks (LANs) throughout a company or a campus.** LANs allow PCs to share programs, data files, and printers and other devices. Novell, the maker of NetWare, thrived as corporate data managers realized that networks of PCs could exchange information more cheaply than the previous generation of mainframes and midrange computers. The biggest challenge to NetWare has been Windows NT, Windows 2000, and now Windows XP.

- **Windows NT and 2000:** Windows 95 and 98 can be used to link PCs in small networks in homes and offices. However, something more

---

Survival Tip

Underlined Letters in Dialog Boxes

The underlined letters you see in certain dialog boxes are shortcuts you can perform on the keyboard without using the mouse to select the option you want. Usually you can press the *Alt* key plus the underlined letter to perform the desired operation.

powerful was needed to run the huge networks linking a variety of computers—PCs, workstations, mainframes—used by many companies, universities, and other organizations, which previously were served principally by Unix and NetWare operating systems. **_Windows NT_ (the NT stands for New Technology), later upgraded to _Windows 2000_, is Microsoft's multitasking operating system designed to run on network servers. It allows multiple users to share resources such as data and programs.**

When it first appeared, in 1993, the system came in two versions. The *Windows NT Workstation* version enabled graphic artists, engineers, and others using stand-alone workstations to do intensive computing at their desks. The *Windows NT Server* version was designed to benefit multiple users tied together in networks. In early 2000, Microsoft rolled out its updated version, *Windows 2000,* to replace Windows NT version 4.0.

The choice of the name Windows 2000 is somewhat unfortunate. This is not the successor to Windows 95 and 98 for home and non-network use. If you're interested in arcade-style games, for instance, 2000 won't work, since it was specifically designed for businesses. (Windows Millennium was designed for home users.)

- **Windows XP:** **_Windows XP_ is Microsoft's newest OS; it combines elements of Windows 2000 and Windows Me and has a new GUI.** It has improved stability and increased driver and hardware support. It has also added new features such as built-in instant messaging, centralized shopping managers, and music, video, and photography managers. (We describe XP in more detail in the Practical Action Box, p. 100.)

- **Unix and Solaris— first to exploit the Internet:** Unix (pronounced "*Yu*-niks") was developed at AT&T's Bell Laboratories in 1969 as an operating system for minicomputers. Today **_Unix_ is a multitasking operating system for multiple users that has built-in networking capability and versions that can run on all kinds of computers.** It is used mostly on workstations and servers. Government agencies, universities, research institutions, large corporations, and banks all use Unix for everything from designing airplane parts to currency trading. Unix is also used for website management. Indeed, the developers of the Internet built their communications system around Unix because it has the ability to keep large systems (with hundreds of processors) churning out transactions day in and day out for years without fail.

Sun Microsystems' *Solaris* is a super-reliable version of Unix that seems to be most popular for handling large e-commerce servers and large websites. Solaris 8 can handle servers with as many as 64

CLICK-ALONG 3-2
The latest on Windows XP

Survival Tip

XP Installation

Every time you install or reinstall Windows XP, you will have to get Microsoft's permission to activate it. You can do this over the Internet or via the phone.

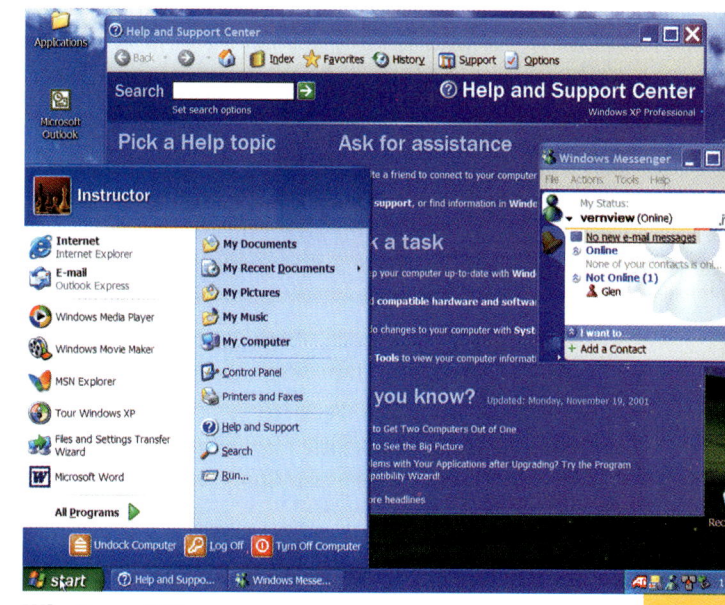

Windows XP screen

microprocessors, compared with 32 for Windows 2000. In addition, eight computers can be clustered together to work as one, compared with four for Windows 2000.[3] Sun is also an application service provider, offering over-the-Net application software StarOffice, a Linux office applications suite.

- **Linux—software built by a community:** It began in 1991 when programmer Linus Torvalds, a graduate student in Finland, posted his free Linux operating system on the Internet. Linux (pronounced "*Linn*-uks") is the rising star of network software. **Linux is a free version of Unix, and its continual improvements result from the efforts of tens of thousands of volunteer programmers.** Whereas Windows is Microsoft's proprietary product, Linux is **open-source software— meaning any programmer can download it from the Internet for free and modify it with suggested improvements.** The only qualification is that changes can't be copyrighted; they must be made available to all and remain in the public domain. From these beginnings, Linux has attained cult-like status. "What makes Linux different is that it's part of the Internet culture," says an IBM general manager. "It's essentially being built by a community."[4] (The People's Republic of China announced in July 2000 that it was adopting Linux as a national standard for operating systems because it feared being dominated by the OS of a company of a foreign power—namely, Microsoft.)

Linus Torvalds

If Linux belongs to everyone, how do companies like Red Hat Software—a company that bases its business on Linux—make money? Their strategy is to give away the software but then sell services and support. Red Hat, for example, makes available an inexpensive ($149) application-software package that offers word processing, spreadsheets, and the like.

Because it was built for use on the Internet, Linux is more reliable than Windows for online applications. Hence, it is better suited to run websites and e-commerce software. Its real growth, however, may come as it reaches outward to other applications. IBM, Red Hat, and 45 other companies have formed the Embedded Linux Consortium, which envisions a time when microchips running Linux are built into everything from Internet-linked microwave ovens and refrigerators to wireless phones and TV set-top cable boxes.[5]

Linux screen

Survival Tip

Ready for Linux?

Linux is an alternative to Windows, and it runs on the same PC hardware. However, Linux may be too difficult for many nontechnical users. Today Linux is still mainly for servers and techies.

Operating Systems for Handhelds: Palm OS & Windows CE/Pocket PC

Maybe you're not one of the millions of owners of a handheld computer or personal digital assistant (PDA). But perhaps you've seen people poking through calendars and address books, beeping through games, or (in the latest versions) checking e-mail or the Web on these palm-size devices. Handhelds have gained headway in the corporate world and on college campuses. Because of their small size, they rely on specialized operating systems, including the Palm OS and Windows CE/Pocket PC.

2000 Share of Server (Non-Mainframe) Market[6]

Netware 17%

Windows NT/2000 41%

Unix 13%

Linux 27%

Other 2%

Visor handheld computer

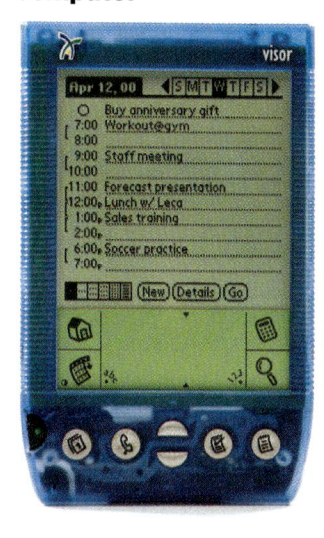

- **Palm OS—the dominant OS for handhelds:** In 1994, Jeff Hawkins took blocks of mahogany and plywood into his garage and emerged with a prototype for the PalmPilot. Two years later, Hawkins and business brain Donna Dubinsky pulled one of the most successful new-product launches in history. Today the company, Palm Computing, sells the popular Palm Vx, VII, and Palm m500. However, Hawkins and Dubinsky left to form another company, Handspring, whose product, Visor, acts like a Palm but costs less and has an expansion slot that can transform the device into a cell phone, MP3 music player, or two-way pager.

  The **Palm OS, which runs the Palm and Visor, is the dominant operating system for handhelds** and is being licensed to competitors, such as IBM and Nokia. With a Palm OS, you have access to thousands of application programs over the Internet, many of them free. (For instance, one $10 program, PayUp, lets you track 19 people's dinner orders and calculate an exact bill for each, including tax and tip.)[7]

- **Windows CE/Pocket PC—Microsoft Windows for handhelds:** In 1996, Microsoft released **Windows CE**, now known as **Pocket PC, a slimmed-down version of Windows for handheld computing devices,** such as those made by Casio, Compaq, and Hewlett-Packard. Windows CE has some of the familiar Windows look and feel and includes mobile versions of word processing, spreadsheet, e-mail, web browsing, and other software.

CONCEPT CHECK

What are the principal desktop operating systems and some of their features?

What are the common network operating systems and some of their features?

What are the leading operating systems for handhelds?

Sending impressions overseas

Japanese and American members of the Mormon church journey via wagon train from Nebraska to Utah, re-enacting a trip taken by Mormons to escape religious persecution more than 150 years ago. Using a solar-powered laptop with a wireless modem, Osamu Sekiguchi of Tokyo posts his impressions of the journey to a Japanese website.

BOOKMARK IT!

PRACTICAL ACTION BOX
Should You Upgrade to Windows XP or Mac OS X?[a]

Let's assume you already have a personal computer, either a Windows PC or a Macintosh. In 2001, Microsoft Corp. introduced Windows XP and Apple Computer introduced Mac OS X ("Ten"). Both are new operating systems that represent major improvements, perhaps the most important being that they (Windows especially) are less inclined to crash.

The question for you to consider: Are these good enough so that you should (1) stay with the computer and operating system you already have, (2) upgrade your present computer to the new operating system, or (3) buy a new computer that already has the new operating system installed?

Microsoft Windows XP

The foremost problem with using Windows 95, 98, and Me (Millennium)—the operating systems favored by students and home users—is that they often crash. That is, whatever is on your screen simply freezes up, and you have to shut down the computer and reboot, losing any unsaved work in the process. Windows 2000, which is the business version, is less prone to crashes. However, most nonbusiness users haven't been using 2000 since it doesn't support as many devices and applications as the home versions of Windows.

Windows XP, which is based on the Windows 2000 core, is said to be less crash-prone than previous Windows versions. This, if true, is an important feature, but here are some others to be aware of to help you decide about upgrading.

Hardware Requirements: Windows XP is available in two versions—Professional (for businesses, at about $200) and Home Edition (at about $100). XP also requires 128 megabytes or more of main memory (RAM) and a 2-gigabyte hard-disk drive. Although it may run with a 300-megahertz Pentium II chip, it performs best with a Pentium III or equally fast Celeron or AMD chip. When you install XP, it will let you know if any components of your computer, such as memory or graphics cards, won't work with it so that you'll have to upgrade them as well.

If you're thinking about installing XP on your old PC, therefore, make sure you have a reasonably current computer system (purchased in the last 18 months) and enough memory. But it's better to simply buy a brand-new PC system with XP already preloaded.

Running Older Application Programs: XP can run older application programs by emulating Windows 95 or other previous versions. It also can handle hundreds of games and other applications that Windows 2000 cannot. Moreover, XP will scan your system to see whether driver updates and software patches are needed, which can be downloaded from Microsoft's website.

Look and Feel: The look of XP is less cluttered than previous Windows versions. When you click on Start, a window appears with, on the left, icons for Internet Explorer 6, e-mail, Media Player 8, and CD recording and video editing programs, plus programs you use most often. On the right are folders such as My Documents, My Music, and My Pictures, which are used to store files. Rarely used notification icons are hidden. A large number of utilities are available for hardware and network monitoring. Software to keep out hackers is included.

"Product Activation"—The Anti-Piracy Component: In the past, you might have bought one copy of a new Windows operating system and then shared it with all the PCs in your household. Now, however, you'll have to buy a separate, full-price copy for each machine. This will be enforced by Microsoft's "product activation" system, which requires you to let Microsoft create and store a profile of every computer on which you install Windows XP. (The company has long educated corporations about installing a separate copy of Windows for each machine, but until now it has not made a point of doing this for home users.)

If you don't allow them to do this, your copy of Windows XP will stop working after 30 days. It might also stop working at any time you make significant changes to your PC's hardware configuration. In addition, if you try to install your copy of XP on another machine, you'll be asked to activate again—and since it probably won't be allowed, you'll need to purchase an additional copy. Finally, if your PC malfunctions and you have to reinstall XP, you'll need to contact Microsoft and explain the situation.

Optional Feature—Getting Added Microsoft Services in Exchange for Personal Information: In order to use certain features of XP, such as instant messaging, you must sign on to Passport, which collects personal information about you. According to *Consumer Reports* magazine, privacy-protection advocates say Passport requires users to disclose far too much personal information.

Macintosh OS X

Although the Mac OS X ($129) won't run on any Windows computer, it will run on any Mac with a G3 or G4 processor and 128 megabytes of main memory (RAM).

The beauty of OS X is that, in the words of *PC Magazine,* it "retains the extremely user-friendly interface for which the Mac is famous" but has "an extremely stylish computing environment that is far different from previous Mac operating systems."

For instance, OS X has an onscreen organizer called the Dock, a window at the bottom of the screen that is similar to the Windows taskbar, which displays an icon for any file, folder, or program you want to run. When you double-click on an icon, it bounces up and down until the folder or program opens. When you click to close a file, it is sucked down into the dock with a vacuum cleaner effect.

In a departure from previous Mac operating systems, OS X rides on top of a variant of Unix. This provides more stability (it's less apt to crash), but it also means that there's not as much application software available for it as before. Until more applications are written, Mac users may have to use Classic, a version of OS 9 that is included with OS X, to run their existing software.

Macintosh G4 system
G-4 tower with a USB hub on top. The hub allows more than two USB peripherals to be attached to the system unit. (The back of the unit has only two USB ports.) The monitor is a 17-inch flat screen, and the keyboard and the mouse plug into the screen. An external Zip drive and an external floppy drive are between the tower and the screen. The top slot on the tower is the CD/DVD drive; the bottom slot (bay) is empty.

3.2 Application Software: Getting Started

KEY QUESTIONS
What are five ways of obtaining application software, tools available to help you learn to use software, three common types of files, and the types of software?

As we have stated, *application software* is software that has been developed to solve a particular problem, to perform useful work on specific tasks, or to provide entertainment. New microcomputers are usually equipped not only with system software but also with some application software.

Application Software: For Sale, for Free, or for Rent?

At one time, just about everyone paid for PC application software. You bought it as part of a computer or in a software store, or you downloaded it

PANEL 3.9
Choices among application software

Types	Definition
Commercial software	Copyrighted. If you don't pay for it, you can be prosecuted
Public-domain software	Not copyrighted. You can copy it for free without fear of legal prosecution
Shareware	Copyrighted. Available free, but you should pay to continue using it
Freeware	Copyrighted. Available free
Rentalware	Copyrighted. Lease for a fee

online with a credit card charge. Now this business model may be changing. The viable alternatives are *commercial software, public-domain software, shareware, freeware,* and *rentalware.* (See ● Panel 3.9.) We also discuss *pirated software.*

- **Commercial software:** *Commercial software,* also called *proprietary software* or *packaged software,* is software that's offered for sale, such as Microsoft Word or Office 2000. Although such software may not show up on the bill of sale when you buy a new PC, you've paid for it as part of the purchase. And, most likely, whenever you order a new game or other commercial program, you'll have to pay for it. This software is copyrighted. **A _copyright_ is the exclusive legal right that prohibits copying of intellectual property without the permission of the copyright holder.**

  Software manufacturers don't sell you their software; rather, they sell you a license to become an authorized user of it. What's the difference? In paying for a **_software license_, you sign a contract in which you agree not to make copies of the software to give away or for resale.** That is, you have bought only the company's permission to use the software and not the software itself. This legal nicety allows the company to retain its rights to the program and limits the way its customers can use it. The small print in the licensing agreement usually allows you to make one copy (backup copy or archival copy) for your own use. (Each software company has a different license; there is no industry standard.)

  Every year or so, software developers find ways to enhance their products and put forth new versions or new releases. A *version* is a major upgrade in a software product, traditionally indicated by numbers such as 1.0, 2.0, 3.0. More recently, other notations have been used. After 1995, for awhile Microsoft labeled its Windows and Office software versions by year instead of by number, as in Microsoft's Office 97, Office 2000, and so forth. However, its latest software version is Office XP. A *release,* which now may be called an "add" or "addition," is a minor upgrade. Often this is indicated by a change in number after the decimal point. (For instance, 3.0 may become 3.1, 3.11, 3.2, and so on.) Some releases are now also indicated by the year in which they are marketed. And, unfortunately, some releases are not clearly indicated at all. (These are "patches," which may be downloaded from the software maker's website.)

- **Public-domain software:** **_Public-domain software_ is not protected by copyright and thus may be duplicated by anyone at will.** Public domain programs—usually developed at taxpayer expense by government agencies—have been donated to the public by their creators. They are often available through sites on the Internet. You can duplicate public domain software without fear of legal prosecution.

- Shareware: **_Shareware_ is copyrighted software that is distributed free of charge but requires users to make a monetary contribution to continue using it.** Shareware is distributed primarily through the Internet, but because it is copyrighted, you cannot use it to develop your own program that would compete with the original product.

- Freeware: **_Freeware_ is copyrighted software that is distributed free of charge,** today most often over the Internet. Why would any software creator let his or her product go for free? Sometimes developers want to see how users respond, so that they can make improvements in a later version. Sometimes it is to further some scholarly or humanitarian purpose—for instance, to create a standard for software on which people are apt to agree. (Linux is such a program.) In its most recent form, freeware is made available by companies trying to make money some other way—actually, by attracting viewers to their advertising. (The web browsers Internet Explorer and Netscape Navigator are of this type.) Freeware developers generally retain all rights to their programs; technically, you are not supposed to duplicate and distribute them further.

- Rentalware: **_Rentalware_ is software that users lease for a fee.** This is the concept behind application services providers, firms that lease software over the Internet (described in Section 3.7). Users download programs whenever they are needed.

- Pirated software: **_Pirated software_ is software obtained illegally,** as when you get a floppy disk from a friend who has made an illicit copy, of, say, a commercial video game. Sometimes pirated software can be downloaded off the Internet. Sometimes it is sold in retail outlets in foreign countries. If you buy such software, not only do the original copyright owners not get paid for their creative work but you risk getting inferior goods and, worse, picking up a *virus,* a deviant program that can corrupt or destroy your computer's programs or data. (We discuss viruses in Chapter 8.)

Ethics

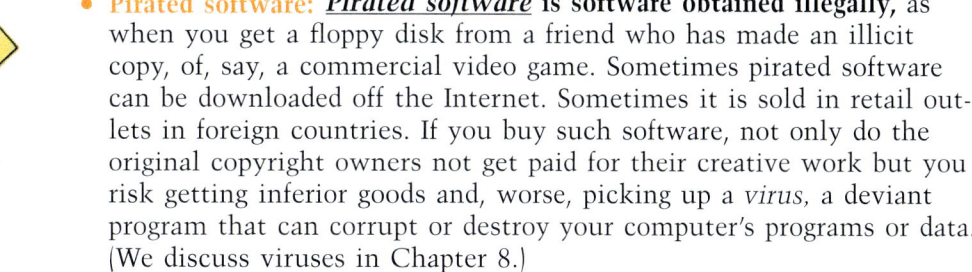

Occasionally, companies or individuals need software written specifically for them, to meet unique needs. This software is called *custom software,* and it's created by software engineers and programmers. (See the appendix for more information on programming.)

Tutorials & Documentation

How are you going to learn a given software program? Most commercial packages come with tutorials and documentation.

- Tutorials: A **_tutorial_ is an instruction book or program that helps you learn to use the product by taking you through a prescribed series of steps.** For instance, our publisher offers several how-to books, known as the *Advantage Series,* that enable you to learn different kinds of software. Tutorials may also form part of the software package.

- Documentation: **_Documentation_ is all information that describes a product to users, including a user guide or reference manual that provides a narrative and graphical description of a program.** While documentation may be print-based, today it is usually available on CD-ROM, as well as via the Internet. Documentation may be instructional, but features and functions are usually grouped by category for reference purposes. For example, in word processing documentation, all features related to printing are grouped together so you can easily look them up.

Files of Data—& the Usefulness of Importing & Exporting

There is only one reason for having application software: to take raw data and manipulate it into useful files of information. A _**file**_ **is a named collection of (1) data or (2) a program that exists in a computer's secondary storage,** such as floppy disk, hard disk, or CD/DVD.

Three well-known types of data files are as follows:

- Document files: _**Document files**_ **are created by word processing programs and consist of documents such as reports, letters, memos, and term papers.**
- Worksheet files: _**Worksheet files**_ **are created by electronic spreadsheets and usually consist of collections of numerical data such as budgets, sales forecasts, and schedules.**
- Database files: _**Database files**_ **are created by database management programs and consist of organized data that can be analyzed and displayed in various useful ways.** Examples are student names and addresses that can be displayed according to age, grade-point average, or home state.

Other common types of files (such as graphics, audio, and video files) are discussed in Chapter 7.

It's useful to know that often files can be exchanged—that is, _imported_ and _exported_—between programs.

- Importing: _**Importing**_ **is defined as getting data from another source and then converting it into a format compatible with the program in which you are currently working.** For example, you might write a letter in your word processing program and include in it—that is, import—a column of numbers from your spreadsheet program.
- Exporting: _**Exporting**_ **is defined as transforming data into a format that can be used in another program and then transmitting it.** For example, you might work up a list of names and addresses in your database program, then send it—export it—to a document you wrote in your word processing program.

The Types of Software

Software can be classified in many ways—for entertainment, personal, education/reference, productivity, and specialized uses. _(See_ ● _Panel 3.10.)_

CONCEPT CHECK

Distinguish among the following kinds of software: commercial, public-domain, shareware, freeware, rentalware, and custom.

Describe tutorials and documentation, and discuss three common types of files.

Distinguish importing from exporting.

In the rest of this chapter we will discuss types of _**productivity software**_—**such as word processing, spreadsheets, and database managers—whose purpose is to make users more productive at particular tasks.** Some productivity software comes in the form of an _**office suite,**_ **which bundles several applications together into a single large package.** Microsoft Office 2000, for example, includes (among other things) Word, Excel, and Access—word processing, spreadsheet, and database programs, respectively. Corel offers similar programs. Other productivity software, such as Lotus Notes, is sold as

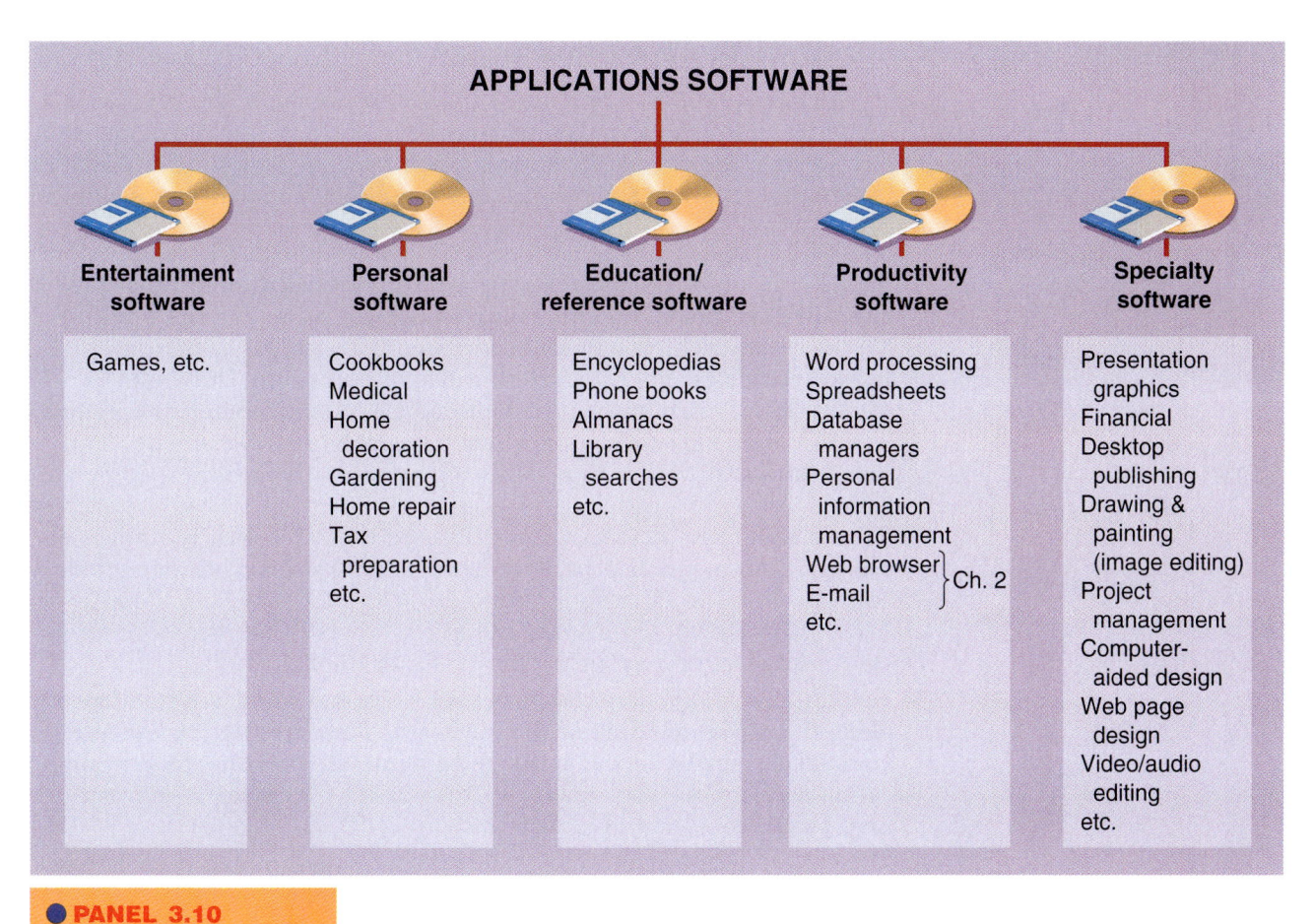

APPLICATIONS SOFTWARE

Entertainment software	Personal software	Education/reference software	Productivity software	Specialty software
Games, etc.	Cookbooks Medical Home decoration Gardening Home repair Tax preparation etc.	Encyclopedias Phone books Almanacs Library searches etc.	Word processing Spreadsheets Database managers Personal information management Web browser ⎱ Ch. 2 E-mail ⎰ etc.	Presentation graphics Financial Desktop publishing Drawing & painting (image editing) Project management Computer-aided design Web page design Video/audio editing etc.

● **PANEL 3.10**
Types of software

groupware—online software that allows several people to collaborate on the same project.

We will now consider the three most important types of productivity software: word processing, spreadsheet, and database software (including personal information managers). We will then discuss more specialized software: presentation graphics, financial, desktop-publishing, drawing and painting, project management, computer-aided design, and image/video/audio editing software.

3.3 Word Processing

KEY QUESTION
What can you do with word processing software that you can't do with pencil and paper?

After a long and productive life, the typewriter has gone to its reward. Indeed, it is practically as difficult today to get a manual typewriter repaired as to find a blacksmith. Word processing software offers a much-improved way of dealing with documents.

Word processing software allows you to use computers to create, edit, format, print, and store text material, among other things. The best-known word processing program is probably Microsoft Word but there are others such as Corel WordPerfect and the word processing components of Lotus Smart Suite and Sun Microsystems' StarOffice. Word processing software allows users to maneuver through a document and *delete, insert,* and *replace* text, the principal correction activities. It also offers such additional features as *creating, editing, formatting, printing,* and *saving.*

Features of the Keyboard

Besides the mouse, the principal tool of word processing is the keyboard. As well as letter, number, and punctuation keys and often a calculator-style numeric keypad, computer keyboards have special-purpose and function keys. *(See* ● *Panel 3.11.)* Sometimes, for the sake of convenience, strings of keystrokes are combined in combinations called *macros*.

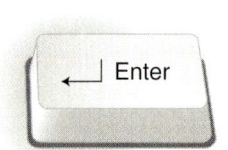

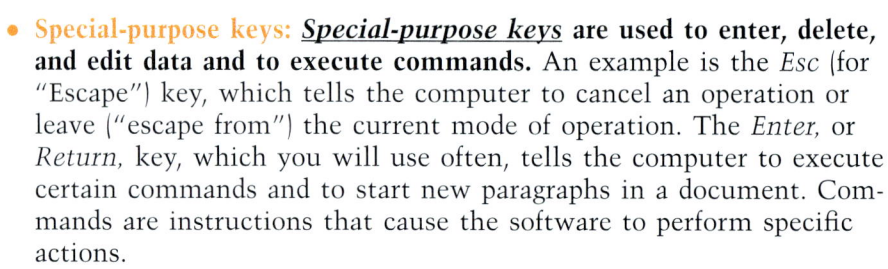

- **Special-purpose keys:** *Special-purpose keys* **are used to enter, delete, and edit data and to execute commands.** An example is the *Esc* (for "Escape") key, which tells the computer to cancel an operation or leave ("escape from") the current mode of operation. The *Enter*, or *Return*, key, which you will use often, tells the computer to execute certain commands and to start new paragraphs in a document. Commands are instructions that cause the software to perform specific actions.

    Special-purpose keys are generally used the same way regardless of the application software package being used. Most keyboards include the following special-purpose keys: *Esc, Ctrl, Alt, Del, Ins, Home, End, PgUp, PgDn, Num Lock,* and a few others. (*Ctrl* means Control, *Del* means Delete, *Ins* means Insert, for example.)

- **Function keys:** *Function keys*, **labeled F1, F2, and so on, are positioned along the top or left side of the keyboard. They are used to execute commands specific to the software being used.** For example, one application software package may use *F6* to exit a file, whereas another may use *F6* to underline a word.

- **Macros:** Sometimes you may wish to reduce the number of keystrokes required to execute a command. To do this, you use a macro.

● **PANEL 3.11**
Common keyboard layout

Escape Key
You can press **Esc** to quit a task you are performing.

Caps Lock and Shift Keys
These keys let you enter text in uppercase (ABC) and lowercase (abc) letters.

Press **Caps Lock** to change the case of all letters you type. Press the key again to return to the original case.

Press **Shift** in combination with another key to type an uppercase letter.

Ctrl and Alt Keys
You can use the **Ctrl** or **Alt** key in combination with another key to perform a specific task. For example, in some programs, you can press **Ctrl** and **S** to save a document.

Function Keys
These keys let you quickly perform specific tasks. For example, in many programs you can press **F1** to display help information.

Windows Key
You can press the **Windows** key to quickly display the Start menu when using many Windows operating systems.

Spacebar
You can press the **Spacebar** to insert a blank space.

A *macro* is a single keystroke or command—or a series of keystrokes or commands—used to automatically issue a longer, predetermined series of keystrokes or commands. Thus, you can consolidate several activities into only one or two keystrokes. The user names the macro and stores the corresponding command sequence; once this is done, the macro can be used repeatedly. (To set up a macro, pull down the Help menu and type in "macro.")

Although many people have no need for macros, individuals who find themselves continually repeating complicated patterns of keystrokes say they are quite useful.

Creating Documents

Creating a document means entering text using the keyboard. Word processing software has three features that affect this process—the *cursor, scrolling,* and *word wrap.*

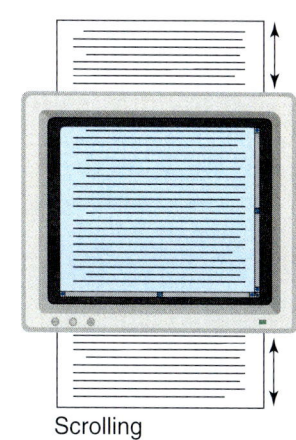

- Cursor: The *cursor* is the movable symbol on the display screen that shows you where you may next enter data or commands. The symbol is often a blinking rectangle or an I-beam. You can move the cursor on the screen using the keyboard's directional arrow keys or a mouse. The point where the cursor is located is called the *insertion point.*

- Scrolling: *Scrolling* means moving quickly upward, downward, or sideways through the text or other screen display. A standard

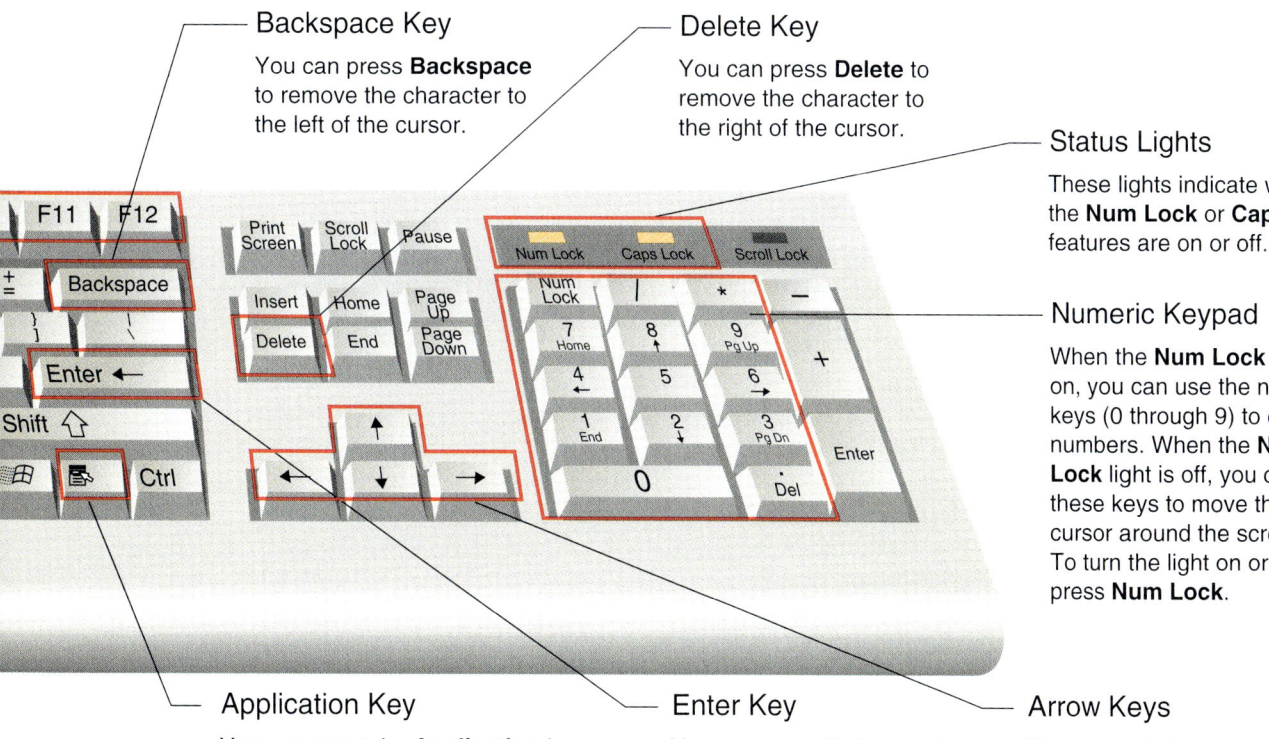

Cursor

Scrolling

Backspace Key
You can press **Backspace** to remove the character to the left of the cursor.

Delete Key
You can press **Delete** to remove the character to the right of the cursor.

Status Lights
These lights indicate whether the **Num Lock** or **Caps Lock** features are on or off.

Numeric Keypad
When the **Num Lock** light is on, you can use the number keys (0 through 9) to enter numbers. When the **Num Lock** light is off, you can use these keys to move the cursor around the screen. To turn the light on or off, press **Num Lock**.

Application Key
You can press the **Application** key to quickly display the shortcut menu for an item on your screen. Shortcut menus display a list of commands commonly used to complete a task related to the current activity.

Enter Key
You can press **Enter** to tell the computer to carry out a task. In a word processing program, press this key to start a new paragraph.

Arrow Keys
These keys let you move the cursor around the screen.

computer screen displays only 20–22 lines of standard-size text. Of course, most documents are longer than that. Using the directional arrow keys, or the mouse and a scroll bar located at the side of the screen, you can move ("scroll") through the display screen and into the text above and below it.

- **Word wrap:** *Word wrap* automatically continues text on the next line when you reach the right margin. That is, the text "wraps around" to the next line. You don't have to hit a "carriage return" key or Enter key, as you do with a typewriter.

To help you organize term papers and reports, the *outline feature* puts tags on various headings to show the hierarchy of heads—for example, main head, subhead, and sub-subhead. The basics of word processing are shown in the accompanying illustration. *(See ● Panel 3.12.)*

Editing Documents

Editing is the act of making alterations in the content of your document. Some features of editing are *insert and delete, undelete, find and replace, cut/copy and paste, spelling checker, grammar checker,* and *thesaurus.* Some of these commands are in the Edit pull-down menu and icons on the toolbar.

- **Insert and delete:** *Inserting* is the act of adding to the document. Simply place the cursor wherever you want to add text and start typing; the existing characters will be pushed along. If you want to write over (replace) text as you write, press the Insert key before typing. When you're finished typing, press the Insert key again to exit Insert mode.

    Deleting is the act of removing text, usually using the *Delete* or *Backspace* keys.

    The *Undo command* allows you to change your mind and restore text that you have deleted. Some word processing programs offer as many as 100 layers of "undo," so that users who delete several paragraphs of text, but then change their minds, can reinstate the material.

- **Find and replace:** The *Find,* or *Search, command* allows you to find any word, phrase, or number that exists in your document. The *Replace command* allows you to automatically replace it with something else.

- **Cut/Copy and paste:** Typewriter users who wanted to move a paragraph or block of text from one place to another in a manuscript used scissors and glue to "cut and paste." With word processing, it takes only a few keystrokes. You select (highlight) the portion of text you want to copy or move. Then you use the *Copy* or *Cut command* to move it to the *clipboard,* a special holding area in the computer's memory. From there, you can "paste," or transfer, the material to any point (indicated with the cursor) in the existing document or in a new document.

- **Spelling checker:** Most word processors have a **spelling checker, which tests for incorrectly spelled words.** As you type, the spelling checker indicates (perhaps with a squiggly line) words that aren't in its dictionary and thus may be misspelled. *(See ● Panel 3.13, page 110.)* Special add-on dictionaries are available for medical, engineering, and legal terms. In addition, programs such as Microsoft Word have an Auto Correct function that automatically fixes such common mistakes as transposed letters—replacing "teh" with "the," for instance.

Survival Tip

When Several Word Documents Are Open

You can write with several Word documents open simultaneously. To go ("toggle") back and forth, hold down *Ctrl* and press *F6.* To go backward, press *Ctrl, Shift,* and press *F6.* To display several documents at once, go to the Window menu and select Arrange All. You can cut and paste text from one document to another.

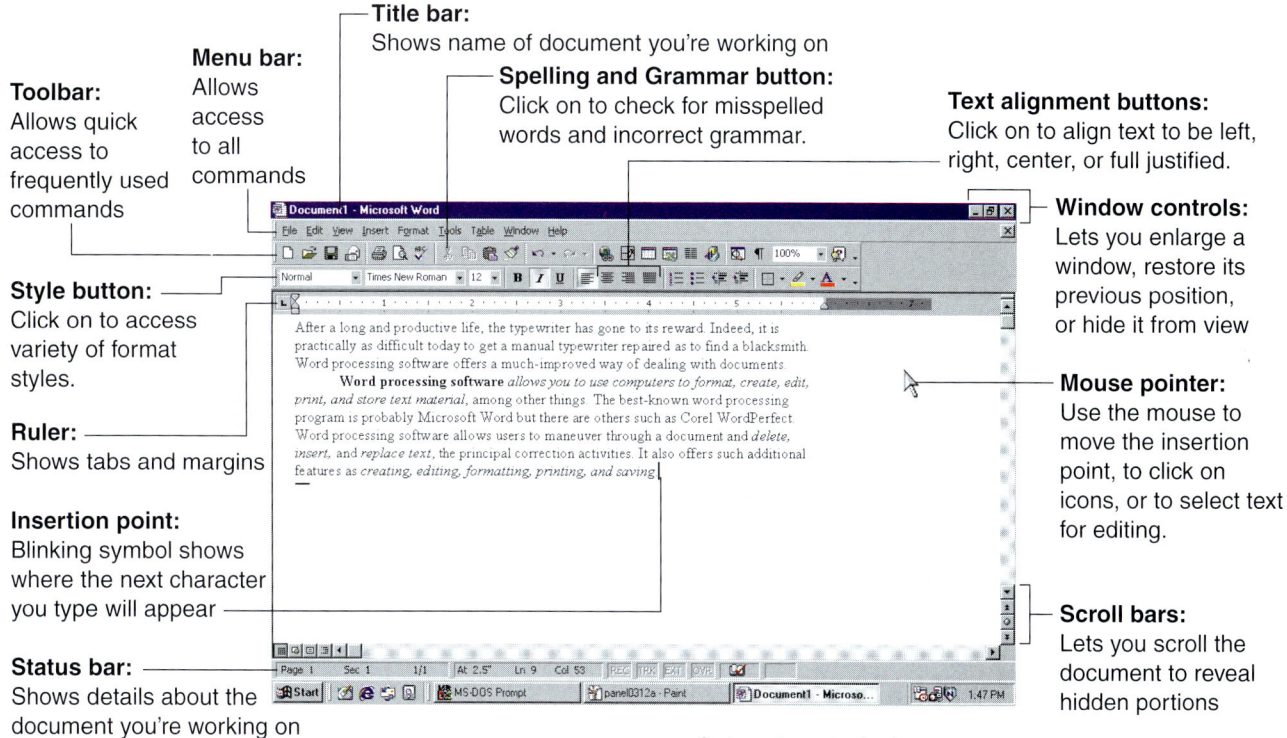

Toolbar:
Allows quick access to frequently used commands

Menu bar:
Allows access to all commands

Title bar:
Shows name of document you're working on

Spelling and Grammar button:
Click on to check for misspelled words and incorrect grammar.

Text alignment buttons:
Click on to align text to be left, right, center, or full justified.

Style button:
Click on to access variety of format styles.

Ruler:
Shows tabs and margins

Window controls:
Lets you enlarge a window, restore its previous position, or hide it from view

Mouse pointer:
Use the mouse to move the insertion point, to click on icons, or to select text for editing.

Insertion point:
Blinking symbol shows where the next character you type will appear

Scroll bars:
Lets you scroll the document to reveal hidden portions

Status bar:
Shows details about the document you're working on

Outline feature:
Enables you to view headings in your document

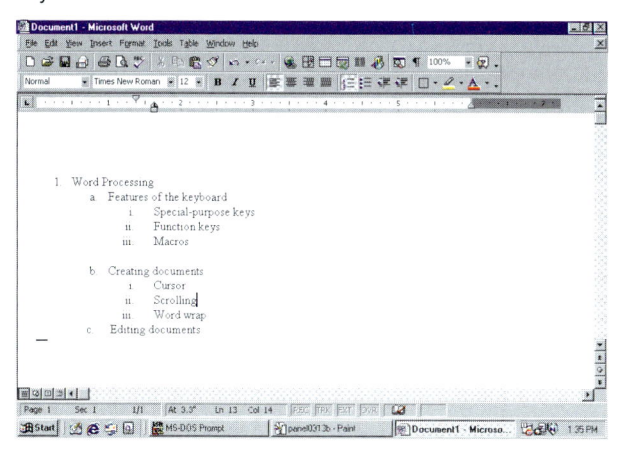

Cut-and-paste feature:
Enables you to move blocks of text. First highlight the text. On Edit menu, select Cut option. Then, on Edit Menu, select Paste option. (You can also use the icons in the toolbar.)

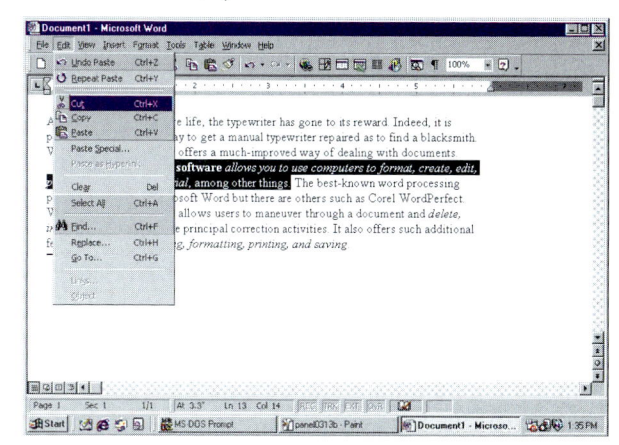

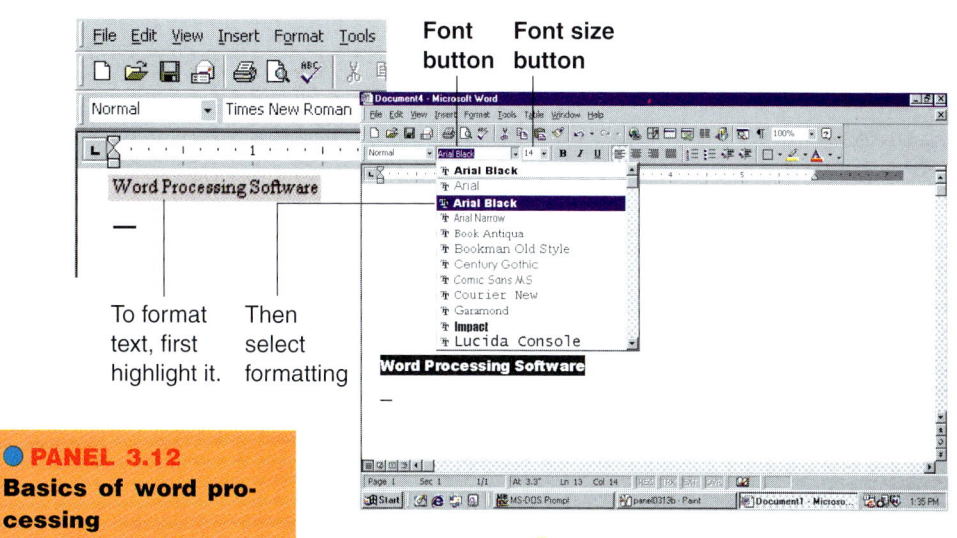

Font button   Font size button

To format text, first highlight it.   Then select formatting

Formatting feature:
Enables you to change type font and size. First highlight the text. Then click next to Font button for pull-down menu of fonts. Click next to Font Size button for menu of type sizes.

● **PANEL 3.12**
Basics of word processing

Software

Red wavy underline: Indicates spelling checker doesn't recognize the word. You have two options.

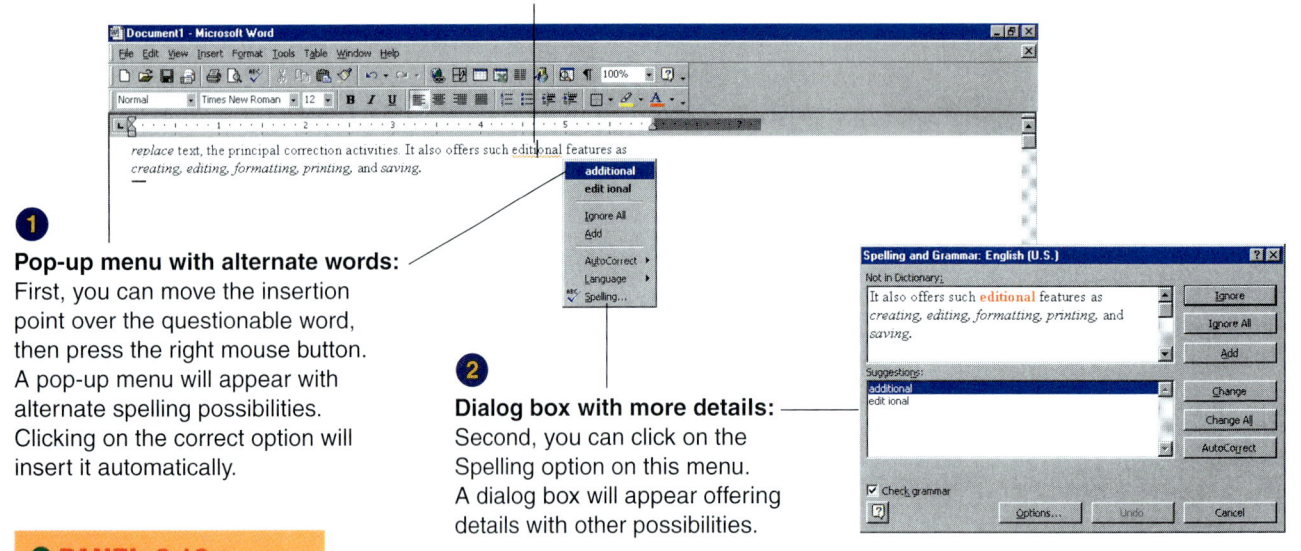

① Pop-up menu with alternate words: First, you can move the insertion point over the questionable word, then press the right mouse button. A pop-up menu will appear with alternate spelling possibilities. Clicking on the correct option will insert it automatically.

② Dialog box with more details: Second, you can click on the Spelling option on this menu. A dialog box will appear offering details with other possibilities.

● **PANEL 3.13**
Spelling checker
How a word processing program checks for misspelled words and offers alternatives

- **Grammar checker:** A *grammar checker* **highlights poor grammar, wordiness, incomplete sentences, and awkward phrases.** The grammar checker won't fix things automatically, but it will flag (perhaps with a different color of squiggly line) possible incorrect word usage and sentence structure. *(See ● Panel 3.14.)*
- **Thesaurus:** If you find yourself stuck for the right word while you're writing, you can call up an on-screen *thesaurus*, **which will present you with the appropriate word or alternative words.**

Formatting Documents with the Help of Templates & Wizards

In the context of word processing, *formatting* **means determining the appearance of a document.** To this end, word processing programs provide two helpful devices—templates and wizards. A *template* **is a preformatted document that provides basic tools for shaping a final document**—the text, layout, and style for a letter, for example. A *wizard* **answers your questions and uses the answers to lay out and format a document.** In Word, you can use the Memo Wizard to create professional-looking memos or the Résumé Wizard to create a résumé.

Among the many aspects of formatting are the following:

- **Font:** You can decide what *font*—**typeface and type size**—you wish to use. For instance, you can specify whether it should be Arial, Courier, or Times New Roman. You can indicate whether the text should be, say, 10 points or 12 points in size and the headings should be 14 points or 16 points. (There are 72 points in an inch.) You can specify what parts of it should be underlined, *italic,* or **boldface.**
- **Spacing and columns:** You can choose whether you want the lines to be *single-spaced* or *double-spaced* (or something else). You can specify

10 point Times Roman

14 point Arial Black

16 point Courier

60
(60 point Arial)

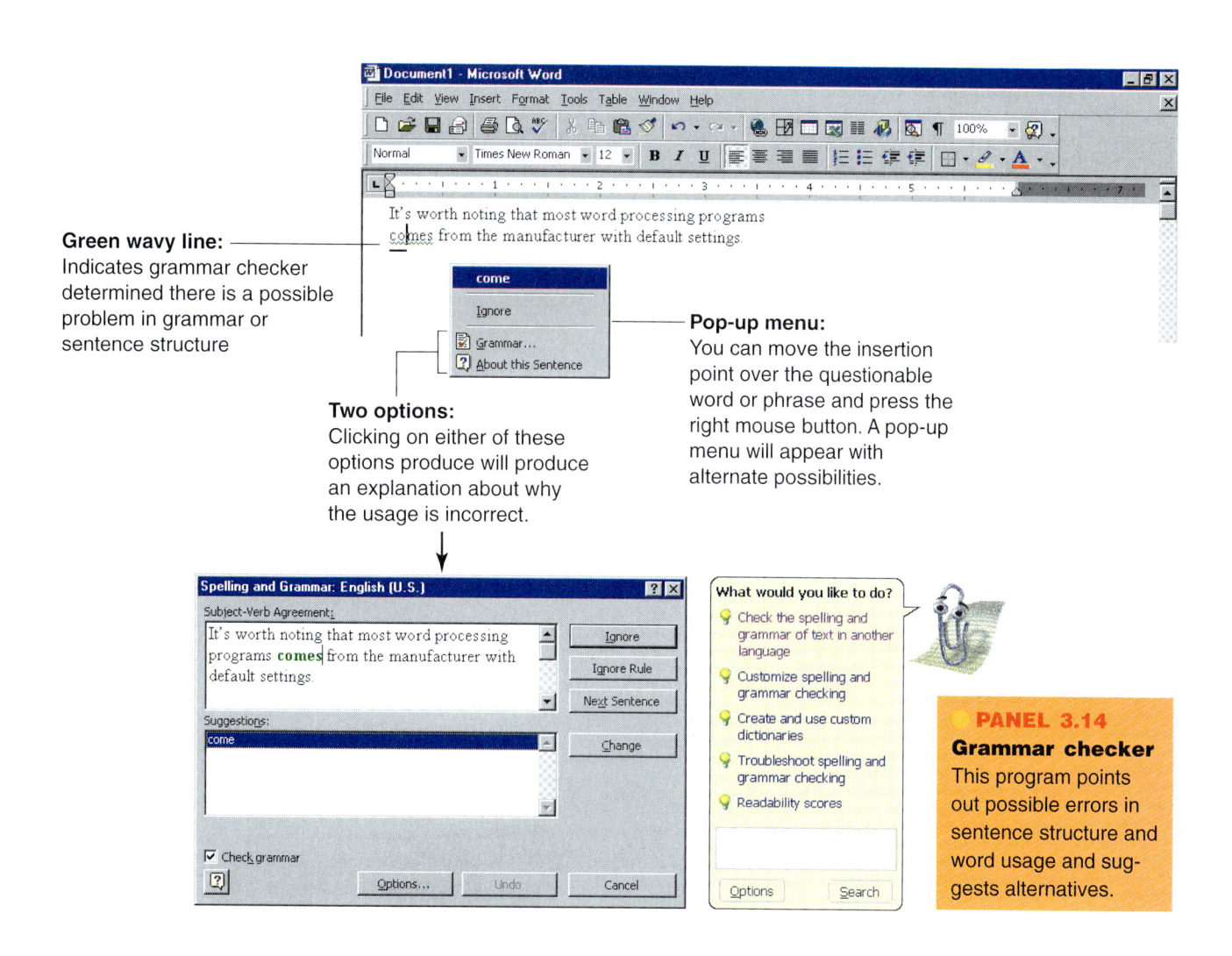

Green wavy line: Indicates grammar checker determined there is a possible problem in grammar or sentence structure

Pop-up menu: You can move the insertion point over the questionable word or phrase and press the right mouse button. A pop-up menu will appear with alternate possibilities.

Two options: Clicking on either of these options produce will produce an explanation about why the usage is incorrect.

● **PANEL 3.14**

Grammar checker This program points out possible errors in sentence structure and word usage and suggests alternatives.

Left-justified

Justified

Centered

Right-justified

whether you want text to be *one column* (like this page), *two columns* (like many magazines and books), or *several columns* (like newspapers).

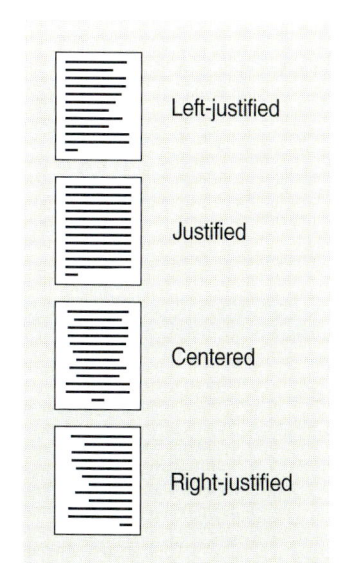

- **Margins and justification:** You can indicate the dimensions of the *margins*—left, right, top, and bottom—around the text. You can specify the text *justification*—how the letters and words are spaced in each line. To *justify* means to align text evenly between left and right margins, as in most newspaper columns. To *left-justify* means to align text evenly on the left. (Left-justified text has a "ragged-right" margin, as do many business letters and this paragraph.) *Centering* centers each text line in the available white space.

- **Pages, headers, footers:** You can indicate page numbers and headers or footers. A *header* is common text (such as a date or document name) printed at the top of every page. A *footer* is the same thing printed at the bottom of every page.

- **Other formatting:** You can specify *borders* or other decorative lines, *shading, tables,* and *footnotes.* You can even import *graphics* or drawings from files in other software programs, including *clip art*—collections of ready-made pictures and illustrations available online or on CD-ROM disks.

It's worth noting that word processing programs (and indeed most forms of application software) come from the manufacturer with *default settings.* **Default settings are the settings automatically used by a program unless the user specifies otherwise, thereby overriding them.** Thus, for example, a word

> **Survival Tip**
>
> **Office XP**
>
> Microsoft's newest version of Word is part of Office XP. This version does *not* work with Microsoft's system software Windows 95; you must have a later version of the system software—Windows 98, Windows Me, Windows XP. However, among other new features, Office XP has new data recovery tools to help you recover lost work. And it supports multi-user collaboration.

processing program may automatically prepare a document single-spaced, left-justified, with 1-inch right and left margins, unless you alter these default settings.

Printing, Faxing, or E-Mailing Documents

Most word processing software gives you several options for printing. For example, you can print *several copies* of a document. You can print *individual pages* or a *range of pages*. You can even preview a document before printing it out. *Previewing (print previewing)* means viewing a document on screen to see what it will look like in printed form before it's printed. Whole pages are displayed in reduced size.

You can also send your document off to someone else by fax or e-mail attachment, if your computer has the appropriate communications link.

Saving Documents

<u>*Saving*</u> **means storing, or preserving, a document as an electronic file permanently**—on floppy disk or hard disk, for example. Saving is a feature of nearly all application software. Having the document stored in electronic form spares you the tiresome chore of retyping it from scratch whenever you want to make changes. You need only retrieve it from the storage medium and make the changes you want. Then you can print it out again.

> ~~Four score and~~ Eighty-seven years ago, our fathers <u>and mothers</u> brought forth on this continent a new nation

Tracking Changes & Inserting Comments

What if you have written an important document and have asked other people to edit it? Word processing software allows editing changes to be tracked by highlighting them, underlining additions, and crossing out deletions. Each person working on the document can choose a different color so that you can tell who's done what. And anyone can insert hidden questions or comments that become visible when you pass the mouse pointer over yellow-highlighted words or punctuation. An edited document can be printed out showing all the changes, as well as a list of comments keyed to the text by numbers. Or it can be printed out "clean," showing the edited text in its new form, without the changes.

Web Document Creation

Most word processing programs allow you to automatically format your documents into HTML (see Chapter 2, p. 51) so they can be used on the Web.

CONCEPT CHECK

What are the important features of the keyboard?

Describe the role of the cursor, scrolling, and word wrap in creating documents.

What word processing features are available to help edit documents?

What assistance is available to help you format documents, and what aspects of formatting should be of concern to you?

3.4 Spreadsheets

KEY QUESTION

What can you do with an electronic spreadsheet that you can't do with pencil and paper and a standard calculator?

What is a spreadsheet? Traditionally, it was simply a grid of rows and columns, printed on special light-green paper, that was used to produce financial projections and reports. A person making up a spreadsheet spent long days and weekends at the office penciling tiny numbers into countless tiny rectangles. When one figure changed, all others numbers on the spreadsheet had to be recomputed. Ultimately, there might be wastebaskets full of jettisoned worksheets.

In the late 1970s, Daniel Bricklin was a student at the Harvard Business School. One day he was staring at columns of numbers on a blackboard when he got the idea for computerizing the spreadsheet. He created the first electronic spreadsheet, now called simply a spreadsheet. **The _spreadsheet_ allows users to create tables and financial schedules by entering data and formulas into rows and columns arranged as a grid on a display screen.** Before long, the electronic spreadsheet was the most popular small-business program. Unfortunately for Bricklin, his version (called VisiCalc) was quickly surpassed by others. Today the principal spreadsheets are Microsoft Excel, Corel Quattro Pro, and Lotus 1-2-3.

You can put your checkbook register on a spreadsheet, and then use it to compute totals and compare income and expenses from one month to the next. In addition, within a spreadsheet file you may have workbooks containing worksheets. A *worksheet* is a single table. A *workbook* is a collection of related worksheets. Thus, within your Microsoft Excel spreadsheet file, you might have one workbook headed *Checkbook,* which would contain worksheets with the history of your checking account for the years 2001, 2002, and so on. You might have another workbook headed *Credit cards,* containing worksheets for each year.

The Basics: How Spreadsheets Work

A spreadsheet is arranged as follows. (See ● *Panel 3.15, next page.*)

- **How a spreadsheet is organized—column headings, row headings, and labels:** In the worksheet's frame area (work area), lettered *column headings* appear across the top ("A" is the name of the first column, "B" the second, and so on). Numbered *row headings* appear down the left side ("1" is the name of the first row, "2" the second, and so forth). *Labels* are any descriptive text, such as APRIL, RENT, or GROSS SALES. You use your computer's keyboard to type in the various headings and labels.

- **Where columns and rows meet—cells, cell addresses, ranges, and values:** **A _cell_ is the place where a row and a column intersect; its position is called a cell address.** For example, "A1" is the cell address for the top left cell, where column A and row 1 intersect. A *range* is a group of adjacent cells—for example, A1 to A5. **A number or date entered in a cell is called a _value_.** The values are the actual numbers used in the spreadsheet—dollars, percentages, grade points, temperatures, or whatever. Headings, labels, and formulas also go into cells. A *cell pointer,* or *spreadsheet cursor,* indicates where data is to be entered. The cell pointer can be moved around like a cursor in a word processing program.

- **Why the spreadsheet has become so popular—formulas, functions, recalculation, and what-if analysis:** Now we come to the reason the electronic spreadsheet has taken offices by storm. **_Formulas_ are instructions for calculations.** For example, a formula might be @SUM(A5:A15), meaning *Sum* (that is, add) *all the numbers in the cells with cell addresses A5 through A15.*

Spreadsheet
A farmer uses a notebook computer with spreadsheet software to record data about pigs, such as age, weight, and eating patterns.

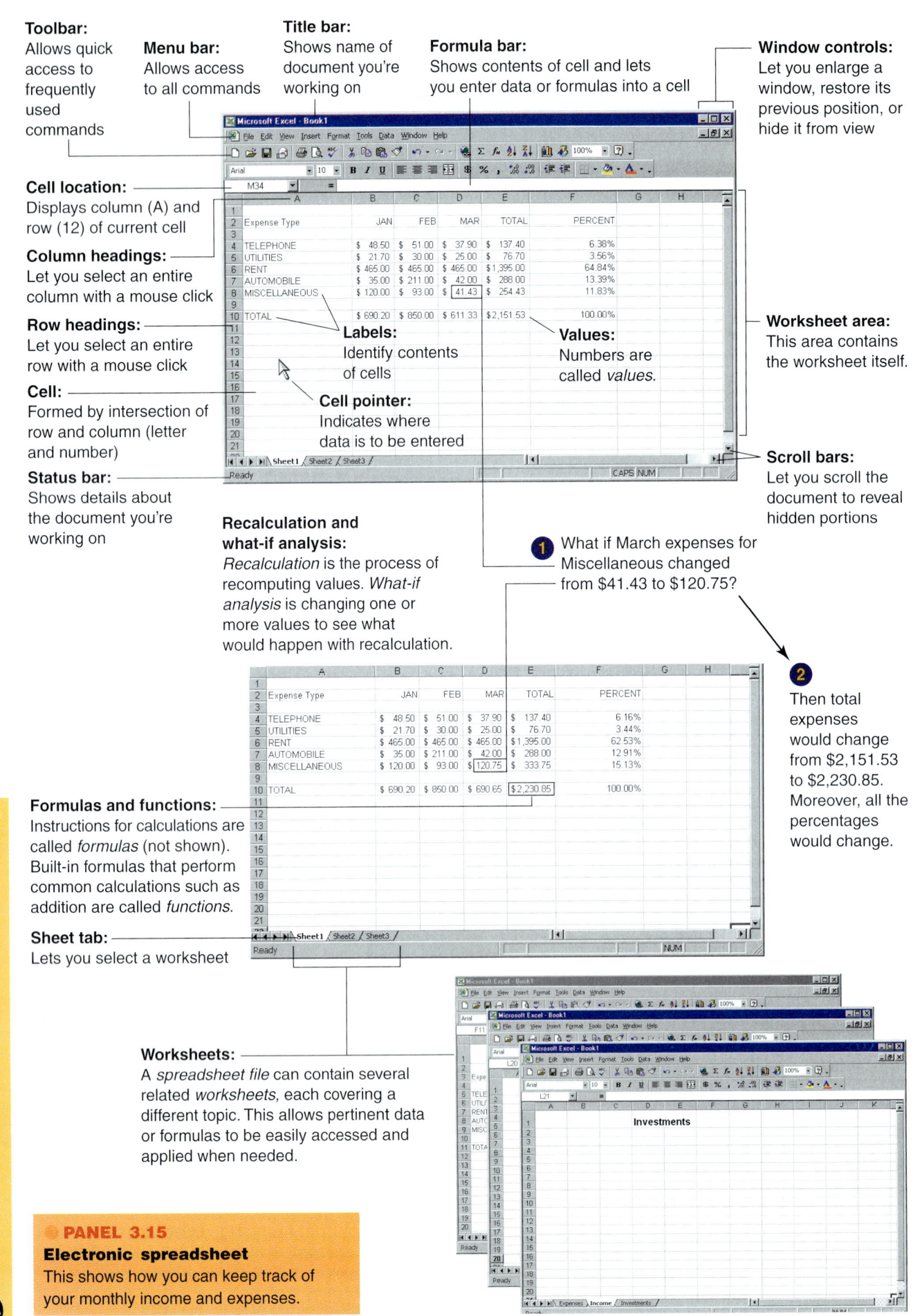

Toolbar: Allows quick access to frequently used commands

Menu bar: Allows access to all commands

Title bar: Shows name of document you're working on

Formula bar: Shows contents of cell and lets you enter data or formulas into a cell

Window controls: Let you enlarge a window, restore its previous position, or hide it from view

Cell location: Displays column (A) and row (12) of current cell

Column headings: Let you select an entire column with a mouse click

Row headings: Let you select an entire row with a mouse click

Cell: Formed by intersection of row and column (letter and number)

Status bar: Shows details about the document you're working on

Labels: Identify contents of cells

Cell pointer: Indicates where data is to be entered

Values: Numbers are called *values*.

Worksheet area: This area contains the worksheet itself.

Scroll bars: Let you scroll the document to reveal hidden portions

Recalculation and what-if analysis: *Recalculation* is the process of recomputing values. *What-if analysis* is changing one or more values to see what would happen with recalculation.

① What if March expenses for Miscellaneous changed from $41.43 to $120.75?

② Then total expenses would change from $2,151.53 to $2,230.85. Moreover, all the percentages would change.

Formulas and functions: Instructions for calculations are called *formulas* (not shown). Built-in formulas that perform common calculations such as addition are called *functions*.

Sheet tab: Lets you select a worksheet

Worksheets: A *spreadsheet file* can contain several related *worksheets*, each covering a different topic. This allows pertinent data or formulas to be easily accessed and applied when needed.

● **PANEL 3.15**
Electronic spreadsheet
This shows how you can keep track of your monthly income and expenses.

Chapter 3

114

Functions are built-in formulas that perform common calculations. For instance, a function might average a range of numbers or round off a number to two decimal places.

After the values have been entered into the worksheet, the formulas and functions can be used to calculate outcomes. However, what was revolutionary about the electronic spreadsheet was its ability to easily do recalculation. **_Recalculation_ is the process of recomputing values,** either as an ongoing process as data is entered or afterward, with the press of a key. With this simple feature, the hours of mind-numbing work required to manually rework paper spreadsheets became a thing of the past.

The recalculation feature has opened up whole new possibilities for decision making. In particular, **_what-if analysis_ allows the user to see how changing one or more numbers changes the outcome of the recalculation.** That is, you can create a worksheet, putting in formulas and numbers, and then ask, "What would happen if we change that detail?"—and immediately see the effect on the bottom line.

- **Using worksheet templates—pre-arranged forms for specific tasks:** You may find that your spreadsheet software makes worksheet templates available for specific tasks. *Worksheet templates* are forms containing formats and formulas custom-designed for particular kinds of work. Examples are templates for calculating loan payments, tracking travel expenses, monitoring personal budgets, and keeping track of time worked on projects. Templates are also available for a variety of business needs—providing sales quotations, invoicing customers, creating purchase orders, and writing a business plan.

Analytical Graphics: Creating Charts

You can use spreadsheet packages to create analytical graphics, or charts. **_Analytical graphics_, or business graphics, are graphical forms that make numeric data easier to analyze** than when it is organized as rows and columns of numbers. Whether viewed on a monitor or printed out, analytical graphics help make sales figures, economic trends, and the like easier to comprehend and analyze.

The principal examples of analytical graphics are *bar charts, line graphs,* and *pie charts.* (See ● Panel 3.16) If you have a color printer, these charts can appear in color. In addition, they can be displayed or printed out so that they look three-dimensional. Spreadsheets can even be linked to more exciting graphics, such as digitized maps.

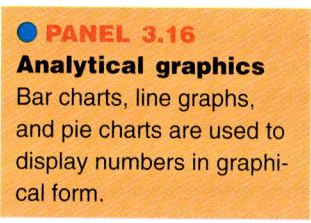

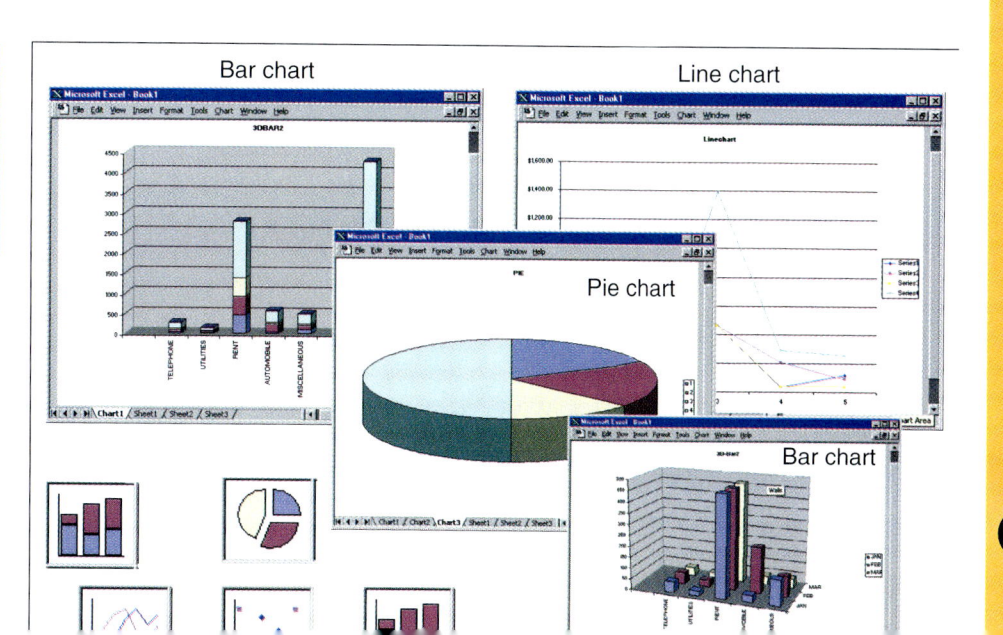

● **PANEL 3.16**
Analytical graphics
Bar charts, line graphs, and pie charts are used to display numbers in graphical form.

> **CONCEPT CHECK**
>
> What is a spreadsheet? A worksheet? A workbook?
>
> What are the components of a spreadsheet?
>
> What is the significance of recalculation and what-if analysis?
>
> What's useful about worksheet templates and analytical graphics?

3.5 Database Software

KEY QUESTIONS
What is database software, and what is personal information management software?

In its most general sense, a database is any electronically stored collection of data in a computer system. In its more specific sense, **a _database_ is a collection of interrelated files** in a computer system. These computer-based files are organized according to their common elements, so that they can be retrieved easily. (Databases are covered in detail in Chapter 7.) Sometimes called a *database manager* or *database management system (DBMS)*, **_database software_ is a program that sets up and controls the structure of a database and access to the data.**

The Benefits of Database Software

When data is stored in separate files, the same data will be repeated in many files. In the old days, each college administrative office—registrar, financial aid, housing, and so on—might have a separate file on you. Thus, there was *redundancy*—your address, for example, was repeated over and over. The advantage of database software is that data is not in separate files. Rather, it is *integrated*. Thus, your address need only be listed once, and all the separate administrative offices will have access to the same information. For that reason, information in databases is considered to have more *integrity*. That is, the information is more likely to be accurate and up to date.

Databases are a lot more interesting than they used to be. Once they included only text. Now they can also include pictures, sound, and animation. It's likely, for instance, that your personnel record in a future company database will include a picture of you and perhaps even a clip of your voice. If you go looking for a house to buy, you will be able to view a real estate agent's database of video clips of homes and properties without leaving the realtor's office.

Today the principal microcomputer database programs are Microsoft Access, Corel Paradox, and Lotus Approach. (In larger systems, Oracle is a major player.)

The Basics: How Databases Work

Let's consider some basic features of databases:

- **How a relational database is organized—tables, records, and fields:** The most widely used form of database, especially on PCs, is the **_relational database_, in which data is organized into related tables.** Each *table* contains rows and columns; the rows are called records, and the columns are called fields. An example of a record is a person's address—name, street address, city, and so on. An example of a field is that person's last name; another field would be that person's first name, a third field would be that person's street address, and so on. (See ● Panel 3.17.)

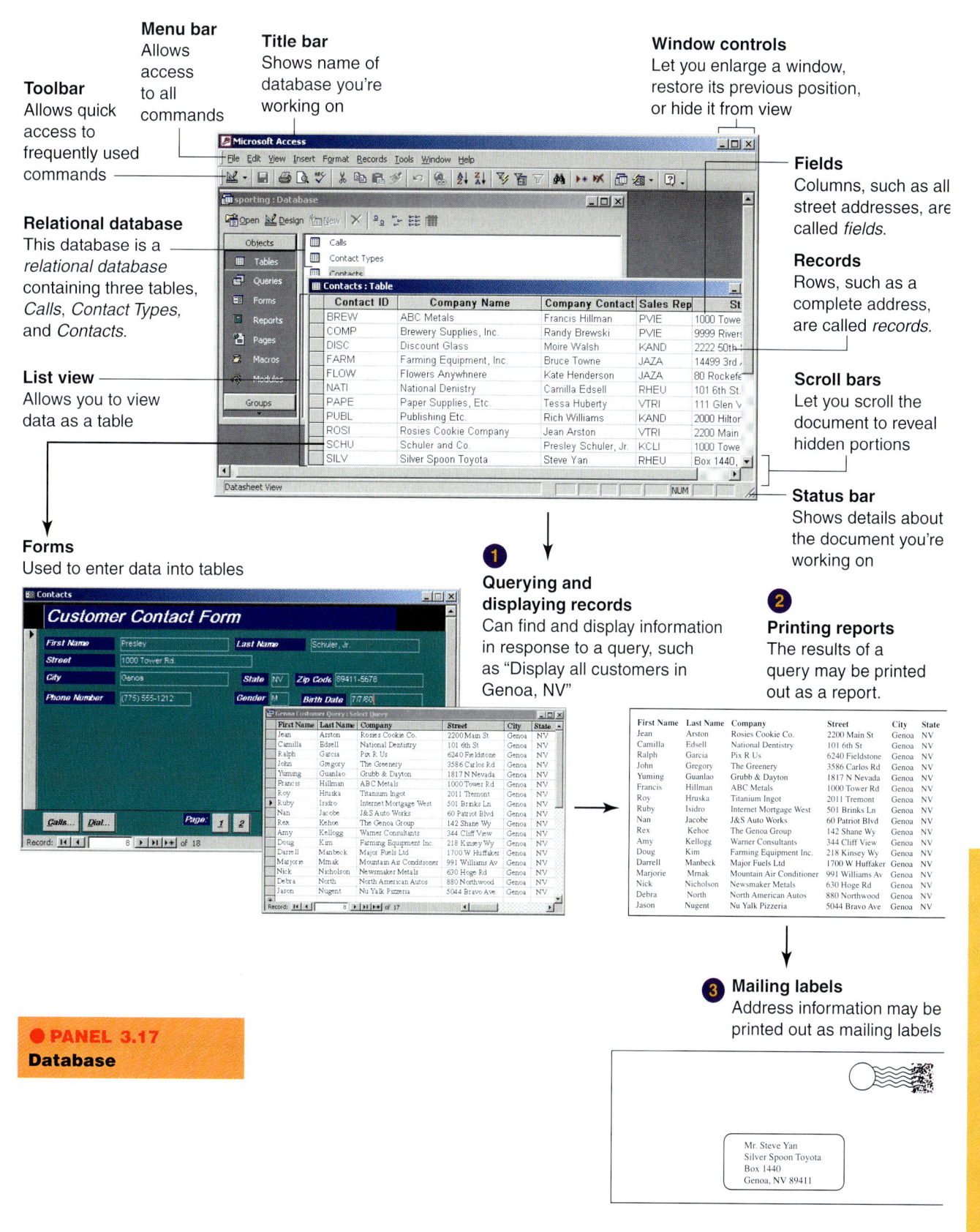

Toolbar
Allows quick access to frequently used commands

Menu bar
Allows access to all commands

Title bar
Shows name of database you're working on

Window controls
Let you enlarge a window, restore its previous position, or hide it from view

Fields
Columns, such as all street addresses, are called *fields*.

Relational database
This database is a *relational database* containing three tables, *Calls*, *Contact Types*, and *Contacts*.

Records
Rows, such as a complete address, are called *records*.

List view
Allows you to view data as a table

Scroll bars
Let you scroll the document to reveal hidden portions

Status bar
Shows details about the document you're working on

Forms
Used to enter data into tables

① Querying and displaying records
Can find and display information in response to a query, such as "Display all customers in Genoa, NV"

② Printing reports
The results of a query may be printed out as a report.

③ Mailing labels
Address information may be printed out as mailing labels

● **PANEL 3.17**
Database

Just as a spreadsheet may include a workbook with several worksheets, so a relational database might include a database with several tables. For instance, if you're running a small company, you might have one database headed *Employees*, containing three tables—*Addresses, Payroll,* and *Benefits.* You might have another database headed *Customers,* with *Addresses, Orders,* and *Invoices* tables.

- **How various records can be linked—the key field:** The records within the various tables in a database are linked by a **_key field_, a field that can be used as a common identifier because it is unique.** The most frequent key field used in the United States is the Social Security number, but any unique identifier could be used, such as employee number.

- **Finding what you want—querying and displaying records:** The beauty of database software is that you can locate records quickly. For example, several offices at your college may need access to your records, but for different reasons: the registrar, financial aid, student housing, and so on. Any of these offices can *query records—locate and display records*—by calling them up on a computer screen for viewing and updating. Thus, if you move, your address field will need to be corrected for all relevant offices of the college. A person making a search might make the query, "Display the address of [your name]." Once a record is displayed, the address field can be changed. Thereafter, any office calling up your file will see the new address.

- **Sorting and analyzing records and applying formulas:** With database software you can easily find and change the order of records in a table. Normally, records are displayed in a database in the same order in which they are entered—for instance, by the date a person registered to attend college. However, all these records can be *sorted* in different ways—arranged alphabetically, numerically, geographically, or in some other order. For example, they can be rearranged by state, by age, or by Social Security number.

  In addition, database programs contain built-in mathematical formulas so that you can analyze data. This feature can be used, for example, to find the grade-point averages for students in different majors or in different classes.

- **Putting search results to use—saving, formatting, printing, copying, or transmitting:** Once you've queried, sorted, and analyzed the records and fields, you can simply save them to your hard disk or to a floppy disk. You can format them in different ways, altering headings and type styles. You can print them out on paper as reports, such as an employee list with up-to-date addresses and phone numbers. A common use is to print out the results as names and addresses on mailing labels—adhesive-backed stickers that can be run through your printer, then stuck on envelopes. You can use the copy command to copy your search results and then paste them into a paper produced on your word processor. You can also cut and paste data into an e-mail message or make the data an attachment file to an e-mail, so that it can be transmitted to someone else.

Personal Information Managers

Pretend you are sitting at a desk in an old-fashioned office. You have a calendar, a Rolodex-type address file, and a notepad. Most of these items could also be found on a student's desk. How would a computer and software improve on this arrangement?

Many people find ready uses for specialized types of database software known as personal information managers. **A _personal information manager (PIM)_ is software to help you keep track of and manage information you use**

● **PANEL 3.18**

Personal information manager

This shows the calendar available with Microsoft Outlook.

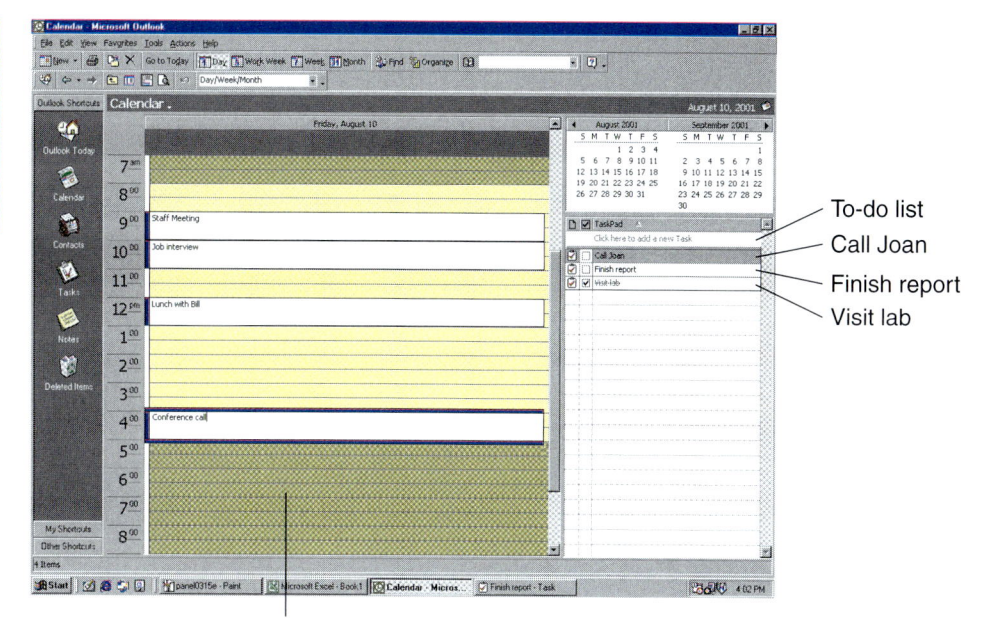

— To-do list
— Call Joan
— Finish report
— Visit lab

Appointment calendar

on a daily basis, such as addresses, telephone numbers, appointments, to-do lists, and miscellaneous notes. Some programs feature phone dialers, outliners (for roughing out ideas in outline form), and ticklers (or reminders). With a PIM, you can key in notes in any way you like and then retrieve them later based on any of the words you typed.

Popular PIMs are Microsoft Outlook, Lotus Organizer, and Act. Microsoft Outlook, for example, has sections labeled Inbox, Calendar, Contacts, Tasks (to-do list), Journal (to record interactions with people), Notes (scratchpad), and Files. *(See ● Panel 3.18.)*

CONCEPT CHECK

What is a database, and what are its benefits?

Describe the basic features of a relational database.

How might a PIM help you?

3.6 Specialty Software

KEY QUESTIONS

What are the principal uses of specialty software such as presentation graphics, financial, desktop publishing, drawing and painting, project management, computer-aided design, and video/audio editing software?

After learning some of the productivity software just described, you may wish to become familiar with more specialized programs. For example, you might first learn word processing and then move on to desktop publishing, or first learn spreadsheets and then learn personal-finance software. We will consider the following kinds of software, although they are but a handful of the thousands of specialized programs available: *presentation graphics, financial, desktop-publishing, drawing and painting, project management*, and *computer-aided design* software.

Presentation Graphics Software

You may already be accustomed to seeing presentation graphics because many college instructors now use such software to accompany their lectures. <u>**Presentation graphics software**</u> uses graphics, animation, sound, and data or

information to make visual presentations. Well-known presentation graphics packages include Microsoft PowerPoint, Corel Presentations, and Lotus Freelance Graphics. *(See* ● *Panel 3.19.)*

Visual presentations are commonly called *slide shows*, although they can consist not only of 35-mm slides but also of paper copies, overhead transparencies, video, animation, and sound. Presentation graphics packages often come with slide sorters, which group together a dozen or so slides in miniature. The person making the presentation can use a mouse or keyboard to bring the slides up for viewing or even start a self-running electronic slide show. You can also use a projection system from the computer itself.

Let's examine the process of using presentation software.

● **PANEL 3.19**
Presentation graphics
Microsoft PowerPoint 2000 helps you prepare and make visual presentations.

1 Outline View
This view helps you organize the content of your material in standard outline form.

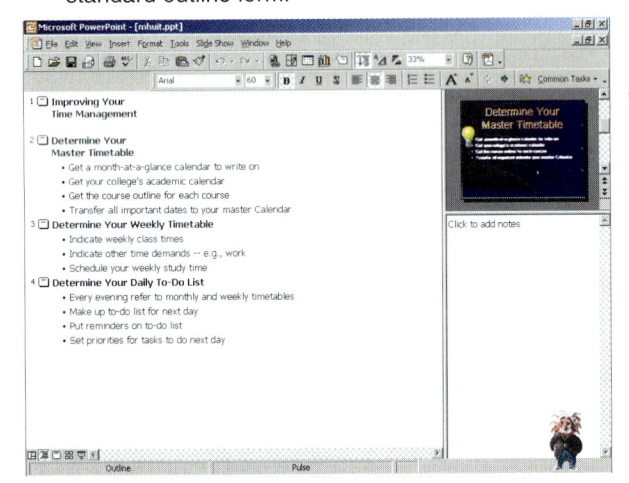

2 Dressing up your presentation
PowerPoint offers professional design templates of text format, background, and borders. You place your text for each slide into one of these templates. You can also import a graphic from the clip art that comes with the program.

3 Slide View
This view allows you to see what a single slide will look like. You can use this view to edit the content and looks of each slide.

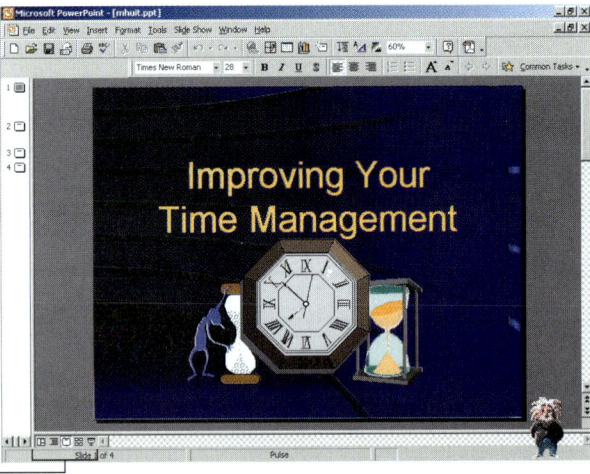

4 Notes Page View
This view displays a small version of the slide plus the notes you will be using as speaker notes.

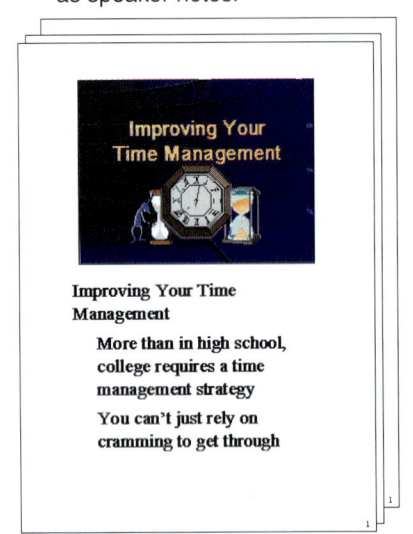

View icons
Clicking on these offers different views: *Slide, Outline, Slide Sorter, Notes Page,* and *Slide Show*

5 Slide Sorter View
This view displays miniatures of each slide, enabling you to adjust the order of your presentation.

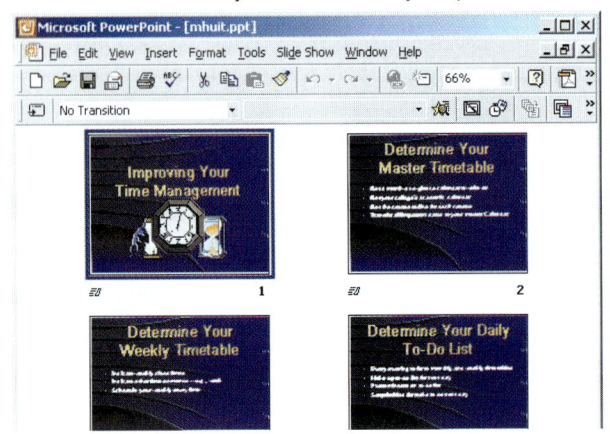

- **Using templates to get started:** Just as word processing programs offer templates for faxes, business letters, and the like, presentation-graphics programs offer templates to help you organize your presentation, whether it's for a roomful of people or over the Internet. Templates are of two types: design and content. *Design templates* offer formats, layouts, background patterns, and color schemes that can apply to general forms of content material. *Content templates* offer formats for specific subjects; for instance, PowerPoint offers templates for "Selling Your Ideas," "Facilitating a Meeting," and "Motivating a Team." The software offers wizards that walk you through the process of filling in the template.

- **Getting assistance on content development and organization:** To provide assistance as you're building your presentation, PowerPoint displays three windows on your screen at the same time—the *Outline View*, the *Slide View*, and the *Notes Page View*. This enables you to add new slides, create and edit the text on the slides, and create notes (to use as lecture or speech notes) while developing your presentation.

    The *Outline View* helps you organize the content of your material in standard outline form. The text you enter into the outline is automatically formatted into slides according to the template you selected. If you wish, you can pull in (import) your outline from a word processing document. The *Slide View* helps you see what a single slide will look like. The outline text appears as slide titles and subtitles in subordinate order. The *Notes Page View* displays the notes you will be using as speaker notes. It includes a small version of the slide.

    Two other views are helpful in organizing and practicing. The *Slide Sorter View* allows you to view a number of slides (4 to 12 or more) at once, so you can see how to order and reorder them. The *Slide Show View* presents the slides in the order in which your audience will view them, so you can practice your presentation.

- **Dressing up your presentation:** Presentation software makes it easy to dress up each visual page ("slide") with artwork by pulling in clip art from other sources. Although presentations may make use of some basic analytical graphics—bar, line, and pie charts—they usually look much more sophisticated. For instance, they may utilize different texture (speckled, solid, cross-hatched), color, and three-dimensionality. In addition, you can add audio clips, special visual effects (such as blinking text), animation, and video clips.

CONCEPT CHECK

What are the benefits of presentation graphics software?

What kind of help is available with a presentation graphics package?

Financial Software

Financial software is a growing category that ranges from personal-finance managers to entry-level accounting programs to business financial-management packages.

Consider the first of these, which you may find particularly useful. **Personal-finance managers let you keep track of income and expenses, write checks, do online banking, and plan financial goals.** Such programs don't promise to make you rich, but they can help you manage your money. They

PANEL 3.20
Quicken financial software

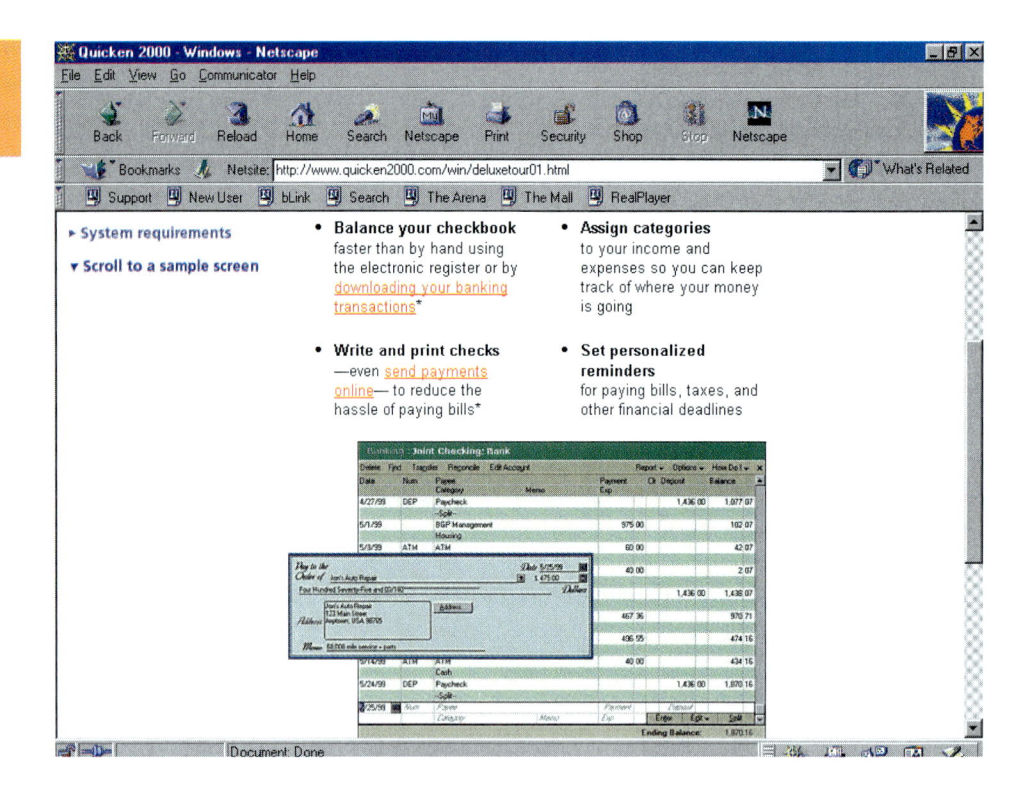

may even get you out of trouble. Many personal-finance programs, such as Quicken (see ● Panel 3.20.) Microsoft Money, include a calendar and a calculator, but the principal features are the following:

- **Tracking of income and expenses:** The programs allow you to set up various account categories for recording income and expenses, including credit card expenses.
- **Checkbook management:** All programs feature checkbook management, with an on-screen check writing form and check register that look like the ones in your checkbook. Checks can be purchased to use with your computer printer.
- **Reporting:** All programs compare your actual expenses with your budgeted expenses. Some will compare this year's expenses to last year's.
- **Income tax:** All programs offer tax categories, for indicating types of income and expenses that are important when you're filing your tax return.
- **Other:** Some of the more versatile personal-finance programs also offer financial-planning and portfolio-management features.

Besides personal-finance managers, financial software includes small business accounting and tax software programs, which provide virtually all the forms you need for filing income taxes. Tax programs such as TaxCut and Turbo Tax make complex calculations, check for mistakes, and even unearth deductions you didn't know existed. Tax programs can be linked to personal finance software to form an integrated tool.

Many financial software programs may be used in all kinds of enterprises. For instance, accounting software automates bookkeeping tasks, while payroll software keeps records of employee hours and produces reports for tax purposes.

Some programs go beyond financial management and tax and accounting management. For example, Business Plan Pro, Management Pro, and Performance Now can help you set up your own business from scratch.

Finally, there are investment software packages, such as StreetSmart from Charles Schwab and Online Xpress from Fidelity, as well as various retirement planning programs.

> **CONCEPT CHECK**
>
> What is financial software?
>
> What functions does financial software perform?

Desktop Publishing

When Margaret Trejo, then 36, was laid off from her job because her boss couldn't meet the payroll, she was stunned. "Nothing like that had ever happened to me before," she said later. "But I knew it wasn't a reflection on my work. And I saw it as an opportunity."[8]

Trejo became head of Trejo Production, a successful desktop-publishing company in Princeton, New Jersey, using Macintosh equipment to produce scores of books, brochures, and newsletters. A few years later she was making twice what she ever made in management positions.

Not everyone can set up a successful desktop-publishing business, because many complex layouts require experience, skill, and knowledge of graphic design. Indeed, use of these programs by nonprofessional users can lead to rather unprofessional-looking results. Nevertheless, the availability of microcomputers and reasonably inexpensive software has opened up a career area formerly reserved for professional typographers and printers.

Desktop publishing (DTP) involves mixing text and graphics to produce high-quality output for commercial printing, using a microcomputer and mouse, scanner, laser or ink-jet printer, and DTP software. Often the printer is used primarily to get an advance look before the completed job is sent to a typesetter for even higher-quality output. Professional DTP programs are QuarkXPress, Adobe InDesign, and Adobe PageMaker. Microsoft Publisher is a "low-end," consumer-oriented DTP package. Some word processing programs, such as Word and WordPerfect, also have many DTP features, though at nowhere near the level of the specialized DTP packages.

Desktop publishing has the following characteristics:

- **Mix of text with graphics:** Desktop-publishing software allows you to precisely manage and merge text with graphics. As you lay out a page on-screen, you can make the text "flow," liquid-like, around graphics such as photographs. You can resize art, silhouette it, change the colors, change the texture, flip it upside down, and make it look like a photo negative.

- **Varied type and layout styles:** As do word processing programs, DTP programs provide a variety of fonts, or typestyles, from readable Times Roman to staid Tribune to wild Jester and Scribble. Additional fonts can be purchased on disk or downloaded online. You can also create all kinds of rules, borders, columns, and page numbering styles.

- **Use of files from other programs:** It's usually not efficient to do word processing, drawing, and painting with the DTP software. As a rule, text is composed on a word processor, artwork is created with drawing and painting software, and photographs are input using a scanner and then modified and stored using image-editing software. Prefabricated art to illustrate DTP documents may be obtained from disks containing clip art, or "canned" images. The DTP program is used to integrate all these files. You can look at your work on the display screen as one page or as two facing pages (in reduced size). Then you can see it again after it has been printed out. (See ● *Panel 3.21, next page.*)

① Text created with word processing software.

② Art created with drawing or painting software.

③ Images scanned to disk by a scanner.

④ The files created in Steps ①, ②, ③ are imported into a DTP document.

⑤ DTP software is used to make up pages.

⑥ A black-and-white or color printer, usually a laser printer, prints out the pages.

● **PANEL 3.21**
How desktop publishing uses other files

Vector image

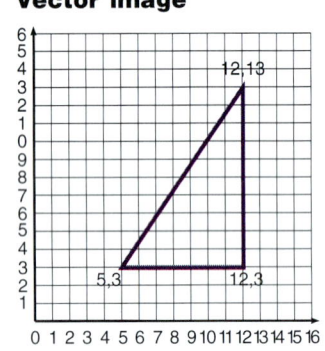

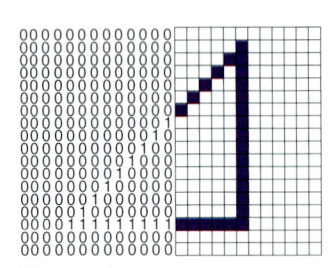

Raster image

Drawing & Painting Programs

John Ennis was trained in realistic oil painting, and for years he used his skill creating illustrations for book covers and dust jackets. Now he "paints" using a computer, software, and mouse. The greatest advantage, he says, is that if "I do a brush stroke in oil and it's not right, I have to take a rag and wipe it off. With the computer, I just hit the 'undo' command."[9]

It may be no surprise to learn that commercial artists and fine artists have begun to abandon the paintbox and pen and ink for software versions of palettes, brushes, and pens. The surprise, however, is that an artist can use mouse and pen-like stylus to create computer-generated art as good as that achievable with conventional artist's tools. More surprising, even *nonartists* can produce good-looking work with these programs.

There are two types of computer art programs, also called *illustration software*: drawing and painting.

- **Drawing programs:** A **_drawing program_** is graphics software that allows users to design and illustrate objects and products. Some drawing programs are CorelDRAW, Adobe Illustrator, Macromedia Freehand, and Sketcher.

    Drawing programs create *vector images*—images created from mathematical calculations.

- **Painting programs:** **_Painting programs_** are graphics programs that allow users to simulate painting on screen. A mouse or a tablet stylus is used to simulate a paintbrush. The program allows you to select "brush" sizes, as well as colors from a color palette. Examples of painting programs are MetaCreations' Painter 3D, Adobe Photoshop, Corel PhotoPaint, and JASC's PaintShop Pro.

    Painting programs produce *raster images* made up of little dots.

Painting software is also called *image-editing software* because it allows you to retouch photographs, adjust the contrast and the colors, and add special effects, such as shadows.

Web Page Design/Authoring Software

As we mentioned in Chapter 2, web page design software is used to create web pages with sophisticated multimedia features. A few of these packages are easy enough for even beginners to use. A few of the best-known are Macromedia Dreamweaver, Macromedia Flash, Adobe GoLive, Adobe PageMill, and Microsoft FrontPage.

Video/Audio Editing Software

Video editing software allows you to modify a section of video, which is called a *clip.* For example, you can add special effects to clips or reorder the clips. Video editing software usually includes audio editing features that allow you, for example, to clean up background noise or emphasize certain sound qualitites. Popular video/audio editing software packages are Adobe Premier and click2learn.com Digital Video Producer.

Survival Tip

Want to Learn How to Use Flash?

www.macromedia.com/support/training
www.shockwave.com
www.fmctraining.com
www.learnit.com
www.Webisode-academy.com

CONCEPT CHECK

What can you do with desktop publishing software?

What are some characteristics of DTP software?

What are some features of drawing programs? Of painting programs?

What can you do with video/audio editing software?

Project Management Software

As we have seen, a personal information manager (PIM) can help you schedule your appointments and do some planning. That is, it can help you manage your own life. But what if you need to manage the lives of others in order to accomplish a full-blown project, such as steering a political campaign or handling a nationwide road tour for a band? Strictly defined, a *project* is a one-time operation involving several tasks and multiple resources that must be organized toward completing a specific goal within a given period of time. The project can be small, such as an advertising campaign for an in-house advertising department, or large, such as construction of an office tower or a jetliner.

Project management software is a program used to plan and schedule the people, costs, and resources required to complete a project on time. For instance, the associate producer on a feature film might use such software to keep track of the locations, cast and crew, materials, dollars, and schedules needed to complete the picture on time and within budget. The software would show the scheduled beginning and ending dates for a particular task—such as shooting all scenes on a certain set—and then the date that task was actually completed. Examples of project management software are Harvard Project Manager, Microsoft Project, Suretrack Project Manager, and ManagerPro.

Computer-Aided Design

Computers have long been used in engineering design. **Computer-aided design (CAD) programs** are intended for the design of products, structures,

civil engineering drawings, and maps. CAD programs, which are available for microcomputers, help architects design buildings and workspaces and help engineers design cars, planes, electronic devices, roadways, bridges, and subdivisions. CAD and drawing programs are similar. However, CAD programs provide precise dimensioning and positioning of the elements being drawn, so that they can be transferred later to computer-aided manufacturing (CAM) programs. Also, CAD programs lack the special effects for illustrations that come with drawing programs. One advantage of CAD software is that the product can be drawn in three dimensions and then rotated on the screen so the designer can see all sides. *(See • Panel 3.22.)* Examples of CAD programs for beginners are Autosketch and CorelCAD.

A variant of CAD is *CADD,* for *computer-aided design and drafting,* software that helps people do drafting. CADD programs include symbols (points, circles, straight lines, and arcs) that help the user put together graphic elements, such as the floor plan of a house. An example is Autodesk's Auto-CAD.

CAD/CAM (computer-aided design/computer-aided manufacturing) software allows products designed with CAD to be input into an automated manufacturing system that makes the products. For example, CAD/CAM systems brought a whirlwind of enhanced creativity and efficiency to the fashion industry. Some CAD systems, says one writer, "allow designers to electronically drape digital-generated mannequins in flowing gowns or tailored suits that don't exist, or twist imaginary threads into yarns, yarns into weaves, weaves into sweaters without once touching needle to garment."[7] The designs and specifications are then input into CAM systems that enable robot pattern-cutters to automatically cut thousands of patterns from fabric with only minimal waste. Whereas previously the fashion industry worked about a year in advance of delivery, CAD/CAM has cut that time to 8 months—a competitive edge for a field that feeds on fads.

CONCEPT CHECK

Describe what project management software can do.

What is the purpose of CAD software? CAD/CAM software?

• PANEL 3.22
CAD
Example of computer-aided design

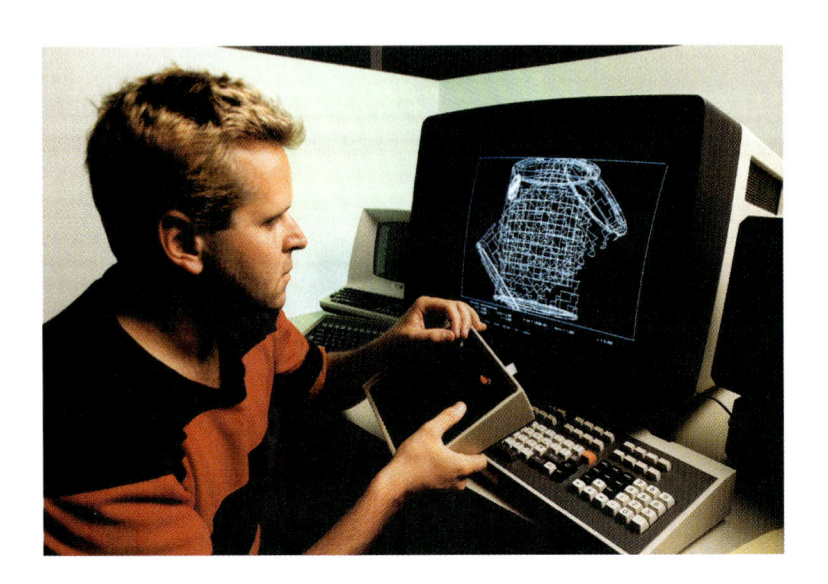

3.7 Online Software & Application Software Providers: Turning Point for the Software Industry?

KEY QUESTION
What are some recent trends in online software?

Everything else seems to be going on the Internet. Why not word processors, spreadsheets, and other software? Indeed, even now, if you're tired of using a paper calendar or datebook, you can go online to jump.com, yahoo.com, or when.com and start using a Web calendar to keep track of appointments, deadlines, and birthdays. (Cautionary note: If, however, there comes a day when the website doesn't come up on your computer screen, as happened to *Time* magazine technology writer Anita Hamilton, you'll be back to jotting down your appointments on Post-it notes.)[10] Other web-based software has also been offered free for some time—browsers, e-mail, and address books, for example—in an attempt to keep users coming back to a particular portal such as AltaVista or Yahoo!

But for businesses in particular—and ultimately probably for you—it does make sense to rent software rather than buy it. Let's take a look at this concept.

Online Software & the Application Service Provider

Every month you pay the phone bill, the electric bill, maybe the ISP (Internet service provider) bill. What about paying an ASP—an application software provider—as well as an ISP? Or even getting both ISP and ASP for free?

An *ASP (application service provider)* is a firm that leases software over the Internet to customers. You no longer have to buy software in a store, in shrink-wrapped packages. Instead, you can simply download a particular program when you need it, for as long as you need it.

Could this be a major turning point for computers and communications? Some experts think ASPs will rock the software industry. Although the ASP market is still in its infancy, one estimate is that the total ASP market will grow from just several hundred million dollars today to one that will exceed $4.5 billion in 2003.[11] Among the purveyors of web-based software are Damango, Atomz.com, Halfbrain.com, Mi8, myWebOS, Sun Microsystems, and Visto.com. (See ● Panel 3.23, next page.) Mi8, for example, offers Microsoft Office and Microsoft Outlook, as well as support services for Palm and Blackberry devices. The ASP called myWebOS offers a free word processor, e-mail, and calendar; free spreadsheet and presentation software is in the works. Sun Microsystems *(www.sun.com)* offers StarOffice. Microsoft leases its software products on a monthly subscription basis.

For more information on ASPs, check the ASP guide at *www.cnet.com*. Clients of ASPs include the U.S. Department of the Interior, Hershey Foods, the Baltimore Sun Co., and the Cystic Fibrosis Foundation.

Network Computers Revisited: "Thin Clients" versus "Fat Clients"

We can link the growth in ASPs to the concept of a *network computer*, proposed a few years ago by Larry Ellison, CEO of database-software maker Oracle Corp. **A *network computer* is an inexpensive, stripped-down computer that connects people to networks and runs applications tied to servers.**

Out of this grew a distinction between "thin clients" and "fat clients." A *client*, you'll recall from Chapter 1, is one part of a client/server network. The *server* is the central computer that holds collections of data and programs; the clients are the PCs, workstations, and other computers on the network that use the data and programs from the server.

A *fat client* is a regular computer, perhaps a PC, that is on a network. Often it contains software with a great many features. Such "bloatware" is the result of software makers constantly trying to top themselves when they

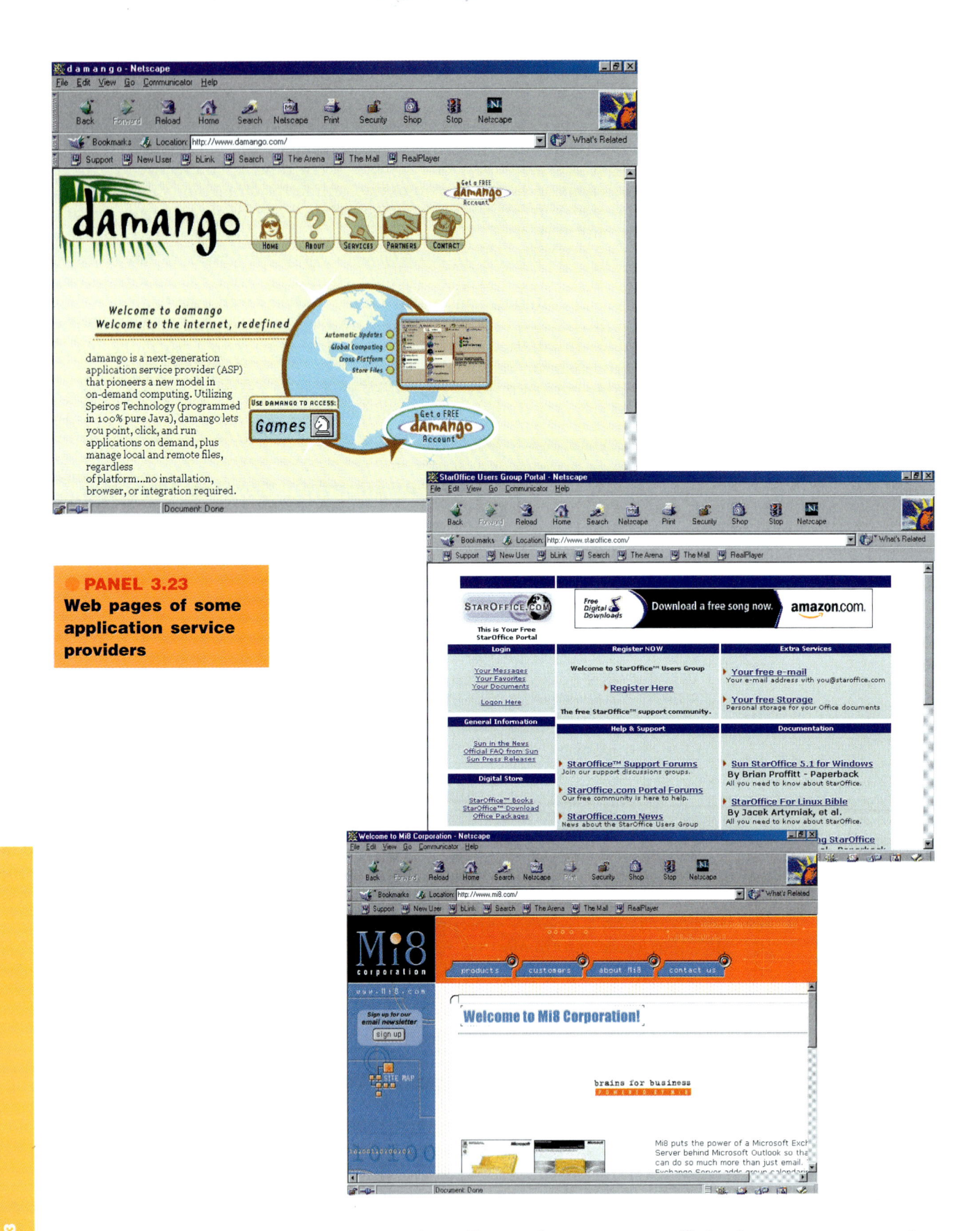

● **PANEL 3.23**
Web pages of some application service providers

issue new versions. To run these programs efficiently, a computer requires lots of main memory and hard-disk storage capacity, as well as a powerful microprocessor. A *thin client,* by contrast, is a network computer—a stripped-down computer without hefty microprocessors or even much storage—that is supposed to operate as an inexpensive terminal tied to a server. (See ● Panel 3.24.)

Contrary to Ellison's expectations, the widespread appearance of network computers/thin clients has been delayed because the prices of regular, fat-client PCs dropped sharply, below $1000. But in the meantime, another movement advanced the cause of online software—the development of enterprise resource planning software.

From ERP to ASP: The Evolution of "Rentalware"

About 10 years ago, businesses began to adopt client/server arrangements. The tasks once performed by mainframes and minicomputers were being divided between desktop computers and servers. To ease this transition, companies relied on business and accounting programs called ERP software. **_ERP (enterprise resource planning) software_ consists of large client/server software applications that help companies organize and operate their businesses.** The makers of such software include SAP, Oracle, and PeopleSoft, as well as Microsoft.

ERP is expensive—a corporation might well spend $30 million on such a system. And the risk is all on the buyer. Even if the system doesn't work as well as it should, the buyer will be inclined to keep it, because it represents such a huge investment.

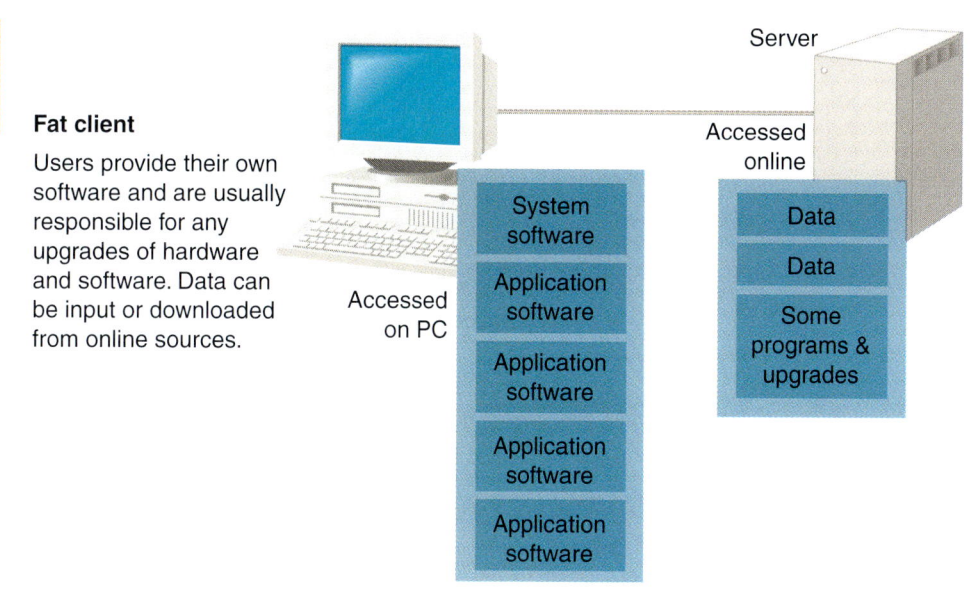

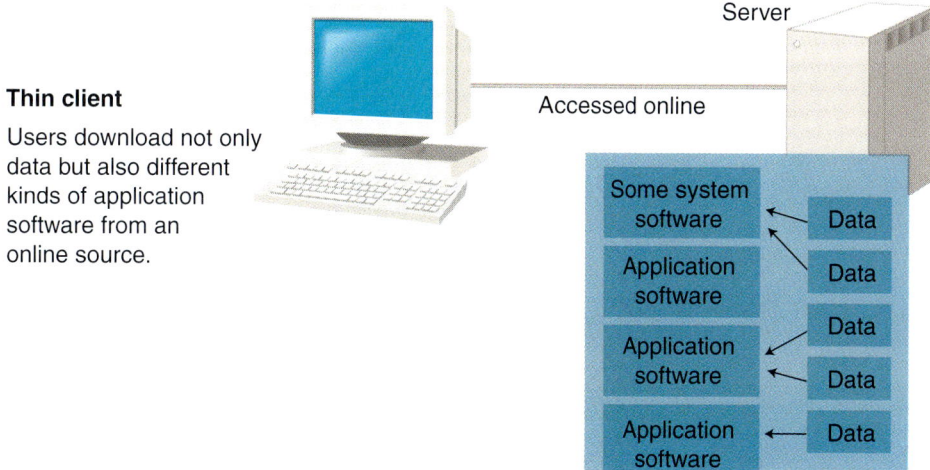

● **PANEL 3.24**
Old and new ways of getting software

Fat client
Users provide their own software and are usually responsible for any upgrades of hardware and software. Data can be input or downloaded from online sources.

Thin client
Users download not only data but also different kinds of application software from an online source.

Enter the ASP model. Instead of buying an ERP system, a company can "rent" the same thing, and the installation and management of all the equipment and software becomes someone else's headache.[12] "The Internet creates an opportunity to change the way people manage information technology," says Gary Bloom, an executive with Oracle Corp. "You will buy software as a service, just the way you buy telephone service today."[13]

CONCEPT CHECK

What is an application software provider?

Explain the concept of thin clients versus fat clients.

What is enterprise resource software?

Summary

analytical graphics (p. 115, KQ 3.4) Also called *business graphics;* graphical forms that make numeric data easier to analyze than when it is organized as rows and columns of numbers. The principal examples of analytical graphics are bar charts, line graphs, and pie charts. Why it's important: *Whether viewed on a monitor or printed out, analytical graphics help make sales figures, economic trends, and the like easier to comprehend and analyze.*

ASP (application service provider) (p. 127, KQ 3.7) Firm that leases software over the Internet. Why it's important: *You no longer have to buy software in a store, in shrink-wrapped packages. Instead, you can simply download a particular program when you need it, for as long as you need it.*

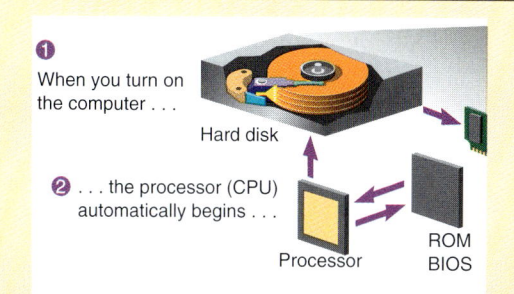

booting (p. 85, KQ 3.1) Loading an operating system into a computer's main memory. Why it's important: *Without booting, computers could not operate. The programs responsible for booting are stored permanently in the computer's electronic circuitry. When you turn on the machine, programs called diagnostic routines test the main memory, the central processing unit, and other parts of the system to make sure they are running properly. Next, BIOS (for basic input/output system) programs are copied to main memory and help the computer interpret keyboard characters or transmit characters to the display screen or to a diskette. Then the boot program obtains the operating system, usually from hard disk, and loads it into the computer's main memory, where it remains until you turn the computer off.*

cell (p. 113, KQ 3.4) Place where a row and a column intersect in a spreadsheet worksheet; its position is called a *cell address.* Why it's important: *The cell is the smallest working unit in a spreadsheet. Data and formulas are entered into cells. Cell addresses provide location references for spreadsheet users.*

computer-aided design (CAD) programs (p. 125, KQ 3.6) Programs intended for the design of products, structures, civil engineering drawings, and maps. Why it's important: *CAD programs, which are available for microcomputers, help architects design buildings and workspaces and help engineers design cars, planes, electronic devices, roadways, bridges, and subdivisions. While similar to drawing programs, CAD programs provide precise dimensioning and positioning of the elements being drawn, so that they can be transferred later to computer-aided manufacturing programs; in addition, they lack special effects for illustrations. One advantage of CAD software is that three-dimensional drawings can be rotated on screen, so the designer can see all sides of the product.*

computer-aided design/computer-aided manufacturing (CAD/CAM) software (p. 126, KQ 3.6) Programs allowing products designed with CAD to be input into an automated manufacturing system that makes the products. Why it's important: *CAM systems have greatly enhanced efficiency in many industries.*

copyright (p. 102, KQ 3.2) Exclusive legal right that prohibits copying of intellectual property without the permission of the copyright holder. Why it's important: *Copyright law aims to prevent people from taking credit for and profiting from other people's work.*

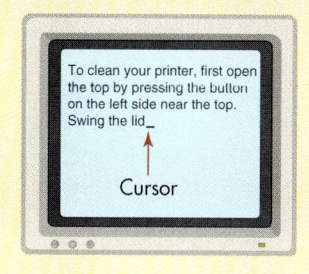

cursor (p. 107, KQ 3.3) Movable symbol on the display screen that shows where you may next enter data or commands. The symbol is often a blinking rectangle or an I-beam. You can move the cursor on the screen using the keyboard's directional arrow keys or a mouse. The point where the cursor is located is called the *insertion point.* Why it's important: *All application software packages use cursors to show the current work location on the screen.*

database (p. 116, KQ 3.5) Collection of interrelated files in a computer system. These computer-based files are organized according to their common elements, so that they can be retrieved easily. *Why it's important:* Businesses and organizations build databases to help them keep track of and manage their affairs. In addition, online database services put enormous resources at the user's disposal.

database file (p. 104, KQ 3.2) File created by database management programs; it consists of organized data that can be analyzed and displayed in various useful ways. *Why it's important:* Database files make up a database.

database software (p. 116, KQ 3.5) Application software that sets up and controls the structure of a database and access to the data. *Why it's important:* Database software allows users to organize and manage huge amounts of data.

default settings (p. 111, KQ 3.3) Settings automatically used by a program unless the user specifies otherwise, thereby overriding them. *Why it's important:* Users need to know how to change default settings in order to customize documents.

desktop (p. 88, KQ 3.1) The operating system's main interface screen. *Why it's important:* The desktop displays pictures (icons) that provide quick access to programs and information.

desktop publishing (DTP) (p. 123, KQ 3.6) Application software and hardware system that involves mixing text and graphics to produce high-quality output for commercial printing, using a microcomputer and mouse, scanner, laser or ink-jet printer, and DTP software (such as QuarkXPress and PageMaker or, at a more consumer-oriented level, Microsoft Publisher). Often the printer is used primarily to get an advance look before the completed job is sent to a typesetter for even higher-quality output. Some word processing programs, such as Word and WordPerfect, have rudimentary DTP features. *Why it's important:* Desktop publishing has reduced the number of steps, the time, and the money required to produce professional-looking printed projects.

device drivers (p. 87, KQ 3.1) Specialized software programs—usually components of system software—that allow input and output devices to communicate with the rest of the computer system. *Why it's important:* Drivers are needed so that the computer's operating system can recognize and run peripheral hardware.

document file (p. 104, KQ 3.2) File created by word processing programs; it consists of documents such as reports, letters, memos, and term papers. *Why it's important:* Document files are probably the most common type of file users deal with.

documentation (p. 103, KQ 3.2) All information that describes a product to users, including a user guide or reference manual that provides a narrative and graphical description of a program. While documentation may be print-based, today it is usually available on CD-ROM, as well as via the Internet. *Why it's important:* Documentation helps users learn software commands and use of function keys, solve problems, and find information about system specifications.

DOS (Disk Operating System) (p. 95, KQ 3.1) Original operating system produced by Microsoft, with a hard-to-use command-driven user interface. Its initial 1982 version was designed to run on the IBM PC as PC-DOS. Later Microsoft licensed the same system to other computer makers as MS-DOS. *Why it's important:* DOS used to be the most common microcomputer operating system, and it is still used on many microcomputers. Today the most popular operating systems use GUIs.

drawing program (p. 124, KQ 3.6) Graphics software that allows users to design and illustrate objects and products. *Why it's important:* Drawing programs are vector-based and are best used for straightforward illustrations based on geometric shapes.

ERP (enterprise resource planning) software (p. 129, KQ 3.7) Large client/server software applications that help companies organize and operate their businesses. *Why it's important:* ERP software coordinates a company's entire business and moves data speedily from one department to another.

exporting (p. 104, KQ 3.2) Transforming data into a format that can be used in another program and then transmitting it. *Why it's important:* Users need to know how to export many types of files.

file (p. 104, KQ 3.2) A named collection of data or a program that exists in a computer's secondary storage, such as on a floppy disk, hard disk, or CD-ROM disk. Why it's important: *Dealing with files is an inescapable part of working with computers. Users need to be familiar with the different types of files.*

financial software (p. 121, KQ 3.6) Applications software that ranges from personal-finance managers to entry-level accounting programs to business financial-management packages. Why it's important: *Financial software provides users with powerful management tools (personal-finance managers) as well as small business programs. Moreover, tax programs provide virtually all the forms needed for filing income taxes, make complex calculations, check for mistakes, and even unearth deductions you didn't know existed. Tax programs can also be integrated with personal finance software to form an integrated tool. Accounting software automates bookkeeping tasks, while payroll software keeps records of employee hours and produces reports for tax purposes. Some programs allow users to set up a business from scratch. Financial software also includes investment software packages and various retirement planning programs.*

fly-out menu (p. 91, KQ 3.1) Menu that seems to explode out to the right. Why it's important: *Menus make software easier to use.*

font (p. 110, KQ 3.3) A particular typeface and type size. Why it's important: *Fonts influence the appearance and effectiveness of documents, brochures, and other publications.*

formatting (initializing) a disk (p. 86, KQ 3.1) Process of preparing a floppy disk so it can store data or programs. Why it's important: *Different computers take disks with different formats; thus, you can't run a "Formatted IBM" disk on a Macintosh or a "Formatted Macintosh" disk on an IBM-compatible PC.*

formatting (p. 110, KQ 3.3) In word processing and desktop publishing, determining the appearance of a document. Why it's important: *The document format should match its users' needs. Ways to format a document include using different fonts, boldface, italics, variable spacing, columns, and margins.*

formulas (p. 113, KQ 3.4) In a spreadsheet, instructions for calculations entered into designated cells. Why it's important: *When spreadsheet users change data in one cell, all the cells linked to it by formulas automatically recalculate their values.*

freeware (p. 103, KQ 3.2) Copyrighted software that is distributed free of charge, today most often over the Internet. Why it's important: *Freeware saves users money.*

function keys (p. 106, KQ 3.3) Keys labeled F1, F2, and so on, positioned along the top or left side of the keyboard. Why it's important: *They are used to execute commands specific to the software being used.*

grammar checker (p. 110, KQ 3.3) Word processing feature that highlights poor grammar, wordiness, incomplete sentences, and awkward phrases. The grammar checker won't fix things automatically, but it will flag (perhaps with a different color of squiggly line) possible incorrect word usage and sentence structure. Why it's important: *Grammar checkers help users produce better-written documents.*

graphical user interface (GUI) (p. 88, KQ 3.1) User interface in which icons and commands from menus may be selected by means of a mouse or keystrokes. Why it's important: *GUIs are easier to use than command-driven interfaces.*

groupware (p. 105, KQ 3.2) Online software that allows several people to collaborate on the same project. Why it's important: *Groupware improves productivity by keeping users continually notified about what their colleagues are thinking and doing.*

Help command (p. 94, KQ 3.1) Command generating a table of contents, an index, and a search feature that can help users locate answers to questions about the software. Why it's important: *Help features provide a built-in electronic instruction manual.*

icons (p. 88, KQ 3.1) Small pictorial figures that represent programs, data files, or procedures. Why it's important: *Icons have simplified the use of software. The feature represented by the icon can be activated by clicking on the icon.*

importing (p. 104, KQ 3.2) Getting data from another source and then converting it into a format compatible with the program in which you are currently working. *Why it's important: Users will often have to import files.*

key field (p. 118, KQ 3.5) Field that can be used as a common identifier because it is unique. The most frequent key field used in the United States is the Social Security number, but any unique identifier could be used, such as an employee number. *Why it's important: Key fields are needed to identify and retrieve specific items in a database.*

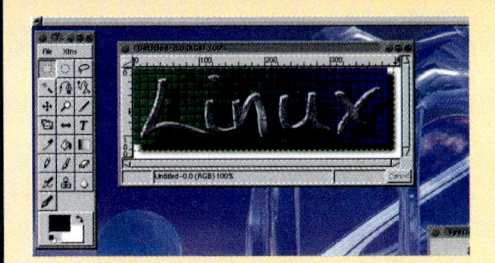

Linux (p. 98, KQ 3.1) Free version of Unix, supported by efforts of thousands of volunteer programmers. *Why it's important: Linux is an inexpensive, open-source operating system useful for online applications and to PC users who have to maintain a web server or a network server.*

Macintosh operating system (Mac OS) (p. 95, KQ 3.1) System software that runs only on Apple Macintosh computers. *Why it's important: Although Macs are not as common as PCs, many people believe they are easier to use. Macs are often used for graphics and desktop publishing.*

macro (p. 107, KQ 3.3) Also called a keyboard shortcut; a single keystroke or command—or a series of keystrokes or commands—used to automatically issue a longer, predetermined series of keystrokes or commands. *Why it's important: Users can consolidate several activities into only one or two keystrokes. The user names the macro and stores the corresponding command sequence; once this is done, the macro can be used repeatedly.*

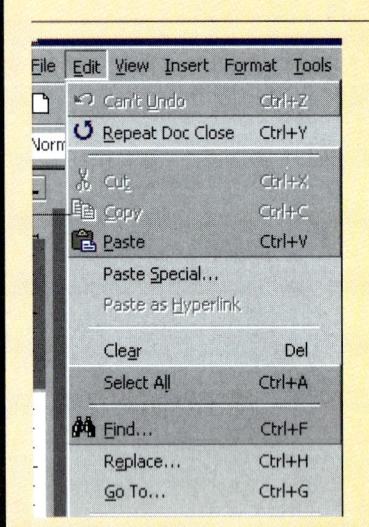

menu (p. 91, KQ 3.1) Displayed list of options—such as commands—to choose from. *Why it's important: Menus are a feature of GUIs that make software easier to use.*

menu bar (p. 93, KQ 3.1) Display at top of screen below the title bar. *Why it's important: It contains the names of pull-down menus.*

Microsoft Pocket PC (p. 99, KQ 3.1) Operating system for handhelds that is simpler and less cluttered than CE and looks and feels a lot less like desktop Windows. *Why it's important: Pocket PC offers pocket versions of Word and Excel that let users read standard word processing and spreadsheet files sent as e-mail attachments from their PCs.*

Microsoft Windows CE (p. 99, KQ 3.1) Greatly slimmed-down version of Windows 95 for handheld computing devices, such as those made by Casio, Compaq, and Hewlett-Packard. Windows CE had some of the familiar Windows look and feel and included rudimentary word processing, spreadsheet, e-mail, web browsing, and other software. *Why it's important: Windows CE was Microsoft's first attempt to modify its Windows desktop operating system for use with handhelds. It has been succeeded by the Pocket PC system.*

Microsoft Windows Millennium Edition (WinMe) (p. 96, KQ 3.1) Successor to Windows 95 and 98, operating system designed to support desktop and portable computers. *Why it's important: Considered a boon to multimedia users because of its ability to handle still pictures, digital video, and audio files. Also claims to have reduced the problem of frequent "crashes."*

Microsoft Windows 95/98 (p. 96, KQ 3.1) Operating system for desktop and portable microcomputers, supporting the most hardware and the most application software. *Why it's important: Windows has become the most common system software used on microcomputers.*

Microsoft Windows NT/2000 (p. 97, KQ 3.1) Multitasking operating system designed to run on network servers. *Why it's important: It allows multi-ple users to share resources such as data and programs.*

Microsoft Windows XP (p. 97, KQ 3.1) Microsoft's newest OS; it combines elements of Windows 2000 and Windows Me and has a new GUI. *Why it's important: With this new version, Microsoft will finally be giving up the last of the Windows software carried forward from the aging DOS programming technology.*

multitasking (p. 86, KQ 3.1) Feature of OS that allows the execution of two or more programs by one user concurrently on the same computer with one CPU. For instance, you might write a report on your computer with one program while another plays a music CD. *Why it's important: Multitasking allows the computer to switch rapidly back and forth among different tasks. The user is generally unaware of the switching process.*

NetWare (p. 96, KQ 3.1) Long-popular network operating system for coordinating micro-computer-based local area networks (LANs) throughout an organization. *Why it's important: LANs allow PCs to share programs, data files, and printers and other devices.*

Thin client
Users download not only data but also different kinds of application software from an online source.

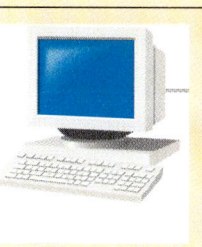

network computer (p. 127, KQ 3.7) Stripped-down computer that connects people to networks and runs applications tied to servers. *Why it's important: It's less expensive than a personal computer, and, since the software is stored on a server, the user theoretically has fewer software headaches.*

office suite (p. 104, KQ 3.2) A single large software package that bundles several applications together. *Why it's important: Office suites cost less than do the applications purchased separately.*

open-source software (p. 98, KQ 3.1) Software that any programmer can down-load from the Internet free and modify with suggested improvements. The only qual-ification is that changes can't be copyrighted; they must be made available to all and remain in the public domain. *Why it's important: Because this software is not proprietary, any programmer can make improvements, which can result in better-quality software.*

operating system (OS) (p. 85, KQ 3.1) Master system of programs that manage the basic operations of the computer. *Why it's important: These programs provide resource management services of many kinds. In particular, they handle the control and use of hardware resources, including disk space, memory, CPU time allocation, and peripheral devices. The operating system allows users to concentrate on their own tasks or applications rather than on the complexities of managing the computer.*

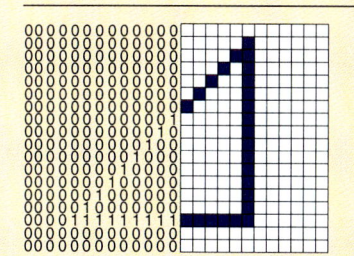

painting program (p. 124, KQ 3.6) Graphics program that allows users to simulate painting on screen. A mouse or a tablet stylus is used to simulate a paintbrush. The program allows you to select "brush" sizes, as well as colors from a color palette. *Why it's important: Painting programs, which produce raster images made up of little dots, are good for creating art with soft edges and many colors.*

Palm OS (p. 99, KQ 3.1) Operating system for the Palm and Visor; the dominant operating system for handhelds. *Why it's important: Because it is not a Windows derivative but was specifically designed for handhelds, Palm OS is a smoother-running operating system.*

personal-finance manager (p. 121, KQ 3.6) Application software that lets you keep track of income and expenses, write checks, do online banking, and plan financial goals. *Why it's important: Personal-finance software can help people manage their money more effectively.*

personal information manager (PIM) (p. 118, KQ 3.5) Software to help you keep track of and manage information you use on a daily basis, such as addresses, telephone numbers, appointments, to-do lists, and miscellaneous notes. Some programs feature phone dialers, outliners (for roughing out ideas in outline form), and ticklers (or reminders). *Why it's important: PIMs can help users better organize and manage daily business activities.*

pirated software (p. 103, KQ 3.2) Software obtained illegally. *Why it's important: Not only copyright holders do not get paid. but buyers may pick up inferior goods or even damage their systems..*

platform (p. 95, KQ 3.1) Particular processor model and operating system on which a computer system is based. *Why it's important: Generally, software written for one platform will not run on any other. Users should be aware that there are Mac platforms (Apple Macintosh) and Windows platforms, or "PC platforms" (for personal computers such as Dell, Compaq, Gateway, Hewlett-Packard, or IBM that run Microsoft Windows). Sometimes the latter are called "Wintel platforms," for "Windows + Intel," because they often combine the Windows operating system with the Intel processor chip.*

pointer (p. 87, KQ 3.1) Indicator that usually appears as an arrow, although it changes shape depending on the application. The mouse is used to move the pointer to a particular place on the display screen or to point to little symbols, or icons. *Why it's important: It is often easier to manipulate the pointer on the screen by means of the mouse than to type commands on a keyboard.*

pop-up menu (p. 92, KQ 3.1) List of command options that can "pop up" anywhere on the screen. In contrast to pull-down or pull-up menus, pop-up menus are not connected to a toolbar. *Why it's important: Pop-up menus make programs easier to use.*

presentation graphics software (p. 119, KQ 3.6) Software that uses graphics, animation, sound, and data or information to make visual presentations. Why it's important: *Presentation graphics software provides a means to produce sophisticated graphics.*

productivity software (p. 104, KQ 3.2) Application software such as word processing, spreadsheets, and database managers. Why it's important: *Productivity software makes users more productive at particular tasks.*

project management software (p. 125, KQ 3.6) Program used to plan and schedule the people, costs, and resources required to complete a project on time. Why it's important: *Project management software increases the ease and speed of planning and managing complex projects.*

public-domain software (p. 102, KQ 3.2) Software, often available on the Internet, that is not protected by copyright and thus may be duplicated by anyone at will. Why it's important: *Public domain software offers lots of software options to users who may not be able to afford much commercial software. Users may download such software from the Internet free and make as many copies as they wish.*

pull-down menu (p. 91, KQ 3.1) Also called a *drop-down menu;* list of options that pulls down from the menu bar at the top of the screen. Why it's important: *Like other menu-based and GUI features, pull-down menus make software easier to use.*

pull-up menu (p. 92, KQ 3.1) List of options that pulls up from the menu bar at the bottom of the screen. Why it's important: *See pull-down menu.*

recalculation (p. 115, KQ 3.4) Recomputing values in a spreadsheet, either as an ongoing process as data is entered or afterward, with the press of a key. Why it's important: *With this simple feature, the hours of mind-numbing work required to manually rework paper spreadsheets became a thing of the past.*

relational database (p. 116, KQ 3.5) Database in which data is organized into related tables. Each table contains rows and columns; the rows are called records, and the columns are called fields. An example of a record is a person's address—name, street address, city, and so on. An example of a field is that person's last name; another field would be that person's first name, a third field would be that person's street address, and so on. Why it's important: *The relational database is a common type of database.*

rentalware (p. 103, KQ 3.2) Software that users lease for a fee. Why it's important: *This is the concept behind application services providers (ASPs).*

Rollover: When you roll your mouse pointer over an icon or graphic, a small box with text appears that briefly explains its function.

rollover (p. 88, KQ 3.1) Icon feature in which a small text box explaining the icon's function appears when you roll the mouse pointer over the icon. A rollover may also produce an animated graphic. Why it's important: *The rollover gives the user an immediate explanation of an icon's meaning.*

saving (p. 112, KQ 3.3) Storing, or preserving, a document as an electronic file permanently—on diskette, hard disk, or CD-ROM, for example. Why it's important: *Saving is a feature of nearly all application software. Having the document stored in electronic form spares users the tiresome chore of retyping it from scratch whenever they want to make changes. Users need only retrieve it from the storage medium and make the changes, then resave it and print it out again.*

scrolling (p. 107, KQ 3.3) Moving quickly upward, downward, or sideways through the text or other screen display. Why it's important: *A standard computer screen displays only 20–22 lines of standard-size text; however, most documents are longer than that. Using the directional arrow keys, or the mouse and a scroll bar located at the side of the screen, users can move ("scroll") through the display screen and into the text above and below it.*

shareware (p. 103, KQ 3.2) Copyrighted software that is distributed free of charge but requires users to make a monetary contribution in order to continue using it. Shareware is distributed primarily through the Internet. Because it is copyrighted, you cannot use it to develop your own program that would compete with the original product. Why it's important: *Like public domain software and freeware, shareware offers an inexpensive way to obtain new software.*

software license (p. 102, KQ 3.2) Contract by which users agree not to make copies of software to give away or for resale. Why it's important: *Software manufacturers don't sell people software; they sell them licenses to become authorized users of the software.*

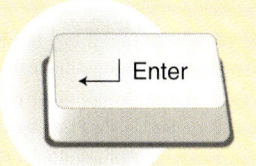

special-purpose keys (p. 106, KQ 3.3) Keys used to enter, delete, and edit data and to execute commands. For example, the Esc (for "Escape") key tells the computer to cancel an operation or leave ("escape from") the current mode of operation. The Enter, or Return, key tells the computer to execute certain commands and to start new paragraphs in a document. Why it's important: *Special-purpose keys are essential to the use of software.*

spelling checker (p. 108, KQ 3.3) Word processing feature that tests for incorrectly spelled words. As you type, the spelling checker indicates (perhaps with a squiggly line) words that aren't in its dictionary and thus may be misspelled. Special add-on dictionaries are available for medical, engineering, and legal terms. Why it's important: *Spelling checkers help users prepare accurate documents.*

spreadsheet (p. 113, KQ 3.4) Application software that allows users to create tables and financial schedules by entering data and formulas into rows and columns arranged as a grid on a display screen. Why it's important: *When data is changed in one cell, values in other cells in the spreadsheet are automatically recalculated.*

supervisor (p. 85, KQ 3.1) Also called *kernel;* the central component of the operating system that manages the CPU. Why it's important: *The supervisor remains in main memory while the computer is running. As well as managing the CPU, it directs other nonresident programs to perform tasks that support application programs.*

template (p. 110, KQ 3.3) In word processing, a preformatted document that provides basic tools for shaping a final document—the text, layout, and style for a letter, for example. Why it's important: *Templates make it very easy for users to prepare professional-looking documents, because most of the preparatory formatting is done.*

thesaurus (p. 110, KQ 3.3) Word processing feature that will present you with the appropriate word or alternative words. Why it's important: *The thesaurus feature helps users prepare well-written documents.*

title bar (p. 93, KQ 3.1) Identifier at the top of the Windows screen. Why it's important: *The title bar contains the name of the window—often the application and the name of the file you're in.*

toolbar (p. 93, KQ 3.1) Bar across the top of the display window. It displays menus and icons representing frequently used options or commands. Why it's important: *Toolbars make it easier to identify and execute commands.*

tutorial (p. 103, KQ 3.2) Instruction book or program that helps you learn to use the product by taking you through a prescribed series of steps. Why it's important: *Tutorials enable users to practice using new software in a graduated fashion and learn the software in an effective manner.*

Unix (p. 97, KQ 3.1) Multitasking operating system for multiple users that has built-in networking capability and versions that can run on all kinds of computers. Why it's important: *Government agencies, universities, research institutions, large corporations, and banks all use Unix for everything from designing airplane parts to currency trading. Unix is also used for website management. The developers of the Internet built their communication system around Unix because it has the ability to keep large systems (with hundreds of processors) churning out transactions day in and day out for years without fail.*

user interface (p. 87, KQ 3.1) User-controllable display screen that allows you to communicate, or interact, with your computer. Why it's important: *The interface determines the ease of use of hardware and software. The most common user interface is the graphical user interface (GUI).*

utility programs (p. 87, KQ 3.1) Also known as *service programs;* system software component that performs tasks related to the control and allocation of computer resources. Why it's important: *Utility programs enhance existing functions or provide services not supplied by other system software programs. Most computers come with built-in utilities as part of the system software.*

value (p. 113, KQ 3.4) A number or date entered in a spreadsheet cell. Why it's important: *Values are the actual numbers used in the spreadsheet—dollars, percentages, grade points, temperatures, or whatever.*

what-if analysis (p. 115, KQ 3.4) Spreadsheet feature that employs the recalculation feature to investigate how changing one or more numbers changes the outcome of the calculation. Why it's important: *Users can create a worksheet, putting in formulas and numbers, and then ask, "What would happen if we change that detail?"—and immediately see the effect.*

window (p. 93, KQ 3.1) Rectangular frame on the computer display screen. Through this frame you can view a file of data—such as a document, spreadsheet, or database—or an application program. Why it's important: *Using windows, users can display at the same time portions of several documents and/or programs on the screen.*

wizard (p. 110, KQ 3.3) Word processing software feature that answers your questions and uses the answers to lay out and format a document. Why it's important: *Wizards make it easy to prepare professional-looking memos, faxes, resumes, and other documents.*

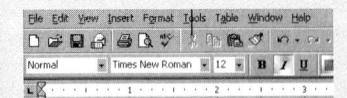

word processing software (p. 105, KQ 3.3) Application software that allows you to use computers to format, create, edit, print, and store text material, among other things. Why it's important: *Word processing software allows users to maneuver through a document and delete, insert, and replace text, the principal correction activities. It also offers such additional features as creating, editing, formatting, printing, and saving.*

worksheet file (p. 104, KQ 3.2) File created by electronic spreadsheets; it consists of a collection of (usually) numerical data such as budgets, sales forecasts, and schedules. Why it's important: *Worksheet files are one common type of file users will have to deal with.*

Chapter Review

stage 1 LEARNING MEMORIZATION

"I can recognize and recall information."

Self-Test Questions

1. _____ enables the computer to perform essential operating tasks.
2. The _____ is the user-controllable display screen that allows you to communicate, or interact, with your computer.
3. _____ is the term for programs designed to perform specific tasks for the user.
4. _____ is the activity of moving upward or downward through the text or other screen display.
5. Name four editing features offered by word processing programs: _____, _____, _____, _____.
6. In a spreadsheet, the place where a row and a column intersect is called a _____.
7. A(n) _____ is a keyboard shortcut used to automatically issue a longer, predetermined series of keystrokes or commands.
8. The _____ is the movable symbol on the display screen that shows you where you may next enter data or commands.
9. Records in a database are linked by a _____, which can be used as a common identifier because it is unique.
10. _____ involves mixing text and graphics to produce high-quality output for commercial printing.

Multiple-Choice Questions

1. Which of the following are functions of the operating system?
   a. file management
   b. CPU management
   c. task management
   d. booting
   e. all of the above

2. Which of the following is not an advantage of using database software?
   a. integrated data
   b. improved data integrity
   c. lack of structure
   d. elimination of data redundancy

3. Which of the following is not a type of menu?
   a. fly-out menu
   b. pop-in menu
   c. pop-out menu
   d. pull-down menu
   e. pull-out menu

4. Which of the following is not a feature of word processing software?
   a. spelling checker
   b. cell address
   c. formatting
   d. cut and paste
   e. find and replace

True/False Questions

T F 1. Spreadsheet software enables you to perform what-if calculations.

T F 2. *Font* refers to a preformatted document that provides basic tools for shaping the final document.

T F 3. Rentalware is software that users lease for a fee.

T F 4. Public-domain software is protected by copyright and so is offered for sale by license only.

T F 5. The records within the various tables in a database are linked by a key field.

stage 2 LEARNING: COMPREHENSION

"I can recall information in my own terms and explain them to a friend."

Short-Answer Questions

1. Briefly define *booting*.
2. What is the difference between a command-driven interface and a graphical user interface (GUI)?
3. Why can't you run your computer without system software?
4. Why is multitasking useful?
5. What is a *device driver*?
6. What is a *utility program*?
7. What is a *platform*?
8. What are the three components of system software? What is the basic function of each?
9. What is *importing*? *exporting*?

Concept Mapping

On a separate sheet of paper, draw a concept map, or visual diagram, linking concepts. Show how the following terms are related.

application software	menu
booting	multitasking
CAD/CAM	operating system
database	productivity software
device driver	spreadsheet
DOS	system software
file	toolbar
DOS	Unix
GUI	utilities
Linux	Windows
Mac OS	

stage 3 LEARNING: APPLYING, ANALYZING, SYNTHESIZING, EVALUATING

"I can apply what I've learned, relate these ideas to other concepts, build on other knowledge, and use all these thinking skills to form a judgment."

Knowledge in Action

1. If you were in the market for a new microcomputer today, what application software would you want to use on it? Why? Where would you get it?
2. How do you think you could use desktop publishing at home? For personal items? Family occasions? Holidays? What else? What hardware and software would you have to buy?

Web Exercises

1. Several websites include libraries of shareware programs. Visit the *www.download.cnet.com* site, click on the Windows shareware icon, and identify three shareware programs that interest you. State the name of each program, the operating system it runs on, and its capabilities. Also, describe the contribution you must make to receive technical support.

2. Did your computer come with a Windows Startup disk and have you misplaced it? If your computer crashes, you'll need this disk to reinstall the operating system. This exercise shows you how to create your own Startup disk. Insert a blank disk in your floppy disk drive. From your Windows desktop, click on Start, Settings, and then Control Panel. Now click on Add/Remove. Click on the tab Startup Disk, then click on the Create Disk button.

    After the disk is created, label it *Startup Disk for Windows* and write the date on the disk. Also note the version of Windows you are using. Store the disk somewhere safe. You'll never know when you might need it.

    To learn the benefits of having a startup disk, visit

    www.microsoft.com/windows98/usingwindows/ maintaining/articles/004Apr/Startup.asp.

3. Microsoft offers "patches," or updates, for its Windows OS. Go to *www.microsoft.com* and search for the list of updates. What kinds of problems do these updates fix?

4. Some people are fascinated by the error message commonly referred to as The Blue Screen of Death (Doom) (BSOD). If you run a search on the Internet, you can find websites that sell T-shirts with the BSOD image on it, photo galleries of public terminals displaying the BSOD, fictional stories of BSOD attacks, and various other forms of entertainment based on the infamous error message.

    To prevent a BSOD attack, keep an eye on how your system is using resources:

    a. Right-click on the My Computer icon on the Windows desktop, click Properties, and then choose the Performance tab. Next to "system resources" you will see the amount of free resources available as a percentage. Try to keep that value above 18%. If your resources dip below that level, save all your work and reboot.

    b. You can also view this information from the System Information file by using the Start menu in this sequence: Start, Programs, Accessories, System Tools, System Information.

    Do a search on the Web to find users' hypotheses of why the BSOD occurs, and find more methods to avoid it.

    Below are some humorous BSOD sites:

    www.daimyo.org/bsod

    http://pla-netx.com/linebackn/news/bsod.html

    http://members.atlasfl.com/sally/stuff/ cuswallpaper.html

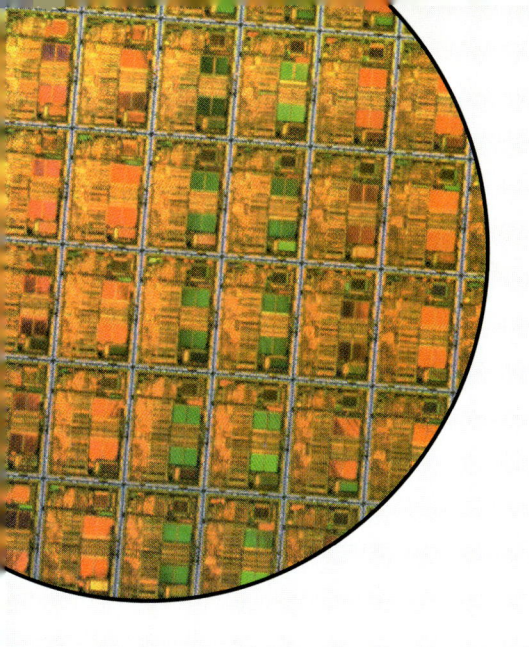

Chapter 4

Hardware: The CPU & Storage

How to Buy a Multimedia Computer System

Key Questions
You should be able to answer the following questions.

- **4.1** **Microchips, Miniaturization, & Mobility** What are the differences between transistors, integrated circuits, chips, and microprocessors?
- **4.2** **The System Unit** How is data represented in a computer; what are the components of the system cabinet; what are processing speeds; how do the processor and memory work; and what are some important ports, buses, and cards?
- **4.3** **Secondary Storage** What are the features of floppy disks, hard disks, optical disks, magnetic tape, smart cards, and online secondary storage?

T he microprocessor "is the most important invention of the 20th century," says Michael Malone, author of *The Microprocessor: A Biography*.[1]

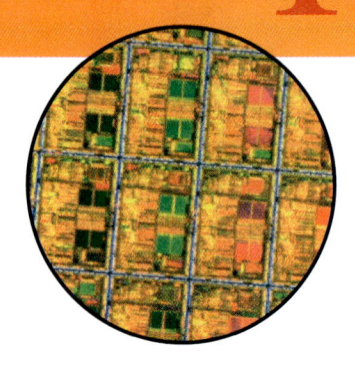

Quite a bold claim, considering the incredible products that have issued forth during those 100 years. More important than the airplane? More than television? More than atomic energy?

According to Malone, the case for the exalted status of this thumbnail-size information-processing device is demonstrated, first, by its pervasiveness in the important machines in our lives, from computers to transportation. Second, "The microprocessor is, intrinsically, something special," he says. "Just as [the human being] is an animal, yet transcends that state, so too the microprocessor is a silicon chip, but more." Why? Because it can be programmed to recognize and respond to patterns in the environment, as humans do. Malone writes:

> Implant [a microprocessor] into a traditional machine—say an automobile engine or refrigerator—and suddenly that machine for the first time can learn, it can adapt to its environment, respond to changing conditions, become more efficient, more responsive to the unique needs of its user.[2]

4.1 Microchips, Miniaturization, & Mobility

KEY QUESTION
What are the differences between transistors, integrated circuits, chips, and microprocessors?

The microprocessor has presented us with gifts that we may only barely appreciate—*portability* and *mobility* in electronic devices.

In 1955, for instance, portability was exemplified by the ads showing a young woman holding a Zenith television set over the caption: IT DOESN'T TAKE A MUSCLE MAN TO MOVE THIS LIGHTWEIGHT TV. That "lightweight" TV weighed a hefty 45 pounds. Today, by contrast, there is a handheld Casio color TV weighing a mere 6.2 ounces.

Had the transistor not arrived, as it did in 1947, the Age of Portability and consequent mobility would never have happened. To us a "portable" telephone might have meant the 40-pound backpack radio-phones carried by some American GIs through World War II, rather than the 6-ounce shirt-pocket cellular models available today.

From Vacuum Tubes to Transistors to Microchips

Old-time radios used vacuum tubes—small lightbulb-size electronic tubes with glowing filaments. One computer to use these tubes, the ENIAC, which was switched on in 1946, employed 18,000 of them. Unfortunately, a tube failure occurred on average once every 7 minutes. Since it took more than 15 minutes to find and replace the faulty tube, it was difficult to get any useful computing work done. Moreover, the ENIAC was enormous, occupying 1500 square feet and weighing 30 tons.

The transistor changed all that. **A *transistor* is essentially a tiny electrically operated switch, or gate, that can alternate between "on" and "off" many millions of times per second.** The first transistors were one-hundredth the size of a vacuum tube, needed no warmup time, consumed less energy, and were faster and more reliable. (See ● Panel 4.1.) Moreover, they marked the beginning of a process of miniaturization that has not ended yet. In 1960 one transistor fit into an area about a half-centimeter square. This was sufficient to permit Zenith, for instance, to market a transistor radio weighing about 1 pound (convenient, they advertised, for "pocket or purse"). Today

● **PANEL 4.1**
Shrinking components
The lightbulb-size 1940s vacuum tube was replaced in the 1950s by a transistor one-hundredth its size. Today's transistors are much smaller, being microscopic in size.

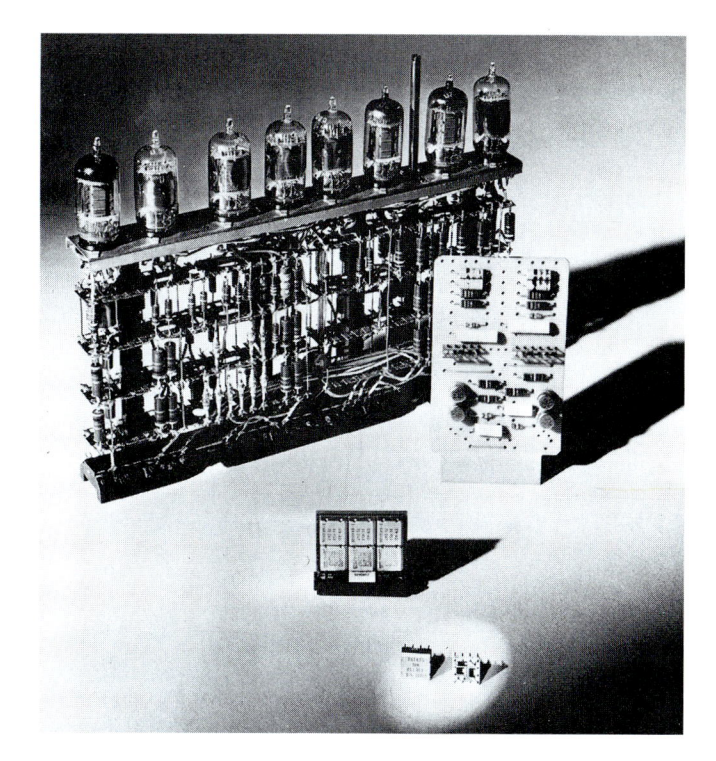

more than 3 million transistors can be squeezed into a half centimeter, and a Sony headset radio, for example, weighs only 6.2 ounces.

In the old days, transistors were made individually and then formed into an electronic circuit with the use of wires and solder. Today transistors are part of an <u>*integrated circuit*</u>—**an entire electronic circuit, including wires, formed on a single "chip," or piece, of special material, usually silicon,** as part of a single manufacturing process. An integrated circuit embodies what is called *solid-state technology*. <u>**Solid state**</u> **means that the electrons are traveling through solid material**—in this case, silicon. They do not travel through a vacuum, as was the case with the old radio vacuum tubes.

What is silicon, and why use it? <u>**Silicon**</u> **is an element that is widely found in clay and sand.** It is used not only because its abundance makes it cheap but also because it is a *semiconductor*. **A <u>semiconductor</u> is material whose electrical properties are intermediate between a good conductor of electricity and a nonconductor of electricity.** (An example of a good conductor of electricity is copper in household wiring; an example of a nonconductor is the plastic sheath around that wiring.) Because it is only a semiconductor, silicon has partial resistance to electricity. As a result, highly conducting materials can be overlaid on the silicon to create the electronic circuitry of the integrated circuit. (See ● Panel 4.2, next page.)

A <u>*chip*</u>, or microchip, **is a tiny piece of silicon that contains millions of microminiature electronic components, mainly transistors.** Chip manufacture requires very clean environments, which is why chip manufacturing workers appear to be dressed for a surgical operation. Such workers must also be highly skilled, which is why chip makers are not found everywhere in the world.

Miniaturization Miracles: Microchips, Microprocessors, & Micromachines

Microchips—"industrial rice," as the Japanese call them—are responsible for the miniaturization that has revolutionized consumer electronics, computers, and communications. They store and process data in all the electronic gadgetry we've become accustomed to—from microwave ovens to videogame controllers to music synthesizers to cameras to automobile fuel-injection systems to pagers to satellites.

● **PANEL 4.2**
Making of a chip
How microscopic circuitry is put onto silicon

1. A large drawing of the electrical circuitry is made; it looks something like the map of a train yard. The drawing is photographically reduced hundreds of times, to microscopic size.

2. That reduced photograph is then duplicated many times so that, like a sheet of postage stamps, there are multiple copies of the same image or circuit.

3. That sheet of multiple copies of the circuit is then printed (in a printing process called *photolithography*) and etched onto a round slice of silicon called a *wafer*. Wafers have gone from 4 inches in diameter to 6 inches to 8 inches, and now are moving toward 12 inches; this allows semiconductor manufacturers to produce more chips at lower cost.

Chip designers checking out an enlarged drawing of chip circuits

A wafer imprinted with many microprocessors.

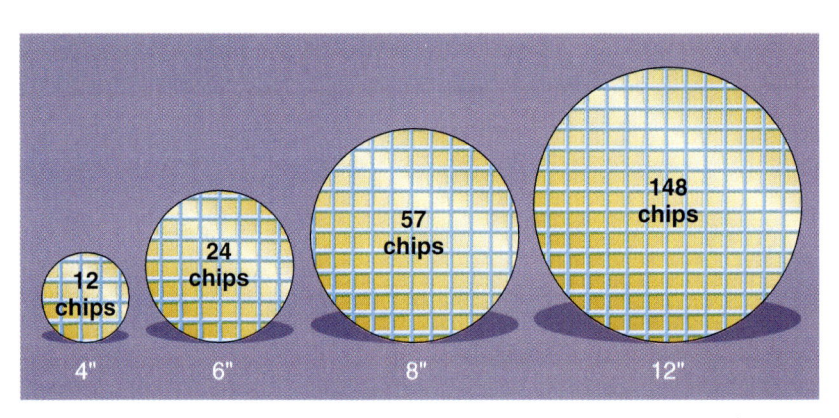

Increasing size of wafers, which means more chips for lower cost. (Each chip measures 20 mm × 20 mm.)

4". 12 chips
6". 24 chips
8". 57 chips
12". 148 chips

4. Subsequent printings of layer after layer of additional circuits produce multilayered and interconnected electronic circuitry built above and below the original silicon surface.

5. Later an automated die-cutting machine cuts the wafer into separate *chips*, which are usually less than 1 centimeter square and about half a millimeter thick. A *chip*, or microchip, is a tiny piece of silicon that contains millions of microminiature electronic circuit components, mainly transistors. An 8-inch silicon wafer will have a grid of nearly 300 chips, each with as many as 5.5 million transistors.

6. After testing, each chip is mounted in a protective frame with extruding metallic pins that provide electrical connections through wires to a computer or other electronic device.

(above) Pentium 4 microprocessor chip mounted in protective frame with pins that can be connected to an electronic device such as a microcomputer.

There are different kinds of microchips—for example, microprocessor, memory, logic, communications, graphics, and math coprocessor chips. We discuss some of these later in this chapter. Perhaps the most important is the microprocessor chip. **A _microprocessor_ ("microscopic processor" or "processor on a chip") is the miniaturized circuitry of a computer processor—the part that processes, or manipulates, data into information.** When modified for use in machines other than computers, microprocessors are called _microcontrollers_, or _embedded computers_ (as discussed in Chapter 1, page 9).

Mobility

Smallness in TVs, phones, radios, camcorders, CD players, and computers is now largely taken for granted. In the 1980s portability, or mobility, meant trading off computing power and convenience in return for smaller size and weight. Today, however, we are getting close to the point where we don't have to give up anything. As a result, experts have predicted that small, powerful, wireless personal electronic devices will transform our lives far more than the personal computer has done so far. "[T]he new generation of machines will be truly personal computers, designed for our mobile lives," wrote one reporter in 1992. "We will read office memos between strokes on the golf course, and answer messages from our children in the middle of business meetings."[3] Today such activities are becoming commonplace.

CONCEPT CHECK

Describe the evolution of the processor.

How has the invention of the microprocessor been important?

Buying an Inexpensive Personal Computer: Understanding Computer Ads

The cost of a PC has declined remarkably. Prices went down 44% over the span of three years, to an average of $844 in early 2000.[4] And if all you want is Internet access, the price is even lower. You can buy a gadget called _Audrey_, a "Netpliance," marketed by 3Com Corp. for people without a home computer. Audrey can be used with any ISP (except, currently, AOL), and it costs $499.

Indeed, Audrey represents a home version of the trend toward "thin-client" computing, mentioned in Chapter 3. Thin clients, which may be had for $350–$700, are simple terminals running a Web browser and small applications downloaded as needed from high-powered network servers.[5] Large organizations are discovering that replacing fat clients (PCs) with thin clients (terminals) produces substantial savings—not only in the upfront costs of the hardware but also in the much larger cost of installing, supporting, and updating the machines over time (as we explain further in Chapter 6).[6]

But even though some people now speak of the "post-PC era," the personal computer isn't dead yet. Since at some point you'll no doubt want to buy or replace a PC, it will help to know how to interpret a typical computer ad. _(See ● Panel 4.3, next page.)_ In this chapter, we explain how.

These days a desktop computer is usually a _multimedia computer_, with sound and graphics capability. As we explained in Chapter 1, the word _multimedia_ means "combination of media," such as the combination of pictures, video, animation, and sound in addition to text. A multimedia computer features such equipment as a fast processor, DVD drive, sound card, graphics card, and speakers, and you may also wish to have headphones and a microphone. You may even wish to add a scanner, sound recorder, and digital camera.

● **PANEL 4.3**
Advertisement for a PC

Great PC Buy!

- 7-Bay Mid-Tower Case
- Intel Pentium III Processor 933 MHz
- 128 MB 133 MHz SDRAM
- 256K Cache
- 2 USB Ports
- 56 Kbps Internal Modem
- 3D AGP Graphics Card (8 MB)
- PCI Wavetable Sound Card
- 3.5" Floppy Drive
- Iomega 250 MB Zip Drive
- 80 GB Ultra ATA 7200 RPM Hard Drive
- 44X Max CD-ROM Drive or CD-R/RW Drive
- 104-Key Keyboard
- Microsoft IntelliMouse
- 17", .25dp Monitor (16" Display)
- HP DeskJet 970Cse Printer

CLICK-ALONG 5-1
More about PC ads

Details of this ad are explained throughout this chapter. See the little magnifying glass:

Let us now go through the parts of a system so that you can understand what you're doing when you buy a new computer. In the remainder of this chapter, we will consider the *system unit* and *storage devices*. In Chapter 5, we look at *input devices* and *output devices*.

4.2 The System Unit

KEY QUESTIONS
How is data represented in a computer; what are the components of the system cabinet; what are processing speeds; how do the processor and memory work; and what are some important ports, buses, and cards?

How is the information in "information processing" in fact processed? The first thing to understand is that computers run on electricity. And what is the most fundamental thing you can say about electricity? Electricity is either *on* or *off*. This two-state situation allows computers to use the *binary system* to represent data and programs.

The Binary System: Using On/Off Electrical States to Represent Data & Instructions

The decimal system that we are accustomed to has 10 digits (0, 1, 2, 3, 4, 5, 6, 7, 8, 9). By contrast, the **binary system has only two digits: 0 and 1.** Thus, in the computer, the 0 can be represented by the electrical current being low power (off) and the 1 by the current being high power (on). All data and program instructions that go into the computer are represented in terms of these binary numbers. (See ● Panel 4.4.)

For example, the letter G is a translation of the electronic signal 01000111, or off-on-off-off-off-on-on-on. When you press the key for G on the computer keyboard, the character is automatically converted into the series of electronic impulses that the computer can recognize. Inside the computer, that character G is represented by a combination of eight *transistors* (as we will describe). Some are closed (representing the 0s), and some are open (representing the 1s).

How many representations of 0s and 1s can be held in a computer or a storage device such as a hard disk? Capacity is denoted by *bits* and *bytes* and multiples thereof:

PANEL 4.4
Binary data representation
How the letters G-R-O-W are represented in one type of low-power/high-power (off/on), 1/0 binary code

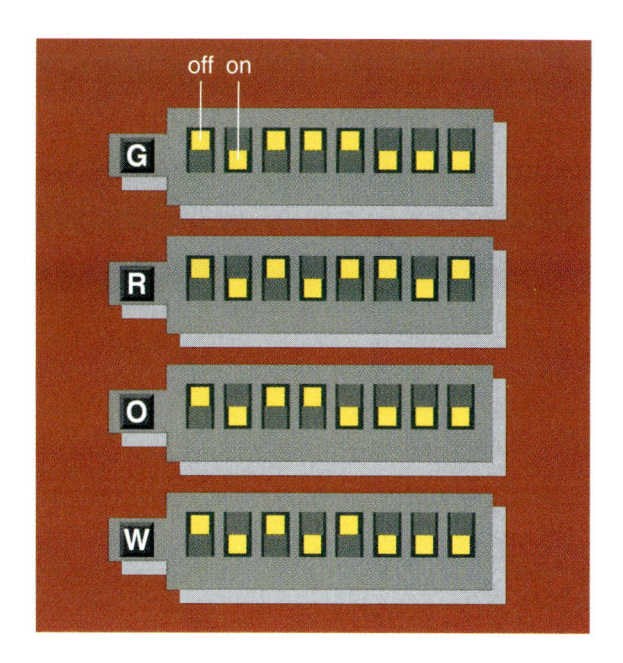

- **Bit:** In the binary system, **each 0 or 1 is called a *bit*, which is short for "binary digit."**
- **Byte:** To represent letters, numbers, or special characters (such as ! or *), bits are combined into groups. **A group of 8 bits is called a *byte*, and a byte represents one character, digit, or other value.** (As we mentioned, in one scheme, 01000111 represents the letter *G*.) The capacity of a computer's memory or of a floppy disk is expressed in numbers of bytes or multiples such as kilobytes and megabytes.
- **Kilobyte:** A *kilobyte (K, KB)* **is about 1000 bytes.** (Actually, it's precisely 1024 bytes, but the figure is commonly rounded.) The kilobyte was a common unit of measure for memory or secondary-storage capacity on older computers. 1 KB equals about 1/2 page of text.
- **Megabyte:** A *megabyte (M, MB)* **is about 1 million bytes** *(1,048,576 bytes).* Measures of microcomputer primary-storage capacity today are expressed in megabytes. 1 MB equals about 500 pages of text.
- **Gigabyte:** A *gigabyte (G, GB)* **is about 1 billion bytes** *(1,073,741,824 bytes).* This measure was formerly used mainly with "big iron" (mainframe) computers, but is typical of the secondary storage (hard disk) capacity of today's microcomputers. 1 GB equals about 500,000 pages of text.
- **Terabyte:** A *terabyte (T, TB)* **represents about 1 trillion bytes** (1,009,511,627,776 bytes). 1 TB equals about 500,000,000 pages of text.
- **Petabyte:** A *petabyte (P, PB)* **represents about 1 quadrillion bytes** (1,048,576 gigabytes).

Letters, numbers, and special characters are represented within a computer system by means of *binary coding schemes.* (See ● *Panel 4.5, next page.*) That is, the off/on 0s and 1s are arranged in such a way that they can be made to represent characters, digits, or other values.

- **ASCII:** Pronounced "*Ask*-ee," **ASCII (American Standard Code for Information Interchange) is the binary code most widely used with microcomputers.** It is also used on many minicomputers.
- **EBCDIC:** Pronounced "*Eb*-see-dick," **EBCDIC (Extended Binary Coded Decimal Interchange Code) is a binary code used with some large computers,** such as mainframes manufactured by IBM and Amdahl.

147

● **PANEL 4.5**
Binary coding schemes: ASCII and EBCDIC

Character	EBCDIC	ASCII	Character	EBCDIC	ASCII
A	1100 0001	0100 0001	N	1101 0101	0100 1110
B	1100 0010	0100 0010	O	1101 0110	0100 1111
C	1100 0011	0100 0011	P	1101 0111	0101 0000
D	1100 0100	0100 0100	Q	1101 1000	0101 0001
E	1100 0101	0100 0101	R	1101 1001	0101 0010
F	1100 0110	0100 0110	S	1110 0010	0101 0011
G	1100 0111	0100 0111	T	1110 0011	0101 0100
H	1100 1000	0100 1000	U	1110 0100	0101 0101
I	1100 1001	0100 1001	V	1110 0101	0101 0110
J	1101 0001	0100 1010	W	1110 0110	0101 0111
K	1101 0010	0100 1011	X	1110 0111	0101 1000
L	1101 0011	0100 1100	Y	1110 1000	0101 1001
M	1101 0100	0100 1101	Z	1110 1001	0101 1010
0	1111 0000	0011 0000	5	1111 0101	0011 0101
1	1111 0001	0011 0001	6	1111 0110	0011 0110
2	1111 0010	0011 0010	7	1111 0111	0011 0111
3	1111 0011	0011 0011	8	1111 1000	0011 1000
4	1111 0100	0011 0100	9	1111 1001	0011 1001
!	0101 1010	0010 0001	;	0101 1110	0011 1011

- **Unicode:** Unlike ASCII, **Unicode** uses two bytes (16 bits) for each character, rather than one byte (8 bits). Instead of the 256 character combinations of ASCII, Unicode can handle 65,536 character combinations. Thus, it allows almost all the written languages of the world to be represented using a single character set.

CONCEPT CHECK

What is the binary system?

Define bits, bytes, kilobytes, megabytes, gigabytes, terabytes, and petabytes.

Distinguish among ASCII, EBCDIC, and Unicode.

The Computer Case: Bays, Buttons, & Boards

The *system unit* houses the motherboard (including the processor chip and memory chips), the power supply, and storage devices. *(See* ● *Panel 4.6.)* In computer ads, the part of the system unit that is the empty box with just the power supply is called the *case* or *system cabinet*.

For today's desktop PC, the system unit may be advertised as something like a "4-bay micro-tower case" or a "7-bay mid-tower case." **A *bay* is a shelf or opening used for the installation of electronic equipment,** generally storage devices such as a hard drive or DVD drive. A computer may come equipped with four or seven bays. (Empty bays are covered by a panel.) A *tower* is a cabinet that is tall, narrow, and deep (so that it can sit on the floor beside or under a table) rather than short, wide, and deep. Originally a tower was considered to be 24 inches high. Micro- and mid-towers may be less than half that size.

A line from the PC ad on page 146

🔍 7-Bay Mid-Tower Case

Survival Tip

Bay Access

Drive bays are the openings in your computer's case into which drives are installed. If the bay is "accessible," it's open to the outside of the PC (tape, floppy, CD/DVD drives). If it's "hidden," it's closed inside the PC case (hard drive).

● **PANEL 4.6**

The system unit
Overhead view of the box, or case. It includes the motherboard, power supply, and storage devices. (The arrangement of the components varies among models.)

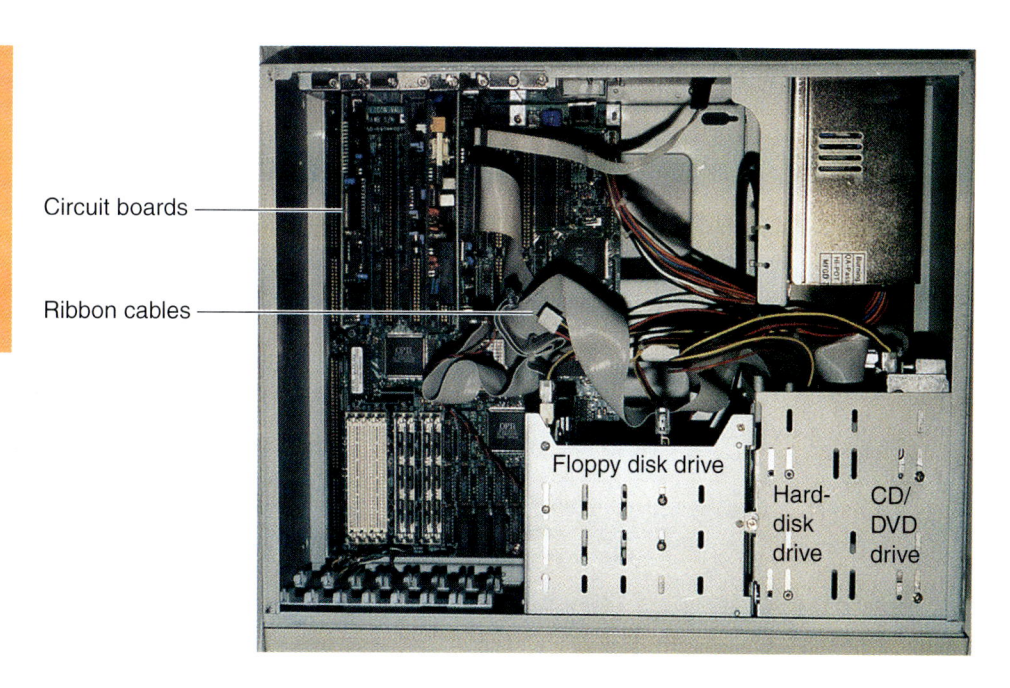

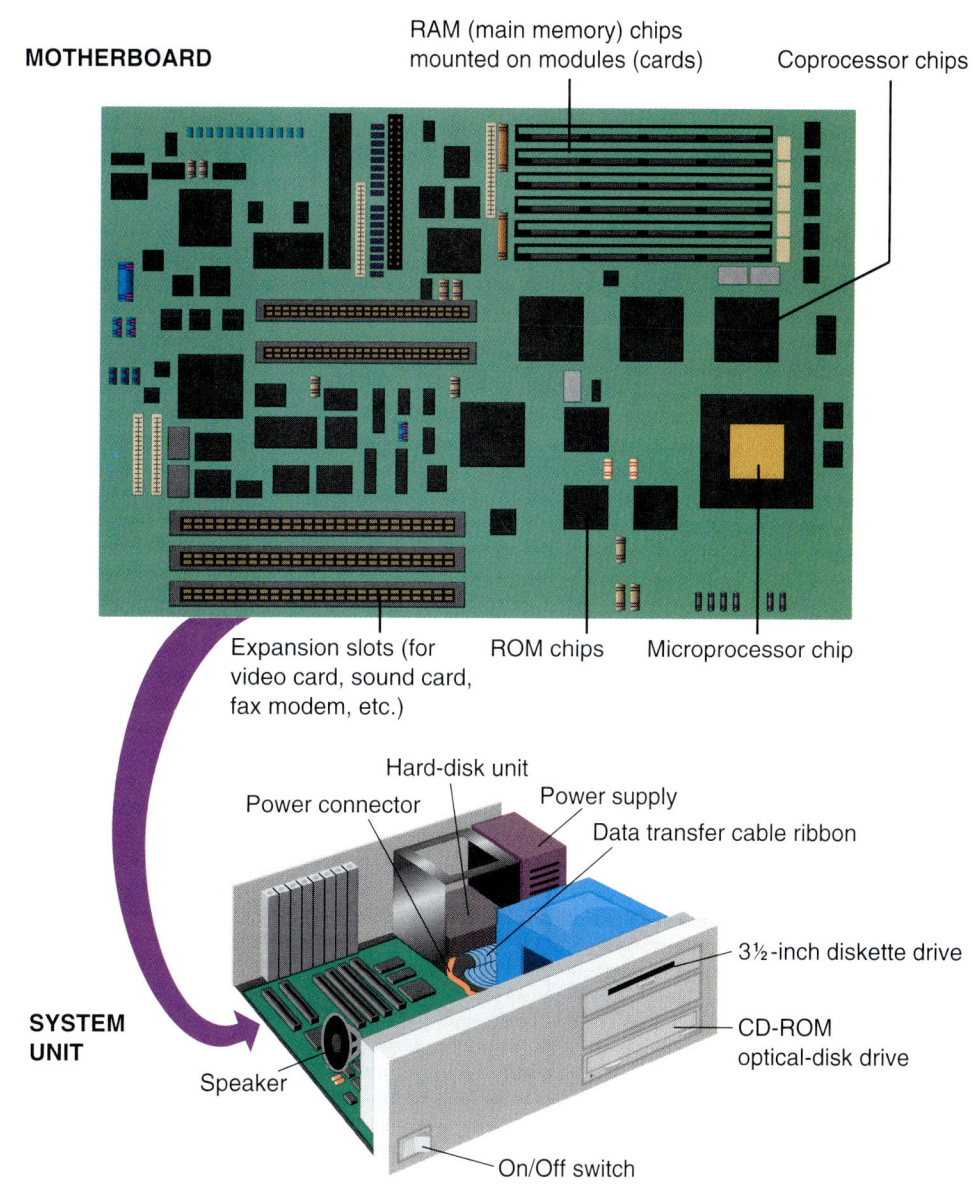

Circuit boards

Ribbon cables

Floppy disk drive

Hard-disk drive

CD/DVD drive

MOTHERBOARD

RAM (main memory) chips mounted on modules (cards)

Coprocessor chips

Expansion slots (for video card, sound card, fax modem, etc.)

ROM chips

Microprocessor chip

SYSTEM UNIT

Power connector

Hard-disk unit

Power supply

Data transfer cable ribbon

3½-inch diskette drive

CD-ROM optical-disk drive

Speaker

On/Off switch

Hardware: The CPU & Storage

149

The number of buttons on the outside of the computer case will vary, but the on/off power switch will appear somewhere, either front or back. There may also be a "sleep" switch; this allows you to suspend operations without terminating them, so that you can conserve electrical power without the need for subsequent "rebooting," or restarting, the computer.

Inside the case—not visible unless you remove the cabinet—are various electrical circuit boards, chief of which is the motherboard, as we'll discuss.

Power Supply

The electricity available from a standard wall outlet is alternating current (AC), but a microcomputer runs on direct current (DC). **The _power supply_ is a device that converts AC to DC to run the computer.** The on/off switch in your computer turns on or shuts off the electricity to the power supply. Because electricity can generate a lot of heat, a fan inside the computer keeps the power supply and other components from becoming too hot.

The Motherboard & the Microprocessor Chip

The _motherboard_, or _system board_, is the main circuit board in the system unit. The motherboard consists of a flat board that fills one side of the case. It contains both soldered, nonremovable components and sockets or slots for components that can be removed—microprocessor chip, RAM chips, and various expansion cards, as we explain later. (See ● Panel 4.7.) Making some components removable allows you to expand or upgrade your system. **_Expansion_ is a way of increasing a computer's capabilities by adding hardware to perform tasks that are beyond the scope of the basic system.** For example, you might want to add video and sound cards. **_Upgrading_ means changing to newer, usually more powerful or sophisticated versions,** such as a more powerful microprocessor or more memory chips.

The most fundamental part of the motherboard is the microprocessor chip. As mentioned, a _microprocessor_ is the miniaturized circuitry of a computer processor. It stores program instructions that process, or manipulate, data into information. The key parts of the microprocessor are transistors. _Transistors_, as we stated, are tiny electronic devices that act as on/off switches, which process the on/off 1/0 bits used to represent data. According to _Moore's_

RAM (main memory) chips mounted on modules (cards)

● **PANEL 4.7**
The motherboard
This main board offers slots or sockets for removable components: microprocessor chip, RAM chips, and various expansion cards.

Microprocessor chip (with CPU)

law, named for legendary Intel cofounder Gordon Moore, the number of transistors that can be packed onto a chip doubles about every 18 months, while the price stays the same. In 1961 a chip had only 4 transistors; in 1971 it had 2300; in 1979 it had 30,000; and in 1997 it had 7.5 million. Current chips have 9.5–43 million transistors.

Two principal "architectures" or designs for microprocessors are CISC and RISC. **_CISC (complex instruction set computing) chips_, which are used mostly in PCs and in conventional mainframes, can support a large number of instructions,** but at relatively low processing speeds. **In _RISC (reduced instruction set computing) chips_, which are used mostly in workstations, a great many seldom-used instructions are eliminated.** As a result, workstations can work up to 10 times faster than most PCs. RISC chips have been used in many Macintosh computers since 1993.

Most personal computers today use microprocessors of two kinds—those based on the model made by Intel and those based on the model made by Motorola.

- **Intel-type chips—for PCs:** About 90% of microcomputers use Intel-type microprocessors. Indeed, the Microsoft Windows operating system is designed to run on Intel chips. As a result, people in the computer industry tend to refer to the Windows/Intel joint powerhouse as *Wintel.*

  Intel-type chips for PCs are made principally by Intel Corp. and Advanced Micro Devices (AMD), but also by Cyrix, DEC, and others. They are used by manufacturers such as Compaq, Dell, Gateway, Hewlett-Packard, and IBM. Since 1993, Intel has marketed its chips under the names *Pentium, Pentium Pro, Pentium MMX, Pentium II, Pentium III, Celeron,* and *Pentium 4* (or *IV*). Currently, the most advanced marketed Intel chip is the P4, with 42 million transistors. Many ads for PCs contain the logo "Intel inside" to show that the systems run an Intel microprocessor.

- **Motorola-type chips—for Macintoshes:** **_Motorola-type chips_ are made by Motorola for Apple Macintosh computers.** Since 1993, Motorola has joined forces with IBM and Apple to produce the PowerPC family of chips. With certain hardware or software configurations, a PowerPC can run PC as well as Mac applications software.

Processing Speeds: From Megahertz to Picoseconds

Often a PC ad will carry a line that says something like "Intel Celeron processor 500 MHz," "Intel Pentium III processor 933 MHz," or "AMD Athlon processor 1-GHz." MHz stands for *megahertz* and GHz for *gigahertz.* These figures indicate how fast the microprocessor can process data and execute program instructions.

Every microprocessor contains a **_system clock_, which controls how fast all the operations within a computer take place.** The system clock uses fixed vibrations from a quartz crystal to deliver a steady stream of digital pulses or "ticks" to the CPU. These ticks are called *cycles.* Faster clock speeds will result in faster processing of data and execution of program instructions, as long as the computer's internal circuits can handle the increased speed.

There are four main ways in which processing speeds are measured:

- **For microcomputers—megahertz and gigahertz:** Microcomputer microprocessor speeds are usually expressed in **_megahertz (MHz)_, a measure of frequency equivalent to 1 million cycles (ticks of the system clock) per second.** The original IBM PC had a clock speed of 4.77 MHz. Today a 933-MHz Pentium III–based microcomputer processes 933 million cycles per second. The latest generation of processors (from

Intel Pentium III
Processor 933 MHz

AMD and Intel) operate in **gigahertz (GHz)**—**a billion cycles per second.** Intel's latest chip, the Pentium 4 (P4), operates at up to 1.9 gigahertz. Some experts predict that advances in microprocessor technology will let Intel produce a 50 GHz CPU in 2010. This kind of power will be necessary to support such functions as true speech interfaces and real-time speech translation.

Intel's mobile processors—for laptops and notebooks—are the Pentium III, which runs at 450 MHz–1 GHz, and the Celeron, which runs at 450–850 MHz.

Since a new high-speed processor can cost many hundred dollars more than the previous generation of chip, experts often recommend that buyers fret less about the speed of the processor (since the work most people do on their PCs doesn't even tax the limits of the current hardware) and more about spending money on extra memory.

- **For workstations, minicomputers, and mainframes—MIPS:** Processing speed can also be measured according to the number of instructions per second that a computer can process. **MIPS stands for millions of instructions per second.** A high-end microcomputer or workstation might perform at 100 MIPS or more, a mainframe at 200–1200 MIPS.

- **For supercomputers—flops:** The abbreviation *flops* **stands for floating-point operations per second.** A floating-point operation is a special kind of mathematical calculation. This measure, used mainly with supercomputers, is expressed as *megaflops* (mflops, or millions of floating-point operations per second), *gigaflops* (gflops, or billions), and *teraflops* (tflops, or trillions). The U.S. supercomputer known as Option Red cranks out 1.34 teraflops. (To put this in perspective, a person able to complete one arithmetic calculation every second would take about 31,000 years to do what Option Red does in a single second.) New supercomputer speeds will be measured in petaflops (1 quadrillion operations per second).

- **For all computers—fractions of a second:** Another way to measure cycle times is in fractions of a second. A microcomputer operates in microseconds, a supercomputer in nanoseconds or picoseconds—thousands or millions of times faster. A *millisecond* is one-thousandth of a second. A *microsecond* is one-millionth of a second. A *nanosecond* is one-billionth of a second. A *picosecond* is one-trillionth of a second.

CONCEPT CHECK

Distinguish expansion from upgrading.

Discuss the most fundamental part of the motherboard, its features, its two principal architectures, and the two principal kinds used in personal computers.

What is the system clock, and what are the various measures of processing speed?

How the Processor or CPU Works: Control Unit, ALU, & Registers

Once upon a time, the processor in a computer was measured in feet. A processing unit in the 1946 ENIAC (which had 20 such processors) was about 2 feet wide and 8 feet high. Today, computers are based on *micro*processors, less than 1 centimeter square. It may be difficult to visualize components so tiny. Yet it is necessary to understand how microprocessors work if you are to grasp what PC advertisers mean when they throw out terms such as "128 MB 133 MHz SDRAM" or "256K Advanced Transfer Cache."

Computer professionals often discuss a computer's word size. ***Word size is the number of bits that the processor may process at any one time.*** The more bits in a word, the faster the computer. A 32-bit computer—that is, one with a 32-bit-word processor—will transfer data within each microprocessor chip in 32-bit chunks or 4 bytes at a time. (Recall there are 8 bits in a byte.) A 64-bit-word computer is faster; it transfers data in 64-bit chunks or 8 bytes at a time.

A processor is also called the *CPU*, and it works hand in hand with other circuits known as *main memory* to carry out processing. **The *CPU (central processing unit)* is the "brain" of the computer; it follows the instructions of the software (program) to manipulate data into information. The CPU consists of two parts—(1) the control unit and (2) the arithmetic/logic unit (ALU), both of which contain registers, or high-speed storage areas** (as we discuss shortly). All are linked by a kind of electronic "roadway" called a *bus*. (See ● Panel 4.8.)

● **PANEL 4.8**
The CPU and main memory
The two main CPU components on a microprocessor are the control unit and the ALU, which contain working storage areas called registers and are linked by a kind of electronic roadway called a *bus*.

CPU

Registers
High-speed storage areas used by control unit and ALU to speed up processing

Control unit
Directs electronic signals between main memory and ALU

Arithmetic/logic unit (ALU)
Performs arithmetic and logical operations

Buses
Electrical data roadways that transmit data within CPU and between CPU and main memory and peripherals

Bus

Main memory
(Random Access Memory, or RAM)

Bus

Expansion slots

- **The control unit—for directing electronic signals:** The *control unit* deciphers each instruction stored in it and then carries out the instruction. It directs the movement of electronic signals between main memory and the arithmetic/logic unit. It also directs these electronic signals between main memory and the input and output devices.

  For every instruction, the control unit carries out four basic operations, known as the machine cycle. In the ***machine cycle***, **the CPU (1) fetches an instruction, (2) decodes the instruction, (3) executes the instruction, and (4) stores the result.**

- **The arithmetic/logic unit—for arithmetic and logical operations:** The ***arithmetic/logic unit (ALU)*** **performs arithmetic operations and logical operations and controls the speed of those operations.**

  As you might guess, *arithmetic operations* are the fundamental math operations: addition, subtraction, multiplication, and division.

  Logical operations are comparisons. That is, the ALU compares two pieces of data to see whether one is equal to (=), greater than (>), or less than (<) the other. (The comparisons can also be combined, as in "greater than or equal to" and "less than or equal to.")

Hardware: The CPU & Storage

153

- **Registers—special high-speed storage areas:** The control unit and the ALU also use registers, special areas that enhance the computer's performance. *Registers* **are high-speed storage areas that temporarily store data during processing.** They may store a program instruction while it is being decoded, store data while it is being processed by the ALU, or store the results of a calculation.
- **Buses—data roadways:** *Buses*, **or bus lines, are electrical data roadways through which bits are transmitted within the CPU and between the CPU and other components of the motherboard.** A bus resembles a multilane highway: The more lanes it has, the faster the bits can be transferred. The old-fashioned 8-bit-word bus of early microprocessors had only eight pathways. Data is transmitted four times faster in a computer with a 32-bit bus, which has 32 pathways, than in a computer with an 8-bit bus. Intel's Pentium chip is a 64-bit processor. Macintosh G4 microcomputers contain buses that are 128 bits, as do some supercomputers. Today there are several principal expansion bus standards, or "architectures," for microcomputers.

    We return to a discussion of buses in a few pages.

How Memory Works: RAM, ROM, CMOS, & Flash

CLICK-ALONG 4-2
All about memory

So far we have described only the kinds of chips known as microprocessors. But other silicon chips called *memory chips* are attached to the motherboard. The four principal types of memory chips are *RAM, ROM, CMOS,* and *flash*.

> 128 MB 133 MHz SDRAM

Survival Tip

Need Info on RAM?

If your system is giving you low-system-resources messages, you may need more memory. Go to these websites for detailed information:

www.crucial.com

www.kingston.com

www.tomshardware.com

- **RAM chips—to temporarily store program instructions and data:** Recall from Chapter 1 that there are two types of storage, primary and secondary. Primary storage is temporary or working storage and is often called *memory* or *main memory*; secondary storage is relatively permanent storage (for example, on floppy disk). *RAM (random access memory) chips* **are for primary storage; they temporarily hold (1) software instructions and (2) data before and after it is processed by the CPU.** Because its contents are temporary, RAM is said to be *volatile*—**the contents are lost when the power goes off or is turned off.** This is why you should *frequently*—every 5–10 minutes, say—transfer (save) your work to a secondary-storage medium such as your hard disk, in case the electricity goes off while you're working. (However, there is one kind of RAM, called flash RAM, that is not temporary, as we'll discuss shortly.)

    Four types of RAM chips are used in personal computers: *DRAM, SDRAM, SRAM,* and *RDRAM*. Pronounced "*dee*-ram," *DRAM (dynamic RAM)* must be constantly refreshed by the CPU or it will lose its contents. The type of dynamic RAM used in most PCs today is *SDRAM (synchronous dynamic RAM)*, which is synchronized by the system clock and is much faster than DRAM. Often in computer ads, the speed of SDRAM is expressed in megahertz. The third type, pronounced "*ess*-ram," *SRAM (static RAM)* is faster than any DRAM and will retain its contents without having to be refreshed by the CPU. *Rambus dynamic RAM,* or *RDRAM,* is faster and more expensive than SDRAM and is the type of memory used with Intel's P4 chip. Microcomputers come with different amounts of RAM, which is usually measured in megabytes. An ad may list "64 MB SDRAM," but you can also get 128 megabytes of RAM. The more RAM you have,

SIMM
Single inline memory module

DIMM
Dual inline memory module

the faster the computer operates, and the better your software performs. Having enough RAM is a critical matter. Before you buy a software package, look at the outside of the box or check the manufacturer's website to see how much RAM is required. Microsoft Office 2000, for instance, states that a minimum of 16 megabytes of RAM is required.

If you're short on memory capacity, you can usually add more RAM chips by plugging them into the motherboard. Chips can be bought single or in so-called *memory modules,* circuit boards that can be plugged into expansion slots on the motherboard. There are two types of such modules: SIMMs and DIMMS, both of which use DRAM chips. A *SIMM (single inline memory module)* has RAM chips on only one side. A *DIMM (dual inline memory module)* has RAM chips on both sides.

- **ROM chips—to store fixed start-up instructions:** Unlike RAM, to which data is constantly being added and removed, **ROM (read-only memory) cannot be written on or erased by the computer user without special equipment. ROM chips contain fixed start-up instructions.** That is, ROM chips are loaded, at the factory, with programs containing special instructions for basic computer operations, such as those that start the computer or put characters on the screen. These chips are nonvolatile; their contents are not lost when power to the computer is turned off.

  In computer terminology, **read means to transfer data from an input source into the computer's memory or CPU.** The opposite is **write—to transfer data from the computer's CPU or memory to an output device.** Thus, with a ROM chip, "read-only" means that the CPU can retrieve programs from the ROM chip but cannot modify or add to those programs. A variation is *PROM (programmable read-only memory),* which is a ROM chip that allows you, the user, to load read-only programs and data. However, this can be done only once.

- **CMOS chips—to store flexible start-up instructions:** Pronounced "see-moss," **CMOS (complementary metal-oxide semiconductor) chips are powered by a battery and thus don't lose their contents when the power is turned off.** CMOS chips contain flexible start-up instructions, such as time, date, and calendar, that must be kept current even when the computer is turned off. Unlike ROM chips, CMOS chips can be reprogrammed, as when you need to change the time for daylight savings time. (Your system software will prompt you to do this.)

- **Flash memory chips—to store flexible programs:** Also a nonvolatile form of memory, **flash memory chips can be erased and reprogrammed more than once** (unlike PROM chips, which can be programmed only once). Flash memory, which doesn't require a battery and which can range from 1 to 64 megabytes in capacity, is used to store programs not only in personal computers but also in pagers, cellphones, MP3 players, Palm organizers, printers, and digital cameras. Flash memory is more expensive than hard disk storage. However, many experts believe that flash memory—a type of *solid state technology*—will eventually replace disk and tape storage.

How Cache Works: Level 1 (Internal) & Level 2 (External)

🔍 256K Cache

Because the CPU runs so much faster than the main system RAM, it ends up waiting for information, which is inefficient. To reduce this effect, we have cache. Pronounced "cash," *cache* temporarily stores instructions and data that the processor is likely to use frequently. Thus, cache speeds up processing.

There are two kinds of cache—Level 1 and Level 2:

- **Level 1 (L1) cache—part of the microprocessor chip:** *Level 1 (L1) cache*, also called internal cache, is built into the processor chip. Ranging from 8 to 256 kilobytes, its capacity is less than that of Level 2 cache, although it operates faster.

- **Level 2 (L2) cache—not part of the microprocessor chip:** This is the kind of cache usually referred to in computer ads. *Level 2 (L2) cache*, also called external cache, resides outside the processor chip and consists of SRAM chips. Capacities range from 64 kilobytes to 2 megabytes. (In Intel ads, L2 is called Advanced Transfer Cache.)

In addition, most current computer operating systems allow for the use of *virtual memory*—that is, some free hard-disk space is used to extend the capacity of RAM. The processor searches for data or program instructions in the following order: first L1, then L2, then RAM, then hard disk (or CD-ROM). In this progression, each kind of memory or storage is slower than its predecessor.

CONCEPT CHECK

Describe the following: word size, CPU, the control unit and the machine cycle, the ALU and arithmetic and logical operations, registers, and buses.

What is RAM and what are four variants?

What is ROM?

Discuss CMOS chips and flash memory chips.

How does L1 cache differ from L2 cache?

Ports & Cables

A *port* **is a connecting socket or jack on the outside of the system unit** *(See* ● *Panel 4.9.)* **into which are plugged different kinds of cables.** A port allows you to plug in a cable to connect a peripheral device, such as a monitor, printer, or modem, so that it can communicate with the computer system.

Ports are of several types. Following are six common ones.

- **Serial ports—for transmitting slow data over long distances: A line connected to a *serial port* will send bits one at a time, one after another,** like cars on a one-lane highway. Because individual bits must follow each other, a serial port is usually used to connect devices that do not require fast transmission of data, such as keyboard, mouse, monitors, and modems. It is also useful for sending data over a long distance. The standard for PC serial ports is the 9-pin or 25-pin RS-232C connector.

- **Parallel ports—for transmitting fast data over short distances: A line connected to a *parallel port* allows 8 bits (1 byte) to be transmitted simultaneously,** like cars on an eight-lane highway. Parallel lines

PC

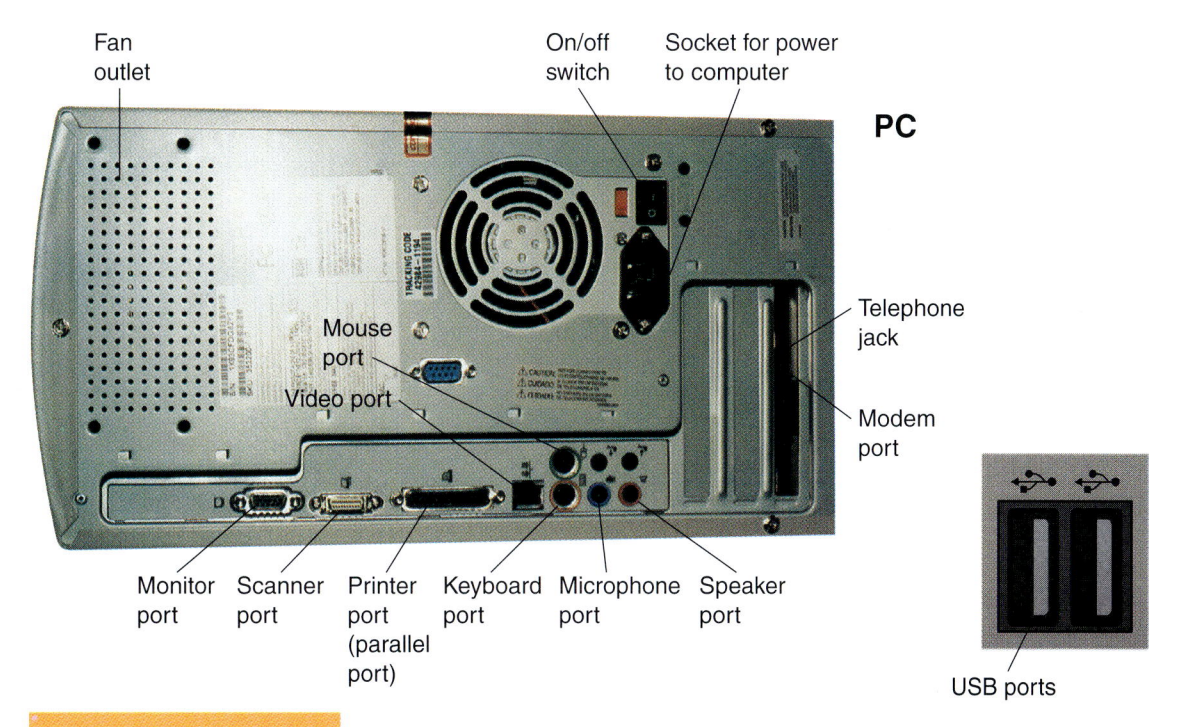

Fan outlet · On/off switch · Socket for power to computer · Mouse port · Video port · Telephone jack · Modem port · Monitor port · Scanner port · Printer port (parallel port) · Keyboard port · Microphone port · Speaker port

USB ports

● **PANEL 4.9**
Ports
The backs of a PC and a Macintosh.

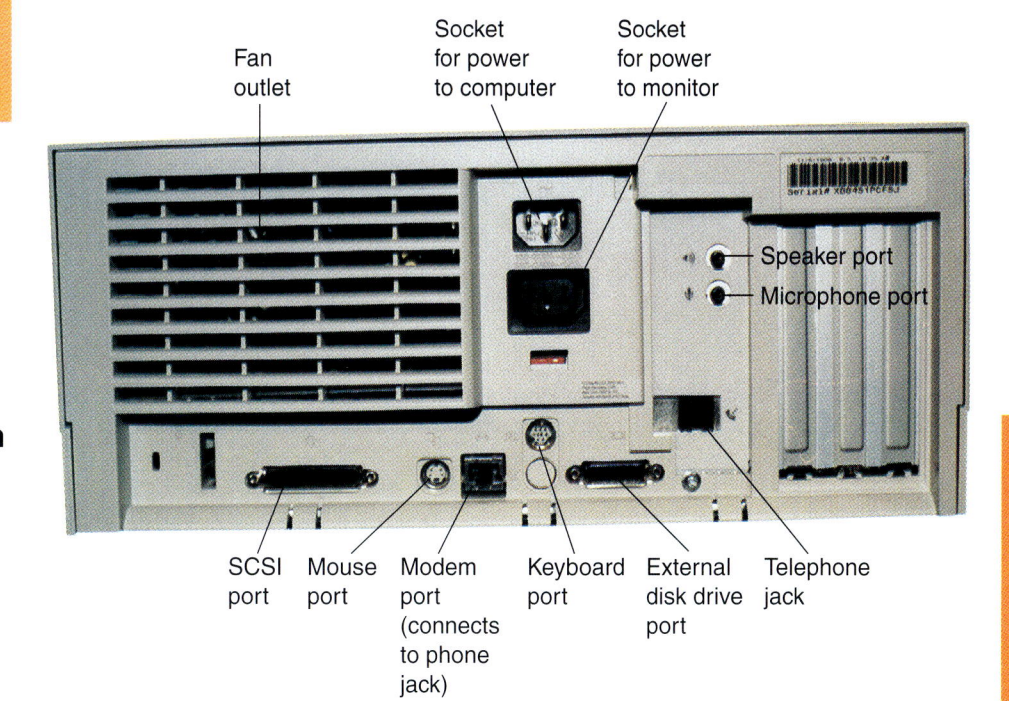

Macintosh

Fan outlet · Socket for power to computer · Socket for power to monitor · Speaker port · Microphone port · SCSI port · Mouse port · Modem port (connects to phone jack) · Keyboard port · External disk drive port · Telephone jack

move information faster than serial lines do, but they can transmit information efficiently only up to 15 feet. Thus, parallel ports are used principally for connecting printers or external disk or magnetic-tape backup storage devices.

- **SCSI ports—for transmitting fast data to up to seven devices in a daisy chain:** Pronounced "scuzzy," a **_SCSI (small computer system interface) port_** allows data to be transmitted in a "daisy chain" to up to 7 devices at speeds (32 bits at a time) higher than those possible with serial and parallel ports. Among the devices that may be connected are external hard-disk drives, CD-ROM drives, scanners, and magnetic-tape backup units. The term *daisy chain* means that several devices are connected in series to each other, so that data for the seventh device, for example, has to go through the other six devices first. Sometimes the equipment on the chain is inside the computer, an internal daisy chain; sometimes it is outside the computer, an external daisy chain.

- **USB ports—for transmitting data to up to 127 devices in a daisy chain:** A <u>USB (universal serial bus) port</u> **can theoretically connect up to 127 peripheral devices daisy-chained to one general-purpose port.** USB ports are useful for peripherals such as digital cameras, digital speakers, scanners, high-speed modems, and joysticks. The so-called USB *hot plug* or *hot swappable* allows such devices to be connected or disconnected even while the PC is running. In addition, USB permits <u>**Plug and Play**</u>, **which allows peripheral devices and expansion cards to be automatically configured while they are being installed.** This avoids the hassle of setting switching and creating special files that plagued earlier users.

    Can you really connect up to 127 devices on a single chain? An Intel engineer did set a world record at an industry trade show before a live audience by connecting 111 peripheral devices to a single USB port on a PC. But respected technology writer Walter Mossberg says that "almost none of the USB peripherals I've seen support this [daisy-chaining] feature."[7] Thus, though most PCs contain only two USB ports, it's worth shopping around to find a model with extra USB connectors.

- **Dedicated ports—for keyboard, mouse, phone, and so on:** So far, we have been considering general-purpose ports, but the back of a computer also has other, *dedicated ports*—ports for special purposes. Among these are the round ports for connecting the keyboard and the mouse. There are also jacks for speakers and microphones and modem-to-telephone jacks. Finally, there is one connector that is not a port at all—the power plug socket, into which you insert the power cord that brings electricity from a wall plug.

- **Infrared ports—for cableless connections over a few feet:** When you use a handheld remote unit to change channels on a TV set, you're using invisible radio waves of the type known as infrared waves. **An <u>infrared port</u> allows a computer to make a cableless connection with infrared-capable devices,** such as some printers. This type of connection requires an unobstructed line of sight between transmitting and receiving ports, and they can be only a few feet apart.

Expandability: Buses & Cards

Today many new microcomputer systems can be expanded. As mentioned earlier, *expansion* is a way of increasing a computer's capabilities by adding hardware to perform tasks that are not part of the basic system. *Upgrading* means changing to a newer, usually more powerful or sophisticated version. (Computer ads often make no distinction between "expansion" and "upgrading." Their main interest is simply to sell you more hardware or software.)

Whether a computer can be expanded depends on its "architecture"—closed or open. *Closed architecture* means a computer has no expansion slots; *open architecture* means it does have expansion slots. (An alternative definition is that closed architecture is a computer design whose specifications are not made freely available by the manufacturer. Thus, other companies cannot create ancillary devices to work with it. With open architecture, the manufacturer shares specifications with outsiders.) **<u>Expansion slots</u> are sockets on the motherboard into which you can plug expansion cards. <u>Expansion cards</u>—also known as expansion boards, adapter cards, interface cards, plug-in boards, controller cards, add-ins, or add-ons—are circuit boards that provide more memory or that control peripheral devices.** *(See* ● *Panel 4.10.)*

Expansion card   Expansion slot

● **PANEL 4.10**
Expandability
How an expansion card fits into an expansion slot.

Common expansion cards connect to the monitor (graphics card), speakers and microphones (sound card), and network (network card), as we'll discuss. Most computers have four to eight expansion slots, some of which may already contain expansion cards included in your initial PC purchase.

Expansion cards are made to connect with different types of buses on the motherboard. (As we mentioned, *buses* are electrical data roadways through which bits are transmitted.) The bus that connects the CPU within itself and to main memory is the *system bus* (also called the *memory bus* or the *frontside bus—FSB*. The bus that connects the CPU with expansion slots on the motherboard and thus with peripheral devices is the *expansion bus*. We already alluded to the universal serial bus (USB), whose purpose, in fact, is to *eliminate* the need for expansion slots and expansion cards, since you can just connect USB devices in a daisy chain outside the system unit. Three expansion buses to be aware of are *ISA, PCI,* and *AGP*.

- **ISA bus—for ordinary low-speed uses:** The <u>ISA (industry standard architecture) bus</u> **is the most widely used expansion bus.** It is also the oldest and, at 8 or 16 bits, the slowest at transmitting data, though it is still used for mice, modem cards, and low-speed network cards.
- **PCI bus—for higher-speed uses:** The <u>PCI (peripheral component interconnect) bus</u> **is a higher-speed bus,** and at 32 or 64 bits wide it is over four times faster than ISA buses. PCI is widely used to connect graphics cards, sound cards, modems, and high-speed network cards.
- **AGP bus—for even higher speeds and 3D graphics:** The <u>AGP (accelerated graphics port) bus</u> **transmits data at even higher speeds and was designed to support video and three-dimensional (3D) graphics.** An AGP bus is twice as fast as a PCI bus.

Among the types of expansion cards are graphics, sound, modem, and network interface cards. A special kind of card is the PC card.

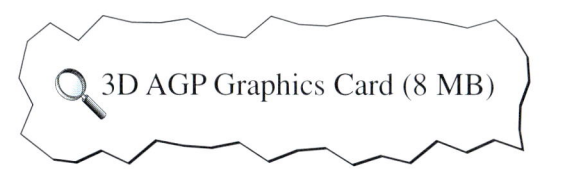

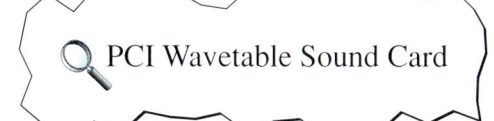

🔍 3D AGP Graphics Card (8 MB)

🔍 PCI Wavetable Sound Card

- **Graphics cards—for monitors:** Graphics cards are included in all PCs. **Also called a video card or video adapter, a <u>graphics card</u> converts signals from the computer into video signals that can be displayed as images on a monitor.** A three-dimensional AGP card is an example of a graphics card. The power of an AGP graphics card is often expressed in megabytes, as in 8, 16, or 32 MB.
- **Sound cards—for speakers and audio output:** **A <u>sound card</u> is used to transmit digital sounds through speakers, microphones, and headsets.** Sound cards come installed on most new PCs. Cards such as PCI wavetable sound cards are used to add music and sound effects to computer video games. *Wavetable synthesis* is a method of creating music based on a wave table, which is a collection of digitized sound samples taken from recordings of actual instruments. The sound samples are then stored on a sound card and are edited and mixed together to produce music. Wavetable synthesis produces higher quality audio output than other sound techniques.

- **Modem cards—for remote communication via phone lines:** Very occasionally you may see a modem that is outside the computer. Most new PCs, however, come with internal modems—modems installed inside as circuit cards. The modem not only sends and receives digital data over telephone lines to and from other computers but can also transmit voice and fax signals.
- **Network interface cards—for remote communication via cable:** A *network interface card* **allows the transmission of data over a cable network,** which connects various computers and other devices such as printers.
- **PC cards—for laptop computers:** Originally called *PCMCIA cards* (for the Personal Computer Memory Card International Association), *PC cards* **are thin, credit-card size (2.1 by 3.4 inches) devices used principally on laptop computers to expand capabilities.** *(See ● Panel 4.11.)* Examples are extra memory (flash RAM—discussed later in the chapter), sound cards, modem, hard disks, and even pagers and cellular communicators. At present there are three sizes for PC cards—I (thin), II (thick), and III (thickest). Type I is used primarily for flash memory cards. Type II, the kind you'll find most often, is used for fax modems and network-interface cards. Type III is for rotating disk devices, such as hard-disk drives, and for wireless communication devices.

CONCEPT CHECK

Distinguish among the following ports: serial, parallel, SCSI, USB, dedicated, and infrared.

Distinguish closed architecture from open architecture.

Define the following three expansion buses: ISA, PCI, AGP.

Why would you need the following cards: graphics, sound, modem, network interface, and PC?

4.3 Secondary Storage

KEY QUESTIONS
What are the features of floppy disks, hard disks, optical disks, magnetic tape, smart cards, and online secondary storage?

You're on a trip with your notebook, or maybe just a cell phone or a personal digital assistant, and you don't have a crucial file of data. Or maybe you need to look up a phone number that you can't get through the phone company's directory assistance. Fortunately, you backed up your data online, using any one of several services (Driveway, FreeDesk.com, I-drive.com, MagicalDesk.com, MyWebOS.com, Visto.com, or X:Drive, for example), and are able to access it through your modem.

● **PANEL 4.11**
PC card
An example of a PC card used in a laptop

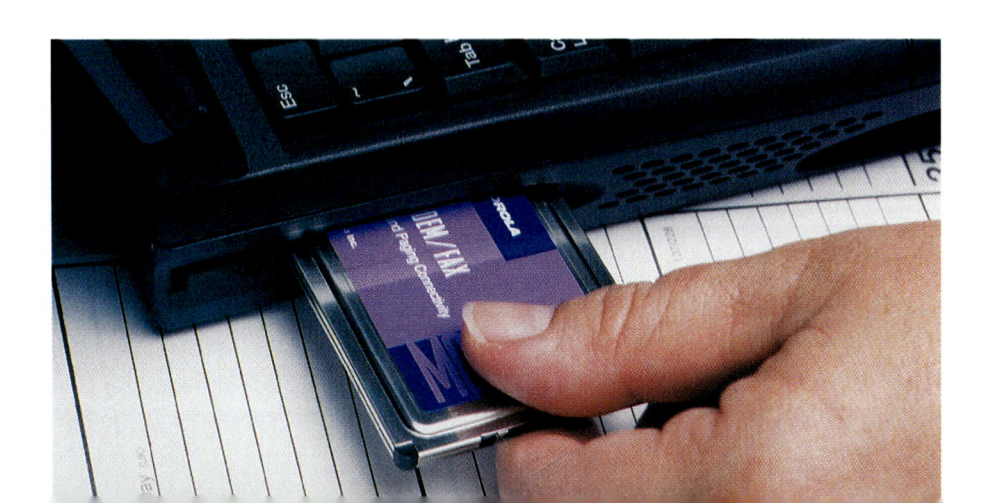

Here is yet another example of how the World Wide Web is offering alternatives to traditional computer functions that once resided within stand-alone machines. We are not, however, fully into the all-online era just yet. Let us consider more traditional forms of <u>**secondary storage hardware**</u>, **devices that permanently hold data and information as well as programs.** We will look at the following types of secondary-storage devices:

- Floppy disks
- Hard disks
- Optical disks
- Magnetic tape
- Smart cards
- Flash memory cards

Finally, we return to online secondary storage, and look at how it actually works.

Floppy Disks

A <u>*floppy disk*</u>, **often called a diskette or simply a disk, is a removable flat piece of mylar plastic packaged in a 3.5-inch plastic case.** Data and programs are stored on the disk's coating by means of magnetized spots, following standard on/off patterns of data representation (such as ASCII). The plastic case protects the mylar disk from being touched by human hands. Originally, when most disks were larger (5.25 inches), the disks actually were "floppy," not rigid; now only the plastic disk inside is flexible or floppy.

Floppy disks are inserted into a floppy-disk drive, a device that holds, spins, reads data from, and writes data to a floppy disk. *Read* means that the data in secondary storage is converted to electronic signals and a copy of that data is transmitted to the computer's memory (RAM). *Write* means that a copy of the electronic information processed by the computer is transferred to secondary storage. Floppy disks have a <u>**write-protect notch**</u>, **which allows you to prevent a diskette from being written to.** In other words, it allows you to protect the data already on the disk. To write-protect, use your thumbnail or the tip of a pen to move the small sliding tab on the lower right side of the disk (viewed from the back), thereby uncovering the square hole. *(See* ● *Panel 4.12, next page.)*

On the diskette, **data is recorded in concentric circles called <u>*tracks*</u>.** Unlike on a vinyl phonograph record, these tracks are neither visible grooves nor a single spiral. Rather, they are closed concentric rings. On a formatted disk each track is divided into <u>**sectors**</u>, **invisible wedge-shaped sections used for storage reference purposes.** When you save data from your computer to a diskette, the data is distributed by tracks and sectors on the disk. That is, the system software uses the point at which a sector intersects a track to reference the data location.

When you insert a floppy disk into the slot (the *drive gate* or *drive door*) in the front of the disk drive, the disk is fixed in place over the spindle of the drive mechanism. The <u>**read/write head**</u> **is used to transfer data between the computer and the disk.** When the disk spins inside its case, the read/write head moves back and forth over the *data access area* on the disk. When the disk is not in the drive, a metal or plastic shutter covers this access area. An access light goes on when the disk is in use. After using the disk, you can retrieve it by pressing an eject button beside the drive. *(See* ● *Panel 4.13, next page.) Note: Do not remove the disk when the access light is on.*

Let's compare the 3.5-inch floppy disk with some 3.5-inch <u>***floppy-disk cartridges***</u>, **or higher-capacity removable disks**—Zip disks, SuperDisks, and HiFD disks:

3.5" Floppy Drive

Survival Tip

Jammed Disk

If your floppy disk is stuck in the drive, try using eyebrow tweezers to pull it out, while simultaneously pressing the eject button.

PANEL 4.12
The parts of a 3.5-inch floppy disk

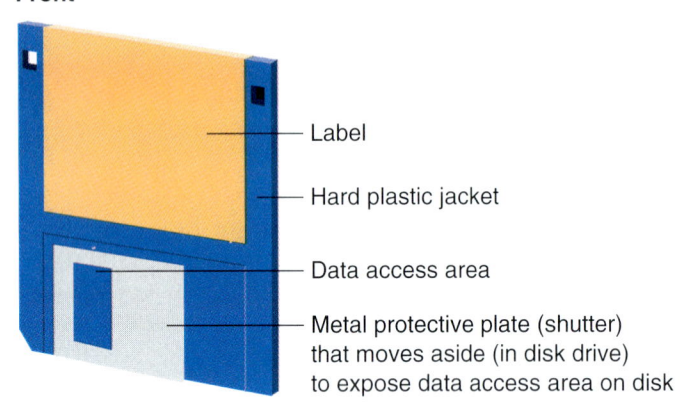

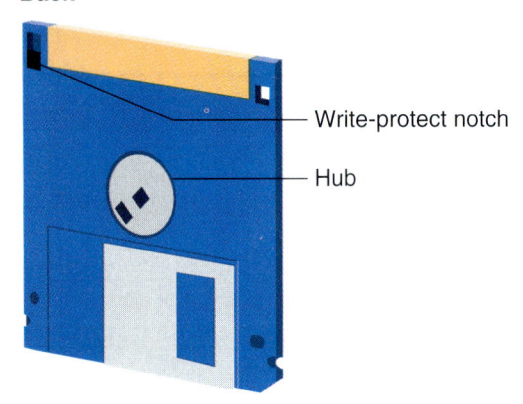

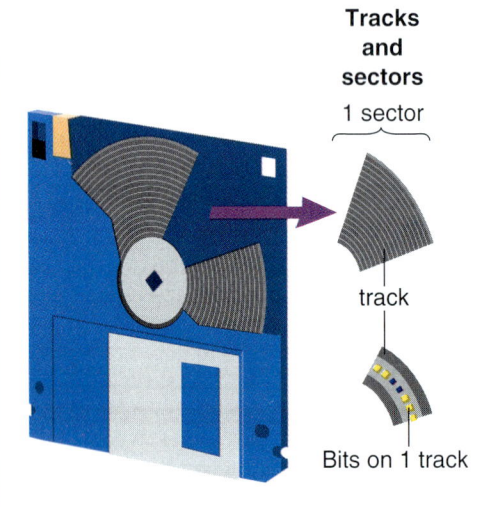

Front
- Label
- Hard plastic jacket
- Data access area
- Metal protective plate (shutter) that moves aside (in disk drive) to expose data access area on disk

Back
- Write-protect notch
- Hub

Tracks and sectors
- 1 sector
- track
- Bits on 1 track

PANEL 4.13
Floppy disk and drive

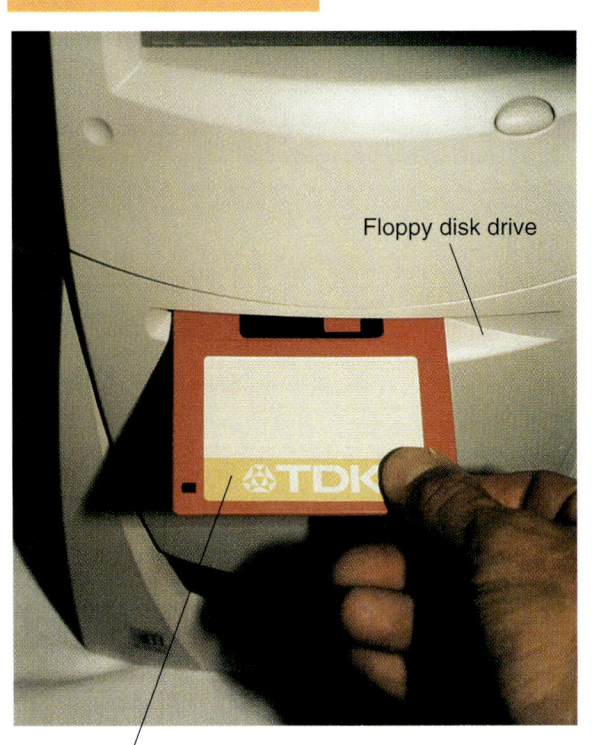

- Floppy disk drive
- Floppy disk

- **3.5-inch floppy disks—1.44 megabytes:** The present-day standard for traditional floppy disks is 1.44 megabytes, the equivalent of 400 typewritten pages. Today's floppy carries the label *2HD*, in which the *2* stands for "double-sided" (it holds data on both sides) and the *HD* stands for "high density" (which means it stores more data than the previous standard—*DD*, for "double density").

  When you buy a box of floppies, be sure to check whether they are "IBM formatted" (for PCs only) or "Macintosh formatted" (for Apple machines). You can also buy "unformatted" disks, which means you have to *format* or *initialize* them yourself—that is, prepare the disks for use so that the operating system can write information on them. Your system software will lead you through the formatting process.

- **Zip disks—100 or 250 megabytes:** Produced by Iomega Corp., **Zip disks** are special disks with a capacity of 100 or 250 megabytes. At 100–250 megabytes, this is at least 70 times the storage capacity of the standard floppy. Among other uses, Zip disks are used to store large spreadsheet files, database files, image files,

Iomega 250 MB Zip Drive

Chapter 4

162

Survival Tip

Backing Up on Zip

The first time you back up your hard disk, it will take a while. Subsequent backups will take less time, because the backup software will archive only the new files. You must be present to back up using Zip disks, because you will have to swap disks. If you plan to back up your computer often, consider tape, which is cheaper than Zip cartridges and which can be set to run automatically, even if you are not there.

multimedia presentation files, and websites. Zip disks require their own Zip disk drives, which may come installed on new computers, although external Zip drives are also available.

- **SuperDisks—120 megabytes:** Produced by Imation, **SuperDisks** are **disks with a capacity of 120 megabytes; the SuperDisk drive can also read standard 1.44-megabyte floppy disks,** which Zip drives cannot do.
- **HiFD disks—200 megabytes:** Made by Sony Corp., **HiFD disks** have a **capacity of 200 megabytes; the disk drive can also read standard 1.44-megabyte floppies.** HiFD disks have 140 times the capacity of today's standard floppy disks.

External Zip drive and disk

Hard Disks

Floppy disks use flexible plastic, but hard disks use metal. **Hard disks** are **thin but rigid metal platters covered with a substance that allows data to be held in the form of magnetized spots.** Hard disks are tightly sealed within an enclosed hard-disk-drive unit to prevent any foreign matter from getting inside. Data may be recorded on both sides of the disk platters. (See • Panel 4.14.)

Hard disks are quite sensitive devices. The read/write head does not actually touch the disk but rather rides on a cushion of air about 0.000001 inch thick. (See • Panel 4.15, next page.) The disk is sealed from impurities within a container, and the whole apparatus is manufactured under sterile conditions. Otherwise, all it would take is a human hair, a dust particle, a fingerprint smudge, or a smoke particle to cause what is called a head crash. A **head crash** happens when the surface of the read/write head or particles on its surface come into contact with the surface of the hard-disk platter, causing the loss of some or all of the data on the disk. A head crash can also

● **PANEL 4.14**
Hard disk
In a microcomputer, the hard disk is enclosed within the system unit. Unlike a floppy disk, it is not accessible. This hard-disk drive is for a notebook computer.

Hardware: The CPU & Storage

● **PANEL 4.15**
Gap between hard disk read/write head and platter
Were the apparatus not sealed, all it would take is a human hair, dust particle, fingerprint, or smoke particle to cause a head crash.

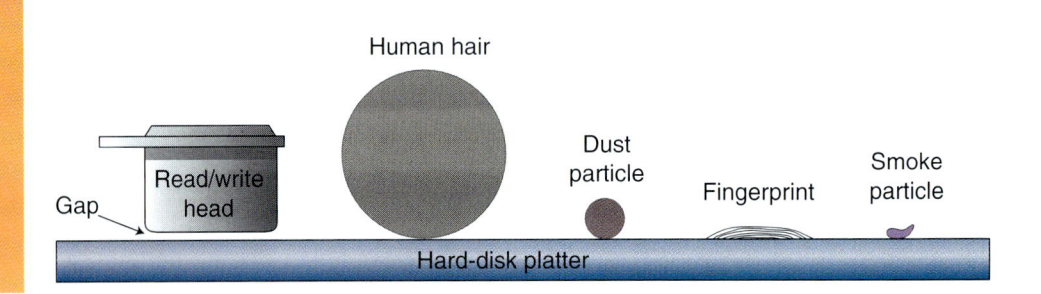

happen when you bump a computer too hard or drop something heavy on the system cabinet. An incident of this sort could, of course, be a disaster if the data has not been backed up. There are firms that specialize in trying to retrieve data from crashed hard disks (for a hefty price), though this cannot always be done.

There are two types of hard disks—nonremovable and removable.

- **Nonremovable hard disks:** A <u>nonremovable hard disk</u>, **also known as a fixed disk, is housed in a microcomputer system unit and is used to store nearly all programs and most data files.** Usually it consists of four 3.5-inch metallic platters sealed inside a drive case the size of a small sandwich, which contains disk platters on a drive spindle, read/write heads mounted on an access arm that moves back and forth, and power connections and circuitry. *(See* ● *Panel 4.16.)* Operation is much the same as for a diskette drive: The read/write heads locate specific instructions or data files according to track or sector.

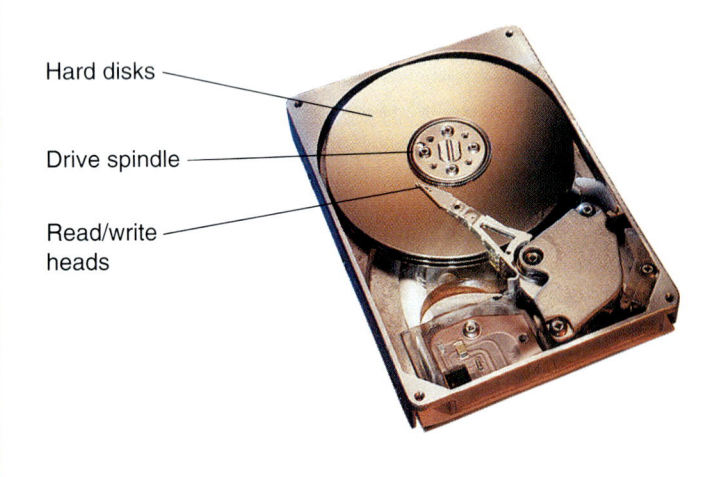

🔍 80 GB Ultra ATA 7200 RPM Hard Drive

● **PANEL 4.16**
Inside a microcomputer's nonremovable hard disk
These platters are installed inside a drive case to prevent contaminants from affecting the read/write heads.

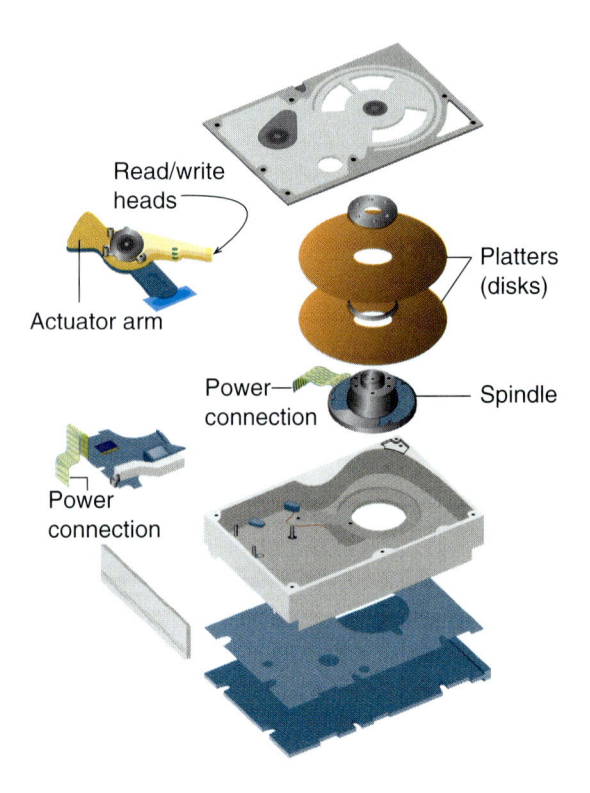

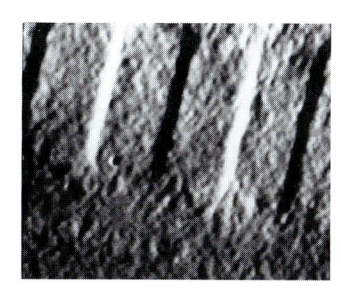

Bits on disk
Magnetic bits on a disk surface, caught by a magnetic force microscope. The dark stripes are 0 bits; the bright stripes are 1 bits.

Microcomputer hard drives with capacities measured in tens of gigabytes—up to 40 gigabytes, according to current ads—are becoming essential because today's programs are so huge. Microsoft Office alone is 500 megabytes. As for speed, hard disks allow faster access to data than floppy disks do, because a hard disk spins many times faster. Computer ads frequently specify speeds in revolutions per minute. A floppy disk drive rotates at only 360 rpm; a 7200-rpm hard drive is going about 300 miles per hour.

In addition, ads may specify the type of **_hard-disk controller_, a special-purpose circuit board that positions the disk and read/write heads and manages the flow of data and instructions to and from the disk.** Popular hard-disk controllers are *Ultra ATA* (or *EIDE*) and *SCSI*. Commonly found on new PCs, *Ultra ATA (advanced technology attachment)* allows fast data transfer and high storage capacity; it is also known as *EIDE (enhanced integrated drive electronics)*. Ultra ATA can support only one or two hard disks. By contrast, *SCSI (small computer system interface)*, pronounced "scuzzy," supports several disk drives as well as other peripheral devices by linking them in a daisy chain of up to seven devices. SCSI controllers are faster and have more storage capacity than EIDE controllers; they are typically found in servers and workstations.

- **Removable hard disks:** **_Removable hard disks_, or hard-disk cartridges, consist of one or two platters enclosed along with read/write heads in a hard plastic case, which is inserted into a microcomputer's cartridge drive.** Typical capacity is 2 gigabytes. Two popular systems are the Iomega's Jaz and SyQuest's SparQ. These cartridges are often used to transport huge files, such as desktop-publishing files with color and graphics and large spreadsheets. They are also frequently used to back up data.

Optical Disks: CDs & DVDs

Everyone who has ever played an audio CD is familiar with optical disks. **An _optical disk_ is a removable disk, usually 4.75 inches in diameter and less than one-twentieth of an inch thick, on which data is written and read through the use of laser beams.** An audio CD holds up to 74 minutes (2 billion bits' worth) of high-fidelity stereo sound. Some optical disks are used strictly for digital data storage, but many are used to distribute multimedia programs that combine text, visuals, and sound.

With an optical disk, there is no mechanical arm, as with floppy disks and hard disks. Instead, a high-power laser beam is used to write data by burning tiny pits or indentations into the surface of a hard plastic disk. To read the data, a low-power laser light scans the disk surface: Pitted areas are not reflected and are interpreted as 0 bits; smooth areas are reflected and are interpreted as 1 bits. *(See ● Panel 4.17, next page.)* Because the pits are so tiny, a great deal more data can be represented than is possible in the same amount of space on a diskette and many hard disks. An optical disk can hold over 4.7 gigabytes of data, the equivalent of 1 million typewritten pages.

Nearly every PC marketed today contains a CD-ROM or DVD-ROM drive, which can also read audio CDs. These, along with their recordable and rewritable variations, are the two principal types of optical-disk technology used with computers. *(See ● Panel 4.18, next page.)*

- **CD-ROM—for reading only:** For microcomputer users, the best-known type of optical disk is the CD-ROM. **_CD-ROM (compact disk read-only memory)_ is an optical-disk format that is used to hold prerecorded text, graphics, and sound.** Like music CDs, a CD-ROM is a read-only disk. *Read-only* means the disk's content is recorded at the time of manufacture and cannot be written on or erased by the

● **PANEL 4.17**
How a laser reads data on an optical disk

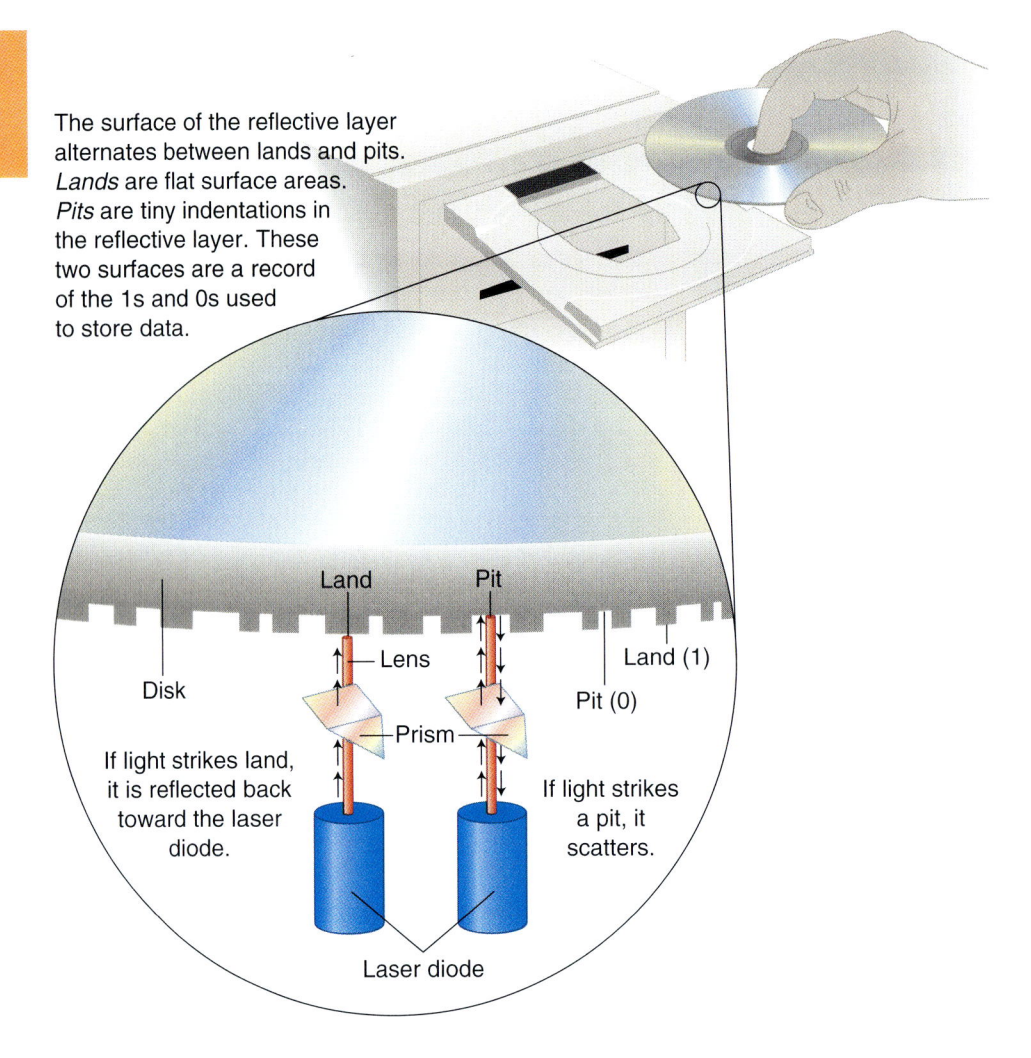

The surface of the reflective layer alternates between lands and pits. *Lands* are flat surface areas. *Pits* are tiny indentations in the reflective layer. These two surfaces are a record of the 1s and 0s used to store data.

user. As the user, you have access only to the data imprinted by the disk's manufacturer. A CD-ROM disk can hold up to 650 megabytes of data, equal to over 300,000 pages of text.

A CD-ROM drive's speed is important because with slower drives, images and sounds may appear choppy. In computer ads, drive speeds are indicated by the symbol "X," as in "44X," which is a high speed. "X" denotes the original data transfer rate of 150 kilobytes per second. (The data transfer rate is the time the drive takes to transmit data to another device. A 44X drive runs at 44 times 150, or 6600 kilobytes (6.6 megabytes) per second. If an ad carries the word "Max," as in "44X Max," this indicates the device's maximum speed. Drives range in speed from 40X to 75X; the faster ones are more expensive.

44X Max CD-ROM Drive

● **PANEL 4.18**
How to use a CD/DVD disk

- **CD-R—for recording on once:** <u>**CD-R (compact disk–recordable) disks**</u> **can be written to only once but can be read many times.** This allows consumers to make their own CD disks, though it's a slow process. (Recording a full disk takes 20–60 minutes.) Also, once recorded, the information cannot be erased. A CD-R can read and write audio CDs and standard CD-ROMs with read speeds of up to 24X and write speeds of up to 8X. CD-R is often used by companies for archiving—that is, to store vast amounts of information. A variant is the Photo CD, an optical disk developed by Kodak that can digitally store photographs taken with an ordinary 35-millimeter camera. Once you've shot a roll of color photographs, you take it for processing to a photo shop, which produces a disk containing your images. You can view the disk on any personal computer with a CD-ROM drive and the right software. Many new computers come equipped with CD-R drives.

- **CD-RW—for rewriting many times:** A <u>**CD-RW (compact disk–rewritable) disk**</u>, **also known as an erasable optical disk, allows users to record and erase data so that the disk can be used over and over again.** Special CD-RW drives and software are required. CD-RW disks are useful for archiving and backing up large amounts of data or work in multimedia production or desktop publishing. The read speed of CD-RW drives is up to 32X, the write speed is up to 8X, and the rewrite speed is up to 4X. CD-RW disks cannot be read by CD-ROM drives.

- **DVD-ROM—the versatile video disk:** A <u>**DVD-ROM (digital versatile disk or digital video disk, with read-only memory)**</u> **is a CD-style disk with extremely high capacity, able to store 4.7–17 gigabytes.** How is this done? Like a CD or CD-ROM, the surface of a DVD contains microscopic pits, which represent the 0s and 1s of digital code that can be read by a laser. The pits on the DVD, however, are much smaller and grouped more closely together than those on a CD, allowing far more information to be represented. Also, the laser beam used focuses on pits roughly half the size of those on current audio CDs. In addition, the DVD format allows for two layers of data-defining pits, not just one. Finally, engineers have succeeded in squeezing more data into fewer pits, principally through data compression.

    Many new computer systems now come with a DVD drive as standard equipment. A great advantage is that these drives can also take standard CD-ROM disks, so that now you can watch DVD movies and play CD-ROMs using just one drive. DVDs have enormous potential to replace CDs for archival storage, mass distribution of software, and entertainment. They not only store far more data but are different in quality from CDs. As one writer points out, "DVDs encompass much more: multiple dialogue tracks and screen formats, and best of all, smashing sound and video."[8] The theater-quality video and sound, of course, is what makes DVD a challenger to videotape as a vehicle for movie rentals. Since its 1997 introduction, DVD technology has gained steadily on videotape players.[9]

    As with CDs, DVDs have their recordable and rewritable variants. <u>**DVD-R (DVD–recordable) disks**</u> **allow one-time recording by the consumer.** Three types of reusable disks are *DVD-RW (DVD–rewritable)*, *DVD-RAM (DVD-random access memory)*, and *DVD+RW (DVD+rewritable)*, all of which can be recorded on and erased many times.

CD-R/RW Drive

Magnetic Tape

Similar to the tape used on an audio tape recorder (but of higher density), *magnetic tape* is thin plastic tape coated with a substance that can be magnetized. Data is represented by magnetized spots (representing 1s) or non-magnetized spots (representing 0s). Today, "mag tape" is used mainly for backup and archiving—that is, for maintaining historical records—where there is no need for quick access.

On large computers, tapes are used on magnetic-tape units or reels, and in cartridges. On microcomputers, tape is used in the form of *tape cartridges*, **modules resembling audio cassettes that contain tape in rectangular, plastic housings.** There are three common types of drive tapes—QIC, DAT, and (most expensive) DLT. *(See ● Panel 4.19.)*

Tapes fell out of favor for a while, supplanted by such products as Iomega's Jaz and Zip drives. However, as hard drives have swelled to multigigabyte size, using Zip disks for backup has become less convenient. Since a single-tape cassette can hold up to 66 gigabytes, tape is looking like a better alternative.

Smart Cards

Today in the United States, most credit cards are old-fashioned magnetic-strip cards. A magnetic-strip card has a strip of magnetically encoded data on its back. The encoded data might include your name, account number, and PIN (personal identification number). Two other kinds of cards, smart cards and optical cards, which hold far more information, are already popular in Europe. Manufacturers are betting they will soon be popular in the United States.

- **Smart cards:** A *smart card* **looks like a credit card but contains a microprocessor embedded in the card.** *Intelligent smart cards* have, in addition to the embedded processor, input, output, and storage capabilities. *Memory cards* have only storage capabilities in addition to the processor. When inserted into a card reader, it transfers data to and from a central computer, and it can store some basic financial records. Smart cards can be used as telephone debit cards. You insert the card into a slot in the phone, wait for a tone, and dial the number. The length of your call is automatically calculated on the chip inside the card, and the corresponding charge is deducted from the balance. Many colleges and universities issue student cards as smart cards.

● **PANEL 4.19**
Magnetic tape cartridge and three types of tapes
Three types of magnetic-tape cartridges and magnetic-tape cartridge drive

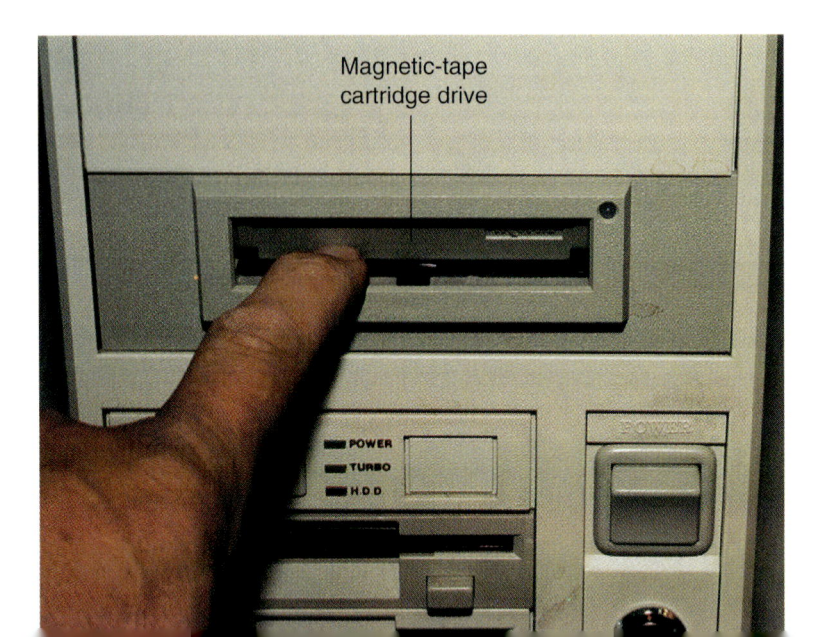

- **Optical cards:** The conventional magnetic-stripe credit card holds the equivalent of a half page of data. The smart card with a microprocessor and memory chip holds the equivalent of 250 pages. The optical card presently holds about 2000 pages of data. Optical cards use the same type of technology as music compact disks but look like silvery credit cards. **_Optical cards_ are plastic, laser-recordable, wallet-type cards used with an optical-card reader.** Because they can cram so much data (6.6 megabytes, as opposed to 16 kilobytes for smart cards) into so little space, they may become popular in the future. For instance, a health card based on an optical card would have room not only for the individual's medical history and health-insurance information but also for digital images, such as electrocardiograms.

Flash Memory Cards

Disk drives, whether for diskettes, hard disks, or CD-ROMs, all involve moving parts—and moving parts can break. Flash memory cards, by contrast, are variations on conventional computer-memory chips, which have no moving parts. **_Flash memory cards_, or flash RAM cards, consist of circuitry on credit-card-size PC cards that can be inserted into slots connecting to the motherboard.** The Memory Stick from Sony and the Secure Digital (SD) card from Panasonic presently hold up to 64 megabytes and are projected to expand to 1 gigabyte of capacity. What makes both the Memory Stick and SD particularly interesting is that they are interoperable between a wide range of computer and electronics devices.

Flash memory cards are not infallible. Their circuits wear out after repeated use, limiting their lifespan. Still, unlike conventional computer memory (RAM or primary storage), flash memory is *nonvolatile*. That is, it retains data even when the power is turned off.

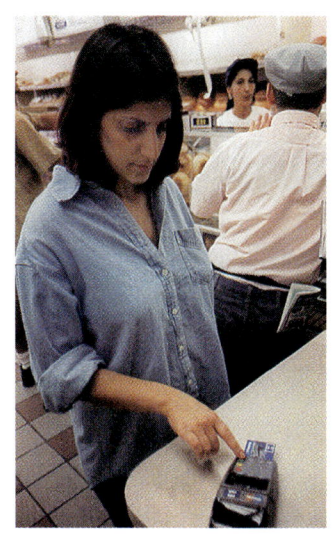

Smart card
Yifat Mabary uses a smart card to make a purchase at the Royale Kosher Bakery in New York City.

Flash memory card

> **CONCEPT CHECK**
>
> What are four types of floppy disks and what are their features?
>
> What are characteristics of hard disks, both nonremovable and removable?
>
> Explain the various types of optical disks—CD-ROM, CD-R, CD-RW, and DVD-ROM.
>
> How is magnetic tape used?
>
> Describe smart cards and optical cards.
>
> Describe flash memory cards.

Online Secondary Storage

If the network computer or thin-client computer actually becomes as popular as its promoters hope, the Internet itself will become, in effect, your hard disk. We described the concept of online storage at the start of this section. The services we mentioned included some application software, such as calendar, address book, and even word processors. Other services, however, simply offer online storage for backup purposes. Examples are @Backup *(www.atbackup.com)*, Connected Online Backup *(www.connected.com)*, Network Associates' Quick Backup to Personal Vault *(www.mcafee.com)*, and Safeguard Interactive *(www.sgii.com)*. Monthly prices are generally in the $10–$15 range. Online backup should be used only for vital files.

BOOKMARK IT!

PRACTICAL ACTION BOX
How to Buy a Notebook

"Selecting a notebook computer is much more complicated than buying a desktop PC," observes *Wall Street Journal* technology writer Walter Mossberg.[a] The reason: notebooks can vary a lot more than the desktop "generic boxes," which tend to be similar, at least within a price class.[b] Trying to choose among the many Windows-based notebooks is a particularly brow-wrinkling experience. (Macintosh notebooks tend to be more straightforward since there are only two models.)

Nevertheless, here are some suggestions:[c]

Purpose What are you going to use your notebook for? You can get a notebook that's essentially a desktop replacement and won't be moved much. If you expect to use the machine a lot in class, in libraries, or on airplanes, however, weight and battery life are important.

Budget & Weight Notebooks range from $1000 to $4000, with high-end brands aimed mainly at business people.

In the $2000 range are the light machines, 3–4 pounds. Designed for mobility, they tend to lack internal disk drives and all the standard ports. The heavy machines (7 pounds and up) in the $3000-plus range generally include all the features, including DVD drives and big screens.

Batteries: The Life–Weight Trade-off A rechargeable lithium-ion battery lasts longer than the nickel–metal hydride battery. Even so, a battery in the less expensive machines will usually run continuously for only about 2¼ hours. (DVD players are particularly voracious consumers of battery power, so it's the rare notebook that will allow you to finish watching a 2-hour movie.)

The trade-off is that heavier machines usually have longer battery life. The lightweight machines tend to get less than 2 hours, and toting extra batteries offsets the weight savings. (A battery can weigh a pound or so.)

Software Many notebooks come with less software than you would get with a typical desktop, though what you get will probably be adequate for most student purposes. On notebook PCs, count on getting Microsoft Works rather than the more powerful Microsoft Office to handle word processing, spreadsheets, databases, and the like.

Keyboards & Pointing Devices The keys on a notebook keyboard are usually the same size as those on a desktop machine, although they can be smaller. However, the up-and-down action feels different, and the keys may feel wobbly. In addition, some keys may be omitted altogether or keys may do double duty or appear in unaccustomed arrangements.

Most notebooks have a small touch-sensitive pad in lieu of a mouse—you drag your finger across the touchpad to move the cursor. Others use a pencil-eraser-size pointing stick in the middle of the keyboard.

Screens If you're not going to carry the notebook around much, go for a big, bright screen. Most people find they are comfortable with a 12- to 14-inch display, measured diagonally, though screens can be as small as 10.4 inches and as large as 15 inches.

The best screens are active-matrix display (TFT). However, some low-priced models have the cheaper passive-matrix screens (HPA, STN, or DSTN), which are harder to read, though you may find you can live with them. XGA screens (1024 × 768 pixels) have a higher resolution than SVGA screens (800 × 600 pixels), but fine detail may not be important to you.

Memory, Speed, & Storage Capacity If you're buying a notebook to complement your desktop, you may be able to get along with reduced memory, slow processor, small hard disk, and no CD-ROM. Otherwise, all these matters become important.

Memory (RAM) is the most important factor in computer performance, even though processor speed is more heavily hyped. Most notebooks have at least 64 megabytes (MB) of memory, but 128 MB is better. A microprocessor running 350–500 megahertz (MHz) or higher is adequate, and more recent models are faster than this.

Sometimes notebooks are referred to as "three-spindle" or "two-spindle" machines. In a three-spindle machine, a hard drive, a floppy-disk drive, and a CD/DVD drive all reside internally (not as external peripherals). A two-spindle machine has a hard drive and space for either a floppy-disk drive or a CD/DVD drive (or a second battery). A hard drive of 6 gigabytes or more is sufficient for most people.

For more information about buying computers, go to *www.zdnet.com/computershopper* and *http://micro.uoregon.edu/buyersguide*.

Summary

AGP (accelerated graphics port) bus (p. 159, KQ 4.2) Bus that transmits data at high speeds; designed to support video and three-dimensional (3-D) graphics. *Why it's important: An AGP bus is twice as fast as a PCI bus.*

arithmetic/logic unit (ALU) (p. 153, KQ 4.2) Part of the CPU that performs arithmetic operations and logical operations and controls the speed of those operations. *Why it's important: Arithmetic operations are the fundamental math operations: addition, subtraction, multiplication, and division. Logical operations are comparisons such as is "equal to," "greater than," or "less than."*

ASCII (American Standard Code for Information Interchange) (p. 147, KQ 4.2) Binary code used with microcomputers. It is also used on many minicomputers. *Why it's important: ASCII is the binary code most widely used in microcomputers.*

bay (p. 148, KQ 4.2) Shelf or opening in the computer case used for the installation of electronic equipment, generally storage devices such as a hard drive or DVD drive. *Why it's important: Bays permit the expansion of system capabilities. A computer may come equipped with four or seven bays.*

binary system (p. 146, KQ 4.2) A two-state system used for data representation in computers; has only two digits—0 and 1. *Why it's important: In the computer, 0 can be represented by electrical current being off and 1 by the current being on. All data and program instructions that go into the computer are represented in terms of these binary numbers.*

bit (p. 147, KQ 4.2) Short for "binary digit," which is either a 0 or a 1 in the binary system of data representation in computer systems. *Why it's important: The bit is the fundamental element of all data and information processed and stored in a computer system.*

bus (p. 154, KQ 4.2) Also called *bus line;* electrical data roadway through which bits are transmitted within the CPU and between the CPU and other components of the motherboard. *Why it's important: A bus resembles a multilane highway: The more lanes it has, the faster the bits can be transferred.*

byte (p. 147, KQ 4.2) Group of 8 bits. *Why it's important: A byte represents one character, digit, or other value. It is the basic unit used to measure the storage capacity of main memory and secondary storage devices (kilobytes and megabytes).*

cache (p. 156, KQ 4.2) Special high-speed memory area on a chip that the CPU can access quickly. It temporarily stores instructions and data that the processor is likely to use frequently. *Why it's important: Cache speeds up processing.*

CD-R (compact disk recordable) disks (p. 167, KQ 4.3) Optical-disk form of secondary storage that can be written to only once but can be read many times. *Why it's important: This format allows consumers to make their own CD disks, though it's a slow process. Once recorded, the information cannot be erased. CD-R is often used by companies for archiving—that is, to store vast amounts of information. A variant is the Photo CD, an optical disk developed by Kodak that can digitally store photographs taken with an ordinary 35-millimeter camera.*

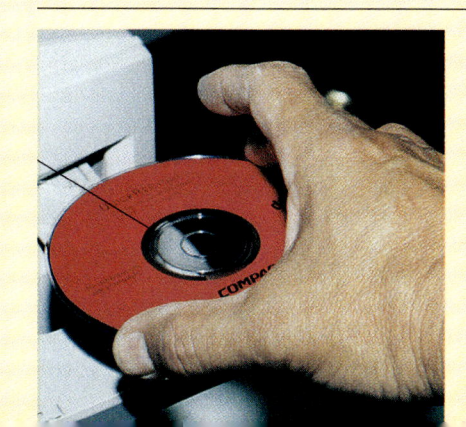

CD-ROM (compact disk read-only memory) (p. 165, KQ 4.3) Optical-disk form of secondary storage that is used to hold prerecorded text, graphics, and sound. *Why it's important: Like music CDs, a CD-ROM is a read-only disk. Read-only means the disk's content is recorded at the time of manufacture and cannot be written on or erased by the user. A CD-ROM disk can hold up to 650 megabytes of data, equal to over 300,000 pages of text.*

CD-RW (compact disk–rewritable) disk (p. 167, KQ 4.3) Also known as *erasable optical disk;* optical-disk form of secondary storage. Users can record and erase data, so that the disk can be used over and over again. Special CD-RW drives and software are required. Why it's important: *CD-RW disks are useful for archiving and backing up large amounts of data or work in multimedia production or desktop publishing; however, they are relatively slow.*

chip (p. 143, KQ 4.1) Also called a *microchip,* or *integrated circuit;* consists of millions of electronic circuits printed on a tiny piece of silicon. Silicon is an element widely found in sand that has desirable electrical (or "semiconducting") properties. Why it's important: *Chips have made possible the development of small computers.*

CISC (complex instruction set computing) chips (p. 151, KQ 4.2) Design that allows a microprocessor to support a large number of instructions. Why it's important: *CISC chips are used mostly in PCs and in conventional mainframes. CISC chips are generally slower than RISC chips.*

CMOS (complementary metal-oxide semiconductor) chips (p. 155, KQ 4.2) Battery-powered chips that don't lose their contents when the power is turned off. Why it's important: *CMOS chips contain flexible start-up instructions, such as time, date, and calendar, that must be kept current even when the computer is turned off. Unlike ROM chips, CMOS chips can be reprogrammed—for example, when you need to change the time for daylight savings time.*

control unit (p. 153, KQ 4.2) Part of the CPU that deciphers each instruction stored in it and then carries out the instruction. Why it's important: *The control unit directs the movement of electronic signals between main memory and the arithmetic/logic unit. It also directs these electronic signals between main memory and the input and output devices.*

CPU (central processing unit) (p. 153, KQ 4.2) The processor; it follows the instructions of the software (program) to manipulate data into information. The CPU consists of two parts—(1) the control unit and (2) the arithmetic/logic unit (ALU), which both contain registers, or high-speed storage areas. All are linked by a kind of electronic "roadway" called a bus. Why it's important: *The CPU is the "brain" of the computer.*

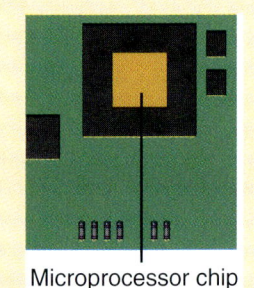

Microprocessor chip

DVD-R (DVD recordable) disks (p. 167, KQ 4.3) DVD disks that allow one-time recording by the consumer. Two types of reusable disks are DVD-RW (DVD rewritable) and DVD-RAM (DVD random access memory), both of which can be recorded on and erased more than once. Why it's important: *Recordable DVD disks offer the user yet another option for storing large amounts of data.*

DVD-ROM (digital versatile disk or digital video disk, with read-only memory) (p. 167, KQ 4.3) CD-type disk with extremely high capacity, able to store 4.7–17 gigabytes. Why it's important: *A powerful and versatile secondary storage medium.*

EBCDIC (Extended Binary Coded Decimal Interchange Code) (p. 147, KQ 4.2) Binary code used with some large computers. Why it's important: *EBCDIC is commonly used in IBM and Amdahl mainframes.*

expansion (p. 150, KQ 4.2) Way of increasing a computer's capabilities by adding hardware to perform tasks that are beyond the scope of the basic system. Why it's important: *Expansion allows users to customize and/or upgrade their computer systems.*

expansion card (p. 158, KQ 4.2) Also known as *expansion board, adapter card, interface card, plug-in board, controller card, add-in,* or *add-on;* circuit board that provides more memory or that controls peripheral devices. Why it's important: *Common expansion cards connect to the monitor (graphics card), speakers and microphones (sound card), and network (network card). Most computers have four to eight expansion slots, some of which may already contain expansion cards included in your initial PC purchase.*

expansion slot (p. 158, KQ 4.2) Socket on the motherboard into which you can plug an expansion card. Why it's important: *See expansion card.*

flash memory cards (p. 169, KQ 4.3) Also known as *flash RAM cards;* form of secondary storage consisting of circuitry on credit-card-size cards that can be inserted into slots connecting to the motherboard. *Why it's important: Flash memory is nonvolatile, so it retains data even when the power is turned off.*

flash memory chips (p. 155, KQ 4.2) Chips that can be erased and reprogrammed more than once (unlike PROM chips, which can be programmed only once). *Why it's important: Flash memory, which can range from 1 to 64 megabytes in capacity, is used to store programs not only in personal computers but also in pagers, cellphones, printers, and digital cameras. Unlike standard RAM chips, flash memory is nonvolatile—data is retained when the power is turned off.*

floppy disk (p. 161, KQ 4.3) Often called a *diskette* or simply a *disk;* removable flat piece of mylar plastic packaged in a 3.5-inch plastic case. Data and programs are stored on the disk's coating by means of magnetized spots, following standard on/off patterns of data representation (such as ASCII). The plastic case protects the mylar disk from being touched by human hands. *Why it's important: Floppy disks are used on all microcomputers.*

floppy-disk cartridges (p. 161, KQ 4.3) High-capacity removable 3.5-inch disks—Zip disks, SuperDisks, and HiFD disks. *Why it's important: These cartridges store more data than regular floppy disks and are just as portable.*

flops (p. 152, KQ 4.2) Stands for *floating-point operations per second.* A floating-point operation is a special kind of mathematical calculation. This measure, used mainly with supercomputers, is expressed as *megaflops* (mflops, or millions of floating-point operations per second), *gigaflops* (gflops, or billions), and *teraflops* (tflops, or trillions). *Why it's important: The measure is used to express the processing speed of supercomputers.*

gigabyte (G, GB) (p. 147, KQ 4.2) Approximately 1 billion bytes (1,073,741,824 bytes); a measure of storage capacity. *Why it's important: This measure was formerly used mainly with "big iron" (mainframe) computers but is typical of the secondary storage (hard disk) capacity of today's microcomputers.*

gigahertz (GHz) (p. 152, KQ 4.2) Measure of speed used for the latest generation of processors: a billion cycles per second. *Why it's important: Since a new high-speed processor can cost many hundred dollars more than the previous generation of chip, experts often recommend that buyers fret less about the speed of the processor (since the work most people do on their PCs doesn't even tax the limits of the current hardware) and more about spending money on extra memory.*

graphics card (p. 159, KQ 4.2) Also called a *video card* or *video adapter;* a graphics card converts signals from the computer into video signals that can be displayed as images on a monitor. *Why it's important: The power of a graphics card, often expressed in megabytes, as in 8, 16, or 32 MB, determines the clarity of the images on the monitor.*

hard disk (p. 163, KQ 4.3) Secondary storage medium; thin but rigid metal platter covered with a substance that allows data to be stored in the form of magnetized spots. Hard disks are tightly sealed within an enclosed hard-disk-drive unit to prevent any foreign matter from getting inside. Data may be recorded on both sides of the disk platters. *Why it's important: Hard disks hold much more data than do floppy disks. All microcomputers use hard disks as their principal storage medium.*

hard-disk controller (p. 165, KQ 4.3) Special-purpose circuit board that positions the disk and read/write heads and manages the flow of data and instructions to and from the disk. *Why it's important: Common PC hard-disk controllers are Ultra ATA (or EIDE) and SCSI.*

head crash (p. 163, KQ 4.3) Name for occurrence in which the surface of the read/write head or particles on its surface come into contact with the surface of the hard-disk platter, causing the loss of some or all of the data on the disk. *Why it's important: Because head crashes are always a possibility, users should always back up data from their hard disks on another storage medium, such as floppy disks.*

HiFD disk (p. 163, KQ 4.3) Disk with a capacity of 200 megabytes, made by Sony Corp. The disk drive can also read standard 1.44-megabyte floppies. Why it's important: *HiFD disks have 140 times the capacity of today's standard floppy disks.*

infrared port (p. 158, KQ 4.2) Port that allows a computer to make a cableless connection with infrared-capable devices, such as some printers. Why it's important: *Infrared ports eliminate the need for cabling.*

Intel-type chip (p. 151, KQ 4.2) Processor chip for PCs; made principally by Intel Corp. and Advanced Micro Devices (AMD), but also by Cyrix, DEC, and others. Why it's important: *These chips are used by manufacturers such as Compaq, Dell, Gateway 2000, Hewlett-Packard, and IBM. Since 1993, Intel has marketed its chips under the names Pentium, Pentium Pro, Pentium MMX, Pentium II, Pentium III, and Pentium 4, Celeron. Many ads for PCs contain the logo "Intel inside" to show that the systems run an Intel microprocessor.*

integrated circuit (p. 143, KQ 4.1). An entire electronic circuit, including wires, formed on a single "chip," or piece, of special material, usually silicon. Why it's important: *In the old days, transistors were made individually and then formed into an electronic circuit with the use of wires and solder. An integrated circuit is formed as part of a single manufacturing process.*

ISA (industry standard architecture) bus (p. 159, KQ 4.2) The most widely used expansion bus. Why it's important: *ISA is also the oldest and, at 8 or 16 bits, the slowest at transmitting data, though it is still used for mouses, modem cards, and low-speed network cards.*

kilobyte (K, KB) (p. 147, KQ 4.2) Approximately 1000 bytes (1024 bytes); a measure of storage capacity. Why it's important: *The kilobyte was a common unit of measure for memory or secondary-storage capacity on older computers.*

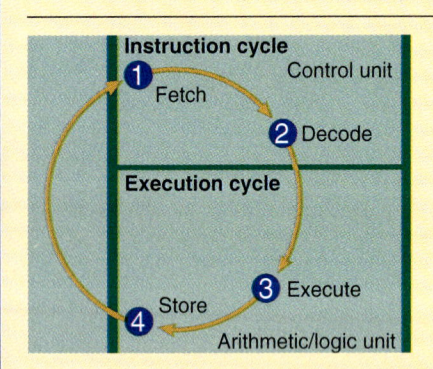

machine cycle (p. 153, KQ 4.2) Series of operations performed by the control unit to execute a single program instruction. It (1) fetches an instruction, (2) decodes the instruction, (3) executes the instruction, and (4) stores the result. Why it's important: *The machine cycle is the essence of computer-based processing.*

magnetic tape (p. 168, KQ 4.3) Thin plastic tape coated with a substance that can be magnetized. Data is represented by magnetized spots (representing 1s) or nonmagnetized spots (representing 0s). Why it's important: *Today, "mag tape" is used mainly for backup and archiving—that is, for maintaining historical records—where there is no need for quick access.*

megabyte (M, MB) (p. 147, KQ 4.2) Approximately 1 million bytes (1,048,576 bytes); measure of storage capacity. Why it's important: *Microcomputer primary-storage capacity is expressed in megabytes.*

megahertz (MHz) (p. 151, KQ 4.2) Measure of microcomputer processing speed, controlled by the system clock. Why it's important: *Generally, the higher the megahertz rate, the faster the computer can process data. A 550-MHz Pentium III–based microcomputer, for example, processes 550 million cycles per second, a 2-gigahertz chip processes 2 billion cycles per second.*

microprocessor (p. 145, KQ 4.1) Miniaturized circuitry of a computer processor. It stores program instructions that process, or manipulate, data into information. The key parts of the microprocessor are transistors. Why it's important: *Microprocessors enabled the development of microcomputers.*

MIPS (p. 152, KQ 4.2) Stands for *millions of instructions per second;* a measure of processing speed. Why it's important: *MIPS are used to measure processing speeds of mainframes, minicomputers, and workstations. A workstation might perform at 100 MIPS or more, a mainframe at 200–1200 MIPS.*

Motorola-type chips (p. 151, KQ 4.2) Microprocessors made by Motorola for Apple Macintosh computers. Why it's important: *Since 1993, Motorola has provided an alternative to the Intel-style chips made for PC microcomputers.*

network interface cards (p. 160, KQ 4.2) Expansion cards that allow the transmission of data over networks. Why it's important: *Installation of a network interface card in one's computer enables one to connect with various computers and other devices such as printers.*

nonremovable hard disk (p. 164, KQ 4.3) Also known as a *fixed disk;* hard disk housed in a microcomputer system unit and used to store nearly all programs and most data files. Usually it consists of four 3.5-inch metallic platters sealed inside a drive case the size of a small sandwich, which contains disk platters on a drive spindle, read/write heads mounted on an access arm that moves back and forth, and power connections and circuitry. Operation is much the same as for a diskette drive: The read/write heads locate specific instructions or data files according to track or sector. Hard disks can also come in removable cartridges. Why it's important: *See hard disk.*

optical card (p. 169, KQ 4.3) Plastic, laser-recordable, wallet-type cards used with an optical-card reader. Why it's important: *Because they can cram so much data (6.6 megabytes) into so little space, they may become popular in the future. For instance, a health card based on an optical card would have room not only for the individual's medical history and health-insurance information but also for digital images, such as electrocardiograms.*

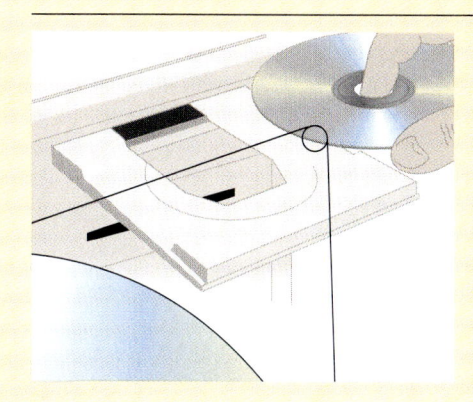

optical disk (p. 165, KQ 4.3) Removable disk, usually 4.75 inches in diameter and less than one-twentieth of an inch thick, on which data is written and read through the use of laser beams. Why it's important: *An audio CD holds up to 74 minutes (2 billion bits' worth) of high-fidelity stereo sound. Some optical disks are used strictly for digital data storage, but many are used to distribute multimedia programs that combine text, visuals, and sound.*

parallel port (p. 156, KQ 4.2) A line connected to a parallel port allows 8 bits (1 byte) to be transmitted simultaneously, like cars on an eight-lane highway. Why it's important: *Parallel lines move information faster than serial lines do. However, because they can transmit information efficiently only up to 15 feet, they are used principally for connecting printers or external disk or magnetic-tape backup storage devices.*

PC card (p. 160, KQ 4.2) Thin, credit-card-size (2.1 by 3.4 inches) hardware device. Why it's important: *PC cards are used principally on laptop computers to expand capabilities.*

PCI (peripheral component interconnect) bus (p. 159, KQ 4.2) High-speed bus; at 32 or 64 bits wide, it is more than four times faster than ISA buses. Why it's important: *PCI is widely used in microcomputers to connect graphics cards, sound cards, modems, and high-speed network cards.*

petabyte (P, PB) (p. 147, KQ 4.2) Approximately 1 quadrillion bytes (1,048,576 gigabytes); measure of storage capacity. Why it's important: *The huge storage capacities of modern databases are now expressed in petabytes.*

Plug and Play (p. 158, KQ 4.2) USB peripheral connection standard that allows peripheral devices and expansion cards to be automatically configured while they are being installed. Why it's important: *Plug and Play avoids the hassle of setting switching and creating special files that plagued earlier users.*

port (p. 156, KQ 4.2) A connecting socket or jack on the outside of the system unit into which are plugged different kinds of cables. Why it's important: *Allows you to plug in a cable to connect a peripheral device, such as a monitor, printer, or modem, so that it can communicate with the computer system.*

power supply (p. 150, KQ 4.2) Device that converts AC to DC to run the computer. Why it's important: *The electricity available from a standard wall outlet is alternating current (AC), but a microcomputer runs on direct current (DC).*

RAM (random access memory) chips (p. 154, KQ 4.2) Also called *primary storage* and *main memory;* these chips temporarily hold software instructions and data before and after it is processed by the CPU. RAM is a volatile form of storage. Why it's important: *RAM is the working memory of the computer. Having enough RAM is critical to your ability to run many software programs.*

read (p. 155, KQ 4.2) To transfer data from an input source into the computer's memory or CPU. *Why it's important:* Reading, along with writing, is an essential computer activity.

read/write head (p. 161, KQ 4.3) Mechanism used to transfer data between the computer and the disk. When the disk spins inside its case, the read/write head moves back and forth over the data access area on the disk. *Why it's important:* The read/write head enables the essential activities of reading and writing data.

registers (p. 154, KQ 4.2) High-speed storage areas that temporarily store data during processing. *Why it's important:* Registers may store a program instruction while it is being decoded, store data while it is being processed by the ALU, or store the results of a calculation.

removable hard disk (p. 165, KQ 4.3) Also called a *hard-disk cartridge*; one or two platters enclosed along with read/write heads in a hard plastic case, which is inserted into a microcomputer's cartridge drive. Typical capacity is 2 gigabytes. Two popular systems are Iomega's Jaz and SyQuest's SparQ. *Why it's important:* These cartridges offer users greater storage capacity than do floppy disks but with the same portability.

RISC (reduced instruction set computing) chips (p. 151, KQ 4.2) Type of chip in which the complexity of the microprocessor is reduced by eliminating many seldom-used instructions, thereby increasing the processing speed. *Why it's important:* RISC chips are used mostly in workstations. As a result, workstations can work up to 10 times faster then most PCs. RISC chips have been used in many Macintosh computers since 1993.

ROM (read-only memory) (p. 155, KQ 4.2) Memory chip that cannot be written on or erased by the computer user without special equipment. *Why it's important:* ROM chips contain fixed start-up instructions. They are loaded, at the factory, with programs containing special instructions for basic computer operations, such as starting the computer or putting characters on the screen. These chips are nonvolatile; their contents are not lost when power to the computer is turned off.

SCSI (small computer system interface) port (p. 157, KQ 4.2) Pronounced "scuzzy," a SCSI allows data to be transmitted in a "daisy chain" to up to 7 devices at speeds (32 bits at a time) higher than those possible with serial and parallel ports. The term *daisy chain* means that several devices are connected in series to each other, so that data for the seventh device, for example, has to go through the other six devices first. *Why it's important:* Enables users to connect external hard-disk drives, CD-ROM drives, scanners, and magnetic-tape backup units.

secondary storage hardware (p. 161, KQ 4.3) Devices that permanently hold data and information as well as programs. *Why it's important:* Secondary storage—as opposed to primary storage—is nonvolatile; that is, saved data and programs are permanent, or remain intact, when the power is turned off.

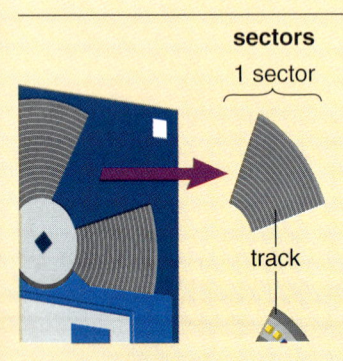

sectors (p. 161, KQ 4.3) Wedge-shaped sections on a formatted diskette used for storage reference purposes. *Why it's important:* When you save data from your computer to a diskette, the data is distributed by tracks and sectors on the disk. That is, the system software uses the point at which a sector intersects a track to reference the data location.

semiconductor (p. 143, KQ 4.1) Material, such as silicon (in combination with other elements), whose electrical properties are intermediate between a good conductor and a nonconductor of electricity. When highly conducting materials are laid on the semiconducting material, an electronic circuit can be created. *Why it's important:* Semiconductors are the materials from which integrated circuits (chips) are made.

serial port (p. 156, KQ 4.2) A line connected to a serial port will send bits one after another, like cars on a one-lane highway. The standard for PC serial ports is the 9-pin or 25-pin RS-232C connector. *Why it's important:* Because individual bits must follow each other, a serial port is usually used to connect devices that do not require fast transmission of data, such as keyboard, mouse, monitors, and modems. It is also useful for sending data over a long distance.

silicon (p. 143, KQ 4.1) An element that is widely found in clay and sand used in the making of solid-state integrated circuits. *Why it's important:* It is used not only because its abundance makes it cheap but also because it is a good semiconductor. As a result, highly conducting materials can be overlaid on the silicon to create the electronic circuitry of the integrated circuit.

smart card (p. 168, KQ 4.3) Looks like a credit card but contains a microprocessor embedded in the card. When inserted into a reader, it transfers data to and from a central computer. Why it's important: *Unlike conventional credit cards, smart cards can hold a fair amount of data and can store some basic financial records. Thus, they are used as telephone debit cards, health cards, and student cards.*

solid-state device (p. 143, KQ 4.1) Electronic component made of solid materials with no moving parts, such as an integrated circuit. Why it's important: *Solid-state integrated circuits are far more reliable, smaller, and less expensive than electronic circuits made from several components.*

sound card (p. 159, KQ 4.2) Used to transmit digital sounds through speakers, microphones, and headsets. Why it's important: *Cards such as PCI wavetable sound cards are used to add music and sound effects to computer video games.*

SuperDisk (p. 163, KQ 4.3) Disks with a capacity of 120 megabytes; produced by Imation. The SuperDisk drive can also read standard 1.44-megabyte floppy disks, which Zip drives cannot do. Why it's important: *See* floppy-disk cartridges.

system clock (p. 151, KQ 4.2) Internal timing device that uses fixed vibrations from a quartz crystal to deliver a steady stream of digital pulses or "ticks" to the CPU. These ticks are called *cycles*. Why it's important: *Faster clock speeds will result in faster processing of data and execution of program instructions, as long as the computer's internal circuits can handle the increased speed.*

tape cartridge (p. 168, KQ 4.3) Module resembling an audio cassette that contains tape in a rectangular plastic housing. There are three common types of drive tapes—QIC, DAT, and (most expensive) DLT. Why it's important: *Tape cartridges are used for secondary storage on microcomputers and also on some large computers. Tape is used mainly for archiving purposes and backup.*

terabyte (T, TB) (p. 147, KQ 4.2) Approximately 1 trillion bytes (1,009,511,627,776 bytes); measure of storage capacity. Why it's important: *The storage capacities of some mainframes and supercomputers are expressed in terabytes.*

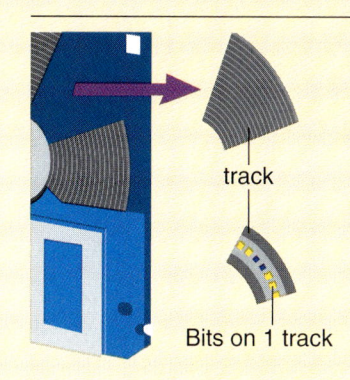

tracks (p. 161, KQ 4.3) The rings on a diskette along which data is recorded. Why it's important: *See* sectors.

transistor (p. 142, KQ 4.1) Tiny electronic device that acts as an on/off switch, switching between on and off millions of times per second. Why it's important: *Transistors are part of the microprocessor.*

Unicode (p. 148, KQ 4.2) Binary coding scheme that uses two bytes (16 bits) for each character, rather than one byte (8 bits). Why it's important: *Instead of the 256 character combinations of ASCII, Unicode can handle 65,536 character combinations. Thus, it allows almost all the written languages of the world to be represented using a single character set.*

upgrading (p. 150, KQ 4.2) Changing to newer, usually more powerful or sophisticated versions, such as a more powerful microprocessor or more memory chips. Why it's important: *Through upgrading, users can improve their computer systems without buying completely new ones.*

USB (universal serial bus) port (p. 158, KQ 4.2) Port that can theoretically connect up to 127 peripheral devices daisy-chained to one general-purpose port. Why it's important: *USB ports are useful for peripherals such as digital cameras, digital speakers, scanners, high-speed modems, and joysticks. The so-called USB hot plug or hot swappable allows such devices to be connected or disconnected even while the PC is running.*

virtual memory (p. 156, KQ 4.2) Type of hard disk space that mimics primary storage (RAM). Why it's important: *When RAM space is limited, virtual memory allows users to run more software at once, provided the computer's CPU and operating system are equipped to use it. The system allocates some free disk space as an extension of RAM; that is, the computer swaps parts of the software program between the hard disk and RAM as needed.*

volatile (p. 154, KQ 4.2) Temporary; the contents of volatile storage media, such as RAM, are lost when the power is turned off. Why it's important: *To avoid data loss, save your work to a secondary-storage medium, such as a hard disk, in case the electricity goes off while you're working.*

word size (p. 153, KQ 4.2) Number of bits that the processor may process at any one time. Why it's important: *The more bits in a word, the faster the computer. A 32-bit computer—that is, one with a 32-bit-word processor—will transfer data within each microprocessor chip in 32-bit chunks, or 4 bytes at a time. A 64-bit computer transfers data in 64-bit chunks, or 8 bytes at a time.*

write (p. 155, KQ 4.2) To transfer data from the computer's CPU or memory to an output device. Why it's important: *See* read.

write-protect notch (p. 161, KQ 4.3) Floppy disk feature that prevents a diskette from being written to. Why it's important: *This feature allows you to protect the data already on the disk. To write-protect, use your thumbnail or the tip of a pen to move the small sliding tab on the lower right side of the disk (viewed from the back), thereby uncovering the square hole.*

Zip disk (p. 162, KQ 4.3) Floppy-disk cartridge with a capacity of 100 or 250 megabytes. At 100 megabytes, this is 70 times the storage capacity of the standard floppy. Why it's important: *Among other uses, Zip disks are used to store large spreadsheet files, database files, image files, multimedia presentation files, and websites. Zip disks require their own Zip disk drives, which may come installed on new computers, although external Zip drives are also available. See also* floppy-disk cartridges.

Chapter Review

stage 1 LEARNING — MEMORIZATION

"I can recognize and recall information."

Self-Test Questions

1. A(n) _____ is about 1000 bytes; a(n) _____ is about 1 million bytes; a(n) _____ is about 1 billion bytes.

2. The _____ is the part of the microprocessor that tells the rest of the computer how to carry out a program's instructions.

3. The process of retrieving data from a storage device is referred to as _____; the process of copying data to a storage device is called _____.

4. To avoid losing data, users should always _____ their files.

5. Formatted diskettes have _____ and _____ that the system software uses to reference data locations.

6. The _____ is often referred to as the "brain" of a computer.

7. The electrical data roadways through which bits are transmitted are called _____.

8. A cable connected to a _____ port sends bits one at a time, one after the other; a cable connected to a _____ port sends 8 bits simultaneously.

9. Part of the disk drive mechanism, the _____, transfers data between the computer and the disk.

10. _____ is the most important factor in computer performance.

Multiple-Choice Questions

1. Which of the following is another term for *primary storage*?
    a. ROM
    b. ALU
    c. CPU
    d. RAM
    e. CD-R

2. Which of the following is not included on a computer's motherboard?
    a. RAM chips
    b. ROM chips
    c. keyboard
    d. microprocessor
    e. expansion slots

3. Which of the following is used to hold data and instructions that will be used shortly by the CPU?
    a. ROM chips
    b. peripheral devices
    c. RAM chips
    d. CD-R
    e. hard disk

4. Which of the following coding schemes is widely used on microcomputers?
    a. EBCDIC
    b. Unicode
    c. ASCII
    d. Microcode
    e. Unix

5. Which of the following is used to measure processing speed in microcomputers?
    a. MIPS
    b. flops
    c. picoseconds
    d. megahertz
    e. millihertz

True/False Questions

T F 1. A bus connects a computer's control unit and ALU.

T F 2. The machine cycle comprises the instruction cycle and the execution cycle.

T F 3. Magnetic tape is the most common secondary storage medium used with microcomputers.

T F 4. Main memory is nonvolatile.

stage 2 LEARNING: COMPREHENSION

"I can recall information in my own terms and explain them to a friend."

Short-Answer Questions

1. What is ASCII, and what do the letters stand for?
2. Why should measures of capacity matter to computer users?
3. What's the difference between RAM and ROM?
4. What is the significance of the term *megahertz*?
5. What is a motherboard? Name at least four components of a motherboard.
6. What advantage does a floppy-disk cartridge have over a regular floppy disk?
7. Why is it important for your computer to be expandable?
8. What would you use a Zip disk for?

Concept Mapping

On a separate sheet of paper, draw a concept map, or visual diagram, linking concepts. Show how the following terms are related.

ALU	floppy disk
ASCII	gigabyte
binary system	hard disk
bit	ISA
bus	machine language
byte	megabyte
cache	microprocessor
CD-ROM	optical disk
chip	RAM
control unit	read/write
CPU	ROM
expansion card	secondary storage
expansion slot	volatile

stage 3 LEARNING: APPLYING, ANALYZING, SYNTHESIZING, EVALUATING

"I can apply what I've learned, relate these ideas to other concepts, build on other knowledge, and use all these thinking skills to form a judgment."

Knowledge in Action

1. If you're using Windows 98, you can easily determine what microprocessor is in your computer and how much RAM it has. To begin, click the Start button in the Windows desktop pull-up menu bar and then choose Settings, Control Panel. Then locate the System icon in the Control Panel window and double-click on the icon.

   The System Properties dialog box will open. It contains four tabs: General, Device Manager, Hardware Profiles, and Performance. The name of your computer's microprocessor will display on the General tab. To see how much RAM is in your computer, click the Performance tab.

2. Visit a local computer store and note the system requirements listed on five software packages. What are the requirements for processor? RAM? operating system? available hard disk space? CD/DVD speed? audio/video cards? Are there any output hardware requirements?

3. Develop a binary system of your own. Use any two objects, states, or conditions, and encode the following statement: *I am a rocket scientist.*

Web Exercises

1. The objective of this project is to introduce you to an online encyclopedia that's dedicated to computer technology. The *www.webopaedia* website is a good resource for deciphering computer ads and clearing up difficult concepts. For practice, visit the site and type "main memory" into the Search text box and then press the enter key. Print out the page that displays. Then locate information on other topics of interest to you.

2. You can customize your own PC through a brand name company such as Dell or Gateway, or you can create your own personal model by choosing each component on your own. Decide which method is best for you.

   Go to the following sites and customize your ideal PC.

   www.dell.com

   www.gateway.com

   www.compaq.com

   www.ibm.com

   www.hpshopping.com

   Then go to

   www.pricewatch.com

   www.computerwarehouse.com

   and see if you could save money by putting your own PC together piece by piece. (This includes purchasing each component separately and verifying compatibility of all components.)

Chapter 5

Hardware: Input & Output

Taking Charge of Computing & Communications

Key Questions
You should be able to answer the following questions.

5.1 **Input & Output** How is input and output hardware used by a computer system?

5.2 **Input Hardware** What are the three categories of input hardware, what devices do they include, and what are their features?

5.3 **Output Hardware** What are the two categories of output hardware, what devices do they include, and what are their features?

The automated teller machine, long a standby for people needing fast cash evenings and weekends, is growing up.

And it is becoming something quite different from its origins.

Across the United States, a new generation of government-sponsored ATMs—better known as *kiosks*—allows people to learn about tourist attractions and bus routes and to get vehicle license plates, garage-sale permits, and information on property taxes. In New York City, kiosks can be used by citizens to pay parking tickets and check for building code violations on their apartment buildings. In San Antonio, they provide information on what animals at the pound are available for adoption. In Seattle, they let commuters at car-ferry terminals view live images of traffic conditions on major highways.

These super-ATMs also have commercial uses. Beyond doling out $20 bills, they also sell stamps, print out checks, and issue movie and plane tickets, for example. At Kmart, kiosks are available to enable customers to order merchandise on the Web that they can't find in the store.[1] Alamo car rental offices have kiosks on which travelers can print out directions, get descriptions of hotel services, and obtain restaurant menus and reviews—in four languages.[2] At Main Stay Suites, an extended-stay hotel chain, guests will not even find front desks. Kiosks have replaced them.[3] Many ATMs or kiosks have been transformed into full-blown multimedia centers, offering ads, coupons, and movie previews.[4]

The ATM or kiosk presents the two faces of the computer that are important to humans: It allows you to *input* data and to *output* information. In this chapter, we discuss what the principal input and output devices are and how you can make use of them.

Kiosk

Electronic ticketing machine at an airport

5.1 Input & Output

KEY QUESTION
How is input and output hardware used by a computer system?

Recall from Chapter 1 that *input* refers to data entered into a computer for processing—for example, from a keyboard or from a file stored on disk. Input includes program instructions that the CPU receives after commands are issued by the user. Commands can be issued by typing keywords, defined by the application program, or pressing certain keyboard keys. Commands can also be issued by choosing menu options or clicking on icons. Finally, input includes user responses—for example, when you reply to a question posed by the application or the operating system, such as *Are you sure you want to put this file in the Recycle Bin?* Output refers to the results of processing—that is, information sent to the screen or the printer or to be stored on disk, or sent to another computer in a network. Some devices combine both input and output functions, examples being not only ATMs but also combination scanner-printer devices.

In this chapter we focus on the common input and output devices used with a computer. (See ● Panel 5.1.) **Input hardware consists of devices that translate data into a form the computer can process.** The people-readable form of the data may be words like those on this page, but the computer-readable form consists of binary 0s and 1s, or off and on electrical signals. **Output hardware consists of devices that translate information processed by the computer into a form that humans can understand.** The computer-processed information consists of 0s and 1s, which need to be translated into words, numbers, sounds, and pictures.

● **PANEL 5.1**
Common input and output devices

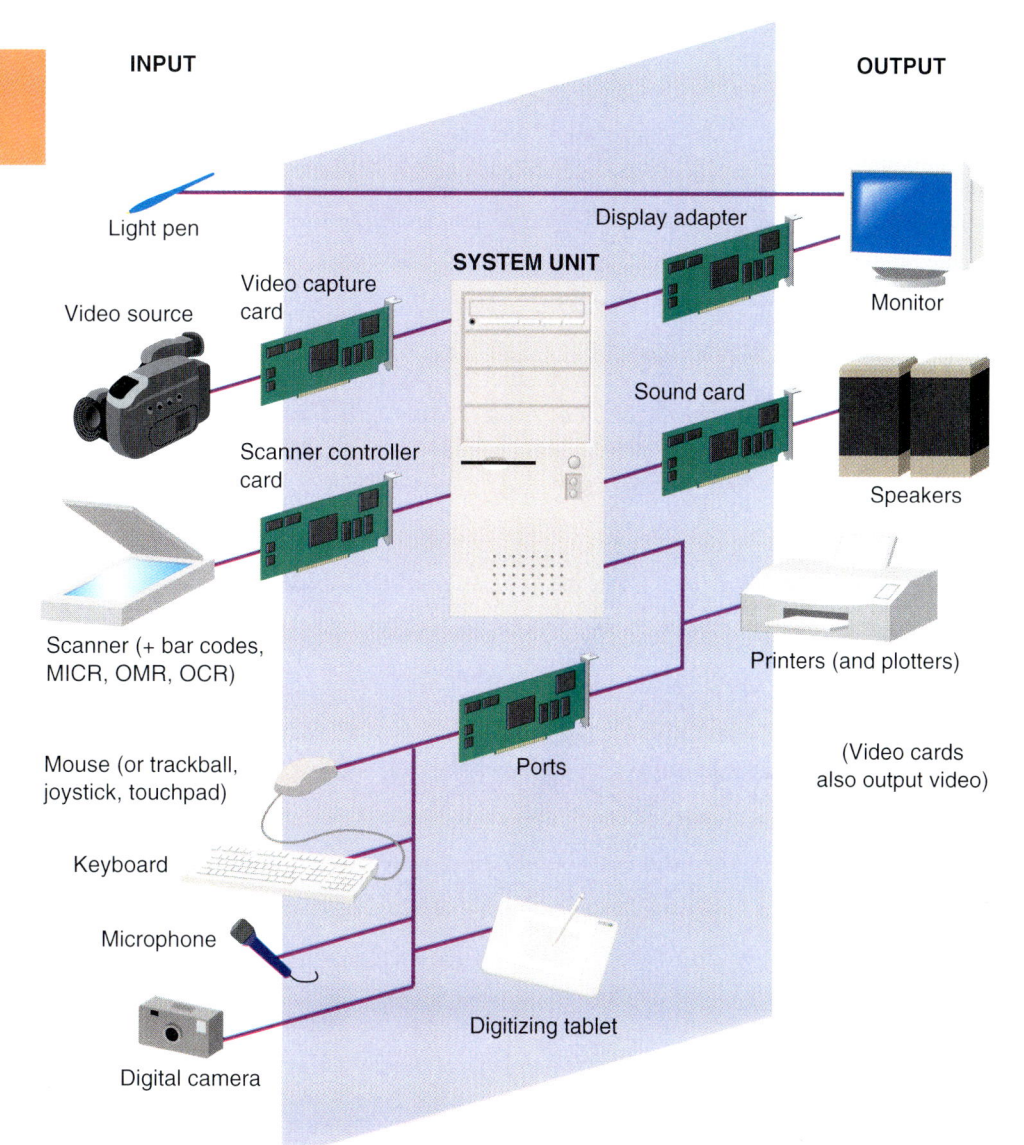

INPUT
- Light pen
- Video source
- Video capture card
- Scanner controller card
- Scanner (+ bar codes, MICR, OMR, OCR)
- Mouse (or trackball, joystick, touchpad)
- Keyboard
- Microphone
- Digital camera
- Digitizing tablet
- Ports

SYSTEM UNIT

OUTPUT
- Display adapter
- Monitor
- Sound card
- Speakers
- Printers (and plotters)
- (Video cards also output video)

Hardware: Input & Output

5.2 Input Hardware

KEY QUESTIONS
What are the three categories of input hardware, what devices do they include, and what are their features?

Input hardware devices are categorized as three types: *keyboards, pointing devices,* and *source data-entry devices.* (See ● Panel 5.2.) Quite often a computer system will combine all three.

Keyboards

A *keyboard* **is a device that converts letters, numbers, and other characters into electrical signals that can be read by the computer's processor.** The keyboard may look like a typewriter keyboard to which some special keys have been added. Alternatively, it may look like the keys on a bank ATM or the keypad of a pocket computer. It may even be a Touch-Tone phone or cable-TV set-top box.

Let's look at *traditional computer keyboards* and various kinds of *specialty keyboards and terminals.*

- **Traditional computer keyboards:** Picking up where we left off with the PC ad presented in Chapter 4 *(Panel 4.3 on page 146),* we see that the seller lists a "104-Key Keyboard." Conventional computer keyboards have all the keys on typewriter keyboards, plus other keys unique to computers. This totals 104–105 keys for desktop computers and 85 keys for laptops. Newer keyboards include extra keys for special activities such as instant Web access, CD/DVD controls, and Windows shortcut keys. The keyboard is built into laptop computers or attached to desktop computers with a cable. Wireless keyboards (which use an infrared signal) are also available.

    The keyboard illustration in Chapter 3 shows keyboard functions. *(Refer back to Panel 3.11, page 106.)*

- **Specialty keyboards and terminals:** Specialty keyboards range from Touch-Tone telephone keypads to keyboards featuring pictures of food for use in fast-food restaurants. Here we will consider *dumb terminals, intelligent terminals,* and *Internet terminals.*

    A *dumb terminal,* **also called a *video display terminal (VDT),* has a display screen and a keyboard and can input and output but not process data.** Usually the output is text only. For instance, airline reservations clerks use these terminals to access a mainframe computer containing flight

Survival Tip

How Do I Use the Print Scrn key?

This key is a holdover from DOS days, before Windows. Then it sent the contents of the screen to the printer. It still can, but in a roundabout way. When you press the key, an image of the screen is placed on the Windows clipboard. Open a blank document in Word or a graphics program and choose Edit, Paste. The image is inserted, and you can print it.

Dumb terminals used by airline reservations clerks

● **PANEL 5.2**
Three types of input devices

Keyboards	Pointing Devices	Source Data-Entry Devices
Traditional computer keyboards	Mice, trackballs, pointing sticks, touchpads	Scanning devices: imaging systems, bar-code readers, mark- and character-recognition devices (MICR, OMR, OCR), fax machines
Specialty keyboards and terminals: dumb terminals, intelligent terminals (ATMs, POS terminals), Internet terminals	Touch screens	Audio-input devices
	Pen-based computer systems, light pens, digitizers (digitizing tablets)	Video-input devices
		Digital cameras
		Voice-recognition systems
		Sensors
		Radio-frequency identification
		Human-biology input devices

information. Dumb terminals cannot perform functions independent of the mainframe to which they are linked.

An **_intelligent terminal_ has its own memory and processor, as well as a display screen and keyboard.** Such a terminal can perform some functions independent of any mainframe to which it is linked. One example is the familiar *automated teller machine (ATM),* a self-service banking machine that is connected through a telephone network to a central computer. Another example is the *point-of-sale (POS) terminal,* used to record purchases at a store's checkout counter.

An **_Internet terminal_ provides access to the Internet.** There are several variants: (1) the *set-top box* or *web terminal,* which displays web pages on a TV set; (2) the *network computer,* a cheap (less than $500), stripped-down computer that connects people to networks; (3) the *online game player,* which not only lets you play games but also connects to the Internet; (4) the full-blown *PC/TV* (or *TV/PC*), which merges the personal computer with the television set; and (5) the *wireless pocket PC* or *personal digital assistant (PDA),* a handheld computer with a tiny keyboard that can do two-way wireless messaging.

Pointing Devices

One of the most natural of all human gestures, the act of pointing, is incorporated in several kinds of input devices. **_Pointing devices_ control the position of the cursor or pointer on the screen.** Pointing devices include the *mouse* and its variants, the *touch screen,* and various forms of *pen input.*

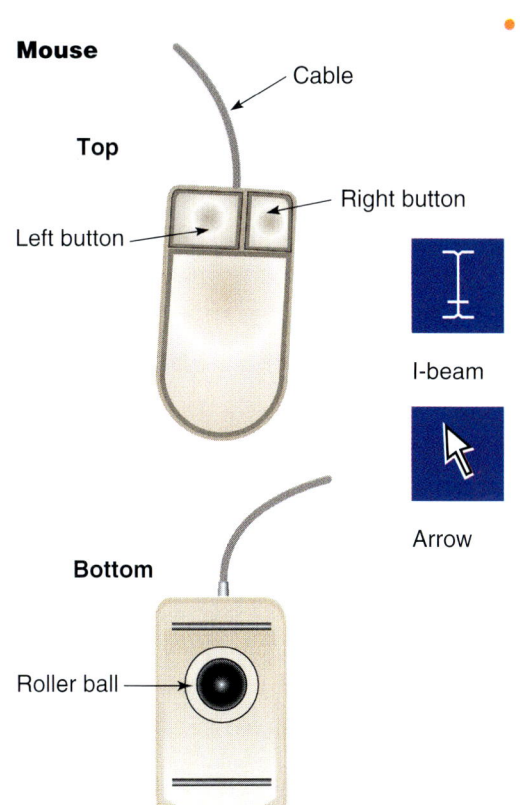

- **The mouse and its variants—trackball, pointing stick, and touchpad:** The principal pointing tool used with microcomputers is the *mouse,* a device that is rolled about on a desktop mouse pad and directs a pointer on the computer's display screen. (The mouse pad is a rectangular rubber/foam pad that provides traction for the mouse ball.) The *mouse pointer*—an arrow, a rectangle, a pointing finger—is the symbol that indicates the position of the mouse on the display screen or that activates icons. When the mouse pointer changes to the shape of an I-beam, it shows the place where text may be inserted or selected for special treatment.

  On the bottom side of the traditional mouse is a ball that translates the mouse movement into digital signals. On the top side are one to five buttons. The first button is used for common functions, such as clicking and dragging. The functions of the other buttons are determined by the software you're using. Microsoft's IntelliMouse is an optical mouse with five programmable buttons, a rotating center wheel, and no moving mechanical parts, which allows the mouse to work on almost any surface. This type of mouse uses light signals to detect mouse movements. Cordless mice are also available.

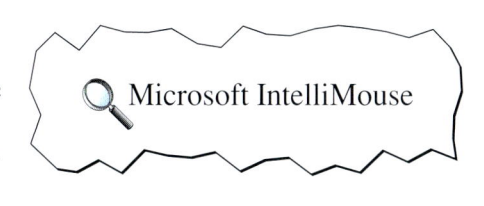

  There are three main variations on the mouse. The **_trackball_ is a movable ball, mounted on top of a stationary device, that can be rotated using your fingers or palm.** In fact, the trackball looks like the mouse turned upside down. Instead of moving the mouse around on the desktop, you move the trackball with the tips of your fingers. A trackball is not as accurate as a mouse, and it requires more frequent cleaning, but it's a good alternative when desktop space is limited.

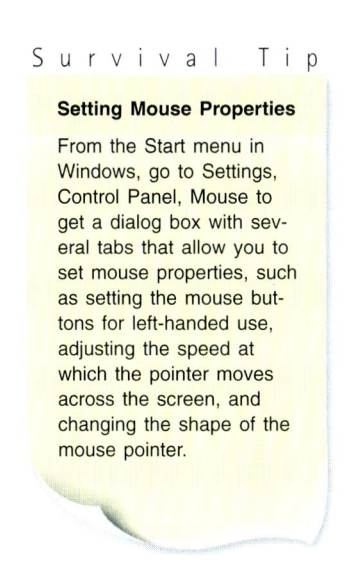

Survival Tip

Setting Mouse Properties

From the Start menu in Windows, go to Settings, Control Panel, Mouse to get a dialog box with several tabs that allow you to set mouse properties, such as setting the mouse buttons for left-handed use, adjusting the speed at which the pointer moves across the screen, and changing the shape of the mouse pointer.

Mouse

Top

Cable
Right button
Left button

Bottom

Roller ball

I-beam

Arrow

Microsoft IntelliMouse

Mouse variants: (top) trackball, (middle) pointing stick, and (bottom) touchpad

A *pointing stick* looks like a pencil eraser protruding from the keyboard between the G, H, and B keys. You move the pointing stick with your forefinger while using your thumb to press buttons located in front of the space bar. (A forerunner of the pointing stick is the *joystick*, which consists of a vertical handle like a gearshift lever mounted on a base with one or two buttons.) Pointing sticks are used on laptop computers.

A *touchpad* is a small, flat surface over which you slide your finger, using the same movements you would with a mouse. The cursor follows the movement of your finger. You "click" by tapping your finger on the pad's surface or by pressing buttons positioned close by the pad. Touchpads are also most often found on laptop computers.

- **Touch screen:** A *touch screen* is a video display screen that has been sensitized to receive input from the touch of a finger. (See ● Panel 5.3.) The screen is covered with a plastic layer, behind which are invisible beams of infrared light. You can input requests for information by pressing on buttons or menus displayed. The answers to your requests are displayed as output in words or pictures on the screen. (There may also be sound.) You find touch screens in kiosks, ATMs, airport tourist directories, hotel TV screens (for guest checkout), department store bridal registries, and campus information kiosks making available everything from lists of coming events to (with proper ID and personal code) student financial-aid records and grades.

- **Pen input:** Some input devices use variations on an electronic pen. Examples are *pen-based systems*, *light pens*, and *digitizers*.

    Pen-based computer systems allow users to enter handwriting and marks onto a computer screen by means of a penlike stylus rather than by typing on a keyboard. (See ● Panel 5.4.) Pen computers use handwriting recognition software that translates handwritten characters made by the pen, or stylus, into data that is usable by the computer. Many handheld computers and PDAs have pen input, as do digital notebooks.

    The *light pen* is a light-sensitive penlike device connected by a wire to the computer terminal. The user brings the pen to a desired point on the display screen and presses the pen button, which identi-

● **PANEL 5.3**
Touch screens
A hotel touch screen provides information about hotel services.

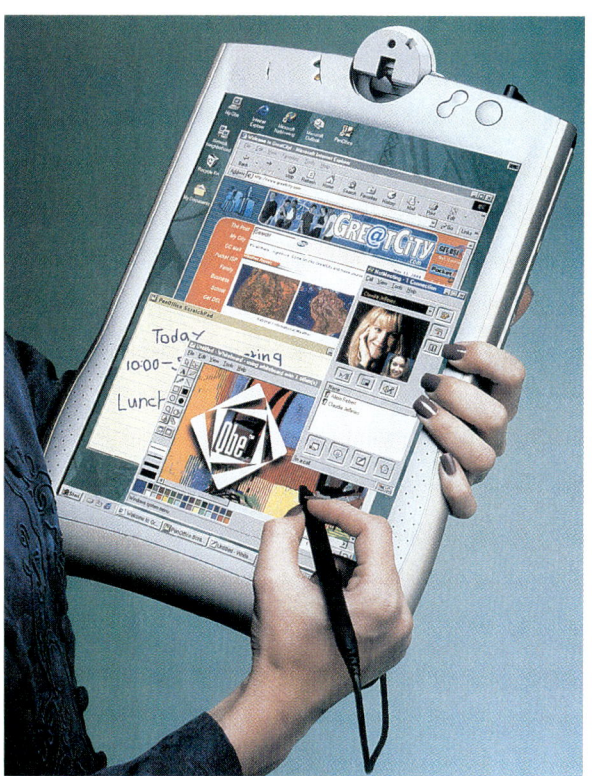

● **PANEL 5.4**
Pen-based computer systems
Handheld computer and digital notebook

fies that screen location to the computer. (See ● Panel 5.5.) Light pens are used by engineers, graphic designers, and illustrators. Some light pens require special monitors.

A *digitizer* **uses a mouselike copying device called a puck, or an electronic pen, which can convert drawings and photos to digital data.** One form of digitizer is the *digitizing tablet*, **an electronic plastic board on which each specific location corresponds to a location on the screen.** When you use the puck or the pen, the tablet converts your movements into digital signals that are input to the computer. Digitizing tablets are often used to make maps and engineering drawings. (See ● Panel 5.6.)

● **PANEL 5.5**
Light pen
This person is using a light pen to input to the computer.

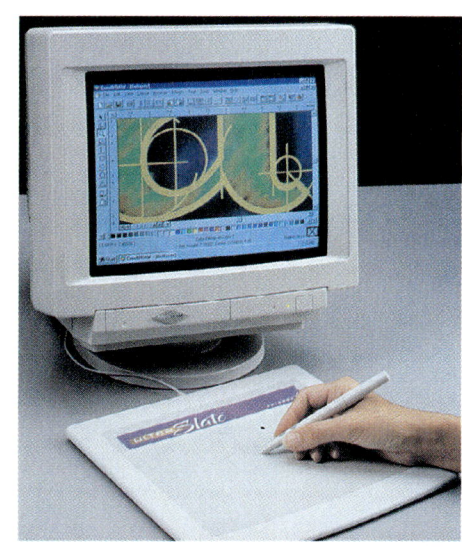

● **PANEL 5.6**
Digitizing tablet
Such tablets are often used in engineering and architectural applications.

Hardware: Input & Output

187

Source Data-Entry Devices

Source data-input devices do not require keystrokes (or require only a few keystrokes) to input data to the computer. Rather data is entered directly from the *source*, without human intervention. **Source data-entry devices create machine-readable data on magnetic media or paper or feed it directly into the computer's processor.** In this section, we cover the following:

1. Scanning devices—imaging systems, bar-code readers, mark- and character-recognition devices, and fax machines
2. Audio-input devices, Web cameras and video input, and photographic input (digital cameras)
3. Voice-recognition systems, sensors, radio-frequency identification devices, and human-biology input devices

- **Scanning devices—imaging systems:** <u>**Scanners**</u> **use light-sensing equipment to translate images of text, drawings, photos, and the like into digital form.** The images can then be processed by a computer, displayed on a monitor, stored on a storage device, or transmitted to another computer. Scanners are similar to photocopy machines except they create electronic files of scanned items instead of paper copies. One type of scanner is the <u>***imaging system***</u>—or image scanner, or graphics scanner—which converts text, drawings, and photographs into digital form that can be stored in a computer system and then manipulated, output, or sent via modem to another computer. (See ● Panel 5.7.) The system scans each image—color or black and white—with light and breaks the image into rows and columns of light and dark dots or color dots, which are then converted to digital code, called a *bitmap*. The more bits in each dot, the more shades of gray and the more colors that can be represented. A 1-bit-per-dot scanner is good enough for simple one-color images, but today's 30-bit to 40-bit scanners are required for the sophisticated scanning needed in, for example, desktop publishing (p. 118). *Resolution* also affects the quality of the scanned image. Resolution is measured in *dots per inch (dpi)*—the number of columns and rows of dots per inch. The higher the number of dots, the clearer and sharper the image. Popular color desktop scanners currently have 600 × 600 dpi, 600 × 1200 dpi, or 1200 ×1200 dpi. Commercial scanners range up to 8000 dpi. (Resolution is covered in more detail in the section on printers.)

● **PANEL 5.7**
Image scanner
A graphics designer scans an image into his desktop publishing system.

Imaging-system technology has led to a whole new art or industry called *electronic imaging*. **Electronic imaging is the software-controlled integration of separate images, using scanners, digital cameras, and advanced graphic computers.** This technology has become an important part of multimedia.

- Scanning devices—bar-code readers: Another scanning device reads **bar codes**, **the vertical zebra-striped marks you see on most manufactured retail products**—everything from candy to cosmetics to comic books. (See ● *Panel 5.8.*) In North America, supermarkets, food manufacturers, and others have agreed to use a bar-code system called the *Universal Product Code.* Other kinds of bar-code systems are used on everything from FedEx packages, to railroad cars, to the jerseys of long-distance runners. **Bar-code readers are photoelectric (optical) scanners that translate the symbols in the bar code into digital code.** In this system, the price of a particular item is set within the store's computer. Once the bar code has been scanned, the corresponding price appears on the salesclerk's point-of-sale terminal and on your receipt. Records of sales from the bar-code readers are input to the store's computer and used for accounting, restocking store inventory, and weeding out products that don't sell well.

- Scanning devices—mark-recognition and character-recognition devices: There are three types of scanning devices that sense marks or characters. They are usually referred to by their abbreviations *MICR*, *OMR*, and *OCR*.

    Magnetic-ink character recognition (MICR) reads the strange-looking numbers printed at the bottom of checks. MICR characters, which are printed with magnetized ink, are read by MICR equipment, producing a digitized signal. The bank's reader/sorter machine employs this signal to sort checks.

    Optical mark recognition (OMR) uses a device that reads pencil marks and converts them into computer-usable form. The best-known example is the OMR technology used to read the College Board Scholastic Aptitude Test (SAT) and the Graduate Record Examination (GRE).

● **PANEL 5.8**
Bar-codes and bar-code reader

```
OCR-A
NUMERIC    0123456789
ALPHA      ABCDEFGHIJ
SYMBOLS    KLMNOPQRST
           UVWXYZ
           >$/-+-#"

OCR-B
NUMERIC    00123456789
ALPHA      ACENPSTVX
SYMBOLS    <+>-¥
```

● **PANEL 5.9**
Optical character recognition
Special typefaces can be read by a scanning device called a wand reader.

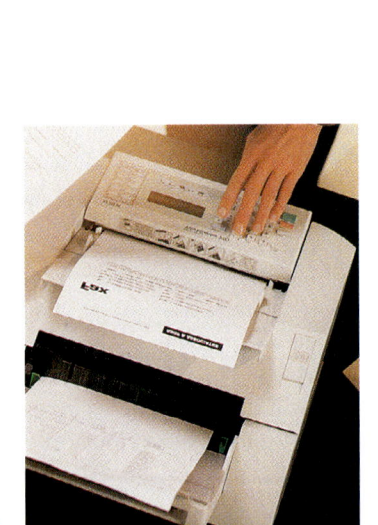

Dedicated fax machine

Fax modem

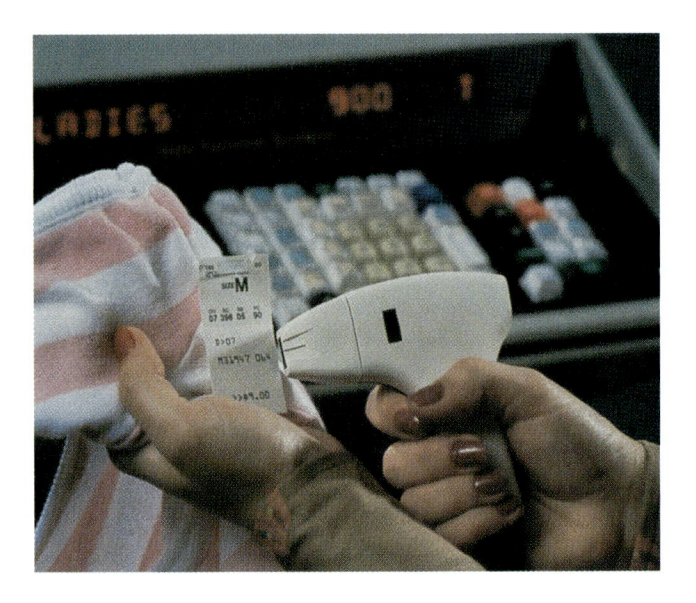

Optical character recognition (OCR) uses a device that reads preprinted characters in a particular font (typeface design) and converts them to digital code. OCR characters appear on utility bills and price tags on department-store merchandise; for example, the wand reader is a common OCR scanning device. *(See ● Panel 5.9.)*

- **Scanning devices—fax machines:** A *fax machine*—or facsimile transmission machine—scans an image and sends it as electronic signals over telephone lines to a receiving fax machine, which prints out the image on paper.

  There are two types of fax machines—*dedicated fax machines* and *fax modems*. **Dedicated fax machines** are specialized devices that do nothing except send and receive fax documents. These are what we usually think of as fax machines. They are found not only in offices and homes but also alongside regular phones in public places such as airports.

  A *fax modem* is installed as a circuit board inside the computer's system cabinet. It is a modem with fax capability that enables you to send signals directly from your computer to someone else's fax machine or computer fax modem. With this device, you don't have to print out the material from your printer and then turn around and run it through the scanner on a fax machine. The fax modem allows you to send information more quickly than if you had to feed it page by page into a machine.

  The fax modem is another feature of mobile computing; it's especially powerful as a receiving device. Fax modems are installed inside portable computers, including pocket PCs and PDAs. If you link up a cellular phone to a fax modem in your portable computer, you can send and receive wireless fax messages no matter where you are in the world.

- **Audio-input devices:** An *audio-input device* records analog sound and translates it for digital storage and processing. An analog sound signal is a continuously variable wave within a certain frequency range. For the computer to process them, these variable waves must be converted to digital 0s and 1s. The principal use of audio-input devices is to produce digital input for multimedia computers.

  An audio signal can be digitized in two ways—by an *audio board* or a *MIDI board*. Analog sound from a cassette player or a microphone goes through a special circuit board called an audio board (or card). An *audio board* is an add-on circuit board in a computer that converts analog sound to digital sound and stores it for further processing and/or plays it back, providing output directly to speakers or an external amplifier. A *MIDI board*—MIDI, pronounced "middie," stands for *Musical Instrument Digital Interface*—provides a standard for the interchange of musical information between musical instruments, synthesizers, and computers.

- **Webcams and video-input cards:** Are you the type who likes to show off for the camera? Maybe, then, you'd like to acquire a *Webcam*—a camera that attaches to a computer to record moving images that can then be posted on a website in real time. You could join the 10,000 other web camera users out there who are hosting such riveting material as a 24-hour view of the aquarium of a turtle named Pixel. Or, like some of those so-called reality-TV programs, you could show your living quarters or messy desk for all of computerland to see.

    As with sound, most film and videotape is in analog form; the signal is a continuously variable wave. For computer use, the signals that come from a VCR or a camcorder must be converted to digital form through a special digitizing card—*a video-capture card*—that is installed in the computer. Two types of video cards are frame-grabber video and full-motion video. *Frame-grabber video cards* can capture and digitize only a single frame at a time. Full-motion video cards can convert analog to digital signals at rates of up to 30 frames per second, giving the effect of a continuously flowing motion picture.

- **Digital cameras:** Digital cameras are particularly interesting because they foreshadow a major change for the entire industry of photography. Instead of using traditional (chemical) film, a **digital camera** uses **a light-sensitive processor chip to capture photographic images in digital form on a small diskette inserted in the camera or on flash-memory chips.** *(See ● Panel 5.10.)* The bits of digital information can then be copied right into a computer's hard disk for manipulation and printing out.

- **Voice-recognition systems:** Can your computer tell whether you want it to "recognize speech" or "wreck a nice beach"? A **voice-recognition system**, **using a microphone (or a telephone) as an input device, converts a person's speech into digital signals by comparing the electrical patterns produced by the speaker's voice with a set of prerecorded patterns stored in the computer.** *(See ● Panel 5.11.)* Programs let you accomplish two tasks: turn spoken dictation into typed text, and issue oral commands (such as "Print file") to control your computer.

Survival Tip

Digital Camera Resource

For more information on digital cameras, go to www.dcresource.com.

● **PANEL 5.10**
Digital camera

● **PANEL 5.11**
Voice input

Hardware: Input & Output

Voice-recognition systems have had to overcome many difficulties, such as different voices, pronunciations, and accents. Recently, however, the systems have measurably improved. Today's programs reach about 98% accuracy at conversational speeds. Two major voice-recognition systems are IBM's Via Voice and L&H's NaturallySpeaking.

Speech-recognition systems are finding many uses. Warehouse workers are able to speed inventory-taking by recording inventory counts verbally. Traders on stock exchanges can communicate their trades by speaking to computers. Radiologists can dictate their interpretations of X-rays directly into transcription machines. Nurses can fill out patient charts by talking to a computer. Drivers can talk to their car radios to change stations. Indeed, for many individuals with disabilities, a computer isn't so much a luxury or a productivity tool as a necessity. It provides freedom of expression, independence, and empowerment.

- **Sensors:** A _sensor_ **is an input device that collects specific data directly from the environment and transmits it to a computer.** Although you are unlikely to see such input devices connected to a PC in an office, they exist all around us, often in nearly invisible form. Sensors can be used to detect all kinds of things: speed, movement, weight, pressure, temperature, humidity, wind, current, fog, gas, smoke, light, shapes, images, and so on.

  Sensors are used to detect the speed and volume of traffic and adjust traffic lights. They are used on mountain highways in wintertime in the Sierra Nevada as weather-sensing devices to tell workers when to roll out snowplows. In California, sensors have been planted along major earthquake fault lines in an experiment to see whether scientists can predict major earth movements. *(See ● Panel 5.12.)* In aviation, sensors are used to detect ice buildup on airplane wings or to alert pilots to sudden changes in wind direction.

- **Radio-frequency identification devices:** **Also known as RF-ID tagging,** _radio-frequency identification technology_ **is based on an identifying tag bearing a microchip that contains specific code numbers. These code numbers are read by the radio waves of a scanner linked to a database.** Drivers with RF-ID tags can breeze through the tollbooths without having to even roll down their windows; the toll is automatically charged to their accounts. Radio-readable ID tags are also used by the Postal Service to monitor the flow of mail; by stores for inventory control and warehousing; and in the railroad industry to keep track of rail cars. They are even injected into dogs and cats, so that veterinarians with the right scanning equipment can identify them if they become separated from their owners.

- **Human biology-input devices:** Characteristics and movements of the human body, when interpreted by sensors, optical scanners, voice recognition, and other technologies, can become forms of input. Two examples are *biometric systems* and *line-of-sight systems.*

  Biometrics is the science of measuring individual body characteristics. Biometric security devices identify a person through a fingerprint, voice intonation, or other biological characteristic. For example, retinal-identification devices use a ray of light to identify the distinctive network of blood vessels at the back of the eyeball.

  Line-of-sight systems enable you to use your eyes to point at the screen. This technology allows some physically disabled users to direct a computer. For example, the Eyegaze System from LC Technologies allows you to operate a computer by focusing on particular areas of a display screen. A camera mounted on the computer analyzes the point of focus of the eye to determine where you are looking. You operate the computer by looking at icons on the screen and "press a key" by looking at one spot for a specified period of time.

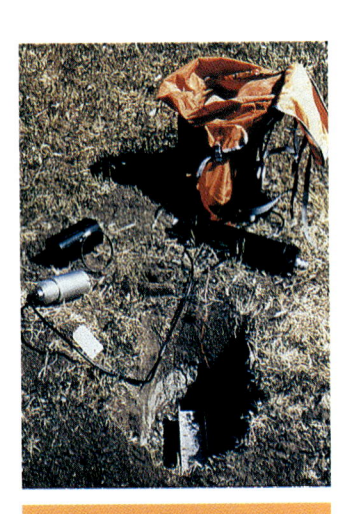

● **PANEL 5.12**
Earthquake sensor
Southern California's advanced seismic-monitoring system uses sensors, computers, and networks to capture data and analyze it. Within minutes of a quake, the system, known as TriNet, produces a map showing regions of intensity and areas that were probably hardest hit, enabling more efficient dispatch of rescue crews.

> **CONCEPT CHECK**
>
> Describe the various types of keyboards.
>
> Discuss the mouse and its variants.
>
> Identify the other types of pointing devices.
>
> Name and characterize the various types of scanning devices.
>
> Describe all the other types of source-data entry devices.

5.3 Output Hardware

KEY QUESTIONS
What are the two categories of output hardware, what devices do they include, and what are their features?

Are we back to old-time radio? Almost. Except that you can call up local programs by downloading them from the Internet. The sound quality isn't even as good as that of AM radio, but no doubt that will improve eventually. Computer output is taking more and more innovative forms and getting better and better.

As mentioned, *output hardware* consists of devices that convert machine-readable information, obtained as the result of processing, into people-readable form. The principal kinds of output are *softcopy* and *hardcopy*. (See ● Panel 5.13.)

- **Softcopy:** <u>Softcopy</u> **is data that is shown on a display screen or is in audio or voice form.** This kind of output is not tangible; it cannot be touched. (Actually, you almost never hear the word "softcopy" used in real life.)
- **Hardcopy:** <u>Hardcopy</u> **is printed output.** The principal examples are printouts, whether text or graphics, from printers. Film, including microfilm and microfiche, is also considered hardcopy output.

There are several types of output devices. In the following three sections, we discuss, first, *softcopy output—display screens;* second, *hardcopy output—printers;* and, third, *other output—sound, voice, animation, and video.*

Softcopy Output: Display Screens

<u>Display screens</u>—also variously called monitors, CRTs, or simply screens—are output devices that show programming instructions and data as they are being input and information after it is processed.

As for television screens, the size of a computer screen is measured diagonally from corner to corner in inches. For desktop microcomputers, the most common sizes are 13, 15, 17, 19, and 21 inches; for

🔍 17", .25dp Monitor (16" Display)

● **PANEL 5.13**
Types of output devices

Softcopy Devices	Hardcopy Devices	Other Devices
CRT display screens	Impact printers: dot-matrix printer	Sound output
Flat-panel display screen (e.g., liquid-crystal display)	Nonimpact printers: laser, ink-jet, thermal	Voice output
		Video output

Monitor screen size	Viewable image area
15 inches	14 inches
17 inches	16 inches
21 inches	20 inches

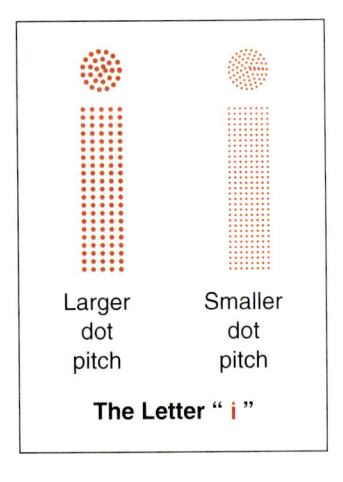

laptop computers, 12.1, 13.3, and 14.1 inches. Increasingly, computer ads state the actual display area, called the *viewable image size (vis)*, which may be an inch or so less. A 15-inch monitor may have a 13.8-inch vis; a 17-inch monitor may have a 16-inch vis.

In deciding which display screen to buy, you will need to consider issues of screen clarity (dot pitch, resolution, and refresh rate), types of monitor (CRT versus flat panel, active-matrix flat panel versus passive-matrix flat panel), and color and resolution standards (SVGA and XGA).

- **Screen clarity—dot pitch, resolution, and refresh rate:** Among the factors affecting screen clarity (often mentioned in ads) are *dot pitch*, *resolution*, and *refresh rate*. These relate to the individual dots known as pixels, which represent the images on the screen. A **pixel**, for "picture element," is the smallest unit on the screen that can be turned on and off or made different shades.

    Dot pitch (dp) is the amount of space between the centers of adjacent pixels; the closer the dots, the crisper the image. For a .28dp monitor, for instance, the dots are 28/100ths of a millimeter apart. Generally, a dot pitch of .28dp will provide clear images.

    Resolution is the image sharpness of a display screen; the more pixels there are per square inch, the finer the level of detail attained. Resolution is expressed in terms of the formula *horizontal pixels × vertical pixels*. Each pixel can be assigned a color or a particular shade of gray. Standard resolutions are 640×480, 800×600, 1024×768, 1280×1024, 1600×1200, and 1920×1440 pixels.

    Refresh rate is the number of times per second that the pixels are recharged so that their glow remains bright. In general, displays are refreshed 45–100 times per second. The higher the refresh rate, the more solid the image looks on the screen—that is, the less it flickers. Refresh rate is measured in *hertz*; a high-quality monitor has a refresh rate of 75 hertz—the screen is redrawn 75 times per second.

- **Two types of monitors—CRT and flat-panel:** Display screens are of two types: *CRT* and *flat-panel*. (See ● Panel 5.14, next page.)

    A **CRT, for cathode-ray tube, is a vacuum tube used as a display screen in a computer or video display terminal.** The same kind of technology is found not only in the screens of desktop computers but also in television sets and flight-information monitors in airports. *Note:* Advertisements for desktop computers often *do not* include a monitor as part of the price of the system. You need to be prepared to spend a few hundred dollars extra for the monitor.

    Compared to CRTs, flat-panel displays are much thinner, weigh less, and consume less power. Thus, they are better for portable computers, although they are available for desktop computers as well. **Flat-panel displays are made up of two plates of glass separated by a layer of a substance in which light is manipulated.** One technology used is **liquid crystal display (LCD), in which molecules of liquid crystal line up in a way that alters their optical properties, creating images on the screen by transmitting or blocking out light.**

    Flat-panel monitors are available for desktop computers. Because they are smaller than CRTs, they fit more easily onto a crowded desk. However, CRTs are considerably cheaper. A new 15-inch CRT costs as little as $100. A flat-panel costs 5–10 times as much.

- **Active-matrix versus passive-matrix flat-panel displays:** Flat-panel screens are either active-matrix or passive-matrix displays, according to where their transistors are located.

● **PANEL 5.14**
CRT (left) versus flat-panel displays

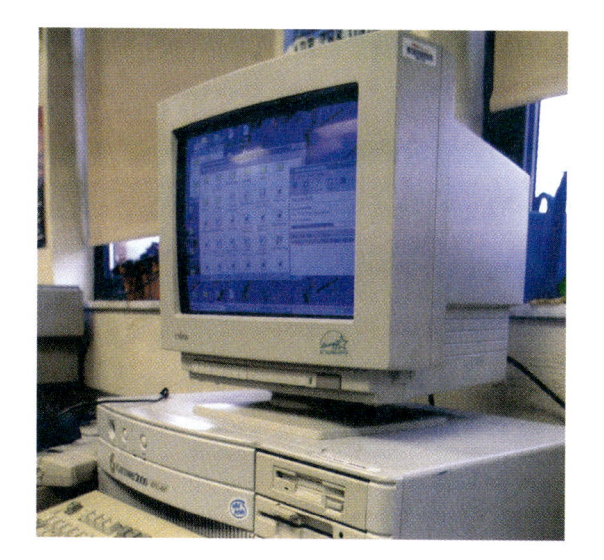

In an *active-matrix display*, also known as TFT (thin-film transistor) display, **each pixel on the screen is controlled by its own transistor.** Active-matrix screens are much brighter and sharper than passive-matrix screens, but they are more complicated and thus more expensive. They also require more power, affecting the battery life in laptop computers.

In a *passive-matrix display*, **a transistor controls a whole row or column of pixels.** Passive matrix provides a sharp image for one-color (monochrome) screens but is more subdued for color. The advantage is that passive-matrix displays are less expensive and use less power than active-matrix displays, but they aren't as clear and bright and can leave "ghosts" when the display changes quickly. Passive-matrix displays go by the abbreviations HPA, STN, or DSTN.

• **Color and resolution standards for monitors—SVGA and XGA:** As we mentioned earlier in the chapter, PCs come with *graphics cards* (also known as *video cards* or *video adapters*) that convert signals from the computer into video signals that can be displayed as images on a monitor. The monitor then separates the video signal into three colors: red, green, and blue signals. Inside the monitor, these three colors combine to make up each individual pixel. Video cards have their own memory, *video RAM*, or *VRAM*, which stores the information about each pixel.

The common color and resolution standards for monitors are *SVGA*, *XGA*, *SXGA*, and *UXGA*. (See ● *Panel 5.15, next page.*)

SVGA (super video graphics array) **supports a resolution of 800 × 600 pixels, or variations, producing 16 million possible simultaneous colors.** SVGA is the most common standard used today with 15-inch monitors.

XGA (extended graphics array) **has a resolution of up to 1024 × 768 pixels, with 65,536 possible colors.** It is used mainly with 17- and 19-inch monitors.

SXGA (super extended graphics array) **has a resolution of 1280 × 1024 pixels.** It is often used with 19- and 21-inch monitors.

UXGA (ultra extended graphics array) **has a resolution of 1600 × 1200 pixels.** It is expected to become more popular with graphic artists, engineering designers, and others using 21-inch monitors.

CLICK-ALONG 5-1
More on monitors

PANEL 5.15
Videographics standards compared for pixels

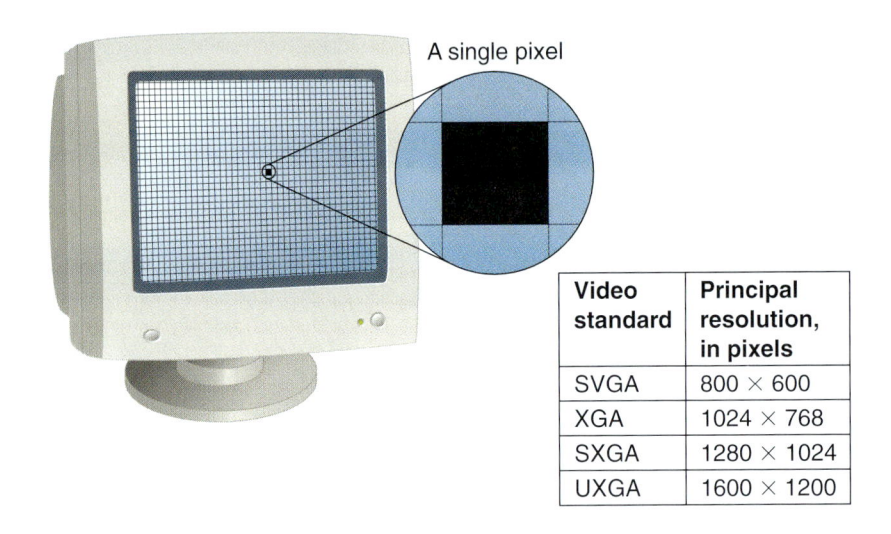

A single pixel

Video standard	Principal resolution, in pixels
SVGA	800 × 600
XGA	1024 × 768
SXGA	1280 × 1024
UXGA	1600 × 1200

Hardcopy Output: Printers

The prices for computer systems in ads often do not include a printer. Thus, you will need to budget an additional $150 to $1000 or more for a printer. A *printer* **is an output device that prints characters, symbols, and perhaps graphics on paper or another hardcopy medium.** The resolution, or quality of sharpness of the image, is indicated by *dpi (dots per inch)*, **which is a measure of the number of dots that are printed in a linear inch.** For microcomputer printers, the resolution is in the range 60–1500 dpi.

HP DeskJet 970Cse Printer

Printers can be separated into two categories, according to whether or not the image produced is formed by physical contact of the print mechanism with the paper. *Impact printers* do have contact with paper; *nonimpact printers* do not. We will also consider plotters and multifunction printers.

- **Impact printers:** An *impact printer* **forms characters or images by striking a mechanism such as a print hammer or wheel against an inked ribbon, leaving an image on paper.** A *dot-matrix printer* **contains a print head of small pins, which strike an inked ribbon against paper, to form characters or images.** Print heads are available with 9, 18, or 24 pins; the 24-pin head offers the best quality. Dot-matrix printers can print *draft quality,* a coarser-looking 72 dpi, or *near-letter-quality (NLQ),* a crisper-looking 144 dpi. The machines print 40–300 characters per second and can handle graphics as well as text. A disadvantage is the noise they produce. Nowadays impact printers are more commonly used with mainframes than with personal computers. Note that dot-matrix printers are the only desktop printers that can use multilayered forms to print "carbon copies."

- **Nonimpact printers:** Nonimpact printers are faster and quieter than impact printers because they have fewer moving parts. *Nonimpact printers* **form characters and images without direct physical contact between the printing mechanism and paper.** Two types of nonimpact printers often used with microcomputers are *laser printers* and *ink-jet printers.* A third kind, the *thermal printer,* is seen less frequently.

Like a dot-matrix printer, a *laser printer* **creates images with dots. However, as in a photocopying machine, these images are produced on a drum, treated with a magnetically charged ink-like toner (powder), and then transferred from drum to paper.** (See ● Panel 5.16.) (Laser printers are also called *page printers,* because they print one page at a time.)

Laser printers run with software called a *page description language (PDL).* This software tells the printer how to lay out the printed page,

Survival Tip

Do I Have to Print the Whole Thing?

If you click on the shortcut icon on the toolbar (the one that looks like a printer), your entire document will be printed. To print only a portion of your document, go to File, Print and check the option you want in the dialog box that appears.

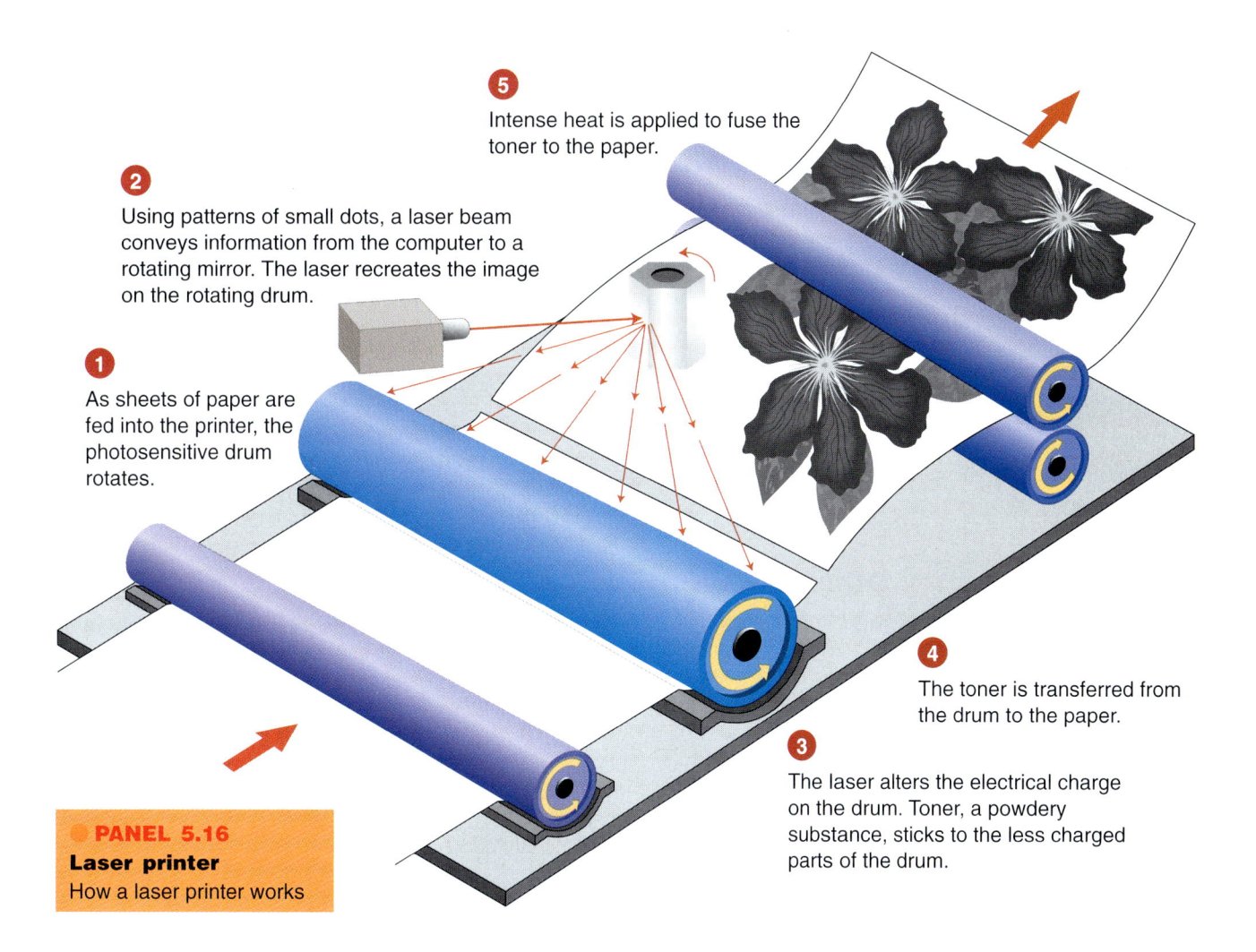

PANEL 5.16
Laser printer
How a laser printer works

1 As sheets of paper are fed into the printer, the photosensitive drum rotates.

2 Using patterns of small dots, a laser beam conveys information from the computer to a rotating mirror. The laser recreates the image on the rotating drum.

3 The laser alters the electrical charge on the drum. Toner, a powdery substance, sticks to the less charged parts of the drum.

4 The toner is transferred from the drum to the paper.

5 Intense heat is applied to fuse the toner to the paper.

and it supports various fonts. A laser printer comes with one of two types of PDL: PostScript (developed by Adobe) or PCL (Printer Control Language, developed by Hewlett-Packard).

There are good reasons that laser printers are among the most common types of nonimpact printer. They produce sharp, crisp images of both text and graphics. They are quiet and fast—able to print 4–32 text-only pages per minute for individual microcomputers and up to 200 pages per minute for mainframes. They can print in different *fonts*—that is, type styles and sizes. The more expensive models can print in different colors. Laser printers have a dpi of 600 to 1200.

Ink-jet printers spray small, electrically charged droplets of ink from four nozzles through holes in a matrix at high speed onto paper. Like laser and dot-matrix printers, ink-jet printers form images with little dots. Ink-jet printers have a dpi of 300 to 1400.

The advantages of ink-jet printers are that they can print in color, are quieter, and are much less expensive than color laser printers. The disadvantages are that they print in a somewhat lower resolution than laser printers and they are slower. Printing a document with high-resolution color graphics may take 10 minutes or more for a single page. Ink-jets, which spray ink onto the page a line at a time, can produce both high-quality black-and-white text and high-quality color graphics. However, if you print a lot of color, you'll find ink-jets much slower and more expensive to operate than laser printers. Moreover, a freshly printed page is apt to smear unless handled carefully. Still, a color ink-jet printer's cost is considerably less than the cost of color laser printers. The rock-bottom price is only about $150.

Hardware: Input & Output

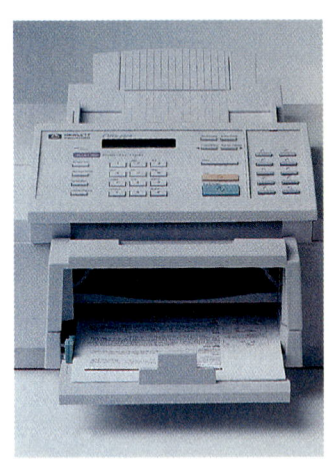

Multifunction device

Thermal printers use colored waxes and heat to produce images by burning dots onto special paper. The colored wax sheets are not required for black-and-white output. However, thermal printers are expensive, and they require expensive paper. For people who want the highest-quality color printing available with a desktop printer, thermal printers are the answer.

- Multifunction printers—printers that do more than print: **Multifunction printers combine several capabilities, such as printing, scanning, copying, and faxing.** Xerox and Hewlett-Packard make machines that combine a photocopier, fax machine, scanner, and laser printer. Multifunction printers take up less space and cost less than the four separate office machines that they replace, but if one component breaks, nothing works.

Specialty printers also exist for such purposes as printing photos and labels, and printing text in Braille. The box opposite gives some questions to consider when you're buying a printer. (See ● Panel 5.17.)

Other Output: Sound, Voice, Animation, & Video

Most PCs are now multimedia computers, capable of outputting not only text and graphics but also sound, voice, and video, as we consider next.

- Sound output: **Sound-output devices produce digitized sounds, ranging from beeps and chirps to music.** To use sound output, you need appropriate software and a sound card. The sound card could be Sound Blaster or, since that brand has become a de facto standard, one that is "Sound Blaster compatible." Well-known brands include Creative Labs, Diamond, and Turtle Beach. The sound card plugs into an expansion slot in your computer; on newer computers, it is integrated with the motherboard. Most computers have simple internal speakers. Many users hook up external speakers for high-quality sound.

- Voice output: **Voice-output devices convert digital data into speech-like sounds.** You hear such forms of voice output on telephones ("Please hang up and dial your call again"), in soft-drink machines, in cars, in toys and games, and recently in mapping software for vehicle-navigation devices. Voice portals read news and other information to users on the go.

- Some uses of speech output are simply frivolous or amusing. You can replace your computer start-up beep with the sound of popular music. But some uses are quite serious. For people with physical challenges, computers with voice output help to level the playing field. Voice output will probably become more important in the future, and it will probably get better.

- Video output: **Video consists of photographic images, which are played at 15–29 frames per second to give the appearance of full motion.** Video is input into a multimedia system using a video camera or VCR and, after editing, is output on a computer's display screen. Because video files can require a great deal of storage—a 3-minute video may require 1 gigabyte of storage—video is often compressed (a topic we later). Good video output requires a powerful processor as well as a video card.

Another form of video output is *videoconferencing*, **in which people in different geographical locations can have a meeting—can see and hear one another—using computers and communications.** Videoconferencing systems range from videophones to group conference rooms with cameras and multimedia equipment to desktop systems with small video cameras, microphones, and speakers.)

● **PANEL 5.17**
Buying a printer

Do I need color, or will black-only do? Are you mainly printing text or will you need to produce color charts and illustrations (and, if so, how often)? If you print lots of black text, consider getting a laser printer. If you might occasionally print color, get an ink-jet that will accept cartridges for both black and color. Unless you are in the publishing or design business, you will probably not need an expensive color laser printer.

Do I have other special output requirements? Do you need to print envelopes or labels? special fonts (type styles)? multiple copies? transparencies or on heavy stock? unusual paper size? Find out if the printer comes with envelope feeders, sheet feeders holding at least 100 sheets, or whatever will meet your requirements.

Is the printer easy to set up? Can you easily put the unit together, plug in the hardware, and adjust the software (the "driver" programs) to make the printer work with your computer?

Is the printer easy to operate? Can you add paper, replace ink/toner cartridges or ribbons, and otherwise operate the printer without much difficulty?

Does the printer provide the speed and quality I want? A regular laser printer prints 4–30 pages per minute (ppm); a color ink-jet printer prints 1–12 ppm. Colors and graphics take longer to print. Are the blacks dark enough and the colors vivid enough?

Will I get a reasonable cost per page? Special paper, ink or toner cartridges (especially color), and ribbons are all ongoing costs. Ink-jet color cartridges, for example, may last 100–500 pages and cost $25–$30 new. Laser toner cartridges last longer but are more expensive. Ribbons for dot-matrix printers are inexpensive. Ask the seller what the cost per page works out to.

Does the manufacturer offer a good warranty and good telephone technical support? Find out if the warranty lasts at least 2 years. See if the printer's manufacturer offers telephone support in case you have technical problems. The best support systems offer toll-free numbers and operate evenings and weekends as well as weekdays.

CLICK-ALONG 5-2
More on printers: photo, label, postage, portable, and 3D printers

CONCEPT CHECK

What is the difference between softcopy and hardcopy output?

What are the different characteristics of display screens?

What is the difference between impact and nonimpact printers?

Identify the characteristics of dot-matrix, laser, ink-jet, thermal, and multifunction printers.

Describe sound, voice, and video output.

BOOKMARK IT!

PRACTICAL ACTION BOX
Good Habits: Protecting Your Computer System, Your Data, & Your Health

Whether you set up a desktop computer and never move it or tote a portable PC from place to place, you need to be concerned about protecting not only your computer but yourself. You don't want your computer to get stolen or zapped by a power surge. You don't want to lose your data. And you certainly don't want to lose your health for computer-related reasons. Here are some tips for taking care of these vital areas.

Guarding Against Hardware Theft & Loss

Portable computers are easy targets for thieves. Obviously, anything conveniently small enough to be slipped into your briefcase or backpack can be slipped into someone else's. Never leave a portable computer unattended in a public place.

It's also possible to simply lose a portable, as in forgetting it's in the overhead-luggage bin in an airplane. To help in its return, use a wide piece of clear tape to tape a card with your name and address to the outside of the machine. You should tape a similar card to the inside also. In addition, scatter a few such cards in the pockets of the carrying case.

Desktop computers are also easily stolen. However, for under $25, you can buy a cable and lock, like those used for bicycles, and secure the computer, monitor, and printer to a work area. If your hardware does get stolen, its recovery may be helped if you have inscribed your driver's license number, Social Security number, or home address on each piece. Some campus and city police departments lend inscribing tools for such purposes. Finally, insurance to cover computer theft or damage is surprisingly cheap. Look for advertisements in computer magazines. (If you have standard tenants' or homeowners' insurance, it may not cover your computer. Ask your insurance agent.)

Guarding Against Heat, Cold, Spills, & Drops

"We dropped 'em, baked 'em, we even froze 'em," proclaimed the *PC Computing* cover, ballyhooing a story about its notebook "torture test."[a]

The magazine put eight notebook computers through durability trials. One approximated putting these machines in a car trunk in the desert heat, another with leaving them outdoors in a Buffalo, New York, winter. A third test simulated sloshing coffee on a keyboard, and a fourth dropped them from desktop height to a carpeted floor. All passed the bake test, but one failed the freeze test. Three completely flunked the coffee-spill test, one other revived, and the rest passed. One that was dropped lost the right side of its display; the others were unharmed. Of the eight, half passed all tests unscathed. In a more recent torture test, nine notebooks survived the heat, cold, and spill tests, but three failed the drop test.

This gives you an idea of how durable computers are. Designed for portability, notebooks may be hardier than desktop machines. Even so, you really don't want to tempt fate by dropping your computer, which could cause your hard-disk drive to fail.

Guarding Against Damage to Software

Systems software and applications software generally come on CD-ROM disks or flexible diskettes. The unbreakable rule is simply this: Copy the original disk, either onto your hard-disk drive or onto another diskette. Then store the original disk in a safe place. If your computer gets stolen or your software destroyed, you can retrieve the original and make another copy.

Protecting Your Data

Computer hardware and commercial software are nearly always replaceable, although perhaps with some expense and difficulty. Data, however, may be major trouble to replace or even irreplaceable. If your hard-disk drive crashes, do you have the same data on a back-up disk? Almost every microcomputer user sooner or later has the experience of accidentally wiping out or losing material and having no copy. This is what makes people true believers in backing up their data—making a duplicate in some form. If you're working on a research paper, for example, it's fairly easy to copy your work onto a floppy disk at the end of your work session.

Floppy disks can be harmed by any number of enemies. These include spills, dirt, heat, moisture, weights, and magnetic fields and magnetized objects. Here are some diskette-maintenance tips:

- Insert the floppy disk *carefully* into the disk drive.
- Don't manipulate the metal shutter on the floppy; it protects the surface of the magnetic material inside.
- Do not place heavy objects on the diskette.
- Do not expose the floppy to excessive heat or light.
- Do not use or place diskettes near a magnetic field, such as a telephone or paper clips stored in magnetic holders. Data can be lost if exposed.
- Do not use alcohol, thinners, or freon to clean the diskette.

Protecting Your Health

More important than any computer system and (probably) any data is your health. What adverse effects might computers cause? As we discussed earlier in the chapter, the most serious are painful hand and wrist injuries, eyestrain and headache, and back and neck pains.

Many people set up their computers in the same way as they would a typewriter. However, the two machines are ergonomically different for various reasons. With a computer, it's important to sit with both feet on the floor, thighs at right angles to your body. The chair should be adjustable and support your lower back. Your forearms should be parallel to the floor. You should look down slightly at the screen. This setup is particularly important if you are going to be sitting at a computer for hours at a stretch.

To avoid wrist and forearm injuries, you should keep your wrists straight and hands relaxed as you type. Instead of putting the keyboard on top of a desk, therefore, you should put it on a low table or in a keyboard drawer under the desk. Otherwise the nerves in your wrists will rub against the sheaths surrounding them, possibly leading to RSI pains. Some experts also suggest using a padded, adjustable wrist rest, attached to the keyboard.

Eyestrain and headaches usually arise because of improper lighting, screen glare, and long shifts staring at the screen. Make sure your windows and lights don't throw a glare on the screen, and that your computer is not framed by an uncovered window. Headaches may also result from too much noise, such as listening for hours to an impact printer printing out.

Back and neck pains occur because furniture is not adjusted correctly or because of heavy computer use. Adjustable furniture and frequent breaks should provide relief here.

Some people worry about emissions of electromagnetic waves and whether they could cause problems in pregnancy or even cause cancer. The best approach is to simply work at an arm's length from computers with CRT-type monitors.

How to set up your computer work area

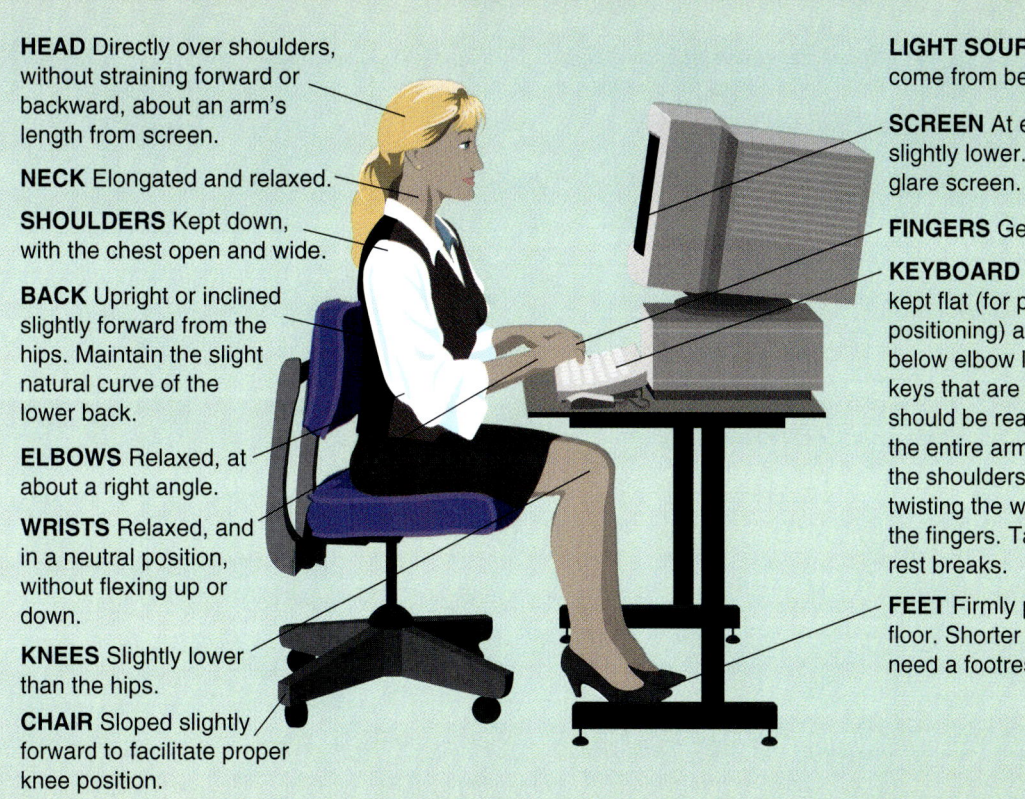

HEAD Directly over shoulders, without straining forward or backward, about an arm's length from screen.

NECK Elongated and relaxed.

SHOULDERS Kept down, with the chest open and wide.

BACK Upright or inclined slightly forward from the hips. Maintain the slight natural curve of the lower back.

ELBOWS Relaxed, at about a right angle.

WRISTS Relaxed, and in a neutral position, without flexing up or down.

KNEES Slightly lower than the hips.

CHAIR Sloped slightly forward to facilitate proper knee position.

LIGHT SOURCE Should come from behind the head.

SCREEN At eye level or slightly lower. Use an anti-glare screen.

FINGERS Gently curved.

KEYBOARD Best when kept flat (for proper wrist positioning) and at or just below elbow level. Computer keys that are far away should be reached by moving the entire arm, starting from the shoulders, rather than by twisting the wrists or straining the fingers. Take frequent rest breaks.

FEET Firmly planted on the floor. Shorter people may need a footrest.

Summary

active-matrix display (p. 195, KQ 5.3) Also known as *TFT (thin-film transistor) display;* flat-panel display in which each pixel on the screen is controlled by its own transistor. Why it's important: *Active-matrix screens are much brighter and sharper than passive-matrix screens, but they are more complicated and thus more expensive. They also require more power, affecting the battery life in laptop computers.*

audio-input device (p. 190, KQ 5.2) Hardware that records analog sound and translates it for digital storage and processing. Why it's important: *Analog sound signals are continuous variable waves within a certain frequency range. For the computer to process them, these variable waves must be converted to digital 0s and 1s. The principal use of audio-input devices is to produce digital input for multimedia computers. An audio signal can be digitized in two ways—by an audio board or a MIDI board.*

bar-code reader (p. 189, KQ 5.2) Photoelectric scanner that translates bar codes into digital codes. Why it's important: *With bar-code readers and the appropriate software system, store clerks can total purchases and produce invoices with increased speed and accuracy; and stores and other businesses can monitor inventory and services with increased efficiency.*

bar codes (p. 189, KQ 5.2) Vertical zebra-striped marks imprinted on most manufactured retail products. Why it's important: *Bar codes provide a convenient means of identifying and tracking items. In North America, supermarkets, food manufacturers, and others have agreed to use a bar-code system called the Universal Product Code. Other kinds of bar-code systems are used on everything from FedEx packages, to railroad cars, to the jerseys of long-distance runners.*

biometrics (p. 192, KQ 5.2) Science of measuring individual body characteristics. Why it's important: *Biometric security devices identify a person through a fingerprint, voice intonation, or other biological characteristic. For example, retinal-identification devices use a ray of light to identify the distinctive network of blood vessels at the back of the eyeball.*

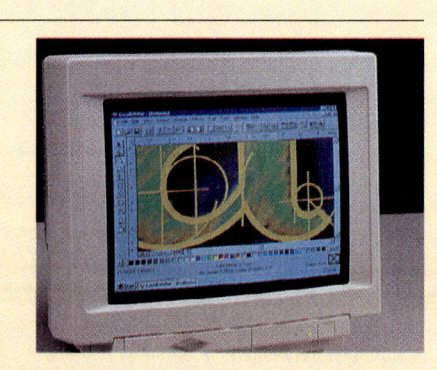

CRT (cathode-ray tube) (p. 194, KQ 5.3) Vacuum tube used as a display screen in a computer or video display terminal. Why it's important: *This technology is found not only in the screens of desktop computers but also in television sets and flight-information monitors in airports.*

dedicated fax machine (p.190, KQ 5.2) Specialized device that does nothing except send and receive fax documents. Why it's important: *Fax machines permit the transmission of text and graphic data over telephone lines quickly and inexpensively. They are found not only in offices and homes but also alongside regular phones in public places such as airports. See also fax modem.*

digital camera (p.191, KQ 5.2) Electronic camera that uses a light-sensitive processor chip to capture photographic images in digital form on a small diskette inserted in the camera or on flash-memory chips. Why it's important: *The bits of digital information—the snapshots you have taken, say—can be copied right onto a computer's hard disk for manipulation and printing out. The environmentally undesirable stage of chemical development required for conventional film is completely eliminated.*

digitizer (p. 187, KQ 5.2) Input unit based on a mouselike copying device called a *puck,* which converts drawings and photos to digital data. Why it's important: *See* digitizing tablet.

digitizing tablet (p. 187, KQ 5.2) One form of digitizer; an electronic plastic board on which each specific location corresponds to a location on the screen. When you use a puck, the tablet converts your movements into digital signals that are input to the computer. Why it's important: *Digitizing tablets are often used to make maps and engineering drawings, as well as to trace drawings.*

display screen (p. 193, KQ 5.3) Also called *monitor, CRT,* or simply *screen*; output device that shows programming instructions and data as they are being input and information after it is processed. Why it's important: *Screens are needed to display soft-copy output.*

dot-matrix printer (p. 196, KQ 5.3) Impact printer with a print head of small pins that strike an inked ribbon against paper to form characters or images. Print heads are available with 9, 18, or 24 pins; the 24-pin head offers the best quality. Why it's important: *Dot-matrix printers are employed much less frequently than laser printers and ink-jet printers but are still useful for some purposes. Dot-matrix printers can print draft quality, a coarser-looking 72 dpi, or near-letter-quality (NLQ), a crisper-looking 144 dpi. The machines print 40–300 characters per second and can handle graphics as well as text. Dot-matrix printers are used much less frequently than laser printers and ink-jet printers.*

dot pitch (dp) (p. 194, KQ 5.3) Amount of space between the centers of adjacent pixels; the closer the dots, the crisper the image. Why it's important: *Dot pitch is one of the measures of display-screen crispness. For a .28dp monitor, for instance, the dots are 28/100ths of a millimeter apart. Generally, a dot pitch of .28dp will provide clear images.*

dpi (dots per inch) (p. 196, KQ 5.3) Measure of the number of dots that are printed in a linear inch. For microcomputer printers, the resolution is in the range 60–1500 dpi. Why it's important: *The higher the dpi, the better the resolution.*

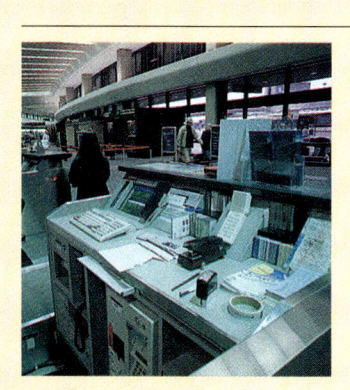

dumb terminal (p. 184, KQ 5.2) Display screen and a keyboard hooked up to a computer system; it can input and output but not process data. Why it's important: *Dumb terminals are used, for example, by airline reservations clerks to access a mainframe computer containing flight information.*

electronic imaging (p. 189, KQ 5.2) Software-controlled integration of separate images, using scanners, digital cameras, and advanced graphic computers. Why it's important: *Electronic imaging is a whole new industry based on imaging-system technology. It has become an important part of multimedia.*

fax machine (p. 190, KQ 5.2). Also called a *facsimile transmission machine;* input device that scans an image and sends it as electronic signals over telephone lines to a receiving fax machine, which prints the image on paper. Why it's important: *See* dedicated fax machine *and* fax modem.

fax modem (p. 190, KQ 5.2) Input device installed as a circuit board inside the computer's system cabinet; a modem with fax capability that enables you to send signals directly from your computer to someone else's fax machine or computer fax modem. Why it's important: *With this device, you don't have to print out the material from your printer and then turn around and run it through the scanner on a fax machine. The fax modem allows you to send information more quickly than if you had to feed it page by page into a machine. Fax modems are installed inside portable computers, including pocket PCs and PDAs. If you can also link up a cellular phone to a fax modem in your portable computer, you can send and receive wireless fax messages no matter where you are in the world.*

flat-panel display (p. 194, KQ 5.3) Display screens that are much thinner, weigh less, and consume less power than CRTs. Flat-panel displays are made up of two plates of glass separated by a layer of a substance in which light is manipulated. *Why it's important:* Flat-panel displays are essential to portable computers, although they are available for desktop computers as well.

hardcopy (p. 193, KQ 5.3) Printed output. *Why it's important:* The principal examples are printouts, whether text or graphics, from printers. Film, including microfilm and microfiche, is also considered hardcopy output.

imaging system (p. 188, KQ 5.2) Also called *image scanner* or *graphics scanner;* type of scanner that converts text, drawings, and photographs into digital form that can be stored in a computer system and then manipulated, output, or sent via modem to another computer. *Why it's important:* An important part of multimedia. For example, the imaging system used in desktop publishing scans in artwork or photos, which can then be positioned within a page of text, using desktop publishing software.

impact printer (p. 196, KQ 5.3) Printer that forms characters or images by striking a mechanism such as a print hammer or wheel against an inked ribbon, leaving an image on paper. *Why it's important:* Nonimpact printers are more commonly used than impact printers, but dot-matrix printers are still used in some businesses.

ink-jet printer (p. 197, KQ 5.3) Printer that sprays small, electrically charged droplets of ink from four nozzles through holes in a matrix at high speed onto paper. Like laser and dot-matrix printers, ink-jet printers form images with little dots. *Why it's important:* Because they produce high-quality images on special paper, ink-jet printers are often used in graphic design and desktop publishing. However, ink-jet printers are slower than laser printers and print at a lower resolution on regular paper.

input hardware (p. 183, KQ 5.1) Devices that translate data into a form the computer can process. *Why it's important:* Without input hardware, computers could not function. The computer-readable form consists of 0s and 1s, represented as off and on electrical signals. Input hardware devices are categorized as three types: keyboards, pointing devices, and source data-entry devices.

intelligent terminal (p. 185, KQ 5.2) Hardware unit with its own memory and processor, as well as a display screen and keyboard, hooked to a larger computer system. *Why it's important:* Such a terminal can perform some functions independent of any mainframe to which it is linked. Examples include the automated teller machine (ATM), a self-service banking machine connected through a telephone network to a central computer, and the point-of-sale (POS) terminal, used to record purchases at a store's customer checkout counter. Recently, many intelligent terminals have been replaced by personal computers.

Internet terminal (p. 185, KQ 5.2) Terminal that provides access to the Internet. There are several variants of Internet terminal: (1) the set-top box or web terminal, which displays web pages on a TV set; (2) the network computer, a cheap, stripped-down computer that connects people to networks; (3) the online game player, which not only lets you play games but also connects to the Internet; (4) the full-blown PC/TV (or TV/PC), which merges the personal computer with the television set; and (5) the wireless pocket PC or personal digital assistant (PDA), a handheld computer with a tiny keyboard that can do two-way wireless messaging. *Why it's important:* In the near future, most likely, Internet terminals will be everywhere.

keyboard (p. 184, KQ 5.2) Input device that converts letters, numbers, and other characters into electrical signals that can be read by the computer's processor. *Why it's important:* Keyboards are the most popular kind of input device.

laser printer (p. 196, KQ 5.3) Nonimpact printer that creates images with dots. As in a photocopying machine, images are produced on a drum, treated with a magnetically charged ink-like toner (powder), and then transferred from drum to paper. *Why it's important:* Laser printers produce much better image quality than do dot-matrix printers and can print in many more colors; they are also quieter. Laser printers, along with page description languages, enabled the development of desktop publishing.

light pen (p. 186, KQ 5.2) Light-sensitive penlike device connected by a wire to the computer terminal. The user brings the pen to a desired point on the display screen and presses the pen button, which identifies that screen location to the computer. *Why it's important:* Light pens are used by engineers, graphic designers, and illustrators.

liquid crystal display (LCD) (p. 194, KQ 5.3) Flat-panel display in which molecules of liquid crystal line up in a way that alters their optical properties, creating images on the screen by transmitting or blocking out light. *Why it's important:* LCD is useful not only for portable computers but also as a display for various electronic devices, such as watches and radios.

magnetic-ink character recognition (MICR) (p. 189, KQ 5.2) Scanning technology that reads magnetized-ink characters printed at the bottom of checks and converts them to digital form. *Why it's important:* MICR technology is used by banks to sort checks.

multifunction printer (p. 198, KQ 5.3) Hardware device that combines several capabilities, such as printing, scanning, copying, and faxing. *Why it's important:* Multifunction printers take up less space and cost less than the four separate office machines that they replace. The downside, however, is that if one component breaks, nothing works.

nonimpact printer (p. 196, KQ 5.3) Printer that forms characters and images without direct physical contact between the printing mechanism and paper. Two types of nonimpact printers often used with microcomputers are laser printers and ink-jet printers. A third kind, the thermal printer, is seen less frequently. *Why it's important:* Nonimpact printers are faster and quieter than impact printers.

```
OCR-A
NUMERIC    0123456789
ALPHA      ABCDEFGHIJ
SYMBOLS    KLMNOPQRST
           UVWXYZ
           >$/-+-#"

OCR-B
NUMERIC    00123456789
ALPHA      ACENPSTVX
SYMBOLS    <+>-¥
```

optical character recognition (OCR) (p. 190, KQ 5.2) Scanning technology that reads special preprinted characters in a particular font (typeface design) and converts them to digital code. *Why it's important:* OCR characters appear on utility bills and price tags on department-store merchandise.

optical mark recognition (OMR) (p. 189, KQ 5.2) Scanning technology that reads pencil marks and converts them into computer-usable form. *Why it's important:* OMR technology is used to read the College Board Scholastic Aptitude Test (SAT) and the Graduate Record Examination (GRE).

output hardware (p. 183, KQ 5.1) Hardware devices that convert machine-readable information, obtained as the result of processing, into people-readable form. The principal kinds of output are softcopy and hardcopy. *Why it's important:* Without output devices, people would have no access to processed data and information.

page description language (p. 196, KQ 5.3) Software that describes the shape and position of characters and graphics to the printer. PostScript and PCL are common page description languages. *Why it's important:* Page description languages are essential to desktop publishing.

passive-matrix display (p. 195, KQ 5.3) Flat-panel display in which a transistor controls a whole row or column of pixels. Passive matrix provides a sharp image for one-color (monochrome) screens but is more subdued for color. *Why it's important:* Passive-matrix displays are less expensive and use less power than active-matrix displays, but they aren't as clear and bright and can leave "ghosts" when the display changes quickly. Passive-matrix displays go by the abbreviations HPA, STN, or DSTN.

pen-based computer system (p. 186, KQ 5.2) Input system that allows users to enter handwriting and marks onto a computer screen by means of a penlike stylus rather than by typing on a keyboard. Pen computers use handwriting recognition software that translates handwritten characters made by the stylus into data that is usable by the computer. *Why it's important:* Many handheld computers and PDAs have pen input, as do digital notebooks.

pixel (p. 194, KQ 5.3) Short for "picture element"; the smallest unit on the screen that can be turned on and off or made different shades. *Why it's important:* Pixels are the building blocks that allow text and graphical images to be displayed on a screen.

pointing devices (p. 185, KQ 5.2) Hardware that controls the position of the cursor or pointer on the screen. It includes the mouse and its variants, the touch screen, and various forms of pen input. *Why it's important:* In many contexts, pointing devices permit quick and convenient data input.

pointing stick (p. 186, KQ 5.2) Pointing device that looks like a pencil eraser protruding from the keyboard between the G, H, and B keys. You move the pointing stick with your forefinger while using your thumb to press buttons located in front of the space bar. Why it's important: *Pointing sticks are used principally in videogames, in computer-aided design systems, and in robots.*

printer (p. 196, KQ 5.3) Output device that prints characters, symbols, and perhaps graphics on paper or another hardcopy medium. Why it's important: *Printers provide one of the principal forms of computer output.*

radio-frequency identification technology (p. 192, KQ 5.2) Also known as RF-ID tagging; a source data-entry technology based on an identifying tag bearing a microchip that contains specific code numbers. These code numbers are read by the radio waves of a scanner linked to a database. Why it's important: *Drivers with RF-ID tags can breeze through tollbooths without having to even roll down their windows; the toll is automatically charged to their accounts. Radio-readable ID tags are also used by the Postal Service to monitor the flow of mail, by stores for inventory control and warehousing, and in the railroad industry to keep track of rail cars.*

refresh rate (p. 194, KQ 5.3) Number of times per second that screen pixels are recharged so that their glow remains bright. In general, displays are refreshed 45–100 times per second. Why it's important: *The higher the refresh rate, the more solid the image looks on the screen—that is, the less it flickers.*

resolution (p. 194, KQ 5.3) Clarity or sharpness of display-screen images; the more pixels there are per square inch, the finer the level of detail attained. Resolution is expressed in terms of the formula *horizontal pixels* × *vertical pixels*. Each pixel can be assigned a color or a particular shade of gray. Standard resolutions are 640 × 480, 800 × 600, 1024 × 768, 1280 × 1024, and 1600 × 1200 pixels. Why it's important: *Users need to know what screen resolution is appropriate for their purposes.*

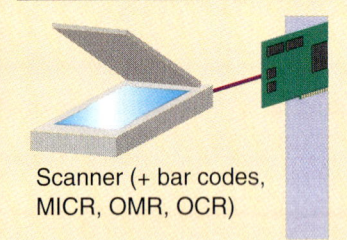

Scanner (+ bar codes, MICR, OMR, OCR)

scanner (p. 188, KQ 5.2) Source-data input device that uses light-sensing equipment to translate images of text, drawings, photos, and the like into digital form. Why it's important: *Scanners simplify the input of complex data. The images can be processed by a computer, displayed on a monitor, stored on a storage device, or communicated to another computer.*

sensor (p. 192, KQ 5.2) Input device that collects specific data directly from the environment and transmits it to a computer. Why it's important: *Although you are unlikely to see such input devices connected to a PC in an office, they exist all around us, often in nearly invisible form. Sensors can be used to detect all kinds of things: speed, movement, weight, pressure, temperature, humidity, wind, current, fog, gas, smoke, light, shapes, images, and so on. In aviation, for example, sensors are used to detect ice buildup on airplane wings and to alert pilots to sudden changes in wind direction.*

softcopy (p. 193, KQ 5.3) Data on a display screen or in audio or voice form. This kind of output is not tangible; it cannot be touched. Why it's important: *This term is used to distinguish nonprinted output from printed (hardcopy) output.*

sound-output devices (p. 198, KQ 5.3) Hardware that produces digitized sounds, ranging from beeps and chirps to music. Why it's important: *To use sound output, you need appropriate software and a sound card. Such devices are used to produce the sound effects when you play a CD-ROM, for example.*

source data-entry devices (p. 188, KQ 5.2) Data-entry devices that create machine-readable data on magnetic media or paper or feed it directly into the computer's processor, without the use of a keyboard. Categories include scanning devices (imaging systems, barcode readers, mark- and character-recognition devices, and fax machines), audio-input devices, video input, and photographic input (digital cameras), voice-recognition systems, sensors, radio-frequency identification devices, and human biology input devices. Why it's important: *Source-data entry devices lessen reliance on keyboards for data entry and can make data entry more accurate.*

SVGA (super video graphics array) (p. 195, KQ 5.3) Graphics board standard that supports a resolution of 800 × 600 pixels, or variations, producing 16 million possible simultaneous colors. Why it's important: *SVGA is the most common standard used today.*

SXGA (super extended graphics array) (p. 195, KQ 5.3) Graphics board standard that supports a resolution of 1280 × 1024 pixels. Why it's important: *SXGA is often used with 19- and 21-inch monitors.*

thermal printer (p. 198, KQ 5.3) Printer that uses colored waxes and heat to produce images by burning dots onto special paper. The colored wax sheets are not required for black-and-white output. Thermal printers are expensive, and they require expensive paper. Why it's important: *For people who want the highest-quality color printing available with a desktop printer, thermal printers are the answer.*

touchpad (p. 186, KQ 5.2) Input device; a small, flat surface over which you slide your finger, using the same movements as you would with a mouse. The cursor follows the movement of your finger. You "click" by tapping your finger on the pad's surface or by pressing buttons positioned close by the pad. Why it's important: *Touchpads let you control the cursor/pointer with your finger, and they require very little space to use. Most laptops have touchpads.*

touch screen (p. 186, KQ 5.2) Video display screen that has been sensitized to receive input from the touch of a finger. The screen is covered with a plastic layer, behind which are invisible beams of infrared light. Why it's important: *You can input requests for information by pressing on buttons or menus displayed. The answers to your requests are displayed as output in words or pictures on the screen. (There may also be sound.) You find touch screens in kiosks, ATMs, airport tourist directories, hotel TV screens (for guest checkout), and cam-pus information kiosks making available everything from lists of coming events to (with proper ID and personal code) student financial-aid records and grades.*

trackball (p. 185, KQ 5.2) Movable ball, mounted on top of a stationary device, that can be rotated using your fingers or palm. It looks like the mouse turned upside down. Instead of moving the mouse around on the desktop, you move the trackball with the tips of your fingers. Why it's important: *Trackballs require less space to use than does a mouse.*

UXGA (ultra extended graphics array) (p. 195, KQ 5.3) Graphics board standard that supports a resolution of 1600 × 1200 pixels. Why it's important: *UXGA is expected to become more popular with graphic artists, engineering designers, and others using 21-inch monitors.*

video (p. 198, KQ 5.3) Output consisting of photographic images played at 15–29 frames per second to give the appearance of full motion. Why it's important: *Video is input into a multimedia system using a video camera or VCR and, after editing, is output on a computer's display screen. Because video files can require a great deal of storage—a 3-minute video may require 1 gigabyte of storage—video is often compressed. Digital video has revolutionized the movie industry, as in the use of special effects.*

videoconferencing (p. 198, KQ 5.3) Form of video output in which people in different geographical locations can have a meeting—can see and hear one another—using computers and communications. Why it's important: *Many organizations use videoconferencing to take the place of face-to-face meetings. Videoconferencing systems range from videophones to group conference rooms with cameras and multimedia equipment to desktop systems with small video cameras, microphones, and speakers.*

voice-output device (p. 198, KQ 5.3) Hardware that converts digital data into speech-like sounds. Why it's important: *You hear such voice output on telephones ("Please hang up and dial your call again"), in soft-drink machines, in cars, in toys and games, and recently in mapping software for vehicle-navigation devices. For people with physical challenges, computers with voice output help to level the playing field.*

voice-recognition system (p. 191, KQ 5.2) Input system that uses a microphone (or a telephone) as an input device and converts a person's speech into digital signals by comparing the electrical patterns produced by the speaker's voice with a set of prerecorded patterns stored in the computer. Why it's important: *Voice-recognition technology is useful in situations where people are unable to use their hands to input data or need their hands free for other purposes.*

XGA (extended graphics array) (p. 195, KQ 5.3) Graphics board display standard with a resolution of up to 1024 × 768 pixels, corresponding to 65,536 possible colors. Why it's important: *XGA offers the most sophisticated standard for color and resolution. It is used mainly on workstation systems, such as those employed by engineering designers.*

Chapter Review

stage 1 LEARNING — MEMORIZATION

"I can recognize and recall information."

Self-Test Questions

1. A(n) _____ terminal is entirely dependent for all its processing activities on the computer system to which it is connected.
2. The two main categories of printer are _____ and _____.
3. A(n) _____ is an input device that is rolled about on a desktop and directs a pointer on the computer's display screen.
4. _____ consists of devices that translate information processed by the computer into a form that humans can understand.
5. _____ is the science of measuring individual body characteristics.
6. CRT is short for _____.
7. LCD is short for _____.
8. A _____ is software that describes the shape and position of characters and graphics to the printer.
9. When people in different geographical locations can have a meeting using computers and communications, it is called _____.

Multiple-Choice Questions

1. Which of the following is not a pointing device?
   a. mouse
   b. touchpad
   c. keyboard
   d. joystick
2. Which of the following is not a source data-entry device?
   a. bar-code reader
   b. sensor
   c. digital camera
   d. scanner
   e. mouse

True/False Questions

T F 1. On a computer screen, the more pixels that appear per square inch, the higher the resolution.
T F 2. Photos taken with a digital camera can be downloaded to a computer's hard disk.
T F 3. *Resolution* is the amount of space between the centers of adjacent pixels.
T F 4. The abbreviation *dpi* stands for *dense pixel intervals*.
T F 5. Pointing devices control the position of the cursor on the screen.

stage 2 LEARNING — COMPREHENSION

"I can recall information in my own terms and explain them to a friend."

Short-Answer Questions

1. What determines how a keyboard's function keys work?
2. What characteristics determine the clarity of a computer screen?
3. Describe two situations in which scanning is useful.
4. What is source-data entry?
5. Why is it important for your computer to be expandable?
6. What is *pixel* short for? What is a pixel?

Concept Mapping

On a separate sheet of paper, draw a concept map, or visual diagram, linking concepts. Show how the following terms are related.

- active matrix
- bar-code reader
- CRT
- digitizer
- display screen
- dot-matrix printer
- dot pitch
- dpi
- dumb
- fax machine/modem
- flat-panel display
- hardcopy
- ink-jet
- input hardware
- intelligent
- keyboard
- laser printer
- LCD
- nonimpact printer
- OCR
- output hardware
- PCL
- PDL
- pixel
- pointing device
- PostScript
- printer
- resolution
- scanner
- source-data entry
- terminal
- touchpad
- touch screen

stage 3 LEARNING: APPLYING, ANALYZING, SYNTHESIZING, EVALUATING

"I can apply what I've learned, relate these ideas to other concepts, build on other knowledge, and use all these thinking skills to form a judgment."

Knowledge in Action

1. Cut out an advertisement from a newspaper or a magazine that features a new microcomputer system. Circle all the terms that are familiar to you now that you have read the first five chapters of this text. Define these terms on a separate sheet of paper. Is this computer expandable? How much does it cost? Is the monitor included in the price? A printer?

2. *Paperless office* is a term that has been around for some time. However, the paperless office has not yet been achieved. Do you think the paperless office is a good idea? Do you think it's possible? Why do you think it has not yet been achieved?

3. Many PC warranties do not cover protection against lightning damage, which is thought to be an "act of God." Does your PC warranty provide coverage for "acts of God"? Read it over to find out.

4. Compare and contrast the pros and cons of different types of monitors. Decide which one is best for you and explain why. Do some research on how each monitor type creates the displayed images.

5. Do you have access to a computer with (a) voice-recognition software and (b) word processing software that determines writing level (such as eighth grade, ninth grade, and so on)? Dictate a few sentences about your day into the microphone. After your speech is encoded into text, use the word processing software to determine what grade level it is appropriate for.

Web Exercises

1. Visit an online shopping site such as *www.yahoo.com* (click on Computers & Internet, then click on Hardware, then click on Printers) and investigate five different types of printers. Note (a) the type of printer, (b) its price, (c) its resolution, and (d) whether the printer is PC or Mac compatible. Which printer would you choose? Why?

2. One particular DVD player received a great deal of attention because of a special feature it had. If you run a search on the Apex AD 600 DVD Player, you'll probably find many search results consisting of websites people have put up in honor of this device. Observe the hit counters on each site to see how many people are looking up information about this player. Two good places to start are *www.dvd-wizards.com* and *www.wired.com/news/technology/0,1282,35365,00.html*. Also, the official Apex website is *www.apexdigitalinc.com*.

3. Concerned about electromagnetic radiation from monitors and other equipment? For a question and answer session on electromagnetic frequencies, go to *http://vitatech.net/q_a.html*. Another site to visit is *http://webnz.com/Authority/1996/electromagneticradiation.html* for information about measuring EMF radiation. Also see *www.has.gov.sg*.

Chapter 6

Telecommunications

Networks & Communications: The "New Story" in Computing

Key Questions
You should be able to answer the following questions.

6.1 **From the Analog to the Digital Age** How do digital and analog data differ, and what does a modem do?

6.2 **The Practical Uses of Communications** What are some offerings of new telecommunications technology?

6.3 **Communications Channels: The Conduits of Communications** What are types of wired and wireless channels and some types of wireless communications?

6.4 **Networks** What are the benefits of networks, and what are their types, components, and variations?

6.5 **Cyberethics: Controversial Material, Censorship, & Privacy Issues** What are important issues in cyberethics?

The essence of all revolution, stated philosopher Hannah Arendt, is the start of a *new story* in human experience.

Before the 1950s, computing devices processed data into information, and communications devices communicated information over distances. The two streams of technology developed pretty much independently, like rails on a railroad track that never merge. Now we have a new story, a revolution.

For us, the new story has been *digital convergence*—the gradual merger of computing and communications into a new information environment, in which *the same information is exchanged among many kinds of equipment, using the language of computers.* (See ● Panel 6.1.) At the same time, there has been a convergence of several important industries—computers, telecommunications, consumer electronics, entertainment, mass media—producing new electronic products that perform multiple functions.

An example is what's been happening in television. *WebTV* consists of a set-top box powered by WebTV Networks, a subsidiary of Microsoft. The box makes a conventional television set function like three different devices: a television with satellite service, an Internet-linked computer, and an enhanced digital video recorder. The result of this convergence of technologies is that, besides watching TV programs, you can also interact with them in real time; for instance, you might participate in *Jeopardy!* along with the contestants you're viewing on the screen. Or you can order a Domino's pizza by pointing and clicking with a remote control. Or you can watch *digital television (DTV)* and the variant called *high-definition television (HDTV)*, which work with digital broadcasting signals to deliver not only crisper images but also more channels and computer connections. HDTV can receive text, graphics, audio, and video, and it can be used as a monitor when you browse the Internet.

● **PANEL 6.1**
Digital convergence—the fusion of computer and communications technologies
Today's new information environment came about gradually from the merger of two separate streams of technological development—computers and communications.

Computer Technology

1621 AD	1642	1833	1843
Slide rule invented (Edmund Gunther)	First mechanical adding machine (Blaise Pascal)	Babbage's difference engine (automatic calculator)	World's first computer programmer, Ada Lovelace, publishes her notes

Communications Technology

1562	1594	1639	1827	1835	1846	1866	1876	1888
First monthly newspaper (Italy)	First magazine (Germany)	First printing press in North America	Photographs on metal plates	Telegraph (first long-distance digital communication system)	High-speed printing	Trans-atlantic telegraph cable laid	Telephone invented	Radio waves identified

6.1 From the Analog to the Digital Age

KEY QUESTIONS
How do digital and analog data differ, and what does a modem do?

Why have the worlds of computers and of telecommunications been so long in coming together? Because *computers are digital, but most of the world has been analog.* Let's take a look at what this means.

The Digital Basis of Computers: Electrical Signals as Discontinuous Bursts

Computers may seem like incredibly complicated devices but, as we saw in Chapter 1, their underlying principle is simple. Because they are based on on/off electrical states, they use the *binary system,* which consists of only two digits—0 and 1. Today **digital specifically refers to communications signals or information represented in a two-stat (binary) way.** More generally, *digital* is usually synonymous with "computer-based."

Digital data consists of data (expressed as 0s and 1s) represented by on/off electrical pulses. These pulses are transmitted in discontinuous bursts rather than (as with analog devices) in continuous waves.

The Analog Basis of Life: Electrical Signals as Continuous Waves

"The shades of a sunset, the flight of a bird, or the voice of a singer would seem to defy the black or white simplicity of binary representation," points out one writer.[1] Indeed, these and most other phenomena of the world are **analog, continuously varying in strength and/or quality.** Sound, light, temperature, and pressure values, for instance, can fall anywhere on a continuum or range. The highs, lows, and in-between states have historically been represented with analog devices rather than in digital form. Examples of analog devices are a speedometer, a thermometer, and a tire-pressure gauge, all of which can measure continuous fluctuations. The electrical signals on a telephone line, for instance, have traditionally been analog-data representations of the original voices.

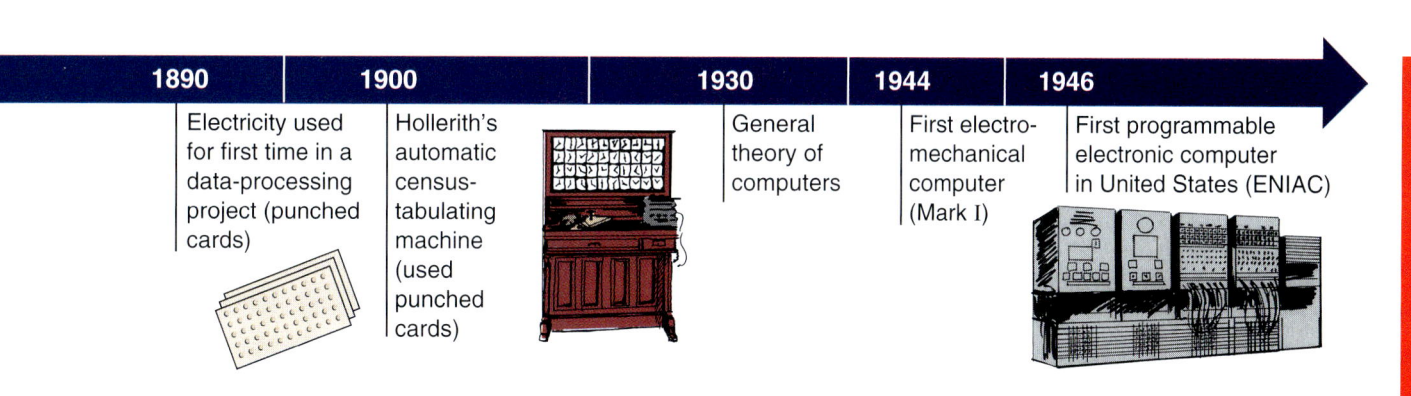

1890	1900	1930	1944	1946
Electricity used for first time in a data-processing project (punched cards)	Hollerith's automatic census-tabulating machine (used punched cards)	General theory of computers	First electro-mechanical computer (Mark I)	First programmable electronic computer in United States (ENIAC)

1894	1895	1912	1915	1928	1939	1946	1947	1948	1950
Edison makes a movie	Marconi develops radio; motion-picture camera invented	Motion pictures become a big business	AT&T long-distance service reaches San Francisco	First TV demonstrated; first sound movie	Commercial TV broadcasting	Color TV demonstrated	Transistor invented	Reel-to-reel tape recorder	Cable TV

Telecommunications

213

Thus, *analog data* is transmitted in a continuous form—a continuous electrical signal in the shape of a wave (called a *carrier wave*). Telephone, radio, television, and cable-TV technologies have long been based on analog data.

Purpose of the Modem: Converting Digital Signals to Analog Signals & Back

To understand the differences between digital and analog transmission, look at a graphic representation of an on/off digital signal emitted from a computer. Like a regular light switch, this signal has only two states—on and off. Compare this with a graphic representation of a wavy analog signal emitted as a signal. The changes in this signal are gradual, as in a dimmer switch, which gradually increases or decreases brightness.

Because telephone lines have traditionally been analog, you need to have a *modem* if your computer is to send communications signals over a telephone line. The modem translates the computer's digital signals into the telephone line's analog signals. The receiving computer also needs a modem to translate the analog signals back into digital signals. *(See ● Panel 6.2.)*

How, in fact, does a modem convert the continuous analog wave to a discontinuous digital pulse that can represent 0s and 1s? The modem can make adjustments to the *frequency*—the number of cycles per second, or the number of times a wave repeats during a specific time interval (the fastness/slowness). Or it can make adjustments to the analog signal's *amplitude*—the height of the wave (the loudness/softness). Thus, in frequency, a slow wave might represent a 0 and a quick wave might represent a 1. In amplitude, a low wave might represent a 0 and a high wave might represent a 1. *(See ● Panel 6.3, page 216.)*

<u>Modem</u> is short for "*mod*ulate/*dem*odulate"; a sending modem modulates digital signals into analog signals for transmission over phone lines. A receiving modem demodulates the analog signals back into digital signals. The

1952	1963	1964	1967	1969	1970	1971	1975	1977	1978
UNIVAC computer correctly predicts election of Eisenhower as U.S. President	BASIC developed at Dartmouth	IBM introduces 360 line of computers	Hand-held calculator	ARPA-Net established, led to Internet	Microprocessor chips come into use; floppy disk introduced for storing data	First pocket calculator	First microcomputer (MITs Altair 8800)	Apple II computer (first personal computer sold in assembled form)	5¼" floppy disk; Atari home videogame

Fusing of computer and communications lines of development

1952	1957	1961	1968	1975	1976	1977	1979	1982
Direct-distance dialing (no need to go through operator); transistor radio introduced	First satellite launched (Russia's Sputnik)	Push-button telephones	Portable video recorders; video cassettes	Flat-screen TV	First wide-scale marketing of TV computer games (Atari)	First interactive cable TV	3-D TV demonstrated	Compact disks; European consortium launches multiple communications satellites

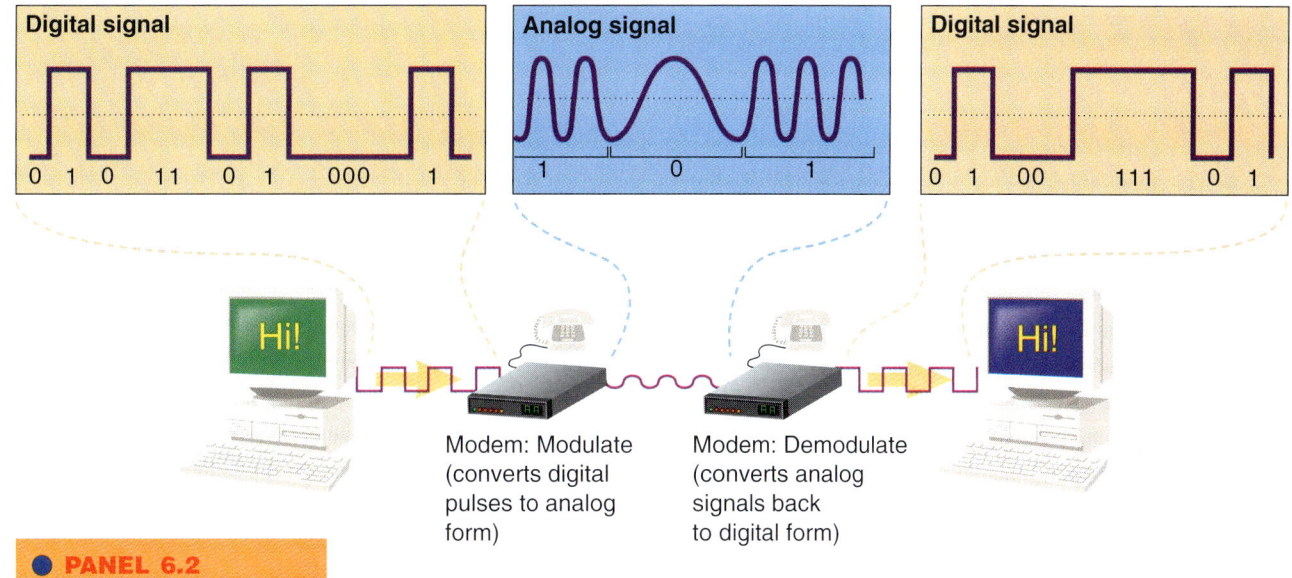

PANEL 6.2
Analog versus digital signals, and the modem
Note that an analog signal represents a continuous electrical signal in the form of a wave. A digital signal is discontinuous, expressed as discrete bursts in on/off electrical pulses.

Modem: Modulate (converts digital pulses to analog form)

Modem: Demodulate (converts analog signals back to digital form)

modem provides a means for computers to communicate with one another using the standard copper-wire telephone network, an analog system that was built to transmit the human voice but not computer signals.

Our concern, however, goes far beyond telephone transmission. How can the analog realities of the world be expressed in digital form? How can light, sounds, colors, temperatures, and other dynamic values be represented so that they can be manipulated by a computer? Let us consider this.

Computer Technology

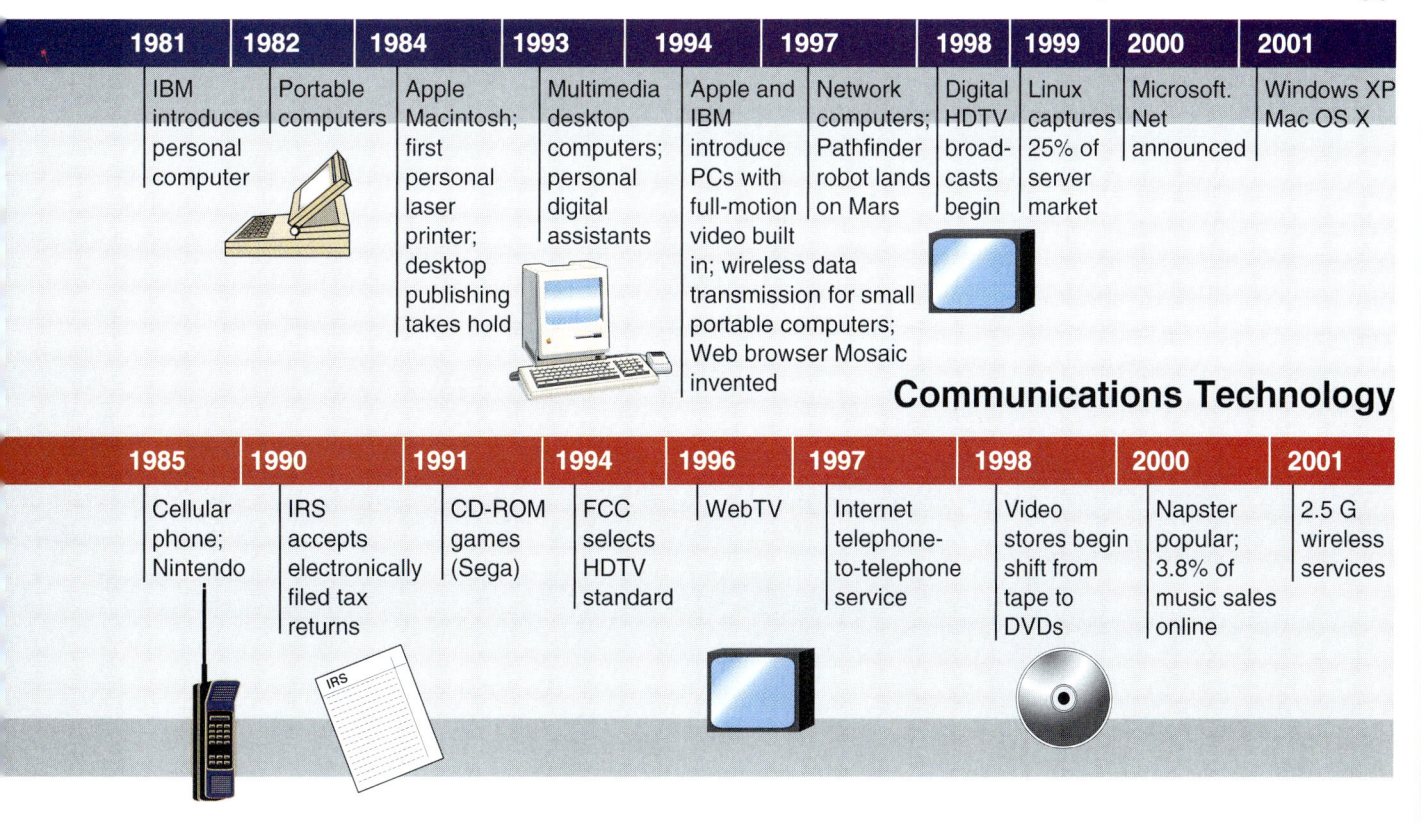

1981	1982	1984	1993	1994	1997	1998	1999	2000	2001
IBM introduces personal computer	Portable computers	Apple Macintosh; first personal laser printer; desktop publishing takes hold	Multimedia desktop computers; personal digital assistants	Apple and IBM introduce PCs with full-motion video built in; wireless data transmission for small portable computers; Web browser Mosaic invented	Network computers; Pathfinder robot lands on Mars	Digital HDTV broadcasts begin	Linux captures 25% of server market	Microsoft .Net announced	Windows XP Mac OS X

Communications Technology

1985	1990	1991	1994	1996	1997	1998	2000	2001
Cellular phone; Nintendo	IRS accepts electronically filed tax returns	CD-ROM games (Sega)	FCC selects HDTV standard	WebTV	Internet telephone-to-telephone service	Video stores begin shift from tape to DVDs	Napster popular; 3.8% of music sales online	2.5 G wireless services

● **PANEL 6.3**
How analog waves are modified to resemble digital pulses

The continuous, even cycle of an analog wave...

... is converted to digital form through *frequency modulation*—the frequency of the cycle increases to represent a 1 and stays the same to represent a 0.

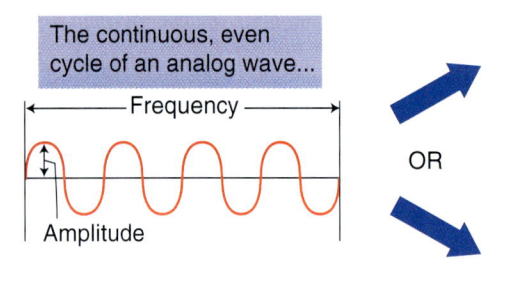

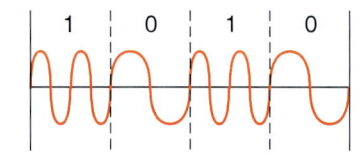

OR

... or is converted to digital form through *amplitude modulation*—the height of the wave is increased to represent a 1 and stays the same to represent a 0.

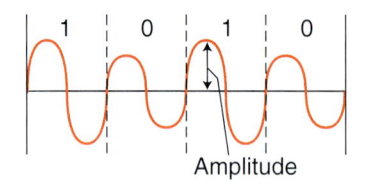

Converting Reality to Digital Form

Suppose you are using an analog tape recorder to record a singer during a performance. The analog process will produce a near duplicate of the sounds. This will include distortions, such as buzzings and clicks, or electronic hums if an amplified guitar is used.

The digital recording process is different. The way in which music is captured for audio CDs does not provide a duplicate of a musical performance. Rather, the digital process uses *representative selections (samples)* to record the sounds. The copy obtained is virtually exact and free from distortion and noise. Computer-based equipment takes samples of sounds at regular intervals—nearly 44,100 times a second. The samples are converted to numbers that the computer then uses to express the sounds. The sample rate of 44,100 times per second and the high precision fool our ears into hearing a smooth, continuous sound. Similarly, for visual material, a computer can take samples of values such as brightness and color. The same is true of other aspects of real-life experience, such as pressure, temperature, and motion.

Are we being cheated out of our experience of "reality" by allowing computers to sample sounds, images, and so on? Actually, people willingly made this compromise years ago, before computers were invented. Movies, for instance, carve up reality into 24 frames a second. Television frames are drawn at 30 lines per second. These processes happen so quickly that our eyes and brains easily jump the visual gaps. Digital processing of analog experience is just one more way of expressing or translating reality.

Turning analog reality into digital form provides tremendous opportunities. One of the most important is that all kinds of multimedia can now be changed into digital form and transmitted as data to all kinds of devices.

Now let us examine the digital world of telecommunications.

CONCEPT CHECK

Distinguish between digital and analog signals.

Explain what a modem does.

6.2 The Practical Uses of Communications

KEY QUESTION
What are some offerings of new telecommunications technology?

Guyana is the lone English-speaking country in South America. In the village of Lethem (population 2000), a collective of women from two native tribes—called the Rupununi Weavers Society—has revived the ancient art of hand-weaving large hammocks. Thanks to donated computer equipment and Internet training for one member, plus free satellite-assisted online access, the collective is selling hammocks around the world through its website *(www.gol.net.gy/rweavers)*. Though it might take 600 hours to make, a hammock could bring as much as $1000—a gigantic sum in this area.[2]

Such is the power of communications connections, or *connectivity*, which give us all instant, around-the-clock information from all over the globe. Let's consider some of the forms this connectivity takes:

- Videoconferencing and videophones
- Workgroup computing and groupware
- Telecommuting and virtual offices
- Home networks
- Information or Internet appliances
- Smart television

Videoconferencing & Videophones: Video/Voice Communication

Videoconferencing, also called teleconferencing, is the use of television video and sound technology as well as computers to enable people in different locations to see, hear, and talk with one another. For a videoconference, people

CLICK-ALONG 6-1
More on creating websites

BOOKMARK IT!

PRACTICAL ACTION BOX
Web-Authoring Tools: How to Create Your Own Simple Website, Easily & for Free

"The full power of the Web will only be unleashed," says technology writer Stephen Wildstrom, "when it's as easy to post information online as it is to write a memo."[a]

We may be getting closer. The World Wide Web is a great way to get information about yourself and your work to co-workers or to potential customers. To do so, however, you need to create a web page and put it online.

Fortunately, some free, easy-to-use web-authoring tools for building simple websites are available. All the following let you create web pages using icons and menus; you don't need to know HTML to get the job done.

- **Online services:** If you're a member of America Online, you can use AOL's Personal Publisher (keyword: *personal publisher*).
- **Browsers:** Netscape's Communicator comes with a web-building program called Composer. The recent releases of Microsoft Explorer also offer web-authoring tools.

- **Word processing software:** Microsoft offers Internet Assistant, which can be used with Word, its word processing program.
- **Desktop publishing software:** If you've taken up desktop publishing, you should know that Adobe PageMaker and Microsoft Publisher come with simple web page editors.

Once you've created your website, you'll need to "publish" it—upload it to a web server for viewing on the Internet. You can get upload instructions from your online service or Internet service provider (ISP). Some ISPs will give you free space on their servers.

Other specific packages for building websites are Macromedia's Dreamweaver, Adobe's PageMill, Soft Quad's Hot Metal Pro, and Microsoft's FrontPage Editor. If you want to create more sophisticated sites, you can try using tools such as Home Page from Claris or Trellix from Trellix Corp.

Videoconferencing
Three online participants discuss an architectural drawing (shown in the right half of the screen).

may go to conference rooms or booths with specially equipped television cameras. Alternatively, videoconferencing equipment can be set up on people's desks, with a camera and microphone to capture the person speaking and a monitor and speakers for the person being spoken to. The *videophone* is a telephone with a TV-like screen and a built-in camera that allows you to see the person you're calling, and vice versa.

Many microcomputers now include video cameras, called *PC cameras* or *webcams*, and network interface cards, which allow users to see people at the same time they communicate on the Internet.

The main difficulty with videoconferencing and videophones is that POTS ("plain old telephone service") equipment based on standard copper wire cannot transmit or receive images very rapidly. Thus, unless you can afford expensive high-speed communications lines, present-day screens will convey a series of jerky, stop-action images of the participants' faces. Today, however, 80% of all corporate videoconferences occur over multiple high-speed ISDN (page 36) phone lines.

Workgroup Computing & Groupware

When microcomputers were first brought into the workplace, they were used simply as another personal-productivity tool, like typewriters or calculators. Gradually, however, companies began to link microcomputers together on a network, usually to share an expensive piece of hardware, such as a laser printer. Then employees found that networks allowed them to share files and databases. Networking using common software also allowed users to buy equipment from different manufacturers—a mix of computers from both Sun Microsystems and Hewlett-Packard, for example. Sharing resources has led to workgroup computing.

In **<u>workgroup computing</u>, also called collaborative computing, teams of co-workers use networks of microcomputers to share information and to cooperate on projects.** Workgroup computing is made possible not only by networks and microcomputers but also by *groupware*. As we stated in Chapter 3, *groupware* is software that allows two or more people on a network to work on the same information at the same time.

In general, groupware, such as Lotus Notes or MS NetMeeting, permits office workers to collaborate with colleagues and to tap into company information through computer networks. It also enables them to link up with crucial contacts outside their organization.

Telecommuting & Virtual Offices

Computers and communications tools have led to telecommuting and virtual offices.

- **Telecommuting: Working at home while in telecommunication with the office is called _telecommuting_.** (A related term is *telework*, which includes not only those who work at least part time from home but also those who work at remote or satellite offices.) There were 21 million workers in 1997 who did some work at home as part of their regular jobs, according to the U.S. Department of Labor. Around 3.6 million worked at home as part of formal telecommuting arrangements for which they were paid.[3]

  Telecommuting can have many benefits. The advantages to society are reduced traffic congestion, energy consumption, and air pollution. What are the advantages to employers? Productivity can increase, it's argued, because telecommuters may experience fewer distractions at home and can work flexible hours. The labor pool can be expanded, because hard-to-get employees don't have to uproot themselves from where they want to live.

  Despite the advantages, however, telecommuting has not become as widespread as was once hoped. Among the reasons: Employees feel isolated, even deserted, or they are afraid that working outside the office will hinder their career advancement. They also find the arrangements blur the line between office and home, straining family life. Employers may feel telecommuting causes resentments among office-bound employees. In addition, with teamwork now more a workplace requirement, managers may worry that telecommuters cannot keep up with the pace of change. Finally, some employers worry that telecommuters create more opportunities for security breeches by hackers or equipment thieves.[4]

- **Virtual offices: The _virtual office_ is an often nonpermanent and mobile office run with computer and communications technology.** Employees work from their homes, cars, and other new work sites. They use pocket pagers, portable computers, fax machines, and various phone and network services to conduct business.

The rise in telecommuting and virtual offices is only part of a larger trend. "Powerful economic forces are turning the whole labor force into an army of freelancers—temps, contingents, and independent contractors and consultants," says business strategy consultant David Kline. The result, he believes, is that "computers, the Net, and telecommuting systems will become as central to the conduct of 21st-century business as the automobile, freeways, and corporate parking lots were to the conduct of mid-20th-century business."[5] The transformation will really gather momentum, some observers feel, when more homes have broadband Internet access (as explained shortly).[6]

Telecommuter
An at-home worker stays connected to his office.

Home Networks

As we shall see, computers linked by telephone lines, cable, or wireless systems are an established component of information technology. Today, however, many new buildings and even homes are built as "network enabled." The new superconnected home or small office is equipped with a *local area network (LAN)*, which allows all the personal computers under the same roof to share peripherals (such as a printer or a fax machine) and a single modem and Internet service.[7] The next development is supposed to be the networking of home appliances, linking stereos, lights, heating systems, phones, and TV sets. Once that has been accomplished, you could walk into your house and give a voice command to turn on the lights or bring up music, for example. Even kitchen appliances would be linked, so that your refrigerator, for instance, could alert a grocery store that you need more milk.[8]

The Information/Internet Appliance

An *information appliance* is a device merging computing capabilities with communications gadgets. Examples include TV set-top boxes, Internet phones, and personal digital assistants. Especially as cable and wireless channels become speedier, these devices will offer the ability to deliver all types of data—text, audio, video, film, still pictures—anywhere at any time.

Smart Television: DTV, HDTV, & SDTV

Today experts differentiate between interactive TV, personalized TV, and Internet TV. *Interactive TV*, which is popular in Europe, lets you interact with the show you're watching, so you can request information about a product or play along with a game show. *Personalized TV* consists of hard-drive–equipped personal video recorders (PVRs), such as TiVo and ReplayTV, that let you not only record shows but also pause, rewind, and replay live TV programs. *Internet TV* lets you read Internet text and web pages on your television set. The foremost example is WebTV; another is Liberate.[9] In the future, interactive, personal, and Internet TV will probably come together in a single box that goes under the umbrella name of digital television. Let's consider this subject.

- **Digital television (DTV):** When most of us tune in our TV sets, we get *analog television*, a system of varying signal amplitude and frequency that represents picture and sound elements. In 1996, however, broadcasters and their government regulator, the Federal Communications Commission (FCC), adopted a standard called **digital television (DTV), which uses a digital signal, or series of 0s and 1s.** DTV is much clearer and less prone to interference than analog TV. (For instance, analog TV has a width-to-height ratio, or aspect ratio, of 4 to 3. One form of digital TV, HDTV, has a ratio of 16 to 9, which is similar to the wide-screen approach used in movies.)

- **High-definition television (HDTV):** If you acquire something advertised as a digital TV, you'll find that it may handle a greater number of channels, but the picture quality won't be much different from traditional analog sets. This is because the cable box often can't communicate with the set. Nevertheless, real DTV is here, in the form known as **high-definition television (HDTV), the high-resolution type of DTV,** which comes in either a 720- or 1080-line mode (compared to 525-line resolution for analog TV). Why don't more people have HDTV sets? The biggest reason is expense—a set may cost $7000, although 95% of the regular televisions purchased in the U.S. retail for under $1500.[10] Another reason is that broadcasters and content producers haven't yet fully backed the standard. Indeed, they might well favor another DTV standard known as SDTV.

- **Standard-definition television (SDTV):** HDTV takes a lot of bandwidth that broadcasters could use instead for **standard-definition television (SDTV), which has a minimum of 480 vertical lines, allowing broadcasters to transmit more information within the HDTV bandwidth.** SDTV would enable broadcasters to effectively multicast their products, on up to as many as six channels instead of one.[11] It also frees up bandwidth for data transmission. Thus, in the future there might be separate channels carrying video, audio, and data.

CLICK-ALONG 6-2
More on the practical uses of communications

CONCEPT CHECK

Explain videoconferencing, workgroup computing, telecommuting and virtual offices, home networks, and information/Internet appliances.

Distinguish among DTV, HDTV, and SDTV.

6.3 Communications Channels: The Conduits of Communications

KEY QUESTION
What are types of wired and wireless channels and some types of wireless communications?

It used to be that two-way individual communications were accomplished mainly in two ways. They were carried by (1) a telephone wire or (2) a wireless method such as shortwave radio. Today there are many kinds of communications channels, although they are still wired or wireless. A **communications channel** is the path—the physical medium—over which information travels in a telecommunications system from its source to its destination. (Channels are also called *links, lines,* or *media.*)

- The electromagnetic spectrum
- Wired communications channels
- Wireless communications channels
- Types of long-distance wireless communications
- Compression and decompression

The Radio Spectrum & Bandwidth

Telephone signals, radar waves, and the invisible commands from a garage-door opener all represent different waves on what is called the electromagnetic spectrum. In the middle of the electromagnetic spectrum is the **radio frequency spectrum**, fields of electrical energy and magnetic energy that carry communications signals. *(See • Panel 6.4, next page.)*

The waves vary according to *frequency*—the number of times a wave repeats, or makes a cycle, in a second. The radio spectrum ranges from low-frequency waves, such as those used for aeronautical and marine navigation equipment, through the medium frequencies for CB radios, cordless phones, and baby monitors, to ultrahigh frequency bands for cell phones and also microwave bands for communications satellites.

A range of frequencies is called a **band** or **bandwidth**. Bandwidth is a measure of the amount of information that can be delivered within a given period of time. For analog signals, bandwidth is expressed in hertz (Hz), or cycles per second. For digital signals, bandwidth is expressed in bits per second (bps). In the United States, certain bands are assigned by the Federal Communications Commission for certain purposes—cell phones within one range, automated teller machines within another, broadcast television within yet another, and so on.

The bandwidth is the difference between the lowest and the highest frequencies transmitted. For example, cellular phones operate within the range 800–900 megahertz—that is, their bandwidth is 100 megahertz. *The wider the bandwidth, the faster data can be transmitted. The narrower the band, the greater the loss of transmission power.* This loss of power must be overcome by using relays or repeaters that rebroadcast the original signal. **Broadband connections** are characterized by very high speed. For instance, the connections that carry broadcast video range in bandwidth from 10 megabits to 30 gigabits per second.

● **PANEL 6.4**

The radio frequency spectrum

The radio frequency spectrum, which carries most communications signals, appears as part of the electromagnetic spectrum.

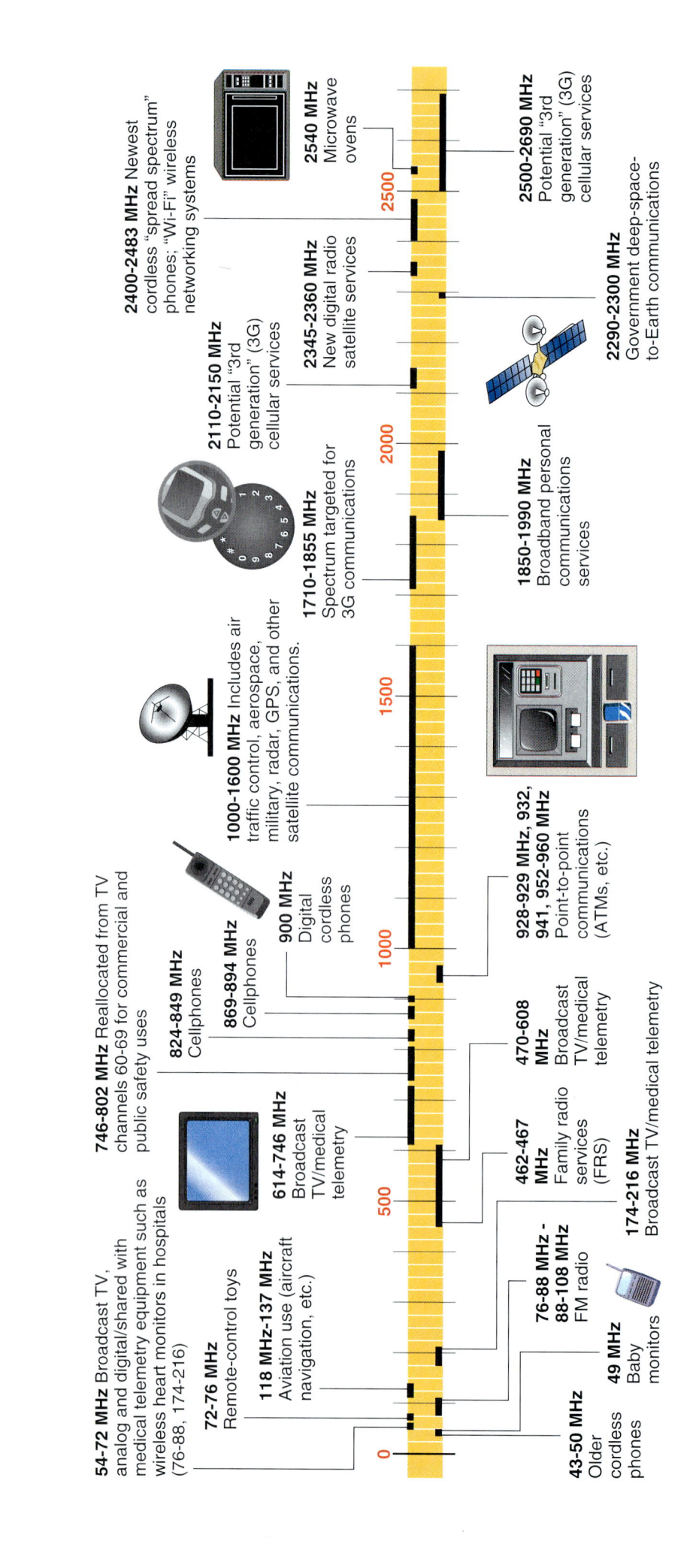

CLICK-ALONG 6-3
More about the electromagnetic spectrum

43-50 MHz Older cordless phones

49 MHz Baby monitors

54-72 MHz Broadcast TV, analog and digital/shared with medical telemetry equipment such as wireless heart monitors in hospitals (76-88, 174-216)

72-76 MHz Remote-control toys

76-88 MHz - 88-108 MHz FM radio

118 MHz-137 MHz Aviation use (aircraft navigation, etc.)

174-216 MHz Broadcast TV/medical telemetry

462-467 MHz Family radio services (FRS)

470-608 MHz Broadcast TV/medical telemetry

614-746 MHz Broadcast TV/medical telemetry

746-802 MHz Reallocated from TV channels 60-69 for commercial and public safety uses

824-849 MHz Cellphones

869-894 MHz Cellphones

900 MHz Digital cordless phones

928-929 MHz, 932, 941, 952-960 MHz Point-to-point communications (ATMs, etc.)

1000-1600 MHz Includes air traffic control, aerospace, military, radar, GPS, and other satellite communications.

1710-1855 MHz Spectrum targeted for 3G communications

1850-1990 MHz Broadband personal communications services

2110-2150 MHz Potential "3rd generation" (3G) cellular services

2290-2300 MHz Government deep-space-to-Earth communications

2345-2360 MHz New digital radio satellite services

2400-2483 MHz Newest cordless "spread spectrum" phones; "Wi-Fi" wireless networking systems

2500-2690 MHz Potential "3rd generation" (3G) cellular services

2540 MHz Microwave ovens

Chapter 6

222

Wired Communications Channels: Transmitting Data by Physical Means

Three types of wired channels are twisted-pair wire (conventional telephone lines), coaxial cable, and fiber-optic cable.

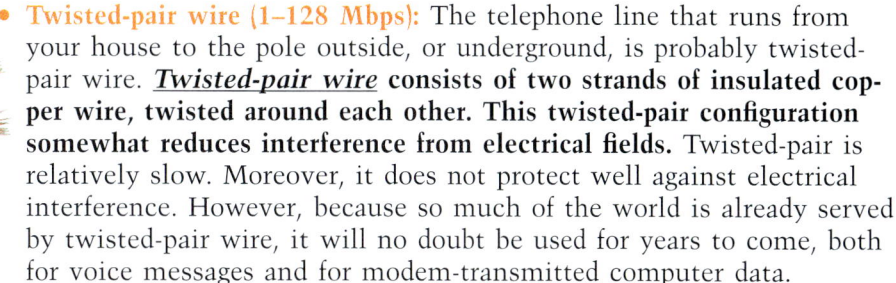

Twisted-pair wire

- **Twisted-pair wire (1–128 Mbps):** The telephone line that runs from your house to the pole outside, or underground, is probably twisted-pair wire. **_Twisted-pair wire_ consists of two strands of insulated copper wire, twisted around each other. This twisted-pair configuration somewhat reduces interference from electrical fields.** Twisted-pair is relatively slow. Moreover, it does not protect well against electrical interference. However, because so much of the world is already served by twisted-pair wire, it will no doubt be used for years to come, both for voice messages and for modem-transmitted computer data.

    The prevalence of twisted-pair wire gives rise to what experts call the "final mile problem." That is, it is relatively easy for telecommunications companies to upgrade the physical connections between cities and even between neighborhoods. But it is expensive for them to replace the "final mile" of twisted-pair wire that connects to individual houses.

Coaxial cable

- **Coaxial cable (up to 200 Mbps):** **_Coaxial cable_, commonly called "co-ax," consists of insulated copper wire wrapped in a solid or braided metal shield, then in an external cover.** Co-ax is widely used for cable television. Thanks to the extra insulation, coaxial cable is much better than twisted-pair wiring at resisting noise. Moreover, it can carry voice and data at a faster rate (up to 200 megabits per second). Often many coaxial cables will be bundled together.

Fiber-optic cable

- **Fiber-optic cable (100 Mbps to 2.4 Gbps):** A _fiber-optic cable_ **consists of dozens or hundreds of thin strands of glass or plastic that transmit pulsating beams of light rather than electricity.** These strands, each as thin as a human hair, can transmit up to 2 billion pulses per second (2 Gbps); each "on" pulse represents one bit. When bundled together, fiber-optic strands in a cable 0.12 inch thick can support a quarter- to a half-million voice conversations at the same time. Moreover, unlike electrical signals, light pulses are not affected by random electromagnetic interference in the environment. Thus, they have much lower error rates than normal telephone wire and cable. In addition, fiber-optic cable is lighter and more durable than twisted-pair and co-ax cable. A final advantage is that it cannot easily be wiretapped, so transmissions are more secure.

The various kinds of wired Internet connections discussed in Chapter 2—dial-up modem, DSL, ISDN, cable modem, T1 lines—are effected through these alternative wired channels.

Wireless Communications Channels: Transmitting Data through the Air

Four types of wireless channels are infrared transmission, broadcast radio, microwave radio, and communications satellite.

- **Infrared transmission (1–4 Mbps):** _Infrared wireless transmission_ **sends data signals using infrared-light waves.** Infrared ports can be found on some laptop computers and printers, as well as wireless mouses. The drawbacks are that _line-of-sight_ communication is required—there must be an unobstructed view between transmitter and receiver—and transmission is confined to short range.
- **Broadcast radio (up to 2 Mbps):** When you tune in to an AM or FM radio station, you are using _broadcast radio_, **a wireless transmission**

Infrared access device

Microwave radio
These dishes are on Midway Island, 1100 miles from Hawaii.

medium that sends data over long distances—between regions, states, or countries. A transmitter is required to send messages and a receiver to receive them; sometimes both sending and receiving functions are combined in a *transceiver.*

In the lower frequencies of the radio spectrum, several broadcast radio bands are reserved not only for conventional AM/FM radio but also for broadcast television, CB (citizens band) radio, ham (amateur) radio, cellular phones, and private radio land mobile services (such as police, fire, and taxi dispatch). Some organizations use specific radio frequencies and networks to support wireless communications. For example, UPC (Universal Product Code) readers are used by grocery-store clerks restocking store shelves to communicate with a main computer so that the store can control inventory levels.

- **Microwave radio (45 Mbps):** <u>Microwave radio</u> **transmits voice and data through the atmosphere as super-high-frequency radio waves called microwaves,** which vibrate at 1 gigahertz (1 billion hertz) per second or higher. These frequencies are used not only to operate microwave ovens but also to transmit messages between ground-based stations and satellite communications systems.

  Nowadays dish- or horn-shaped microwave reflective dishes, which contain transceivers and antennas, are nearly everywhere—on towers, buildings, and hilltops.

  Why, you might wonder, do we have to interfere with nature by putting a microwave dish on top of a mountain? As with infrared waves, microwaves are line-of-sight; they cannot bend around corners or around the earth's curvature, so there must be an unobstructed view between transmitter and receiver. Thus, microwave stations need to be placed within 25–30 miles of each other, with no obstructions in between. The size of the dish varies with the distance (perhaps 2–4 feet in diameter for short distances, 10 feet or more for long distances). A string of microwave relay stations will each receive incoming messages, boost the signal strength, and relay the signal to the next station.

  More than half of today's telephone system uses dish microwave transmission. However, the airwaves are becoming so saturated with microwave signals that future needs will have to be satisfied by other channels, such as satellite systems.

- **Communications satellites:** To avoid some of the limitations of microwave earth stations, communications companies have added microwave "sky stations"—communications satellites. <u>**Communications satellites**</u> **are microwave relay stations in orbit around the earth.** Transmitting a signal from a ground station to a satellite is called *uplinking;* the reverse is called *downlinking.* The delivery process will be slowed if, as is often the case, more than one satellite is required to get the message delivered.

  Satellite systems may occupy one of three zones in space: *GEO, MEO,* and *LEO.*

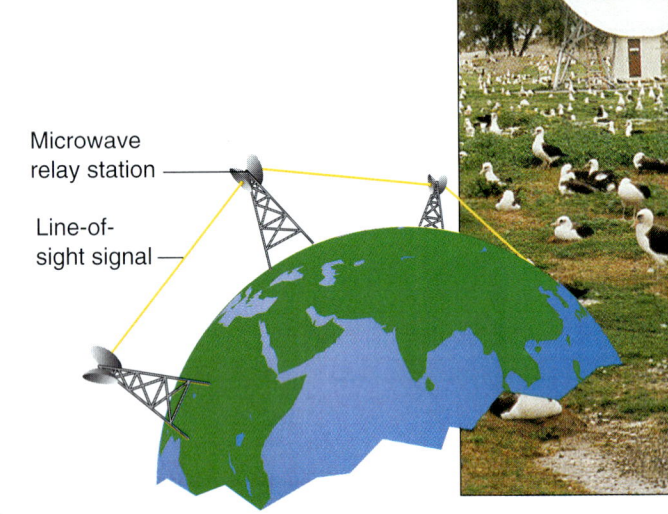

Microwave relay station
Line-of-sight signal

Communications satellite

The highest level, known as *geostationary earth orbit (GEO)*, is 22,300 miles and up and is always directly above the equator. Because the satellites in this orbit travel at the same speed as the earth, they appear to an observer on the ground to be stationary in space—that is, they are *geostationary*. Consequently, microwave earth stations are always able to beam signals to a fixed location above. The orbiting satellite has solar-powered transceivers to receive the signals, amplify them, and retransmit them to another earth station. At this high orbit, fewer satellites are required for global coverage; however, their quarter-second delay makes two-way conversations difficult.

The *medium-earth orbit (MEO)* is 5000–10,000 miles up. It requires more satellites for global coverage than GEO.

The *low-earth orbit (LEO)* is 400–1000 miles up and has no signal delay. LEO satellites may be smaller and are cheaper to launch.

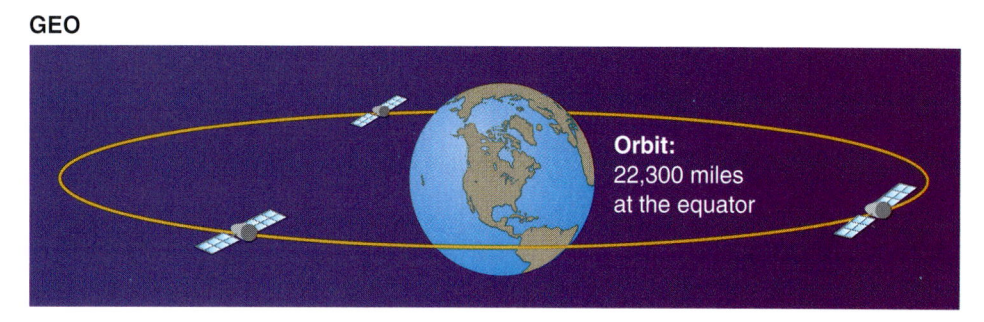

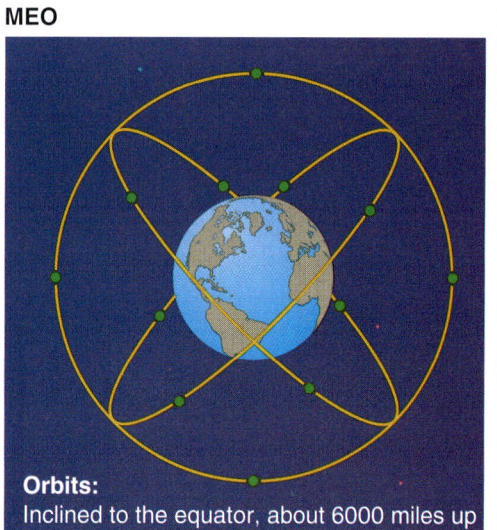

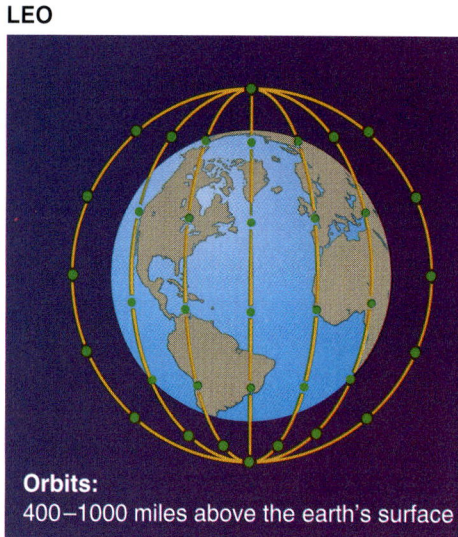

GEO
Orbit: 22,300 miles at the equator

MEO
Orbits: Inclined to the equator, about 6000 miles up

LEO
Orbits: 400–1000 miles above the earth's surface

A GPS car-navigation system

Types of Long-Distance Wireless Communications

There are essentially two ways to move information through the air long-distance on radio frequencies. The first is via *one-way communications*, as typified by the satellite navigation system known as the Global Positioning System, and by most pagers. The second is via *two-way communications*: (1) two-way pagers, (2) analog cellular phones, (3) 2G digital wireless, and (4) 3G digital wireless. (Other wireless methods operate at short distances.)

- **One-way communications—the Global Positioning System:** A $10 billion infrastructure developed by the military in the mid-1980s, the <u>**Global Positioning System (GPS)**</u> **consists of a series of earth-orbiting satellites continuously transmitting timed radio signals that can be used to identify earth locations.** A GPS receiver—handheld or mounted in a vehicle, plane, or boat—can pick up transmissions from any four satellites, interpret the information from each, and pinpoint the receiver's

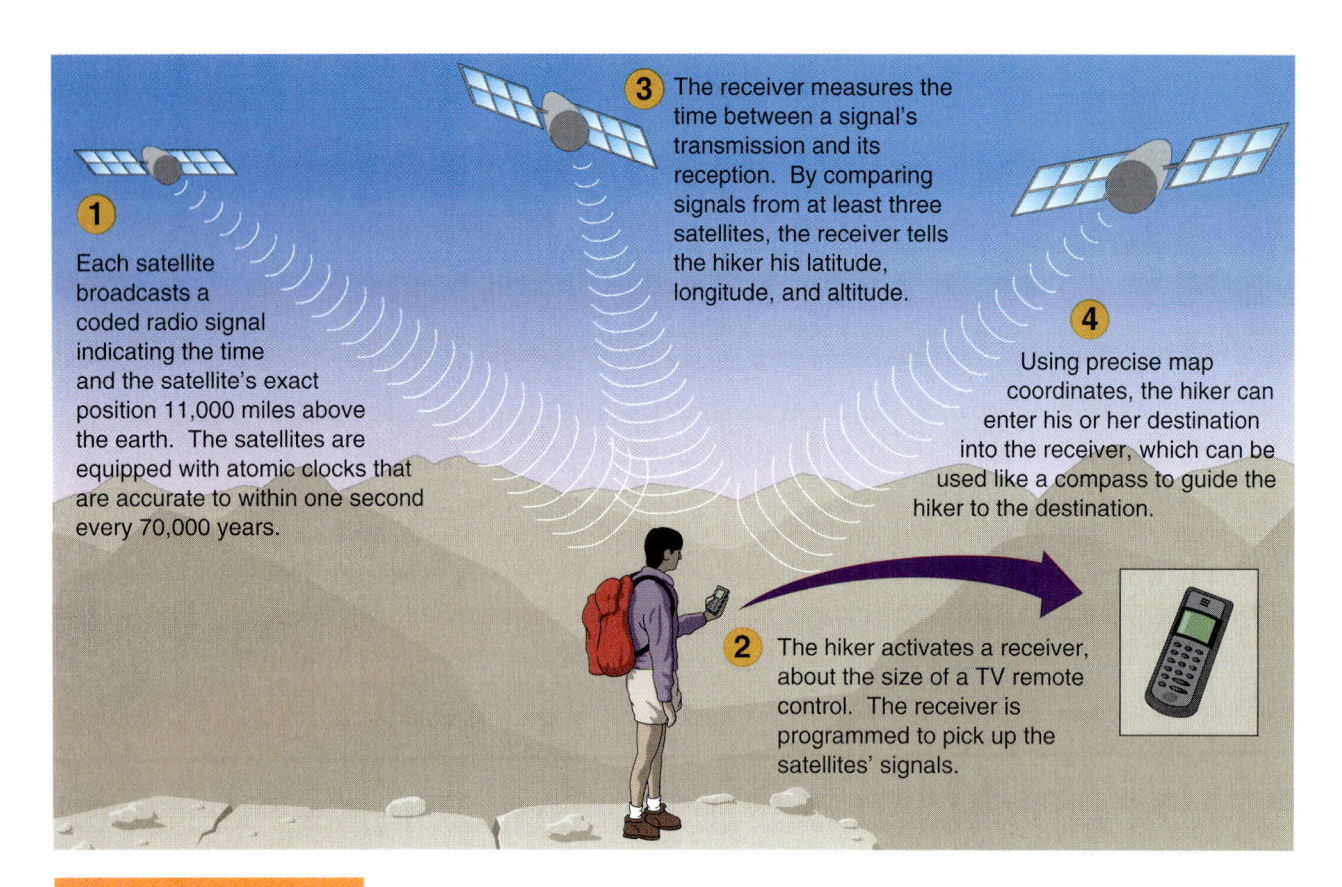

1 Each satellite broadcasts a coded radio signal indicating the time and the satellite's exact position 11,000 miles above the earth. The satellites are equipped with atomic clocks that are accurate to within one second every 70,000 years.

3 The receiver measures the time between a signal's transmission and its reception. By comparing signals from at least three satellites, the receiver tells the hiker his latitude, longitude, and altitude.

4 Using precise map coordinates, the hiker can enter his or her destination into the receiver, which can be used like a compass to guide the hiker to the destination.

2 The hiker activates a receiver, about the size of a TV remote control. The receiver is programmed to pick up the satellites' signals.

● **PANEL 6.5**
GPS receiver: hand-held compass
The Global Positioning System uses 24 satellites, developed for military use, to pinpoint a location on the earth's surface.

Survival Tip

Pagers Help the Deaf

People who are hard of hearing (10% of the population) often use two-way pagers supported by Wynd Communications. The tiny keyboard allows users to type messages that can be e-mailed, faxed, or sent to a text telephone, or TTY device, that allows text messages to be sent over phone lines.

longitude, latitude, and altitude. (See ● Panel 6.5.) Some GPS receivers include map software for finding your way around, as with the navigation systems available with some rental cars.

The system, accurate within 3–60 feet, is used to tell military units carrying special receivers where they are. GPS is used for such civilian activities as tracking trucks and taxis, locating stolen cars, orienting hikers, and aiding in surveying. Some public-transportation systems have installed GPS receivers on buses, where they can tell drivers when they fall behind schedule. GPS has also been used by scientists to keep a satellite watch over a Hawaiian volcano, Mauna Loa, and to capture infinitesimal movements that may be used to predict eruptions.[12]

By January 2003, cellular carriers are required to have E-911 (for Enhanced 911) capability. This enables them to locate, through tiny GPS receivers embedded in users' cellphones, the position of every person making an emergency 911 call.

- **One-way communications—pagers:** Once stereotyped as devices for doctors and drug dealers, pagers are now consumer items. Commonly known as beepers, for the sound they make when activated, **_pagers are simple radio receivers that receive data (but not voice messages) sent from a special radio transmitter._** Often the pager has its own telephone number. When the number is dialed from a phone, the call goes by way of the transmitter straight to the designated pager. Pagers are very efficient for transmitting one-way information—emergency messages, news, prices, stock quotations, delivery-route assignments, even sports news and scores—at low cost to single or multiple receivers.

Pagers do more than beep, transmitting full-blown alphanumeric text (such as four-line, 80-character messages) and other data. Newer ones are mini-answering machines, capable of relaying digitized voice messages.

- **Two-way communications—pagers:** Recent advances have given us _two-way paging_ or _enhanced paging._ Service is provided by carriers like SkyTel and Go.Web, and you can use paging gadgets such as the BlackBerry RIM 957. For instance, in one version, users can send a preprogrammed message or acknowledgment that they have received a message. Another version allows consumers to compose and send e-mail to anyone on the Internet and to other pagers. Typing a message on a tiny keyboard no larger than those on pocket calculators can pose a challenge, however.

- **Two-way communications—first-generation analog cellular services (1G):** <u>Analog cellular phones</u> **are designed primarily for communicating by voice through a system of ground-area cells. Each** <u>cell</u> **is hexagonal in shape, usually 8 miles or less in diameter, and is served by a transmitter-receiving tower.** Communications are handled in the bandwidth of 824–894 megahertz. Calls are directed between cells by a mobile telephone switching office. Movement between cells requires that calls be "handed off" by this switching office. (See ● Panel 6.6.) This technology is known as *1G*, for *first generation*.

    Handing off voice calls between cells poses only minimal problems. However, handing off data transmission (where every bit counts), with the inevitable gaps and pauses on moving from one cell to another, is much more difficult.

- **Two-way communications—second-generation digital wireless services (2G):** <u>Digital wireless services</u>—**which support digital cellphones and personal digital assistants—use a network of cell towers to send voice communications and data over the airwaves in digital form.** Known as *2G*, for *second-generation, technology*, digital cellphones began replacing analog cellphones during the 1990s as telecommunications companies added digital transceivers to their cell towers. 2G technology not only dramatically increased voice clarity; it also allowed the telecoms to cram many more voice calls into the same slice of bandwidth.

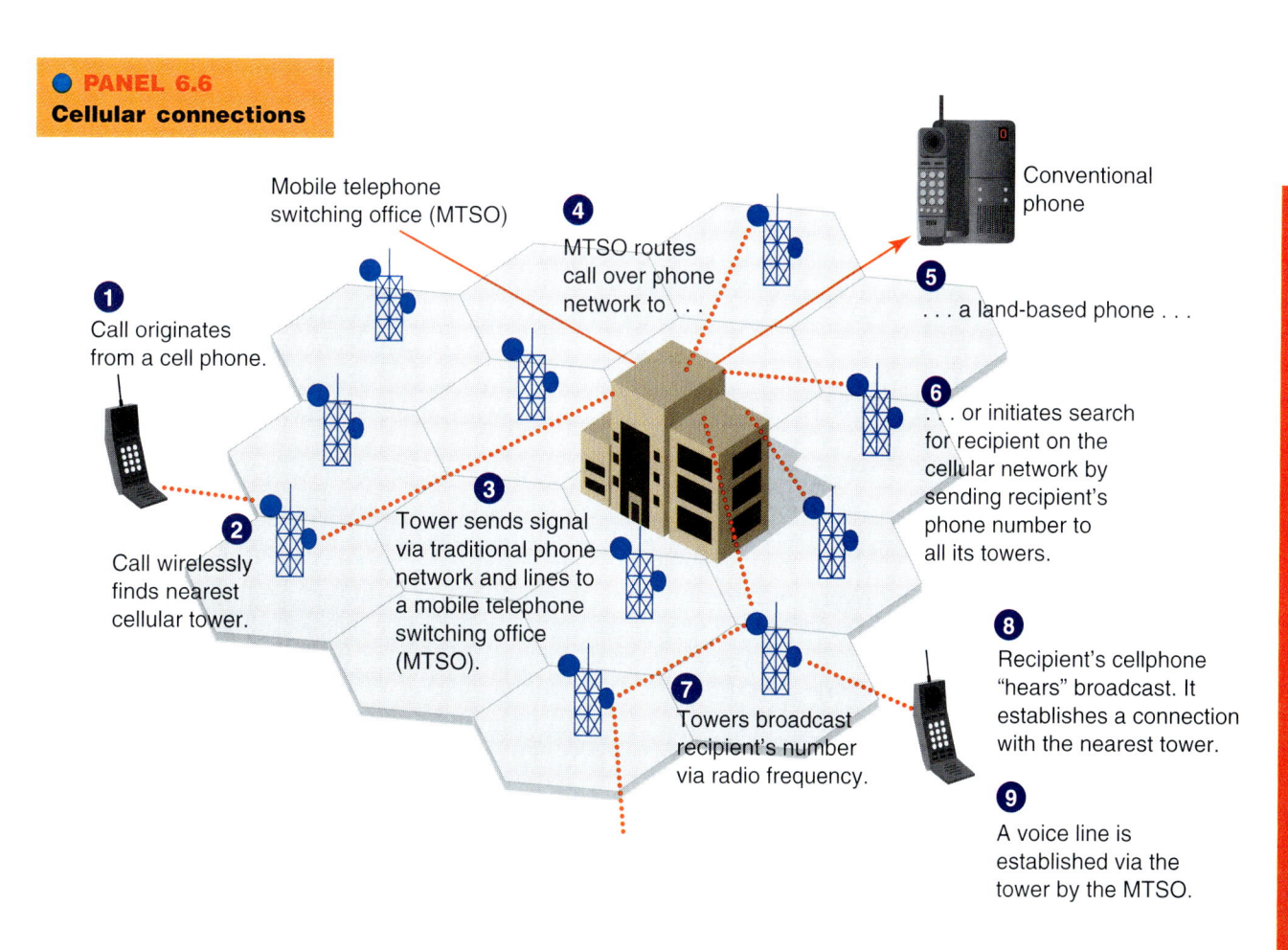

● **PANEL 6.6**
Cellular connections

1 Call originates from a cell phone.

2 Call wirelessly finds nearest cellular tower.

3 Tower sends signal via traditional phone network and lines to a mobile telephone switching office (MTSO).

4 MTSO routes call over phone network to . . .

5 . . . a land-based phone . . .

6 . . . or initiates search for recipient on the cellular network by sending recipient's phone number to all its towers.

7 Towers broadcast recipient's number via radio frequency.

8 Recipient's cellphone "hears" broadcast. It establishes a connection with the nearest tower.

9 A voice line is established via the tower by the MTSO.

Mobile telephone switching office (MTSO)

Conventional phone

Telecommunications

International phones
Many parts of the world adhere to a universal cellphone standard, but not the United States.

Even so, transmission speeds have a maximum of only 14.4 kilobits per second, and even that rate is rarely achieved. This means that today's 2G digital cellphones and wireless devices transmit data so slowly that they are, like the 1G analog devices, best used for talking rather than for Internet access, although they can handle e-mail and web pages in a limited way.

Other countries adhere to a single standard for wireless, but in the United States there are four incompatible digital technologies operating at different frequencies in the electromagnetic spectrum: *TDMA, GSM, CDMA,* and *iDEN*. Thus, there is no true national coverage. In the future, however, it appears that GSM and CDMA will prevail in the United States. GSM has become the standard not only in Western Europe but in many other countries, especially in the Middle East and Asia. Unfortunately, in the United States. GSM is on a different frequency from the GSM phones in the rest of the world, which limits use of an American GSM cellphone in other countries. (There are a few phones on the market that include both U.S. and European-based GSM frequencies.)

- **Two-way communications planned for the future—third-generation broadband wireless digital services (3G):** <u>Broadband wireless digital services</u>, usually referred to as *third-generation (3G) technology*, **are based on the GSM standard, are (like cable modems) "always on," carry data at high speeds (56 kilobits per second up to about 2 megabits per second eventually), and are able to quickly transmit video, still pictures, and music, along with offering better ways to tap into websites than today's 2G wireless systems.** The leaders in 3G cellphone technology are Japan and Europe; the United States lags behind both. Worldwide 3G service is supposed to occur by 2004, but many observers are wondering if it will ever happen.

   The impressive application for 3G is that you can send pictures and data. There is some question, however, whether enough consumers actually see the need to walk (or drive) while watching videos, say, on their cellphones or PDAs. A 2001 survey by Accenture found that only 15% of cellphone users ever make use of data capabilities. Moreover, European telecoms have spent billions of dollars to acquire 3G licenses, yet they need to spend billions more to upgrade the networks and to launch marketing campaigns. Meanwhile, in the United States, there are many problems, not the least of which is that the frequencies needed for 3G are presently reserved for military and security agencies, which won't readily give them up.

- **Two-way communication in the near future—2.5G, a compromise between 2G and 3G:** Because of the uncertainties about 3G, many telecom carriers are taking a wait-and-see attitude. Thus, many have decided to test the waters using 2.5G technology. *2.5 digital wireless services* consist of a minor hardware upgrade that uses the existing GSM standards and same frequencies as 2G networks but offers one important difference: It is "always on." That is, it is instantly available; you don't have to wait 30 seconds after dialing in for the cellphone or mobile device to connect to the network. You wouldn't get the full broadband experience possible with cable modems and ISDN fixed-wired systems. But you would still get data such as headlines, sports scores, and traffic updates, which would come to your mobile wireless device anytime it's on—and on a separate channel from your voice calls. Moreover, your data speeds would be two to three times faster than is now currently possible with 2G devices, which will make using the wireless Web easier.[13]

Short-Range Wireless Communications at 2.4 Gigahertz: Bluetooth, WiFi, & HomeRF

We have discussed the standards for wireless digital communications in the 800–1900 megahertz part of the radiowave spectrum, which are considered long-range waves. Now let us consider those in the 2.4-gigahertz part of the radio spectrum, which are short-range and effective only within several feet of a wireless access point. The 2.4-gigahertz band is available globally for unlicensed, low-power uses and is set aside as an innovation zone where new devices can be tested without the need for a government license.

There are three such short-distance wireless standards—Bluetooth, WiFi, and HomeRF.[14]

- Bluetooth—up to 30 feet: _Bluetooth_ **is a short-range wireless digital standard aimed at linking cellphones, PDAs, computers, and peripherals up to distances of 30 feet.** That is, Bluetooth is principally designed to replace cables connecting PCs to printers and PDAs or wireless phones. The concept is named for the 10th-century Danish king who unified Denmark and Norway. In Windows XP, Microsoft chose not to support the Bluetooth standard.

- WiFi—up to 300 feet: Known formally as 802.11b (for the wireless technical standard specified by the Institute of Electrical and Electronics Engineers, _WiFi_ **is a short-range wireless digital standard aimed at helping machines inside offices to communicate at high speeds and share Internet connections at distances up to 300 feet; it connects to a kind of local area network known as the Ethernet.** The Ethernet standard for wiring computers into local networks has been popular for two decades because it is an "open" standard—no single company controls it. Because of this openness and because it is cheaper than Bluetooth, WiFi may become the more popular standard. It is also 10 times faster than Bluetooth and has about 10 times the range. Some enthusiasts have set up transmitters on rooftops, thereby distributing Internet service throughout their neighborhoods. WiFi is supported by Windows XP and is becoming a standard in most new computers.

- HomeRF—up to 150 feet: _HomeRF_ is a separate, incompatible standard designed to network up to 10 PCs and peripherals as far as 150 feet apart. Many observers think HomeRF will probably not succeed because people will not want to use one wireless card (for WiFi or Bluetooth) at the office and another in one's home PC.

Wireless digital phones are sometimes called "WAP phones." _WAP_—short for _Wireless Application Protocol_—is the main set of communications conventions for connecting wireless uses to the World Wide Web. Both Bluetooth and WiFi use WAP, which requires websites to strip out graphics and shorten stories.

Compression & Decompression: Putting More Data in Less Space

The vast streams of text, audio, and visual information threaten to overwhelm us. The file of a 2-hour movie, for instance, contains so much sound and visual information that, if stored without modification on a standard CD-ROM, it would require 360 disk changes during a single showing. A broadcast of _Oprah_ that presently fits into one conventional, or analog, television channel would require 45 channels if sent in digital language.

To fit more data into less space, we use the mathematical process called compression. _Compression_, or _digital-data compression_, **is a method of removing repetitive elements from a file so that the file requires less storage space and therefore less time to transmit.** Before we use the data, it is decompressed—the repeated patterns are restored. These methods are sometimes referred to as *codec* (for *compression/decompression*) techniques.

- **Lossless versus lossy compression:** There are two principal methods of compressing data—lossless and lossy. In any situation, which of these two techniques is more appropriate will depend on whether data quality or storage space is more critical.

    Lossless compression uses mathematical techniques to replace repetitive patterns of bits with a kind of coded summary. During decompression, the coded summaries are replaced with the original patterns of bits. In this method, the data that comes out is exactly the same as what went in; it has merely been repackaged for purposes of storage or transmission. Lossless techniques are used when it's important that nothing be lost—for instance, for computer data, database records, spreadsheets, and word processing files.

    Lossy compression techniques permanently discard some data during compression. Lossy data compression involves a certain loss of accuracy in exchange for a high degree of compression (to as little as 5% of the original file size). This method of compression is often used for graphics files and sound files. Thus, a lossy codec might discard subtle shades of color or very soft sounds. Most users wouldn't notice the absence of these details.

- **Compression standards—JPEG and MPEG:** Several standards exist for compression, particularly of visual data. Data recorded and compressed in one standard cannot be played back in another. The main reason for the lack of agreement is that different industries have different priorities. What will satisfy the users of still photographs, for instance, will not work for the users of movies.

    As we have seen, lossless compression schemes are used for text and numeric data files, whereas lossy compression schemes are used with graphics and video files. The principal lossy compression schemes are *JPEG* and *MPEG*.

    The leading compression standard for still images is <u>JPEG</u> (pronounced "jay-peg"), which stands for the Joint Photographic Experts Group of the International Standards Organization. The file extension that identifies graphic images files in the JPEG format is *.jpeg* or *.jpg*. In storing and transmitting still photographs, the data must remain of high quality. The JPEG codec looks for a way to squeeze a single image, mainly by eliminating repetitive pixels (picture-element dots) within the image. Higher or lower degrees of JPEG compression may be chosen; greater compression corresponds to greater image loss.

    The leading compression standard for moving images is <u>MPEG</u> ("em-peg"), for Motion Picture Experts Group. The file extension that identifies video (and sound) files compressed in this format is *.mpeg* or *.mpg*. People who work with videos are mainly interested in storing or transmitting an enormous amount of visual information in economical form; preserving details is a secondary consideration. The Motion Picture Experts Group sets standards for weeding out redundancies between neighboring images in a stream of video. Three MPEG standards have been developed for compressing visual information—MPEG-1, MPEG-2, and MPEG-4.

CONCEPT CHECK

What are radio frequency spectrums and bandwidth?

What are wired communications channels? Wireless channels?

Describe how information is moved through the air long-distance—one-way methods and two-way methods.

What is Bluetooth? WiFi? HomeRF?

Discuss compression and decompression and the leading compression standards.

6.4 Networks

KEY QUESTIONS
What are the benefits of networks, and what are their types, components, and variations?

Whether wired, wireless, or both, all the channels we've described can be used singly or in mix-and-match fashion to form networks. A *__network__, or communications network,* **is a system of interconnected computers, telephones, or other communications devices that can communicate with one another and share applications and data.** The tying together of so many communications devices in so many ways is changing the world we live in.

The Benefits of Networks

People and organizations use computers in networks for several reasons. These include the following:

- **Sharing of peripheral devices:** Peripheral devices such as laser printers, disk drives, and scanners are often quite expensive. Consequently, to justify their purchase, management wants to maximize their use. Usually the best way to do this is to connect the peripheral to a network serving several computer users.

- **Sharing of programs and data:** In most organizations, people use the same software and need access to the same information. It is less expensive for a company to buy a separate word processing program that will serve many employees than to buy a separate word processing program for each employee.

    Moreover, if all employees have access to the same data on a shared storage device, the organization can save money and avoid serious problems. If each employee has a separate machine, some employees may update customer addresses, while others remain ignorant of the changes. Updating information on a shared server is much easier than updating every user's individual system.

    Finally, network-linked employees can more easily work together online on shared projects.

- **Better communications:** One of the greatest features of networks is electronic mail. With e-mail, everyone on a network can easily keep others posted about important information.

- **Security of information:** Before networks became commonplace, an individual employee might be the only one with a particular piece of information, which was stored in his or her desktop computer. If the employee was dismissed—or if a fire or flood demolished the office—the company would lose that information. Today such data would be backed up or duplicated on a networked storage device shared by others. (On the other hand, networks that can be accessed remotely present new risks of unauthorized access of data, as we discuss in Chapter 8 in considering hackers.)

- **Access to databases:** Networks enable users to tap into numerous databases, whether private company databases or public databases available online through the Internet.

Types of Networks: WANs, MANs, & LANs

Networks, which consist of various combinations of computers, storage devices, and communications devices, may be divided into three main categories, differing primarily in their geographical range.

- **Wide area network:** A _wide area network (WAN)_ **is a communications network that covers a wide geographical area, such as a country or the world.** Most long-distance and regional Bell telephone companies are WANs. A WAN may use a combination of satellites, fiber-optic cable, microwave, and copper wire connections and link a variety of computers, from mainframes to terminals. *(See ● Panel 6.7.)*

- **Metropolitan area network:** A _metropolitan area network (MAN)_ **is a communications network covering a city or a suburb.** The purpose of a MAN is often to bypass local telephone companies when accessing long-distance services. Many cellular phone systems are MANs.

- **Local area network:** A _local area network (LAN)_ **connects computers and devices in a limited geographical area,** such as one office, one building, or a group of buildings close together (for instance, a college campus). A small LAN in a modest office, or even in a home, might link a file server with a few terminals or PCs and a printer or two. Such small LANs have been called *TANs,* for "tiny area networks."

Most large computer networks have at least one **host computer**, **a mainframe or midsize central computer that controls the network.** The other devices within the network are called nodes. A **node** **is any device that is attached to a network**—for example, a microcomputer, terminal, storage device, or printer.

Networks may be connected together—LANs to MANs and MANs to WANs. A **backbone** **is a high-speed network that connects LANs and MANs to the Internet.**

● **PANEL 6.7**
Wide area network

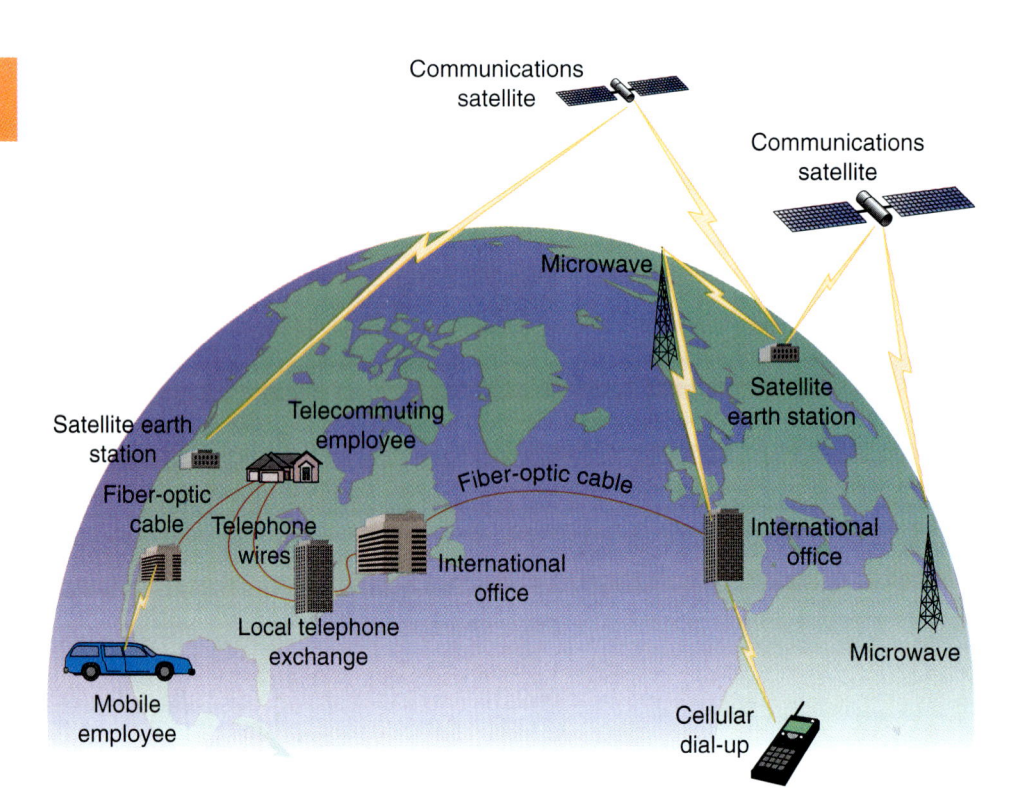

● **PANEL 6.8**

Two types of LANs: client/server and peer-to-peer

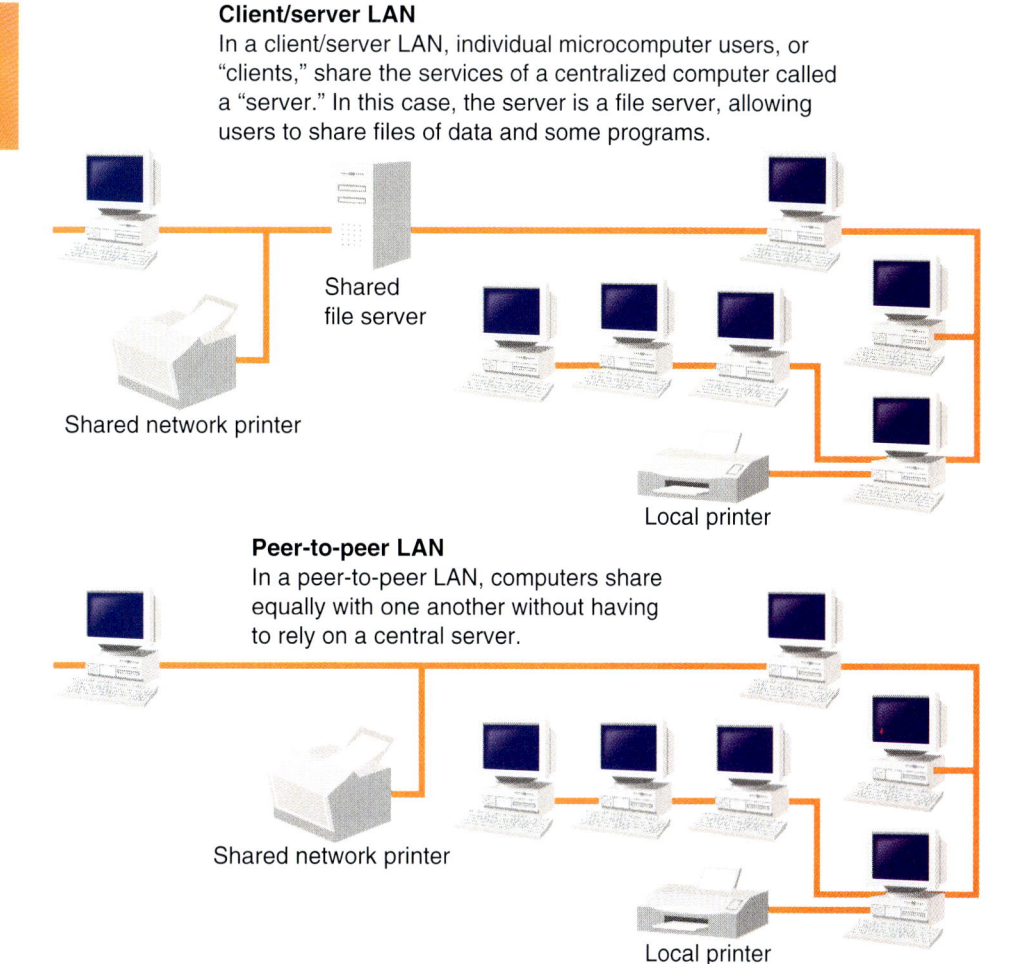

Client/server LAN
In a client/server LAN, individual microcomputer users, or "clients," share the services of a centralized computer called a "server." In this case, the server is a file server, allowing users to share files of data and some programs.

Shared file server

Shared network printer

Local printer

Peer-to-peer LAN
In a peer-to-peer LAN, computers share equally with one another without having to rely on a central server.

Shared network printer

Local printer

Types of LANs: Client-Server & Peer-to-Peer

Local area networks consist of two principal types: client/server and peer-to-peer. *(See ● Panel 6.8.)*

- **Client/server LANs:** A *client/server LAN* consists of clients, which are microcomputers that request data, and servers, which are computers used to supply data. The server is a powerful microcomputer that manages shared devices, such as laser printers. It runs server software for applications such as e-mail and web browsing.

  Different servers may be used to manage different tasks. A *__file server__* **is a computer that acts like a disk drive, storing the programs and data files shared by users on a LAN.** A *database server* is a computer in a LAN that stores data but doesn't store programs. A *print server* controls one or more printers and stores the print-image output from all the microcomputers on the system. *Web servers* contain web pages that can be viewed using a browser. *Mail servers* manage e-mail.

- **Peer-to-peer LANs:** The word *peer* denotes one who is equal in standing with another (as in the phrases "peer pressure" and "jury of one's peers"). In a *__peer-to-peer LAN__*, **all microcomputers on the network communicate directly with one another without relying on a server.** Peer-to-peer networks are less expensive than client-server networks and work effectively for up to 25 computers. Beyond that, they slow down under heavy use. They are appropriate for small networks.

Many LANs mix elements from both client-server and peer-to-peer models.

LAN

At an early-morning flower auction in Alsmeer, Netherlands, buyers bid electronically via a LAN. Within three hours, 17 million flowers will have been snapped up. By noon, the flowers will be on planes jetting toward shops around the world.

Components of a LAN

Local area networks are made up of several standard components.

- **Connection or cabling system:** LANs may use a wired or wireless connection system. Wired connections may be twisted-pair wiring, coaxial cable, or fiber-optic cable. Wireless connections may be infrared, radio-wave transmission, Bluetooth, or WiFi.
- **Microcomputers with network interface cards:** Two or more microcomputers are required, along with network interface cards. As we mentioned in Chapter 5 (p. 160), a *network interface card* enables the computer to send and receive messages over a cable network. The network card can be inserted into an expansion slot in a PC. Alternatively, a network card in a stand-alone box may serve a number of devices. Many new computers come with network cards already installed.
- **Network operating system:** The *network operating system (NOS)* is the system software that manages the activity of a network. The NOS supports access by multiple users and provides for recognition of users based on passwords and terminal identifications. Depending on whether the LAN is client/server or peer-to-peer, the operating system may be stored on the file server, on each microcomputer on the network, or a combination of both.

  Examples of popular NOS software are Novell NetWare, Microsoft Windows NT/2000, Unix, and Linux. Peer-to-peer networking can also be accomplished with Microsoft Windows 95/98/Me/XP and Microsoft Windows for Workgroups.
- **Other shared devices:** Printers, scanners, storage devices, and other peripherals may be added to the network as necessary and shared by all users.
- **Routers, bridges, and gateways:** In principle, a LAN may stand alone. Today, however, it invariably connects to other networks, especially the Internet. Network designers determine the types of hardware and software necessary as interfaces to make these connections. Routers, bridges, and gateways are used for this purpose. (See ● Panel 6.9.)

  A <u>router</u> **is a special computer that directs communicating messages when several networks are connected together.** High-speed routers can serve as part of the Internet backbone, or transmission path, handling the major data traffic.

  A <u>bridge</u> **is an interface used to connect the same types of networks.**

  A <u>gateway</u> **is an interface permitting communication between dissimilar networks**—for instance, between a LAN and a WAN or between two LANs based on different network operating systems or different layouts.

Intranets, Extranets, & Firewalls: Private Internet Networks

Early in the Online Age, businesses discovered the benefits of using the World Wide Web to get information to customers, suppliers, or investors. For example, in the mid-1990s, Federal Express found it could save millions by allowing customers to click through web pages to trace their parcels, instead of having FedEx customer-service agents do it. From there, it was a short step to the application of the same technology inside companies—in internal Internet networks called *intranets*.

PANEL 6.9
Components of a typical LAN

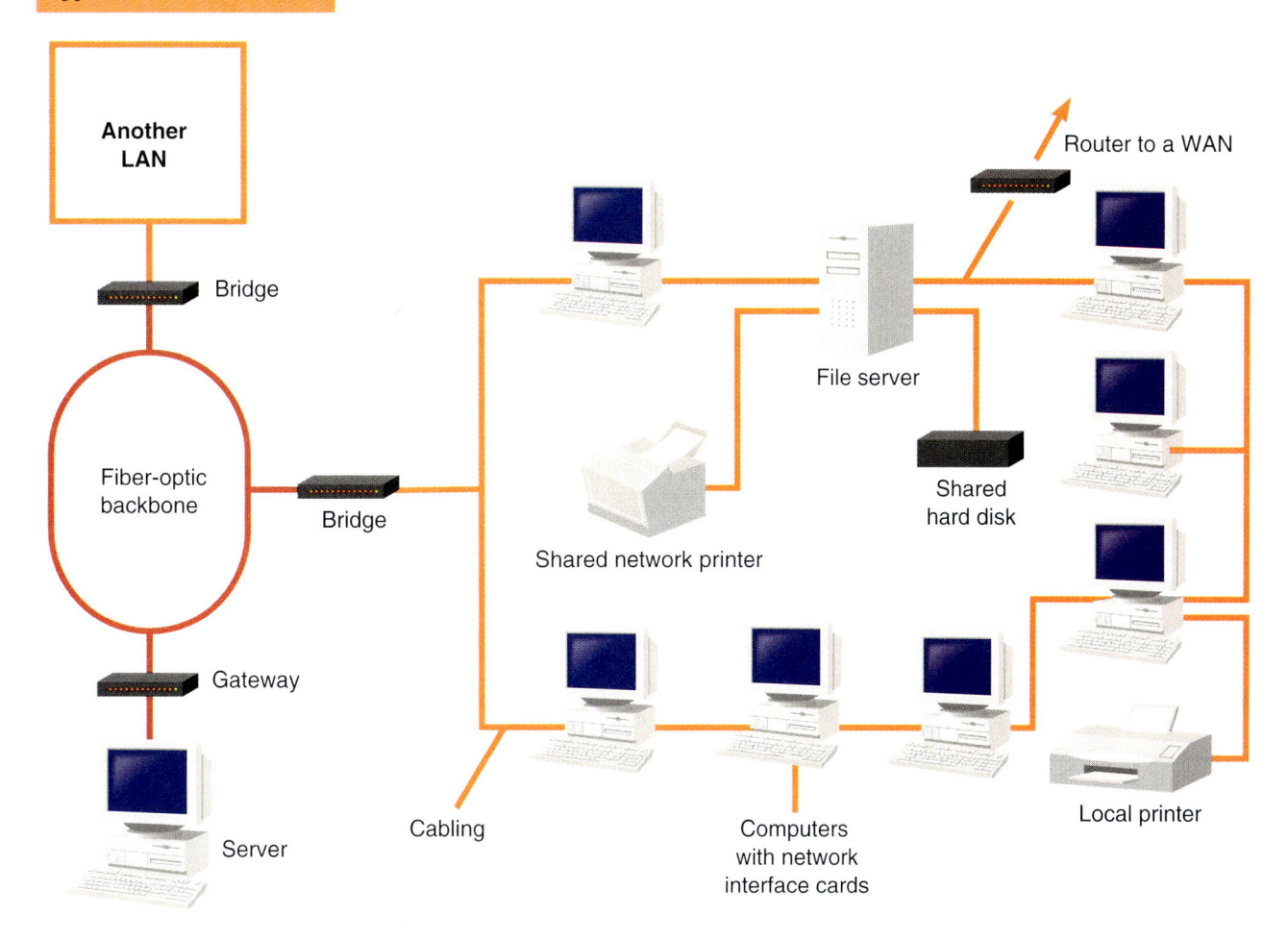

Another LAN

Bridge

Fiber-optic backbone

Bridge

Gateway

Server

Cabling

Shared network printer

File server

Shared hard disk

Computers with network interface cards

Router to a WAN

Local printer

Survival Tip

PC Firewall

Free PC firewall software called Zone Alarm is available at *www.zonelabs.com*. You can go to Gibson Research at *www.grc.com* to run Shields Up! to test your firewall.

- **Intranets—for internal use only:** An *intranet* **is an organization's internal private network that uses the infrastructure and standards of the Internet and the World Wide Web.** When a corporation develops a public website, it is making selected information available to consumers and other interested parties. When it creates an intranet, it enables employees to have quicker access to internal information and to share knowledge so that they can do their jobs better. Information exchanged on intranets may include employee e-mail addresses and telephone numbers, product information, sales data, employee benefit information, and lists of jobs available within the organization.

- **Extranets—for certain outsiders:** Taking intranet technology a few steps further, extranets offer security and controlled access. As we have seen, intranets are internal systems, designed to connect the members of a specific group or a single company. By contrast, *extranets* **are private intranets that connect not only internal personnel but also selected suppliers and other strategic parties.** Extranets have become popular for standard transactions such as purchasing. Ford Motor Company, for instance, has an extranet that connects more than 15,000 Ford dealers worldwide. Called FocalPt, the extranet supports sales and servicing of cars, with the aim of improving service to Ford customers.

- **Firewalls:** Security is essential to an intranet (or even an extranet). Sensitive company data, such as payroll information, must be kept private, by means of a *firewall*. A *firewall* **is a system of hardware and**

Telecommunications

235

software that blocks unauthorized users inside and outside the organization from entering the intranet. The firewall monitors all Internet and other network activity, looking for suspicious data and preventing unauthorized access.

A firewall consists of two parts, a choke and a gate. The *choke* forces all data packets flowing between the Internet and the intranet to pass through a gate. The *gate* regulates the flow between the two networks. It identifies authorized users, searches for viruses, and implements other security measures. Thus, intranet users can gain access to the Internet (including key sites connected by hyperlinks), but outside Internet users cannot enter the intranet.

CONCEPT CHECK

What are the benefits of networks?

Characterize WANs, MANs, and LANs.

What are the differences between client-server and peer-to-peer LANs?

What are the components of a LAN?

What functions do intranets, extranets, and firewalls fulfill?

6.5 Cyberethics: Controversial Material, Censorship, & Privacy Issues

KEY QUESTION
What are important issues in cyberethics?

Communications technology gives us more choices of nearly every sort. It provides us with different ways of working, thinking, and playing. It also presents us with some different moral choices—determining right actions in the digital and online universe. Let's consider some important aspects of "cyberethics"—controversial material and censorship, and matters of privacy.

Controversial Material & Censorship

Ethics

Since computers are simply another way of communicating, there should be no surprise that many people use them to communicate about sex. Yahoo!, the Internet directory company, says that the word "sex" is the most popular search word on the Net.[15] All kinds of online X-rated message boards, chat rooms, and Usenet newsgroups exist. These raise serious issues for parents. Do we want children to have access to sexual conversations, to download hard-core pictures, or to encounter criminals who might try to meet them offline? "Parents should never use [a computer] as an electronic baby sitter," computer columnist Lawrence Magid says. People online are not always what they seem to be, he points out, and a message seemingly from a 12-year-old girl could really be from a 30-year-old man. "Children should be warned never to give out personal information," says Magid, "and to tell their parents if they encounter mail or messages that make them uncomfortable."[16]

What can be done about X-rated materials? Some possibilities:

- **Blocking software:** Some software developers have discovered a golden opportunity in making programs like SurfWatch, Net Nanny, and CYBERsitter. These "blocking" programs screen out objectionable material, typically by identifying certain unapproved keywords in a user's request or comparing the user's request for information against a list of prohibited sites.

- **Browsers with ratings:** Another proposal in the works is browser software that contains built-in ratings for Internet, Usenet, and World Wide Web files. Parents could, for example, choose a browser that has been endorsed by the local school board or the online service provider.
- **The V-chip:** The 1996 Telecommunications Law officially launched the era of the V-chip, a device that will be required equipment in most new television sets. The *V-chip* allows parents to automatically block out programs that have been labeled as high in violence, sex, or other objectionable material.

However, any attempts at restricting the flow of information are hindered by the basic design of the Internet itself, with its strategy of offering different roads to the same place. "If access to information on a computer is blocked by one route," writes the *New York Times*'s Peter Lewis, "a moderately skilled computer user can simply tap into another computer by an alternative route." Lewis cites an Internet axiom attributed to an engineer named John Gilmore: "The Internet interprets censorship as damage and routes around it."[17]

Protecting children
Blocking software can help restrict objectionable material from children.

Privacy

Privacy **is the right of people not to reveal information about themselves.** Technology, however, puts constant pressure on this right.

Consider web cookies, little pieces of data left in your computer by some sites you visit.[18] A *cookie* **is a file that the web server stores on your hard-disk drive when you visit a website.** Thus, unknown to you, a website operator or companies advertising on the site can log your movements within the site. These records provide information that marketers can use to target customers for their products. Other websites can also get access to the cookies and acquire information about you.

There are other intrusions on your privacy. Think your medical records are inviolable? Actually, private medical information is bought and sold freely by various companies since there is no federal law prohibiting it. (And they simply ignore the patchwork of state laws.)

Think the boss can't snoop on your e-mail at work? The law allows employers to "intercept" employee communications if one of the parties involved agrees to the "interception." The party who agrees in this case is the employer. Indeed, employer snooping seems to be widespread.

A great many people are concerned about the loss of their right to privacy. Indeed, one survey found that 80% of the people contacted worried that they had lost "all control" of the personal information being collected and tracked by computers.[19] Although several laws restrain the government's ability to acquire and disseminate information and to listen in on private conversations, there are reasons to be alarmed.

Survival Tip

Cookies

Some websites sell information stored in cookies. If you dislike cookies, you can set your browser to disable them.

CONCEPT CHECK

Discuss some ways to protect children from X-rated material.

What is privacy?

Why should intellectual property copyrights be respected?

Summary

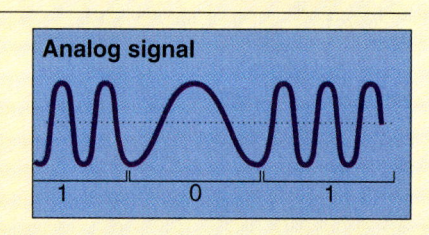

analog (p. 213, KQ 6.1) Continuous and varying in strength and/or quantity. An analog signal is a continuous electrical signal with such variation. Why it's important: *Sound, light, temperature, and pressure values, for instance, can fall anywhere on a continuum or range. The highs, lows, and in-between states have historically been represented with analog devices rather than in digital form.* Examples of analog devices are a speedometer, a thermometer, and a tire-pressure gauge, all of which can measure continuous fluctuations. The electrical signals on a telephone line have traditionally been analog-data representations of the original voices. Telephone, radio, television, and cable-TV technologies have long been based on analog data.

analog cellular phone (p. 227, KQ 6.3) Mobile telephone designed primarily for communicating by voice through a system of ground-area cells. Calls are directed to cells by a mobile telephone switching office (MTSO). Moving between cells requires that calls be "handed off" by the MTSO. Why it's important: *Cellular phone systems allow callers mobility.*

backbone (p. 232, KQ 6.4) High-speed network that connects LANs and MANs to the Internet. Why it's important: *The backbone is an essential part of the Internet.*

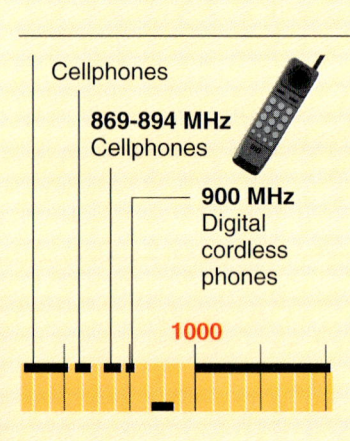

band (p. 221, KQ 6.3) Also called *bandwidth;* range of frequencies serving as a measure of the amount of information that can be delivered within a given period of time. The bandwidth is the difference between the lowest and the highest frequencies transmitted. For analog signals, bandwidth is expressed in hertz (Hz), or cycles per second. For digital signals, bandwidth is expressed in bits per second (bps). In the United States, certain bands are assigned by the Federal Communications Commission (FCC) for certain purposes. Why it's important: *The wider the bandwidth, the faster data can be transmitted. The narrower the band, the greater the loss of transmission power. This loss of power must be overcome by using relays or repeaters that rebroadcast the original \signal.*

Bluetooth (p. 229, KQ 6.3) Short-range wireless digital standard aimed at linking cellpones. Why it's important: *Bluetooth technology can replace cables between PCs and printers and can connect PCs to PDAs and wireless phones.*

bridge (p. 234, KQ 6.4) Interface used to connect the same types of networks. Why it's important: *Similar networks (local area networks) can be joined together to create larger area networks.*

broadband connections (p. 221, KQ 6.3) Connections characterized by very high speed (wide bandwidth). Why it's important: *Broadband connections are necessary for reliable, high-speed Internet hook-ups.*

broadband wireless digital services (p. 228, KQ 6.3) Two-way, third-generation wireless digital services based on an "always-on" standard. Why it's important: *3G technology is faster than 2G technology; it can send pictures and data clearly and quickly.*

broadcast radio (p. 223, KQ 6.3) Wireless transmission medium that sends data over long distances—between regions, states, or countries. A transmitter is required to send messages and a receiver to receive them; sometimes both sending and receiving functions are combined in a transceiver. Why it's important: *In the lower frequencies of the radio spectrum, several broadcast radio bands are reserved not only for conventional AM/FM radio but also for broadcast television, CB (citizens band) radio, ham (amateur) radio, cellular phones, and private radio land mobile services (such as police, fire, and taxi dispatch). Some organizations use specific radio frequencies and networks to support wireless communications.*

coaxial cable (p. 223, KQ 6.3) Commonly called "co-ax"; insulated copper wire wrapped in a solid or braided metal shield, then in an external cover. Why it's important: Co-ax is widely used for cable television. Because of the extra insulation, coaxial cable is much better than twisted-pair wiring at resisting noise. Moreover, it can carry voice and data at a faster rate.

communications channel (p. 221, KQ 6.3) Path over which information travels in a telecommunications system from its source to its destination. Why it's important: Channels may be wired or wireless. Three types of wired channels are twisted-pair wire (conventional telephone lines), coaxial cable, and fiber-optic cable.

communications satellite (p. 224, KQ 6.3) Microwave relay station in orbit around the earth. Why it's important: Transmitting a signal from a ground station to a satellite is called uplinking; the reverse is called downlinking. The delivery process will be slowed if, as is often the case, more than one satellite is required to get the message delivered.

compression (p. 229, KQ 6.3) Also called *digital-data compression*; method of removing repetitive elements from a file so that the file requires less storage space and therefore less time to transmit. Before we use the data, it is decompressed—the repeated patterns are restored. These methods are sometimes referred to as *codec* (for *compression/decompression*) techniques. Why it's important: Many of today's files, with graphics, sound, and video, require huge amounts of storage space; data compression makes the storage and transmission of these files more feasible.

cookie (p. 237, KQ 6.5) A file that the web server stores on a computer user's hard-disk drive when he or she visits a website. Why it's important: Unknown to users, a website operator or a company advertising on the site can log their movements within the site. These records provide information that marketers can use to target customers for their products.

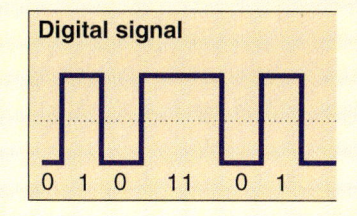

digital (p. 213, KQ 6.1) Represented in a two-state (binary) way. Why it's important: Digital signals are the basis of computer-based communications. "Digital" is usually synonymous with "computer-based."

digital television (DTV) (p. 220, KQ 6.2) A television standard that uses a digital signal, or series of 0s and 1s, rather than the customary analog standard, a system of varying signal amplitude and frequency that represents picture and sound elements. DTV was adopted as a standard in 1996 by television broadcasters and the Federal Communications Commission. Why it's important: DTV is much clearer and less prone to interference than analog TV and is better suited to handling computer and Internet data.

digital wireless services (p. 227, KQ 6.3) Two-way, second-generation (2G) wireless services that support digital cellphones and PDAs. They use a network of cell towers to send voice communications and data over the airwaves in digital form. Why it's important: This technology is a dramatic improvement over analog cellphones. Voice clarity is better, and more calls can be squeezed into the same bandwidth.

extranet (p. 235, KQ 6.4) Private intranet that connects not only internal personnel but also selected suppliers and other strategic parties. Why it's important: Extranets have become popular for standard transactions such as purchasing.

fiber-optic cable (p. 223, KQ 6.3) Cable that consists of dozens or hundreds of thin strands of glass or plastic that transmit pulsating beams of light rather than electricity. Why it's important: These strands, each as thin as a human hair, can transmit up to 2 billion pulses per second (2 Gbps); each "on" pulse represents one bit. When bundled together, fiber-optic strands in a cable 0.12 inch thick can support a quarter-million to a half-million voice conversations at the same time. Moreover, unlike electrical signals, light pulses are not affected by random electromagnetic interference in the environment. Thus, they have much lower error rates than normal telephone wire and cable. In addition, fiber-optic cable is lighter and more durable than twisted-pair and co-ax cable. A final advantage is that it cannot easily be wiretapped, so transmissions are more secure.

file server (p. 233, KQ 6.4) Computer in a client/server network that acts like a disk drive, storing the programs and data files shared by users. Why it's important: A file server enables all users of a LAN to have access to the same programs and data.

firewall (p. 235, KQ 6.4) System of hardware and software that blocks unauthorized users inside and outside the organization from entering the intranet. *Why it's important:* Security is essential to an intranet. A firewall consists of two parts, a choke and a gate. The choke forces all data packets flowing between the Internet and the intranet to pass through a gate. The gate regulates the flow between the two networks. It identifies authorized users, searches for viruses, and implements other security measures. Thus, intranet users can gain access to the Internet (including key sites connected by hyperlinks), but outside Internet users cannot enter the intranet.

gateway (p. 234, KQ 6.4) Interface permitting communication between dissimilar networks. *Why it's important:* Gateways permit communication between a LAN and a WAN or between two LANs based on different network operating systems or different layouts.

Global Positioning System (GPS) (p. 225, KQ 6.3) A series of earth-orbiting satellites continuously transmitting timed radio signals that can be used to identify earth locations. *Why it's important:* A GPS receiver—handheld or mounted in a vehicle, plane, or boat—can pick up transmissions from any four satellites, interpret the information from each, and calculate to within a few hundred feet or less the receiver's longitude, latitude, and altitude. Some GPS receivers include map software for finding your way around, as with the Guidestar system available with some rental cars.

high-definition television (HDTV) (p. 220, KQ 6.2) A high-resolution type of digital television (DTV), which comes in either a 720- or 1080-line mode (compared to 525-line resolution for analog TV). *Why it's important:* The Federal Communications Commission expects that eventually HDTV will supplant analog TV as the dominant digital standard. At present, most consumers consider the HDTV sets available to be too expensive.

host computer (p. 232, KQ 6.4) Mainframe or midsize central computer that controls a network. *Why it's important:* The host is responsible for managing the entire network.

infrared wireless transmission (p. 223, KQ 6.3) Transmission of data signals using infrared-light waves. *Why it's important:* Infrared ports can be found on some laptop computers and printers, as well as wireless mouses. The advantage is that no physical connection is required among devices. The drawbacks are that line-of-sight communication is required—there must be an unobstructed view between transmitter and receiver—and transmission is confined to short range.

intranet (p. 235, KQ 6.4) An organization's internal private network that uses the infrastructure and standards of the Internet and the World Wide Web. *Why it's important:* When an organization creates an intranet, it enables employees to have quicker access to internal information and to share knowledge so that they can do their jobs better. Information exchanged on intranets may include employee e-mail addresses and telephone numbers, product information, sales data, employee benefit information, and lists of jobs available within the organization.

JPEG (p. 230, KQ 6.3) The leading compression standard for still images; it stands for Joint Photographic Experts Group of the International Standards Organization. The file extension that identifies graphic images files in the JPEG format is .jpeg or .jpg. *Why it's important:* JPEG compression is commonly used to store and transmit still graphic images. Higher or lower degrees of JPEG compression may be chosen; greater compression corresponds to greater image loss.

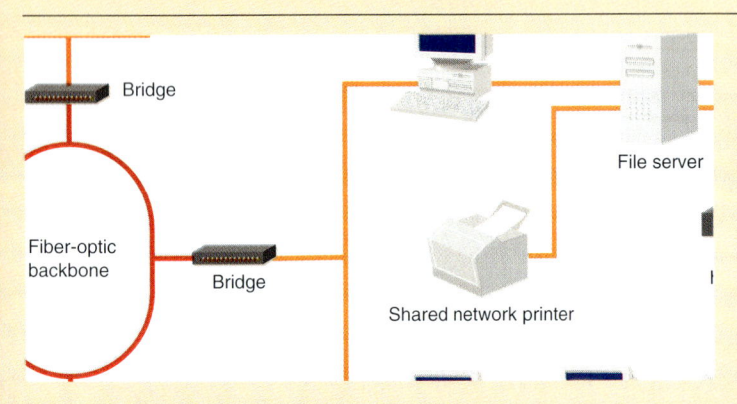

local area network (LAN) (p. 232, KQ 6.4) Network that connects computers and devices in a limited geographical area, such as one office, one building, or a group of buildings close together (for instance, a college campus). *Why it's important:* LANS have replaced large computers for many functions and are considerably less expensive.

metropolitan area network (p. 232, KQ 6.4) Communications network covering a city or a suburb. *Why it's important: The purpose of a MAN is often to bypass local telephone companies when accessing long-distance services. Many cellular phone systems are MANs.*

microwave radio (p. 224, KQ 6.3) Transmission of voice and data through the atmosphere as superhigh-frequency radio waves called *microwaves*. These frequencies are used to transmit messages between ground-based stations and satellite communications systems. *Why it's important: Microwaves are line-of-sight; they cannot bend around corners or around the earth's curvature, so there must be an unobstructed view between transmitter and receiver. Thus, microwave stations need to be placed within 25–30 miles of each other, with no obstructions in between. A string of microwave relay stations will each receive incoming messages, boost the signal strength, and relay the signal to the next station. Nowadays dish- or horn-shaped microwave reflective dishes, which contain transceivers and antennas, are nearly everywhere.*

Modem: Modulate (converts digital pulses to analog form)

Modem: Demodulate (converts analog signals back to digital form)

modem (p. 214, KQ 6.1) Short for "**mo**dulate/**dem**odulate"; device that converts digital signals into a representation of analog form (modulation) to send over phone lines; a receiving modem then converts the analog signal back to a digital signal (demodulation). *Why it's important: The modem provides a means for computers to communicate with one another using the standard copper-wire telephone network, an analog system that was built to transmit the human voice but not computer signals.*

MPEG (p. 230, KQ 6.3) Stands for Motion Picture Experts Group. MPEG is the leading compression standard for video images. The file extension that identifies video (and sound) files compressed in this format is .mpeg or .mpg. *Why it's important: People who work with videos are mainly interested in storing or transmitting an enormous amount of visual information in economical form; preserving details is a secondary consideration. The Motion Picture Experts Group sets standards for weeding out redundancies between neighboring images in a stream of video. Three MPEG standards have been developed for compressing visual information—MPEG-1, MPEG-2, and MPEG-4.*

network (p. 231, KQ 6.4) Also called *communications network;* system of interconnected computers, telephones, or other communications devices that can communicate with one another and share applications and data. *Why it's important: The tying together of so many communications devices in so many ways is changing the world we live in.*

node (p. 232, KQ 6.4) Any device that is attached to a network. *Why it's important: A node may be a microcomputer, terminal, storage device, or peripheral device, any of which enhance the usefulness of the network.*

pager (p. 226, KQ 6.3) Simple radio receiver that receives data, but not voice messages, sent from a special radio transmitter. The pager number is dialed from a phone and travels via the transmitter to the pager. *Why it's important: Pagers have become a common way of receiving notification of phone calls so that the user can return the calls immediately; some pagers can also display messages of up to 80 characters and send preprogrammed messages.*

peer-to-peer LAN (p. 233, KQ 6.4) Type of local area network in which all microcomputers on the network communicate directly with one another without relying on a server. *Why it's important: Peer-to-peer networks are less expensive than client-server networks and work effectively for up to 25 computers. Beyond that, they slow down under heavy use. They are appropriate for small networks.*

privacy (p. 237, KQ 6.5) The right of people not to reveal information about themselves. *Why it's important: Privacy is a basic democratic right. Information technology presents constant threats to this right.*

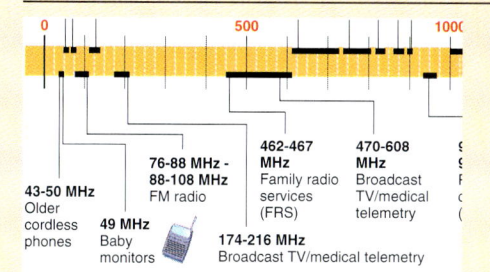

radio frequency spectrum (p. 221, KQ 6.3) The part of the electromagnetic spectrum that carries communications signals. *Why it's important: The radio spectrum ranges from low-frequency waves, such as those used for aeronautical and marine navigation equipment, through the medium frequencies for CB radios, cordless phones, and baby monitors, to ultrahigh frequency bands for cell phones and also microwave bands for communications satellites.*

router (p. 234, KQ 6.4) Special computer that directs communicating messages when several networks are connected together. Why it's important: *High-speed routers can serve as part of the Internet backbone, or transmission path, handling the major data traffic.*

standard-definition television (SDTV) (p. 220, KQ 6.2) A standard of digital television that has a minimum of 480 vertical lines, allowing broadcasters to transmit more information within the HDTV bandwidth. Why it's important: *SDTV would enable broadcasters to effectively multicast their products, on up to as many as six channels instead of one. It also frees up bandwidth for data transmission. Thus, in the future there might be separate channels carrying video, audio, and data.*

telecommuting (p. 219, KQ 6.2) Working at home while in telecommunication with the office. Why it's important: *Telecommuting has many benefits. Examples are reduced traffic congestion, energy consumption, and air pollution, increased productivity, and improved teamwork. A disadvantage is that people may feel isolated.*

twisted-pair wire (p. 223, KQ 6.3) Two strands of insulated copper wire, twisted around each other. Why it's important: *Twisted-pair wire has been the most common channel or medium used for telephone systems. However, it is relatively slow and does not protect well against electrical interference.*

videoconferencing (p. 217, KQ 6.2) Also called *teleconferencing;* use of television video and sound technology as well as computers to enable people in different locations to see, hear, and talk with one another. Why it's important: *Videoconferencing may eliminate the need for some travel for the purpose of meetings and allow people who cannot travel to visit face to face.*

virtual office (p. 219, KQ 6.2) Nonpermanent and mobile office run with computer and communications technology. Why it's important: *Employees work from their homes, cars, and other new work sites, using pagers, portable computers, and cellphones to conduct business. This reduces the office expenses of their employer.*

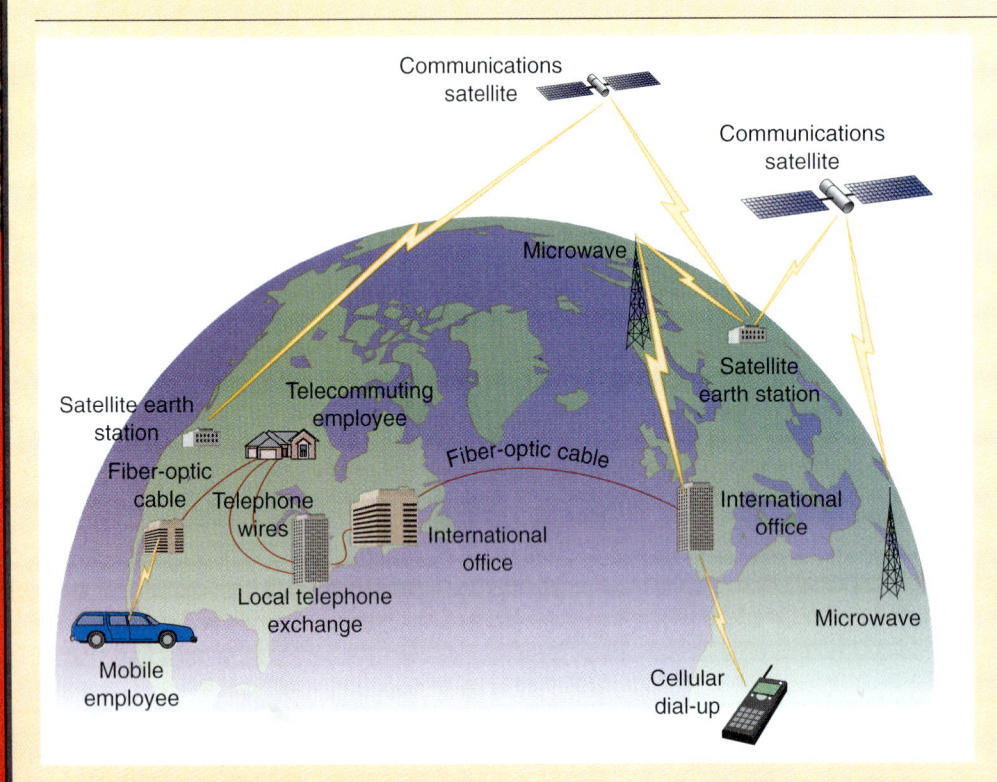

wide area network (WAN) (p. 232, KQ 6.4) Communications network that covers a wide geographical area, such as a country or the world. Why it's important: *Most long-distance and regional Bell telephone companies are WANs. A WAN may use a combination of satellites, fiber-optic cable, microwave, and copper wire connections and link a variety of computers, from mainframes to terminals.*

WiFi (p. 229, KQ 6.3) Short-range wireless digital standard aimed at helping machines inside offices to communicate at high speeds and share Internet connections at distances up to 300 feet. It connects to Ethernet LANs. Why it's important: *WiFi is a standard in most new computers.*

workgroup computing (p. 218, KQ 6.2) Also called *collaborative computing;* technology by which teams of co-workers can use networks of microcomputers to share information and to cooperate on projects. Workgroup computing is made possible not only by networks and microcomputers but also by groupware. Why it's important: *Workgroup computing allows co-workers to collaborate with colleagues, suppliers, and customers and to tap into company information through computer networks.*

Chapter Review

stage 1 LEARNING — MEMORIZATION
"I can recognize and recall information."

Self-Test Questions

1. A(n) _____ converts digital signals into analog signals for transmission over phone lines.

2. A(n) _____ network covers a wide geographical area, such as a state or a country.

3. _____ cable transmits data as pulses of light rather than as electricity.

4. _____ refers to waves continuously varying in strength and/or quantity; _____ refers to communications signals or information in a binary form.

5. _____ is a method of removing repetitive elements from a file so that the file requires less storage space.

6. A(n) _____ is a computer that acts as a disk drive, storing programs and data files shared by users on a LAN.

7. The _____ is the system software that manages the activities of a network.

8. *Modem* is short for _____.

9. The leading compression standard for still images is _____.

10. _____ programs can screen out objectionable material on the Internet.

Multiple-Choice Questions

1. Which of the following best describes the telephone line that is used in most homes today?
   a. coaxial cable
   b. modem cable
   c. twisted-wire pair
   d. fiber-optic cable
   e. LAN

2. Which of the following do local area networks enable?
   a. sharing of peripheral devices
   b. sharing of programs and data
   c. better communications
   d. access to databases
   e. all of the above

3. Which of the following is not a data compression standard/method?
   a. lossless
   b. JPEG
   c. MPEG
   d. lossy
   e. NOS

4. Which of the following is not a type of server?
   a. file server
   b. print server
   c. mail server
   d. disk server
   e. database server

5. Which of the following is not a short-distance wireless standard?
   a. Bluetooth
   b. PDA
   c. WiFi
   d. Home RF

True/False Questions

T F 1. In a LAN, a bridge is used to connect the same type of networks, whereas a gateway is used to enable dissimilar networks to communicate.

T F 2. Frequency and amplitude are two characteristics of analog carrier waves.

T F 3. A range of frequencies is called a *spectrum*.

T F 4. Twisted-pair wire commonly connects residences to external telephone systems.

T F 5. A cookie is an ID number used to visit specific websites.

Telecommunications

Stage 2 LEARNING: COMPREHENSION

"I can recall information in my own terms and explain them to a friend."

Short-Answer Questions

1. What is the difference between an intranet and an extranet?
2. What is workgroup computing?
3. What is the difference between a LAN and a WAN?
4. Why is bandwidth a factor in data transmission?
5. What is a firewall?
6. What do 2G and 3G mean?

Concept Mapping

On a separate sheet of paper, draw a concept map, or visual diagram, linking concepts. Show how the following terms are related.

analog
backbone
band
Bluetooth
bridge
broadband connections
coaxial cable
communications channel
digital
extranet
fiber-optic cable
file server
firewall
gateway
GPS
host computer
Intranet
JPEG
LAN
modem
MPEG
network
node
satellite
spectrum
twisted wire
WAN
WiFi
wireless network

Stage 3 LEARNING: APPLYING, ANALYZING, SYNTHESIZING, EVALUATING

"I can apply what I've learned, relate these ideas to other concepts, build on other knowledge, and use all these thinking skills to form a judgment."

Knowledge in Action

1. Are the computers at your school connected to a network? If so, what kind of network(s)? What types of computers are connected? What hardware and software allows the network to function? What department(s) did you contact to find the information you needed to answer these questions?

2. Using current articles, publications, and/or the Web, research cable modems. Where are they being used? What does a residential user need to hook up to a cable modem system? Do you think you will use a cable modem in the near future? Write a short report.

3. Research the role of the Federal Communications Commission in regulating the communications industry. How are new frequencies opened up for new communications services? How are the frequencies determined? Who gets to use new frequencies?

Web Exercises

1. Compare digital cable and satellite TV in your area. Which offers more channels? Which offers more features? How much do the services cost? Does either allow Internet connectivity? Does either use a telephone line to download the programming listings?

2. Wondering what those astronauts are doing on the International Space Station? NASA has a TV station via the Internet dedicated to providing real-time coverage of what is going on in, and out of, this world. Visit the following website to get the programming schedule:

   www.nasa.gov/ntv

   When you're at the site, read the frequency of transmission information in the second paragraph.

3. Calculate the amount of airborne data transmission that travels through your body. Research the amount of radio and television station broadcast signals in your area, as well as the estimated number of mobile-phone users. Imagine what the world would look like if you could see all the radio-wave signals the way you can see the waves in the ocean.

4. On the Web, go to *www.trimble.com* and work through some of the tutorial on "About GPS Technology." Then write a short report on the applications of a GPS system.

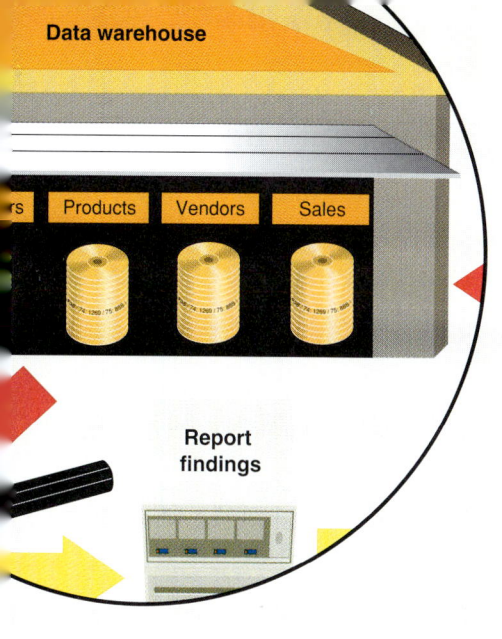

Chapter 7

Files, Databases, & E-Commerce

Digital Engines for the New Economy

Key Questions
You should be able to answer the following questions.

7.1 **Managing Files: Basic Concepts** What are the data storage hierarchy, the key field, types of files, sequential versus direct access, and offline versus online storage?

7.2 **Database Management Systems** What are the benefits of database management systems, and what are four types of database access?

7.3 **Database Models** What are four types of database models?

7.4 **Databases & the New Economy: E-Commerce, Data Mining, & B2B Systems** How are e-commerce, data mining, and business-to-business systems using databases?

7.5 **The Ethics of Using Databases: Concerns about Accuracy & Privacy** What are some ethical concerns about the uses of databases?

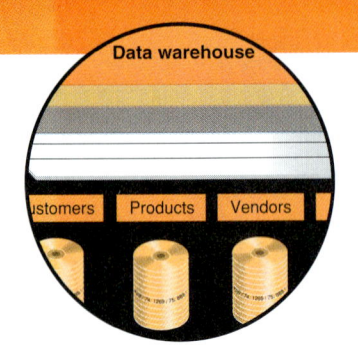

If data exists in one place, it exists in more than one place.

Perhaps this statement could stand as a summary of one of the most important developments of the Digital Age. How does data in one place get to be in more than one place? The answer has to do with databases. A database is not just the computerization of what used to go into manila folders and a filing cabinet. A database is an organized collection of related files, a technology for pulling together facts that allows the slicing and dicing and mixing and matching of data in all kinds of ways. As a result, the arrival of databases—especially when linked to the Internet—has stood many of our business and social institutions on their heads.

Your name and some facts about you can probably be found in scores—if not hundreds—of far-flung databases. How is that data being used? That is a very interesting question. In this chapter, we will examine the importance of databases and how they work.

7.1 Managing Files: Basic Concepts

KEY QUESTIONS
What are the data storage hierarchy, the key field, types of files, sequential versus direct access, and offline versus online storage?

When you create a letter or a spreadsheet or an address file on a computer, you often want to save it for future reference. Let's look at the mechanics of this process.

How Data Is Organized: The Data Storage Hierarchy

Data can be grouped into a hierarchy of categories, each increasingly more complex. The <u>*data storage hierarchy*</u> **consists of the levels of data stored in a computer: bits, bytes (characters), fields, records, files, and databases.** (See ● Panel 7.1.)

Computers, we have said, are based on the principle that electricity may be on or off. Thus, individual items of data are represented by the bits 0 for off and 1 for on. Bits and bytes are the building blocks for representing data, whether it is being processed, stored, or telecommunicated. The computer deals with the bits and bytes; you, however, will need to deal with characters, fields, records, files, and databases.

- **Characters:** A <u>*character*</u> (byte) is a letter, number, or special character. A, B, C, 1, 2, 3, #, $, % are all examples of single characters.
- **Field:** A <u>*field*</u> **is a unit of data consisting of one or more characters (bytes).** An example of a field is your name, your address, or your Social Security number.
- **Record:** A <u>*record*</u> **is a collection of related fields.** An example of a record would be your name *and* address *and* Social Security number.
- **File:** A <u>*file*</u> **is a collection of related records.** An example of a file is data collected on everyone employed in the same department of a company, including all names, addresses, and Social Security numbers. You use files a lot because the file is the collection of data or information that is treated as a unit by the computer.
- **Database:** As we've stated, a <u>*database*</u> **is an organized collection of integrated files.** A company database might include files on all past and current employees in all departments. There would be various files for each employee: payroll, retirement benefits, sales quotas and achievements (if in sales), and so on.

● PANEL 7.1
How data is organized

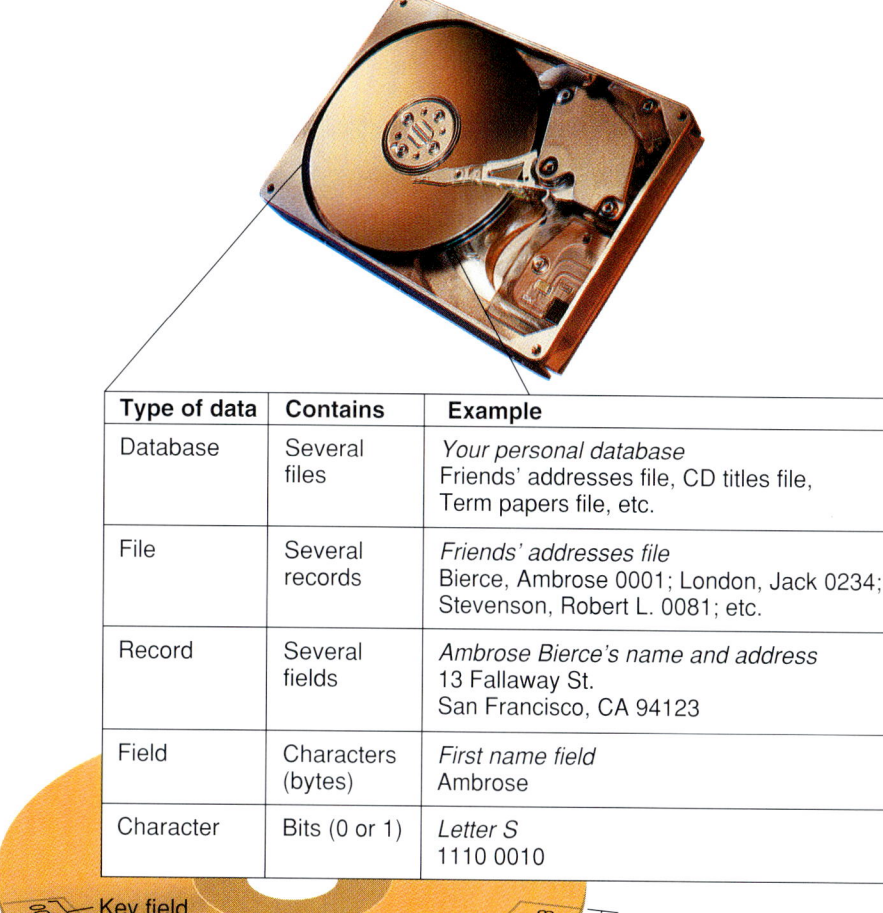

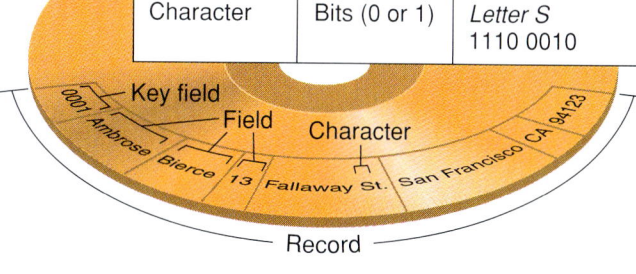

Type of data	Contains	Example
Database	Several files	*Your personal database* Friends' addresses file, CD titles file, Term papers file, etc.
File	Several records	*Friends' addresses file* Bierce, Ambrose 0001; London, Jack 0234; Stevenson, Robert L. 0081; etc.
Record	Several fields	*Ambrose Bierce's name and address* 13 Fallaway St. San Francisco, CA 94123
Field	Characters (bytes)	*First name field* Ambrose
Character	Bits (0 or 1)	*Letter S* 1110 0010

The Key Field

An important concept in data organization is that of the *key field* (see ● Panel 7.1). A **key field is a field that is chosen to uniquely identify a record so that it can be easily retrieved and processed.** The key field is often an identification number, Social Security number, customer account number, or the like. The primary characteristic of the key field is that it is *unique*. Thus, numbers are clearly preferable to names as key fields because there are many people with common names like James Johnson, Susan Williams, Ann Wong, or Roberto Sanchez, whose records might be confused. Student records are often identified by student ID numbers used as key fields.

Types of Files: Program Files, Data Files, & Others

As we said, the *file* is the collection of data or information that is treated as a unit by the computer. **Files are given names—*filenames*.** If you're using a word processing program to write a psychology term paper, you might name it Psychreport.

Filenames also have *extension names*. These extensions of up to three letters are added after a period following the filename—for example, the *.doc* in Psychreport.doc is recognized by Microsoft Word as a "document." Extensions are usually inserted automatically by the application software.

Survival Tip

Guard Your Social Security Number

The SSN is an important key field to records about you. Don't carry it with you. Don't print it on checks. Don't give it out unless really necessary.

Files, Databases, & E-Commerce

247

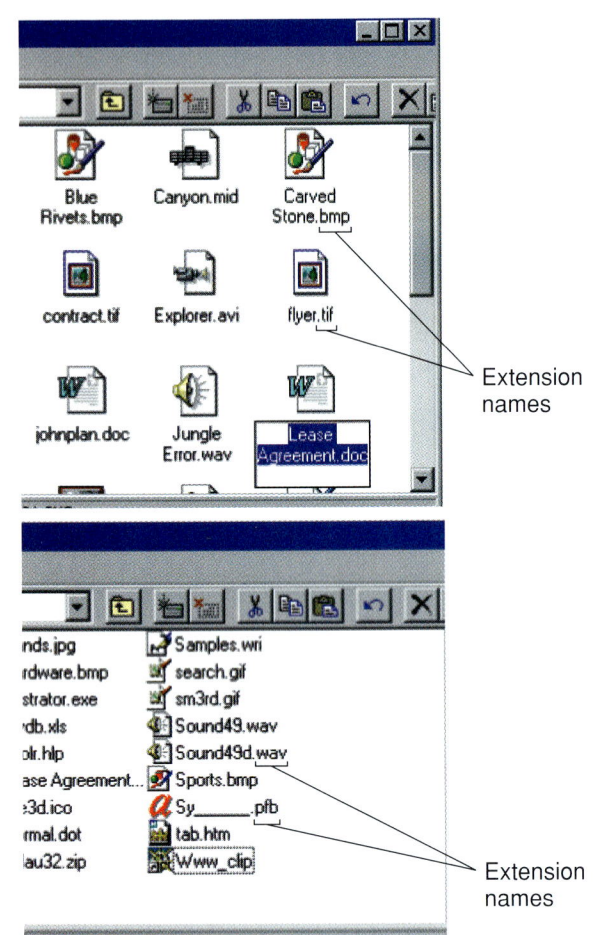

When you look up the filenames listed on your hard drive (on the directory, as we will explain), you will notice a number of extensions, such as *.doc*, *.exe*, and *.com*. There are many kinds of files, but perhaps the two principal ones are *program files* and *data files*.

- **Program files:** **_Program files_ are files containing software instructions.** Examples are word processing or spreadsheet programs, which are made up of several different program files. The two most important are source program files and executable files.

  Source program files contain high-level computer instructions in the original form written by the programmer. Some source program files have the extension of the language in which they are written, such as *.bas* for BASIC, *.pas* for Pascal, or *.jav* for Java. (The appendix has information about programming languages.)

  For the processor to use source program instructions, they must be translated into an *executable file,* which contains the instructions that tell the computer how to perform a particular task. You can identify an executable file by its extension, *.exe.* You use an executable file by running it—as when you select Microsoft Excel from your onscreen menu and run it. (There are some executable files that you cannot run—another computer program causes them to execute. These are identified by such extensions as *.dll, .drv, ocx, .sys,* and *.vbx.*)

- **Data files:** **_Data files_ are files that contain data**—words, numbers, pictures, sounds, and so on. Unlike program files, data files don't instruct the computer to do anything. Rather, data files are there to be acted on by program files. Examples of common extensions in data files are *.txt* (text) and *.xls* (spreadsheets). Certain proprietary software programs have their own extensions, such as *.ppt* for PowerPoint and *.mdb* for Access.

Other common types of files are *ASCII files, image files, audio files, animation/video files,* and *web files.*

- **ASCII files:** ASCII is a common binary coding scheme used to represent data in a computer (page 175). ASCII ("as-key") files are text-only files that contain no graphics and no formatting, such as boldface or italics. This format is used to transfer documents between incompatible computers, such as PC and Macintosh. Such files may use the *.txt* extension.

- **Image (graphic) files:** If ASCII files are for text, *image files* are for digitized graphics, such as art or photographs. They are indicated by such extensions as *.bmp, .gif, .jpg, .pcx, .tif,* and *.wmf.*

- **Audio files:** *Audio files* contain digitized sound and are used for conveying sound in CD-ROM multimedia and over the Internet. They have extensions such as *.wav* and *.mid.*

- **Animation/video files:** *Video files,* used for such purposes as conveying moving images over the Internet, contain digitized video images. Common extensions are *.avi, flc, .fli,* and *.mpg.*

- **Web files:** *Web files* are files carried over the World Wide Web. Their extensions include *.html, .htm,* and *.xml.*

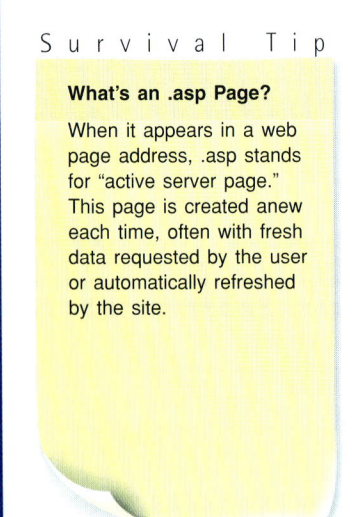

> Survival Tip
>
> **What's an .asp Page?**
>
> When it appears in a web page address, .asp stands for "active server page." This page is created anew each time, often with fresh data requested by the user or automatically refreshed by the site.

Two Types of Data Files: Master File & Transaction File

Among the several types of data files, two are commonly used to update data: a master file and a transaction file.

- **Master file:** The *master file* **is a data file containing relatively permanent records that are generally updated periodically.** An example of a master file would be the address-label file for all students currently enrolled at your college.

- **Transaction file:** The *transaction file* **is a temporary holding file that holds all changes to be made to the master file: additions, deletions, revisions.** For example, in the case of the address labels for your college, a transaction file would hold new names and addresses to be added (because over time new students enroll) and names and addresses to be deleted (because students leave). It would also hold revised names and addresses (because students change their names or move). Each month or so, the master file would be *updated* with the changes called for in the transaction file.

Data Access Methods: Sequential versus Direct Access

The way that a secondary-storage device allows access to the data stored on it affects its speed and its usefulness for certain applications. The two main types of data access are sequential and direct.

- **Sequential storage:** *Sequential storage* **means that data is stored and retrieved in sequence,** such as alphabetically. Tape storage falls in the category of sequential storage. Thus, if you are looking for employee number 8888 on a tape, the computer will have to start with 0001, then go past 0002, 0003, and so on, until it finally comes to 8888. This data access method is less expensive than other methods because it uses magnetic tape, which is cheaper than disks. The disadvantage of sequential file organization is that searching for data is slow.

- **Direct access storage:** *Direct access storage* **means that the computer can go directly to the information you want**—just as a CD player can go directly to a particular track on a music CD. The data is retrieved (accessed) according to a unique data identifier called a *key field*, as we will discuss. It also uses a *file allocation table (FAT)*, a hidden on-disk table that records exactly where the parts of a given file are stored.

  This method of file organization is used with hard disks and other types of disks. It is ideal for applications where there is no fixed pattern to the requests for data—for example, in airline reservation systems or computer-based directory-assistance operations.

  If you need to find specific data, direct file access is much faster than sequential access. However, direct file access is also more expensive, for two reasons: (1) the complexity involved in maintaining a file allocation table and (2) the need to use hard-disk technology, rather than cheaper magnetic tape technology.

Offline versus Online Storage

Whether it's on magnetic tape or on some form of disk, data may be stored either offline or online.

- **Offline:** *Offline storage* **means that data is not directly accessible for processing until the tape or disk it's on has been loaded onto an input device.** That is, the storage is not under the direct, immediate control of the central processing unit.

- **Online:** *Online storage* **means that stored data is randomly (directly) accessible for processing.** That is, storage is under the direct, immediate control of the central processing unit. You need not wait for a tape or disk to be loaded onto an input device.

For processing to be online, the storage must be online and *fast.* This nearly always means storage on disk (direct access storage) rather than magnetic tape (sequential storage).

> **CONCEPT CHECK**
>
> Describe the data storage hierarchy and the concept of a key field.
>
> Distinguish program files from data files.
>
> List other types of files. What are their extension names?
>
> Distinguish sequential from direct access.
>
> How are offline and online storage different?

7.2 Database Management Systems

KEY QUESTIONS
What are the benefits of database management systems, and what are four types of database access?

As we said earlier, a *database* is an organized collection of related (integrated) files. A database may be small, contained entirely within your own personal computer, or it may be massive, available through online connections. Such massive databases are of particular interest because they offer phenomenal resources that until recently were unavailable to most ordinary computer users.

In the 1950s, when commercial use of computers was just beginning, a large organization would have different files for different purposes. For example, a university might have one file for course grades, another for student records, another for tuition billing, and so on. In a corporation, people in the accounting, order-entry, and customer-service departments all had their own separate files. Thus, if an address had to be changed, for example, each file would have to be updated separately. The database files were stored on magnetic tape and had to be accessed in sequence in what was called a *file-processing system.* Later magnetic disk technology came along, allowing any file to be accessed randomly. This permitted the development of new technology and new software: the database management system. A *database management system (DBMS),* **or database manager, consists of programs that control the structure of a database and access to the data.** In a DBMS, an address change need be entered only once, and the updated information is then available in any relevant file.

Four Types of Database Access

Databases can be set up to serve people in several different ways. (See ● Panel 7.2.)

- **Individual databases:** *Individual databases* **are collections of integrated files used by one person.** As we discussed in Chapter 3, microcomputer users can set up their own individual databases using popular database management software; the information is stored on the hard drives of their personal computers. Today the principal database programs are Microsoft Access, Corel Paradox, and Filemaker Pro. Such programs are used, for example, by graduate students to conduct research, by salespeople to keep track of clients, by purchasing agents to monitor orders, by coaches to keep watch on other teams and players, and by

PANEL 7.2
Four types of database access

Database	Description
Individual database	Collection of integrated files used by one person
Shared database	Database shared by users in one organization in one location
Distributed database	Database stored on different computers in different locations connected by a client/server network
Public databank	Compilation of data available to the public

home microcomputer users to maintain holiday card lists, hobby collection inventories, and inventories of valuables, for example.

In addition, types of individual databases known as *personal information managers (PIMs)* (page 118) can help you keep track of and manage information you use on a daily basis, such as addresses, telephone numbers, appointments, to-do lists, and miscellaneous notes. Popular PIMs are Microsoft Outlook, Lotus Organizer, and Act.

- **Shared (company) databases:** A ***shared database*** **or company database is shared by users in one company or organization in one location.** The organization owns the database, which may be stored on a server such as a mainframe. Users are linked to the database via a local area or wide area network; the users access the network through terminals or microcomputers.

  Shared databases, such as those you find when surfing the Web, are the foundation for a great deal of electronic commerce.

- **Distributed databases:** A ***distributed database*** **is stored on different computers in different locations connected by a client/server network.** For example, Cisco Systems, which supplies the vast network that connects computers to the Internet, uses its worldwide intranet to connect its distributed databases, even those located overseas. As a result, it is able to execute what is called a "virtual close"—defined as the ability to close the financial books with a one-hour notice. "By connecting an entire company via intranet, even one with operations in dozens of countries," explains Cisco CEO John Chambers, "what was once done quarterly can now be done anytime."[1]

- **Public databanks:** If you're looking for very specific information, you can use your browser to do a web search, investigating hundreds or thousands of websites. Many of these websites represent ***public databanks***, **compilations of data that are available to the public.**

  Some public databanks are fee-based, such as Dialog Information Services, which offers scientific and technical information, and Dow Jones Interactive Publishing, which provides business information. Certain fee-based public databases are specialized, such as Lexis, which gives lawyers access to local, state, and federal laws, or Nexis, which gives journalists access to published articles in a range of newspapers.

  Other public databanks are free. For instance, the U.S. government provides a great deal of free information, such as economic figures from the Bureau of Labor Statistics. Finally, many public databanks, such as Yahoo! or Amazon.com, are supported by advertising or online sales. The most popular revenue-producing websites—those supported by selling products or ads—are devoted to shopping, news and media, entertainment, online games, sweepstakes and lotteries, travel, finance, health and family, sports, and home and food.[2]

Survival Tip

Some Records Have to Be Hardcopy

You could scan your own birth certificate, will, or car ownership title into your computer to make a digital record. But such records printed off a hard drive aren't legally useful. Some, for instance, need to show a raised seal.

CLICK-ALONG 7-1
Ways of using individual databases

> **CONCEPT CHECK**
>
> What are the advantages of database management systems?
>
> Distinguish among four types of database access.

7.3 Database Models

KEY QUESTION
What are four types of database models?

Just as files can be organized in different ways, so databases can be organized in ways to best fit their use. The four most common arrangements are *hierarchical, network, relational,* and *object-oriented.* Every database and database management system is based on one of these four data organization models.

Hierarchical Database

In a <u>*hierarchical database*</u>, **fields or records are arranged in related groups resembling a family tree, with child (lower-level) records subordinate to parent (higher-level) records.** The parent record at the top of the database is called the *root record.* (See ● Panel 7.3.)

The hierarchical database is the oldest of the four models. It lent itself well to the tape storage systems used by mainframes in the 1970s.[3] It is still used in some types of passenger reservation systems. In hierarchical databases, accessing or updating data is very fast, because the relationships have been predefined. However, because the structure must be defined in advance, it is quite rigid. There may be only one parent per child, and no relationships among the child records are possible. Moreover, adding new fields to database records requires that the entire database be redefined. A new database model was needed to address the problems of data redundancy and complex data relationships.

Network Database

The network database was in part developed to solve some of the problems of the hierarchical database model. A <u>*network database*</u> **is similar to a hierarchical database, but each child record can have more than one parent record.** (See ● Panel 7.4.) Thus, a child record, which in network database terminology is called a *member,* may be reached through more than one parent, which is called an *owner.*

Also used principally with mainframes, the network database is more flexible than the hierarchical arrangement, because different relationships may be established between different branches of data. However, it still requires that the structure be defined in advance, and, as with the hierarchical model, the user must be very familiar with the structure of the database. Moreover, there are limits to the number of possible links among records.

Although the network database was an improvement over the hierarchical database, some people in the database community believed there must be a better way to manage large amounts of data.[4]

> **Survival Tip**
>
> **Music File Sharing**
>
> Napster users swapping their music (MP3) files for free had to go through a central server, which could be blocked. But other kinds of file-sharing software—Gnutella, LimeWire, BearShare, Aimster, Morpheus—connect computers on the Internet directly to one another; users can then search for files on others' PCs.

> **PANEL 7.3**
> **Hierarchical database**
> Example of a cruise ship reservation system

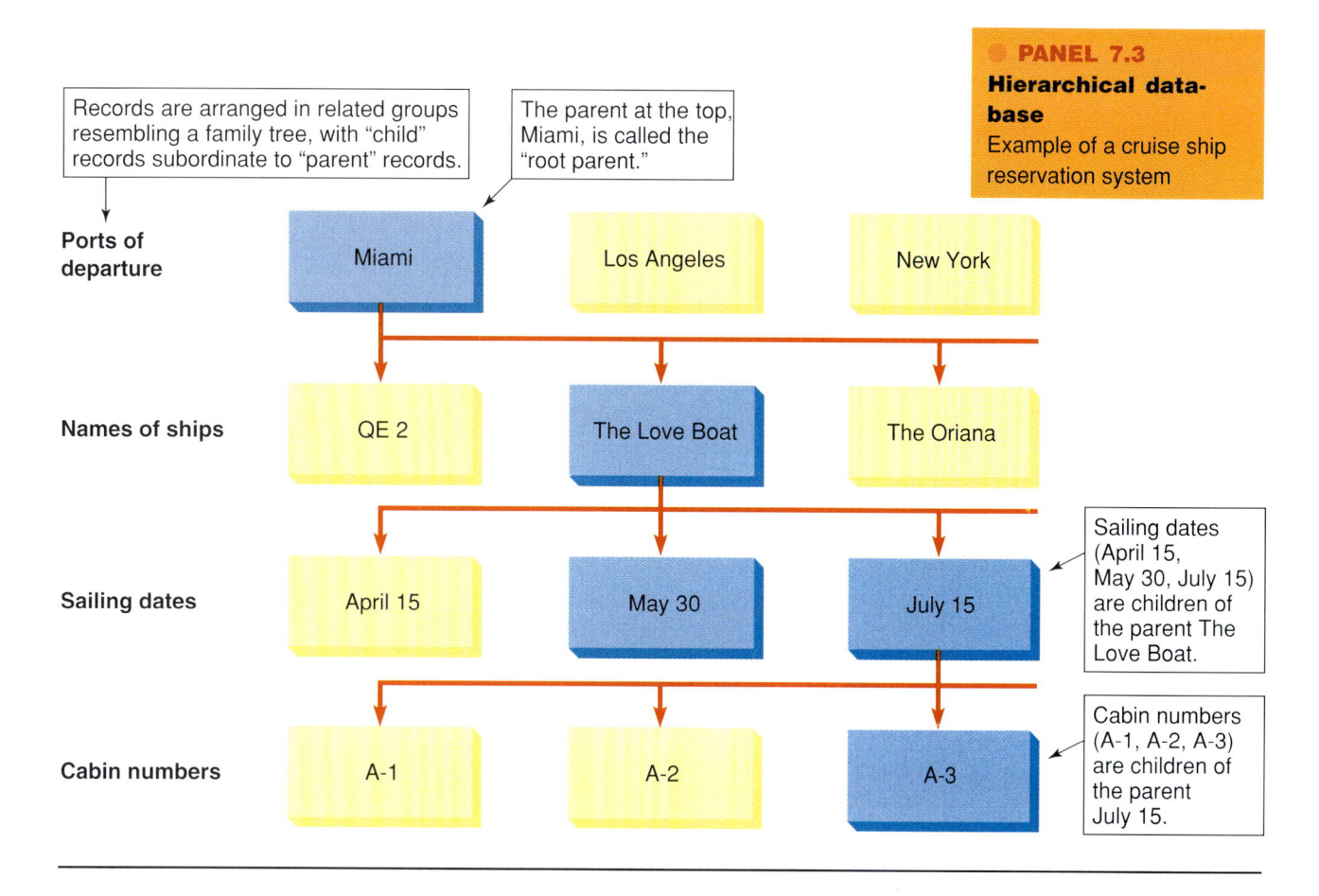

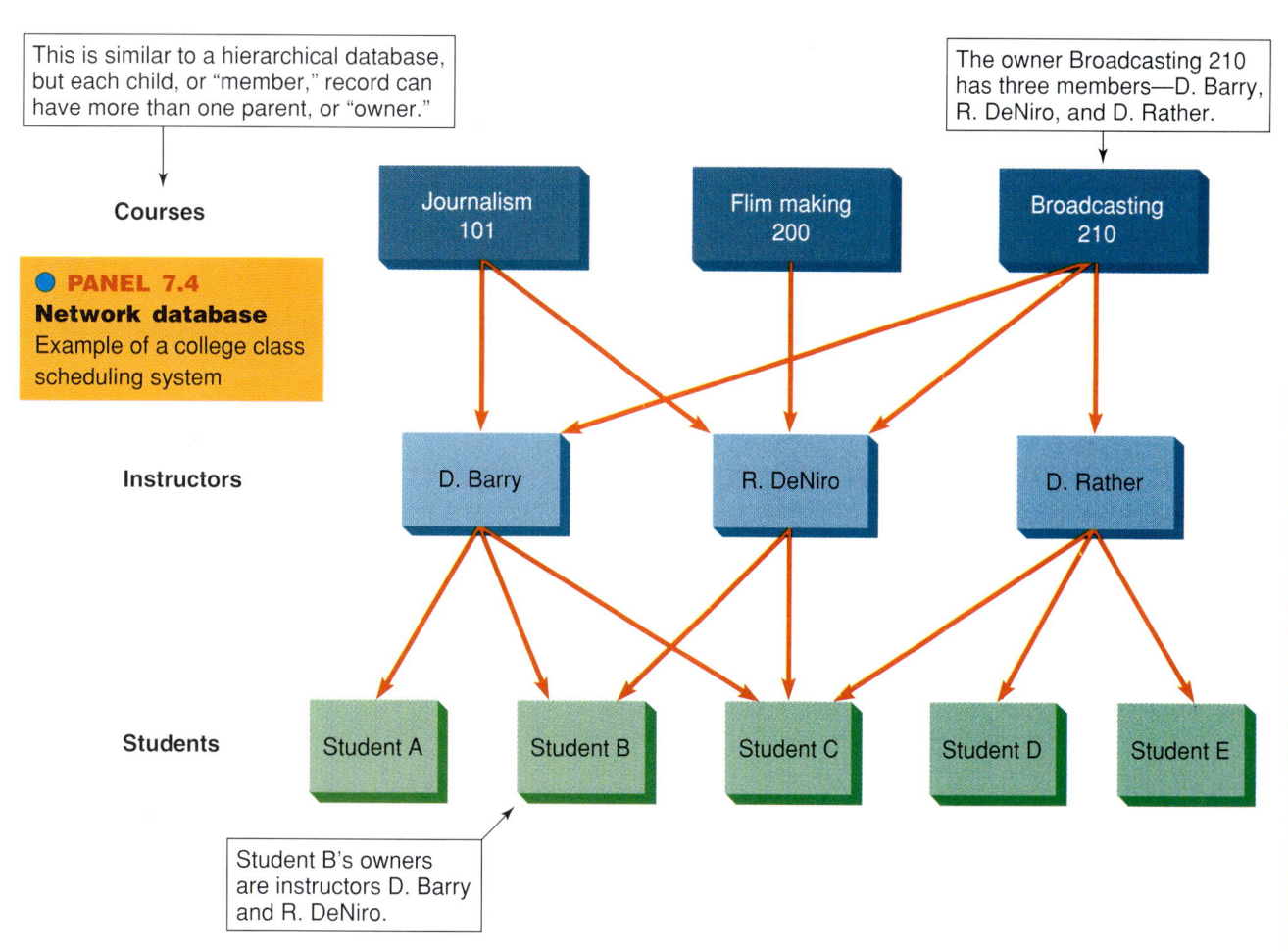

Records are arranged in related groups resembling a family tree, with "child" records subordinate to "parent" records.

The parent at the top, Miami, is called the "root parent."

Ports of departure: Miami, Los Angeles, New York

Names of ships: QE 2, The Love Boat, The Oriana

Sailing dates: April 15, May 30, July 15

Sailing dates (April 15, May 30, July 15) are children of the parent The Love Boat.

Cabin numbers: A-1, A-2, A-3

Cabin numbers (A-1, A-2, A-3) are children of the parent July 15.

This is similar to a hierarchical database, but each child, or "member," record can have more than one parent, or "owner."

The owner Broadcasting 210 has three members—D. Barry, R. DeNiro, and D. Rather.

Courses: Journalism 101, Flim making 200, Broadcasting 210

> **PANEL 7.4**
> **Network database**
> Example of a college class scheduling system

Instructors: D. Barry, R. DeNiro, D. Rather

Students: Student A, Student B, Student C, Student D, Student E

Student B's owners are instructors D. Barry and R. DeNiro.

Files, Databases, & E-Commerce

Relational Database

More flexible than hierarchical and network database models, the ***relational database* relates, or connects, data in different files through the use of a key field, or common data element.** *(See ● Panel 7.5.)* In this arrangement there are no access paths down through a hierarchy. Instead, data elements are stored in different tables made up of rows and columns. In database terminology, the tables are called *relations* (files), the rows are called *tuples* (records), and the columns are called *attributes* (fields). The physical order of the records or fields in a table is completely immaterial. Each record in the table is identified by a field that contains a unique value. These two characteristics allow the data in a relational database to exist independently of the way it is physically stored on the computer. Thus a user is not required to know the physical location of a record in order to retrieve its data—as is the case with hierarchical and network databases.[5]

The relational model has become popular for microcomputer database management programs, such as Paradox, Access, and Visual FoxPro. DB2, Oracle, and Sybase are relational models used on larger systems.

Object-Oriented Database

An ***object-oriented database* uses "objects," software written in small, reusable chunks, as elements within database files.** An *object* consists of (1) data in any form, including graphics, audio, and video, and (2) instructions on the action to be taken on the data.

An object-oriented database can store more types of data than can a relational database. For example, an object-oriented student database might contain each student's photograph, a "sound bite" of his or her voice, and even a short piece of video, in addition to grades and personal data. Moreover, the object would store operations, called *methods,* the programs that objects use to process themselves. For example, these programs might indicate how to calculate the student's grade-point average or how to display or print the student's record. Examples of object-oriented databases are FastObjects, eXcelon, Object Spark, Objectware, Jeevan, and KE Texpress. Many high-tech companies exist that can create custom databases.

> **CONCEPT CHECK**
>
> Describe the four models of database organization.

7.4 Databases & the New Economy: E-Commerce, Data Mining, & B2B Systems

KEY QUESTIONS
How are e-commerce, data mining, and business-to-business systems using databases?

At one time there was a difference between the Old Economy and the New Economy. The first consisted of traditional companies—car makers, pharmaceuticals, retailers, publishers. The second consisted of computer, telecommunications, and Internet companies (AOL, Amazon, eBay, and a raft of "dot-com" firms). Now, however, Old Economy companies have begun to absorb the new Internet-driven technologies, and the differences between the two sectors are dwindling.

One sign of growth is that the number of Internet host computers has been almost doubling every year. But the mushrooming of computer networks and the booming popularity of the World Wide Web are only the most obvious signs of the digital economy. Behind them lies something equally important: the growth of vast stores of information in databases.

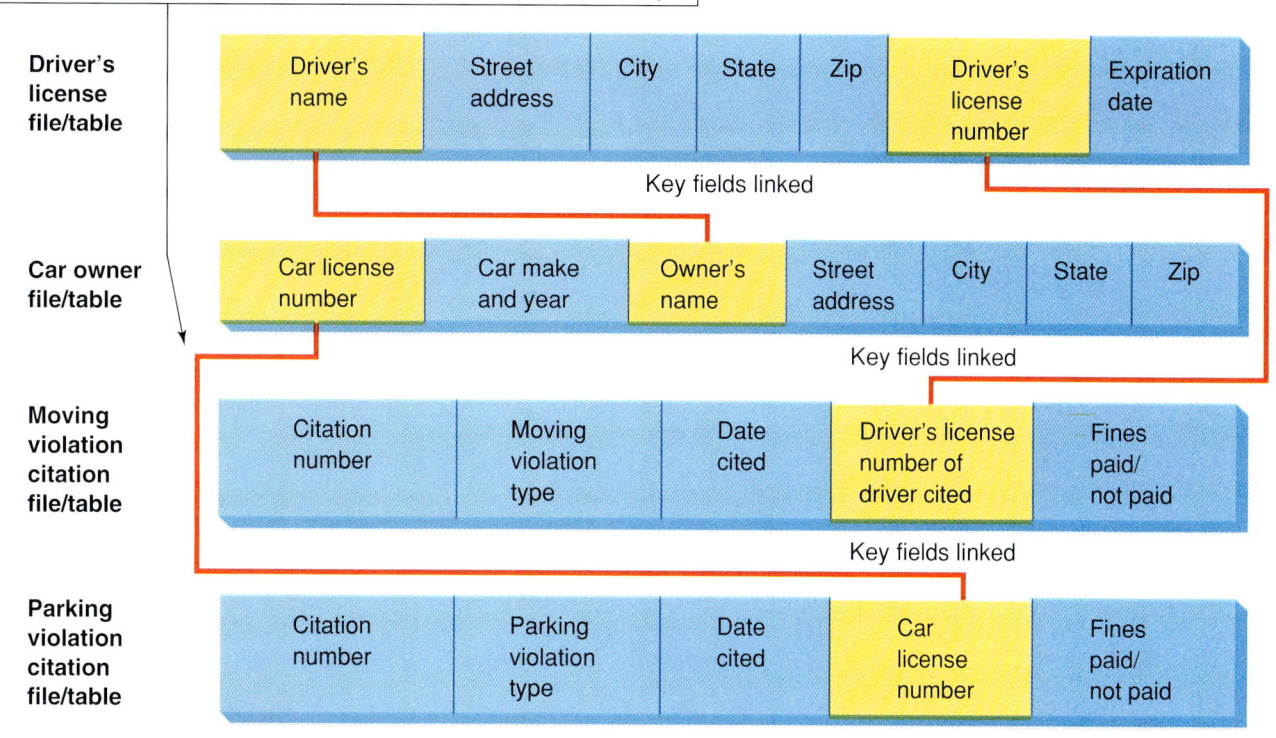

This kind of database relates, or connects, data in different files through the use of a key field, or common data element. The relational database does not require predefined relationships.

| Driver's license file/table | **Driver's name** | Street address | City | State | Zip | **Driver's license number** | Expiration date |

Key fields linked

| Car owner file/table | **Car license number** | Car make and year | **Owner's name** | Street address | City | State | Zip |

Key fields linked

| Moving violation citation file/table | Citation number | Moving violation type | Date cited | **Driver's license number of driver cited** | Fines paid/ not paid |

Key fields linked

| Parking violation citation file/table | Citation number | Parking violation type | Date cited | **Car license number** | Fines paid/ not paid |

● **PANEL 7.5**
Relational database
Example of a state department of motor vehicles database

E-commerce
There seem to be no limits on the uses of e-commerce.

How are databases underpinning the New Economy? Let us consider three aspects: *e-commerce, data mining,* and *business-to-business (B2B) systems.*

E-Commerce

The Internet might have remained a text-based realm, the province of academicians and researchers, had it not been for the creative contributions of Tim Berners-Lee. He was the computer scientist who came up with the coding system (HyperText Markup Language, page 55), linkages, and addressing scheme (URLs) that debuted in 1991 as the graphics-laden and multimedia World Wide Web. "It's hard to overstate the impact of the global system he created," writes *Time* technology writer Joshua Quittner. "He took a powerful communications system [the Internet] that only the elite could use, and turned it into a mass medium."[6]

The arrival of the Web quickly led to **_e-commerce_, or electronic commerce, the buying and selling of products and services through computer networks.** By 2003, total U.S. e-commerce sales to consumers were expected to reach $108 billion, or 6% of consumer retail spending.[7] Indeed, online shopping is growing even faster than the increase in computer use, which has been fueled by the falling price of personal computers. Among the best-known e-firms are bookseller Amazon.com; auction network eBay; and Priceline.com, which lets you name the price you're willing to pay for airline tickets and hotel rooms.

Probably the foremost example of e-commerce is Amazon.com.[8] In 1994, seeing the potential for electronic retailing on the World Wide Web, Jeffrey Bezos left a successful career on Wall Street to launch an online bookstore called Amazon.com. Why the name *Amazon?*

"Earth's biggest river, Earth's biggest bookstore," said Bezos in a 1996 interview. "The Amazon River is ten times as large as the next largest river, which is the Mississippi, in terms of volume of water. Twenty percent of the world's fresh water is in the Amazon River Basin, and we have six times as many titles as the world's largest physical bookstore."[9] A more hard-headed reason is that, according to consumer tests, words starting with "A" show up on search-engine lists first.[10]

Still, Bezos realized that no bookstore with four walls could possibly stock the more than 2.5 million titles that are now active and in print. Moreover, he saw that an online bookstore wouldn't have to make the same investment in retail clerks, store real estate, or warehouse space (in the beginning, Amazon.com ordered books from the publisher *after* it took an order), so it could pass savings along to customers in the form of discounts. In addition, he appreciated that there would be opportunities to obtain demographic information about customers in order to offer personalized services. For example, Amazon could let customers know of books that might be of interest to them. Such personalized attention is difficult for traditional bookstores. Finally, Bezos saw that there could be a good deal of online interaction: Customers could post reviews of books they read and could reach authors by e-mail to provide feedback. All this was made possible on the Web by the recording of information on giant databases.

Amazon.com sold its first book in July 1995 and by early 2000 had 1.1 million customers. In August 2001, 23 million people visited Amazon.com.[11] The firm also expanded into the online retailing of music CDs, toys, electronics, drugs, cosmetics, pet supplies, kitchen appliances, and other goods and into online auctions.

In 2000, Amazon.com's stock plummeted, along with that of many other so-called "dot-com" companies, which had become overvalued. The company made a modest profit in late 2001. But the lasting impact of Amazon's trail blazing is clear from the surge of Old Economy "brick and mortar" companies into the online sector.

Data Mining

A personal database, such as the address list of friends you have on your microcomputer, is generally small. But some databases are almost unimaginably vast, involving records for millions of households and trillions of bytes of data. Some of these activities require the use of so-called massively parallel database computers that cost $1 million or more. "These machines gang together scores or even hundreds of the fastest microprocessors around," says one description, "giving them the oomph to respond in minutes to complex database queries."[12]

These large-scale efforts go under the name *data mining*. **Data mining (DM) is the computer-assisted process of sifting through and analyzing vast amounts of data in order to extract meaning and discover new knowledge.** The purpose of DM is to describe past trends and predict future trends. Thus, data-mining tools might sift through a company's immense collections of customer, marketing, production, and financial data and identify what's worth noting and what's not.

Data mining has come about because companies find that, in today's fierce competitive business environment, they need to turn the gazillions of bytes of raw data at their disposal to new uses for further profitability. However, nonprofit institutions have also found DM methods useful, as in the pursuit of scientific and medical discoveries.

Some applications of data mining:[13]

- **Marketing:** Marketers use DM tools (such as one called Spotlight) to mine point-of-sale databases of retail stores, which contain facts (such as prices, quantities sold, dates of sale) for thousands of products in hundreds of geographic areas. By understanding customer preferences and buying patterns, marketers hope to target consumers' individual needs.

Databases for marketing
Harrah's casinos use data mining to identify gambling patron preferences for marketing and special promotions.

- **Health:** A coach in the U.S. Gymnastics Federation is using a DM system (called IDIS) to discover what long-term factors contribute to an athlete's performance, so as to know what problems to treat early on. A Los Angeles hospital is using the same tool to see what subtle factors affect success and failure in back surgery. Another system helps health-care organizations pinpoint groups whose costs are likely to increase in the near future, so that medical interventions can be made.
- **Science:** DM techniques are being employed to find new patterns in genetic data, molecular structures, global climate changes, and more. For instance, one DM tool (called SKICAT) is being used to catalog more than 50 million galaxies, which will be reduced to a 3-terabyte galaxy catalog.

Clearly, short-term payoffs can be dramatic. One telephone company, for instance, mined its existing billing data to identify 10,000 supposedly "residential" customers who spent more than $1000 a month on their phone bills. When it looked more closely, the company found these customers were really small businesses trying to avoid paying the more expensive business rates for their telephone service.[14]

However, the payoffs in the long term could be truly astonishing. Sifting medical-research data or subatomic-particle information may reveal new treatments for diseases or new insights into the nature of the universe.[15]

Business-to-Business (B2B) Systems

In a _business-to-business (B2B) system_, **a business sells to other businesses, using the Internet or a private network to cut transaction costs and increase efficiencies.** Business-to-business activity is expected to balloon to a $2.8 trillion industry by 2004.[16]

One of the most famous examples of B2B is the auto industry online exchange developed by the big three U.S. automakers—General Motors, Ford, and DaimlerChrysler. The companies are putting their entire system of purchasing, involving more than $250 billion in parts and materials and 60,000 suppliers, on the Internet. Already so-called "reverse auctions," in which suppliers bid to provide the lowest price, have driven down the cost of parts such as tires and window sealers. This system replaces the old-fashioned bureaucratic procurement process built on phone calls and fax machines, and provides substantial cost savings.[17]

Online B2B exchanges have been developed to serve a variety of businesses, from manufacturers of steel and airplanes to convenience stores to olive oil producers.[18] B2B exchanges are expected to help business by moving beyond pricing mechanisms and encompassing product quality, customer support, credit terms, and shipping reliability, which often count for more than price. The name given to this system is the _business web_, or _b-web_, in which suppliers, distributors, customers, and e-commerce service providers use the Internet for communications and transactions. In addition, b-webs are expected to provide extra revenue from ancillary services, such as financing and logistics.[19] (See ● Panel 7.6, next page.)

None of these innovations is possible without databases and the communications lines connecting them.

CLICK-ALONG 7-2
More about databases

CONCEPT CHECK

Describe what e-commerce is.

What is data mining, and how is it used?

What is a B2B system?

● **PANEL 7.6**
B2B Exchanges
B2B exchanges, which draw on data from various databases, act as centralized online markets for buyers and sellers in specific fields, such as car parts or olive oil. Exchanges are expected to evolve into "b-webs," or business webs, encompassing other factors besides price.

Sellers

Buyers

Internet Exchange

The B2B World

Content management

Value-Added Services
Order management

Logistics

Financial

Marketing

Customer service

Shipping

7.5 The Ethics of Using Databases: Concerns about Accuracy & Privacy

KEY QUESTION
What are some ethical concerns about the uses of databases?

Ethics

The enormous capacities of today's storage devices have given photographers, graphics professionals, and others a new tool—the ability to manipulate images at the pixel level. For example, photographers can easily do *morphing*—transforming one image into another. In **morphing, a film or video image is displayed on a computer screen and altered pixel by pixel, or dot by dot. As a result, the image metamorphoses into something else**—a pair of lips into the front of a Toyota, for example, or an owl into a baby.

The ability to manipulate digitized output—images and sounds—has brought a wonderful new tool to art. However, it has created some big new problems in the area of credibility, especially for journalism. How can we know that what we're seeing or hearing is the truth? Consider the following.

Chapter 7

258

Manipulation of Sound

Frank Sinatra's 1994 album *Duets* paired him through technological tricks with singers like Barbra Streisand, Liza Minnelli, and Bono of U2. Sinatra recorded solos in a recording studio. His singing partners, while listening to his taped performance on earphones, dubbed in their own voices. These second voices were recorded not only at different times but often, through distortion-free phone lines, from different places. The illusion in the final recording is that the two singers are standing shoulder to shoulder.

Newspaper columnist William Safire called *Duets* "a series of artistic frauds." Said Safire, "The question raised is this: When a performer's voice and image can not only be edited, echoed, refined, spliced, corrected, and enhanced—but can be transported and combined with others not physically present—what is performance? . . . Enough of additives, plasticity, virtual venality; give me organic entertainment."[20] Some listeners feel that the technology changes the character of a performance for the better. Others, however, think the practice of assembling bits and pieces in a studio drains the music of its essential flow and unity.

Whatever the problems of misrepresentation in art, however, they pale beside those in journalism. What if, for example, a radio station were to edit a stream of digitized sound so as to misrepresent what actually happened?

Manipulation of Photos

When O. J. Simpson was arrested in 1994 on suspicion of murder, the two principal American newsmagazines both ran pictures of him on their covers.[21] *Newsweek* ran the mug shot unmodified, as taken by the Los Angeles Police Department. At *Time*, an artist working with a computer modified the shot with special effects as a "photo-illustration." Simpson's image was darkened so that it still looked like a photo but, some critics said, with a more sinister cast to it.

Should a magazine that reports the news be taking such artistic license? Should *National Geographic* in 1982 have photographically moved two Egyptian pyramids closer together so that they would fit on a vertical cover? Was it even right for *TV Guide* in 1989 to run a cover showing Oprah Winfrey's head placed on Ann-Margret's body? In another case, to show what can be done, a photographer digitally manipulated the famous 1945 photo showing the meeting of the leaders of the wartime Allied powers at Yalta. Joining Stalin, Churchill, and Roosevelt are some startling newcomers: Sylvester Stallone and Groucho Marx. The additions are so seamless that it is impossible to tell the photo has been altered. *(See ● Panel 7.7.)*

● **PANEL 7.7**
Photo manipulation
In this 1945 photo, World War II Allied leaders Joseph Stalin, Winston Churchill, and Franklin Roosevelt are shown from left to right. Digital manipulation has added Sylvester Stallone standing behind Roosevelt and Groucho Marx seated at right.

The potential for abuse is clear. "For 150 years, the photographic image has been viewed as more persuasive than written accounts as a form of 'evidence,'" says one writer. "Now this authenticity is breaking down under the assault of technology."[22] Asks a former photo editor of the *New York Times Magazine*, "What would happen if the photograph appeared to be a straightforward recording of physical reality, but could no longer be relied upon to depict actual people and events?"[23]

Many editors try to distinguish between photos used for commercialism (advertising) versus for journalism, or for feature stories versus for news stories. However, this distinction implies that the integrity of photos is only important for some narrow category of news. In the end, it can be argued, altered photographs pollute the credibility of all of journalism.

Manipulation of Video & Television

The technique of morphing, used in still photos, takes a massive jump when used in movies, videos, and television commercials. Digital image manipulation has had a tremendous impact on filmmaking. Director and digital pioneer Robert Zemeckis *(Death Becomes Her)* compares the new technology to the advent of sound in Hollywood.[24] It can be used to erase jet contrails from the sky in a western and to make digital planes do impossible stunts. It can even be used to add and erase actors.

Films and videotapes are widely thought to accurately represent real scenes (as evidenced by the reaction to the amateur videotape of the Rodney King beating by police in Los Angeles). Thus, the possibility of digital alterations raises some real problems. Videotapes supposed to represent actual events could easily be doctored. Another concern is for film archives: Because digital videotapes suffer no loss in resolution when copied, there are no "generations." Thus, it will be impossible for historians and archivists to tell whether the videotape they're viewing is the real thing or not.[25]

Indeed, it is possible to create virtual images during live television events. These images—such as a Coca-Cola logo in the center of a soccer field—don't exist in reality but millions of viewers see them on their TV screens.[26]

Virtual advertising
The oil company 76 logo doesn't really appear on this wall but on TV looks as though it does.

Not a real person, Ananova is the first "virtual newscaster," an animated figure who presents the latest headlines.

Accuracy & Completeness

Databases—including public data banks such as Nexis/Lexis—can provide you with *more* facts and *faster* facts but not always *better* facts. Penny Williams, professor of broadcast journalism at Buffalo State College in New York and formerly a television anchor and reporter, suggests five limitations to bear in mind when using databases for research:[27]

- **You can't get the whole story:** For some purposes, databases are only a foot in the door. There may be many facts or facets of the topic that are not in a database. Reporters, for instance, find a database is a starting point. It may take intensive investigation to get the rest of the story.
- **It's not the gospel:** Just because you see something on a computer screen doesn't mean it's accurate. Numbers, names, and facts must be verified in other ways.

- **Know the boundaries:** One database service doesn't have it all. For example, you can find full text articles from the *New York Times* on Lexis/Nexis, from *The Wall Street Journal* on Dow Jones News Retrieval, and from the *San Jose Mercury News* on America Online, but no service carries all three.
- **Find the right words:** You have to know which keywords (search words) to use when searching a database for a topic. As Lynn Davis, a professional researcher with ABC News, points out, if you're searching for stories on guns, the keyword "can be guns, it can be firearms, it can be handguns, it can be pistols, it can be assault weapons. If you don't cover your bases, you might miss something."[28]
- **History is limited:** Most public databases, Davis says, have information going back to 1980, and a few into the 1970s, but this poses problems if you're trying to research something that happened or was written about earlier.

Matters of Privacy

Privacy **is the right of people not to reveal information about themselves.** Who you vote for in a voting booth and what you say in a letter sent through the U.S. mail are private matters. However, the ease of pulling together and disseminating information via databases and communications lines has put privacy under extreme pressure.

As you've no doubt discovered, it's no trick at all to get your name on all kinds of mailing lists. Theo Theoklitas, for instance, received applications for credit cards, invitations to join video clubs, and notification of his finalist status in Ed McMahon's $10 million sweepstakes. Theo is a black cat who's been getting mail ever since his owner sent in an application for a rebate on cat food. A whole industry has grown up of professional information gatherers and sellers, who collect personal data and sell it to fund-raisers, direct marketers, and others.

These concerns have led to the enactment of a number of laws to protect individuals from invasion of privacy. *(See • Panel 7.8, next page.)*

Survival Tip

Some Websites about Privacy

The following websites offer ways to guard your privacy:

www.epic.org

www.privacyfoundation.org

www.eff.org

www.ncinet.org/essentials/privacy.html

www.ftc.gov/privacy/index.html

www.junkbusters.com

www.kidsprivacy.org

www.spamfree.org

PANEL 7.8
Privacy laws
Important federal privacy laws

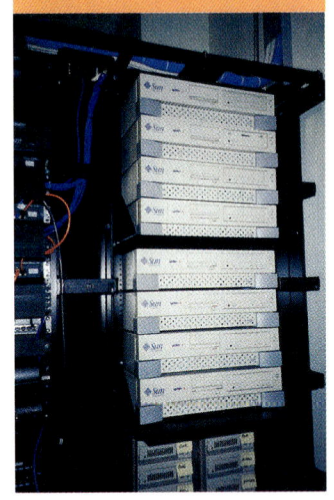

Database servers

Important Federal Privacy Laws

Freedom of Information Act (1970): Gives you the right to look at data concerning you that is stored by the federal government. A drawback is that sometimes a lawsuit is necessary to pry it loose.

Fair Credit Reporting Act (1970): Bars credit agencies from sharing credit information with anyone but authorized customers. Gives you the right to review and correct your records and to be notified of credit investigations for insurance or employment. A drawback is that credit agencies may share information with anyone they reasonably believe has a "legitimate business need." Legitimate is not defined.

Privacy Act (1974): Prohibits federal information collected about you for one purpose from being used for a different purpose. Allows you the right to inspect and correct records. Federal agencies share information anyway.

Family Educational Rights and Privacy Act (1974): Gives students and their parents the right to review, and to challenge and correct, students' school and college records; limits sharing of information in these records.

Right to Financial Privacy Act (1978): Sets strict procedures that federal agencies must follow when seeking to examine customer records in banks; regulates financial industry's use of personal financial records. A drawback is that the law does not cover state and local governments.

Privacy Protection Act (1980): Prohibits agents of federal government from making unannounced searches of press offices if no one there is suspected of a crime.

Cable Communications Policy Act (1984): Restricts cable companies in the collection and sharing of information about their customers.

Computer Fraud and Abuse Act (1984): Allows prosecution for unauthorized access to computers and databases. A drawback is that people with legitimate access can still get into computer systems and create mischief without penalty.

Electronic Communications Privacy Act (1986): Makes eavesdropping on private conversations illegal without a court order.

Computer Security Act (1987): Makes actions that affect the security of computer files and telecommunications illegal.

Computer Matching and Privacy Protection Act (1988): Regulates computer matching of federal data; allows individuals a chance to respond before government takes adverse actions against them. A drawback is that many possible computer matches are not affected, such as those done for law-enforcement or tax reasons.

Video Privacy Protection Act (1988): Prevents retailers from disclosing video-rental records without the customer's consent or a court order.

Telephone Consumer Protection Act (1991): Restricts the activities of telemarketing salespeople.

Cable Act (1992): Extends to cellphone and wireless services the privacy protections of the Cable Communications Policy Act of 1984.

Computer Abuse Amendments Act (1994): Outlaws transmission of harmful computer codes such as worms and viruses.

National Information Infrastructure Protection Act (1996): Provides penalties for trespassing of computer systems, threats made to networks, and theft of information across state lines.

No Electronic Theft (NET) Act (1997): Eliminates legal loophole that enabled people to give away copyrighted material.

Child Online Protection Act (COPA) (1998): Penalties for commercial firms that knowingly distribute materials harmful to minors.

Children's Online Privacy Protection Act (COPPA) (2000): Requires online sites that attract young children to get permission from parents before asking for names, locations, and other details.

BOOKMARK IT!

PRACTICAL ACTION BOX
Preventing Your Identity from Getting Stolen

One day, Kathryn Rambo, 28, of Los Gatos, California, learned that she had a new $35,000 sports utility vehicle listed in her name, along with five credit cards, a $3000 loan, and even an apartment—none of which she'd asked for. "I cannot imagine what would be weirder, or would make you angrier, than having someone pretend to be you, steal all this money, and then leave you to clean up all their mess later," said Rambo, a special-events planner.[a] Added to this was the eerie matter of constantly having to prove that she was, in fact, herself: "I was going around saying, 'I am who I am!'"[b]

Identity Theft: Stealing Your Good Name—and More

Theft of identity (TOI) is a crime in which thieves hijack your very name and identity and use your good credit rating to get cash or to buy things. To begin, all they need is your full name or Social Security number. Using these, they tap into Internet databases and come up with other information—your address, phone number, employer, driver's license number, mother's maiden name, and so on. Then they're off to the races, applying for credit everywhere.

In Rambo's case, someone had used information lifted from her employee-benefits form. The spending spree went on for months, unbeknownst to her. The reason it took so long was that Rambo never saw any bills. They went to the address listed by the impersonator, a woman, who made a few payments to keep creditors at bay while she ran up even more bills. For Rambo, straightening out the mess required months of frustrating phone calls, time off from work, court appearances, and legal expenses.

How Does Identity Theft Start?

Identity theft typically starts in one of several ways:[c]

- **Wallet or purse theft:** There was a time when a thief would steal a wallet or purse, take the cash, and toss everything else. No more. Everything from keys to credit cards can be parlayed into further thefts.

- **Mail theft:** Thieves also consider mailboxes fair game. The mail will yield them bank statements, credit-card statements, new checks, tax forms, and other personal information.

- **Mining the trash:** You might think nothing of throwing away credit-card offers, portions of utility bills, or old cancelled checks. But "dumpster diving" can produce gold for thieves. Credit-card offers, for instance, may have limits of $5000 or so.

- **Telephone solicitation:** Prospective thieves may call you up and pretend to represent a bank, credit-card company, government agency, or the like in an attempt to pry loose essential data about you.

- **Insider access to databases:** You never know who has, or could have, access to databases containing your personnel records, credit records, car-loan applications, bank documents, and so on. This is one of the harder TOI methods to guard against.

What to Do Once Theft Happens

If you're the victim of a physical theft (or even loss), as when your wallet is snatched, you should immediately contact—first by phone, and then in writing—all your credit card companies, other financial institutions, the Department of Motor Vehicles, and any other organization whose cards you use that are now compromised. Be sure to call utility companies—telephone, electricity, and gas; identity thieves can run up enormous phone bills. Also call the local police and your insurance company to report the loss.

It's important to notify financial institutions *within two days of learning of your loss* because then you are legally responsible for only the first $50 of any theft. If you become aware of fraudulent transactions, immediately contact the fraud units of the three major credit bureaus: Equifax, Experian, and TransUnion. (See ● Panel 7.9.)

If your Social Security number has been fraudulently used, alert the Social Security Administration (800-772-1213). It's possible, as a last resort, to have your Social Security Number changed.

If you have a check guarantee card that was stolen, if your checks have been lost, or if a new checking account has been opened in your name, there are two organizations to notify so that payment on any fraudulent checks will be denied. They are Telecheck (800-366-2424) and National Processing Company (800-526-5380).

If your mail has been used for fraudulent purposes or if an identity thief filed a change of address form, look in the phone directory under U.S. Government Postal Service for the local Postal Inspector's office.

● PANEL 7.9
The three major credit bureaus

Equifax	Experian	TransUnion
To check your credit report: 800-685-1111 www.equifax.com	To check your credit report: 800-397-3742 www.experian.com	To check your credit report: 800-888-4213 www.tuc.com

How to Prevent Identity Theft

One of the best ways to keep your finger on the pulse of your financial life is, on a regular basis—once a year, say—to get a copy of your credit report from one or all three of the main credit bureaus. This will show you whether there is any unauthorized activity. Reports cost $8.

In addition, there are some specific measures you can take to guard against personal information getting into the public realm.

- **Check your credit-card billing statements:** If you see some fraudulent charges, report them immediately. If you don't receive your statement, call the creditor first. Then call the post office to see if a change of address has been filed under your name.

- **Treat credit cards and other important papers with respect:** Make a list of your credit cards and other important documents, and the list of numbers to call if you need to report them lost. (You can photocopy the cards front and back, but make sure the numbers are legible.)

  Carry only one or two credit cards at a time. Carry your Social Security card, passport, or birth certificate only when needed.

  Don't dispose of credit card receipts in a public place.

  Don't give out your credit-card numbers or Social Security number over the phone, unless you have some sort of trusted relationship with the party on the other end.

  Tear up credit-card offers before you throw them away.

  Keep tax records and other financial documents in a safe place.

- **Treat passwords with respect:** Memorize passwords and PINs. Don't use your birth date, mother's maiden name, or similar common identifiers, which thieves may be able to guess.

- **Treat checks with respect:** Pick up new checks at the bank. Shred cancelled checks before throwing them away. Don't let merchants write your credit-card number on the check.

- **Watch out for "shoulder surfers" when using phones and ATMs:** When using PINs and passwords at public telephones and automated teller machines, shield your hand so that anyone watching through binoculars or using a video camera—"shoulder surfers"—can't read them.

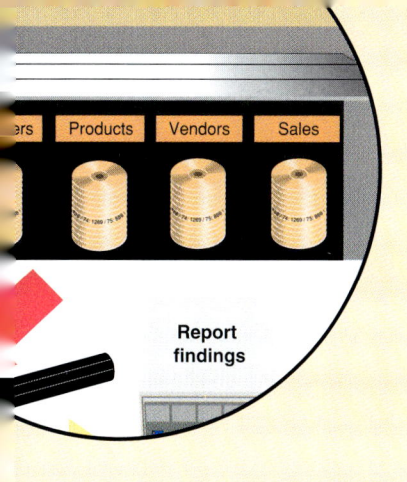

Summary

business-to-business (B2B) system (p. 257, KQ 7.4) Direct sales between businesses, using the Internet or private network to cut transaction costs and increase efficiencies. Why it's important: *Business-to-business activity is expected to balloon to a $1 trillion industry by 2002 and a $2.8 trillion industry by 2004.*

character (p. 246, KQ 7.1) A single letter, number, or special character. Why it's important: *Characters—such as A, B, C, 1, 2, 3, #, $, %—are part of the data storage hierarchy.*

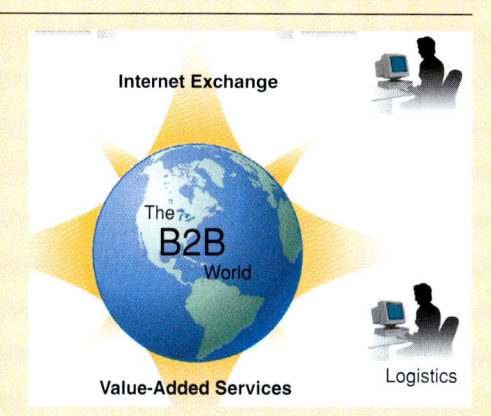

database (p. 246, KQ 7.1) Organized collection of related (integrated) files. Why it's important: *Businesses and organizations build databases to help them keep track of and manage their affairs. In addition, online database services put enormous research resources at the user's disposal.*

database management system (DBMS) (p. 250, KQ 7.2) Also called a *database manager;* software that controls the structure of a database and access to the data. Allows users to manipulate more than one file at a time. Why it's important: *This software enables sharing of data (same information is available to different users); economy of files (several departments can use one file instead of each individually maintaining its own files, thus reducing data redundancy, which in turn reduces the expense of storage media and hardware); data integrity (changes made in the files in one department are automatically made in the files in other departments); security (access to specific information can be limited to selected users).*

data files (p. 248, KQ 7.1) Files that contain data—words, numbers, pictures, sounds, and so on. Why it's important: *Unlike program files, data files don't instruct the computer to do anything. Rather, data files are there to be acted on by program files. Examples of common extensions in data files are .txt (text) and .xls (spreadsheets). Certain proprietary software programs have their own extensions, such as .ppt for PowerPoint and .mdb for Access.*

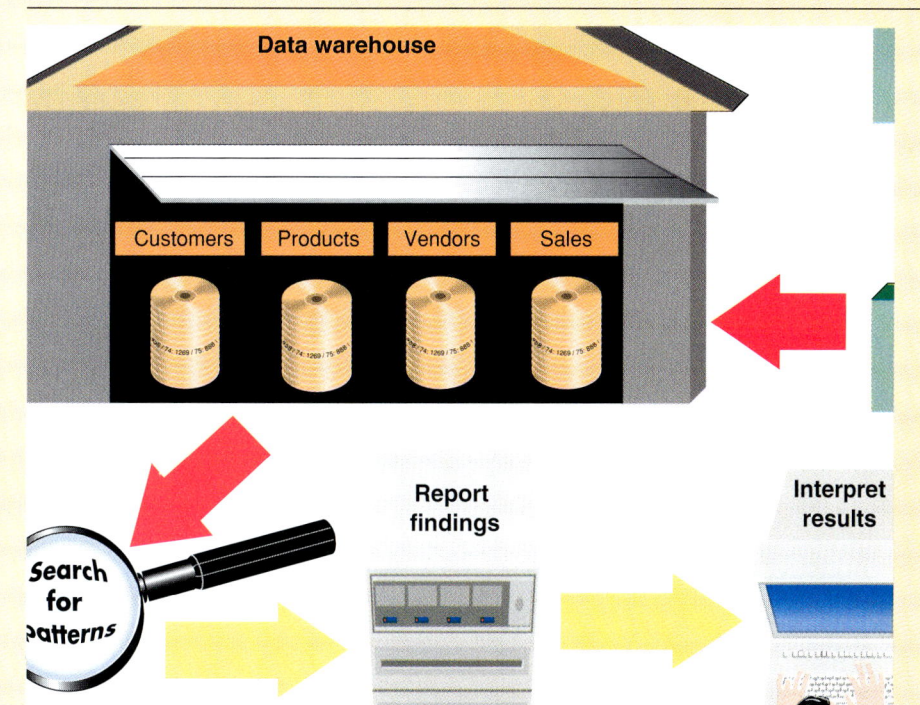

data mining (DM) (p. 256, KQ 7.4) Computer-assisted process of sifting through and analyzing vast amounts of data in order to extract meaning and discover new knowledge. Why it's important: *The purpose of DM is to describe past trends and predict future trends. Thus, data-mining tools might sift through a company's immense collections of customer, marketing, production, and financial data and identify what's worth noting and what's not.*

data storage hierarchy (p. 246, KQ 7.1) The ranked levels of data stored in a computer: bits, bytes (characters), fields, records, files, and databases. Why it's important: *Understanding the data storage hierarchy is necessary to understand how to use a database.*

direct access storage (p. 249, KQ 7.1) Storage system that allows the computer to go directly to the desired information. The data is retrieved (accessed) according to a unique data identifier called a key field. It also uses a file allocation table (FAT), a hidden on-disk table that records exactly where the parts of a given file are stored. *Why it's important:* This method of file organization, used with hard disks and other types of disks, is ideal for applications where there is no fixed pattern to the requests for data—for example, in airline reservation systems or computer-based directory-assistance operations. Direct access storage is much faster than sequential access storage.

distributed database (p. 251, KQ 7.2) Database that is stored on different computers in different locations connected by a client/server network. *Why it's important:* Data need not be centralized in one location.

e-commerce (p. 255, KQ 7.4) Electronic commerce; the buying and selling of products and services through computer networks. *Why it's important:* By 2003, total U.S. e-commerce sales to consumers are expected to reach $108 billion, or 6% of consumer retail spending; online shopping is growing even faster than the increase in computer use, which has been fueled by the falling price of personal computers.

File	Several records	*Friends' addresses file* Bierce, Ambrose 0001; London, Jack 0234; Stevenson, Robert L. 0081; etc.
Record	Several fields	*Ambrose Bierce's name and address* 13 Fallaway St. San Francisco, CA 94123
Field	Characters (bytes)	*First name field* Ambrose
Character	Bits (0 or 1)	*Letter S* 1110 0010

field (p. 246, KQ 7.1) Unit of data consisting of one or more characters (bytes). An example of a field is your name, your address, or your Social Security number. *Why it's important:* A collection of fields makes up a record. Also see key field.

file (p. 246, KQ 7.1) Collection of related records. An example of a file is data collected on everyone employed in the same department of a company, including all names, addresses, and Social Security numbers. *Why it's important:* A file is the collection of data or information that is treated as a unit by the computer; a collection of related files makes up a database.

filename (p. 247, KQ 7.1) The name given to a file. *Why it's important:* Files are given names so that they can be differentiated. Filenames also have extension names. These extensions of up to three letters are added after a period following the filename—for example, the .doc in Psychreport.doc is recognized by Microsoft Word as the extension for "document." Extensions are usually inserted automatically by the application software.

Records are arranged in related groups resembling a family tree, with "child" records subordinate to "parent" records.

Ports of departure	Miami
Names of ships	QE 2
Sailing dates	April 15
Cabin numbers	A-1

hierarchical database (p. 252, KQ 7.3) Database in which fields or records are arranged in related groups resembling a family tree, with child (lower-level) records subordinate to parent (higher-level) records. The parent record at the top of the database is called the *root record*. *Why it's important:* The hierarchical database is one of the common database structures.

individual database (p. 250, KQ 7.2) Collection of integrated files used by one person. *Why it's important:* Microcomputer users can set up their own individual databases using popular database management software; the information is stored on the hard drives of their personal computers. Today the principal database programs are Microsoft Access, Corel Paradox, and Lotus Approach. In addition, types of individual databases known as personal information managers (PIMs) can help users keep track of and manage information used on a daily basis, such as addresses, telephone numbers, appointments, to-do lists, and miscellaneous notes. Popular PIMs are Microsoft Outlook, Lotus Organizer, and Act.

key field (p. 247, KQ 7.1) Field that is chosen to uniquely identify a record so that it can be easily retrieved and processed. The key field is often an identification number, Social Security number, customer account number, or the like. *Why it's important:* The primary characteristic of the key field is that it is unique and thus can be used to identify one specific record.

master file (p. 249, KQ 7.1) Data file containing records that are generally updated periodically. *Why it's important:* Master files contain relatively permanent information used for reference purposes. They are updated through the use of transaction files.

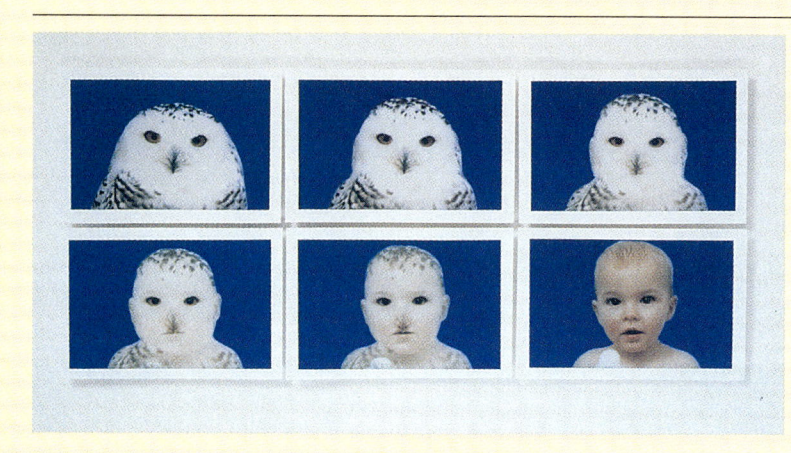

morphing (p. 258, KQ 7.5) Altering a film or video image displayed on a computer screen pixel by pixel, or dot by dot. *Why it's important: Morphing and other techniques of digital manipulation can produce images that misrepresent reality.*

network database (p. 252, KQ 7.3) Database similar in structure to a hierarchical database; however, each child record can have more than one parent record. Thus, a child record, which in network database terminology is called a *member*, may be reached through more than one parent, which is called an *owner*. *Why it's important: The network database is one of the common database structures.*

object-oriented database (p. 254, KQ 7.3) Database that uses "objects," software written in small, reusable chunks, as elements within database files. An object consists of (1) data in any form, including graphics, audio, and video, and (2) instructions on the action to be taken on the data. *Why it's important: A hierarchical or network database might contain only numeric and text data. By contrast, an object-oriented database might also contain photographs, sound bites, and video clips. Moreover, the object would store operations, called* methods, *the programs that objects use to process themselves.*

offline storage (p. 249, KQ 7.1) System in which stored data is not directly accessible for processing until the tape or disk it's on has been loaded onto an input device. *Why it's important: The storage medium and data are not under the direct, immediate control of the central processing unit.*

online storage (p. 250, KQ 7.1) System in which stored data is randomly (directly) accessible for processing. *Why it's important: The storage medium and data is under the direct, immediate control of the central processing unit. There's no need to wait for a tape or disk to be loaded onto an input device.*

privacy (p. 261, KQ 7.5) Right of people not to reveal information about themselves. *Why it's important: The ease of pulling together and disseminating information via databases and communications lines has put the basic democratic right of privacy under extreme pressure.*

program files (p. 248, KQ 7.1) Files containing software instructions. *Why it's important: Contrast data files.*

public databank (p. 251, KQ 7.2) Compilation of data available to the public. *Why it's important: The public databank is one of the basic types of database.*

record (p. 246, KQ 7.1) Collection of related fields. An example of a record would be your name and address and Social Security number. *Why it's important: Related records make up a file.*

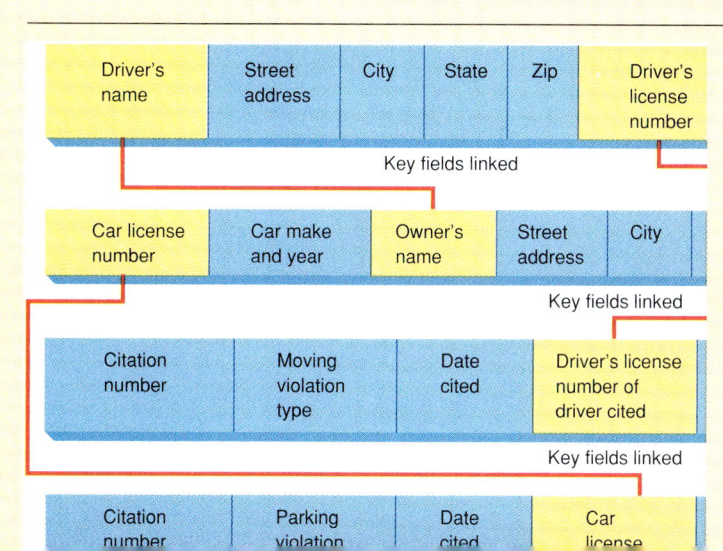

relational database (p. 254, KQ 7.3) Common database structure that relates, or connects, data in different files through the use of a key field, or common data element. In this arrangement there are no access paths down through a hierarchy. Instead, data elements are stored in different tables made up of rows and columns. In database terminology, the tables are called *relations* (files), the rows are called *tuples* (records), and the columns are called *attributes* (fields). All related tables must have a key field that uniquely identifies each row; that is, the key field must be in all tables. *Why it's important: The relational database is one of the common database structures; it is more flexible than hierarchical and network database models.*

sequential storage (p. 249, KQ 7.1) Storage system whereby data is stored and retrieved in sequence, such as alphabetically. Why it's important: *An inexpensive form of storage, sequential storage is the only type of storage provided by tape, which is used mostly for archiving and backup. The disadvantage of sequential file organization is that searching for data is slow. Compare direct access storage.*

shared database (p. 251, KQ 7.2) Also called a *company database;* a database shared by users in one company or organization in one location. The organization owns the database, which may be stored on a server such as a mainframe. Users are linked to the database via a local area or wide area network; the users access the network through terminals or microcomputers. Why it's important: *Shared databases, such as those you find when surfing the Web, are the foundation for a great deal of electronic commerce, particularly B2B commerce.*

transaction file (p. 249, KQ 7.1) Temporary holding file that holds all changes to be made to the master file: additions, deletions, revisions. Why it's important: *The transaction file is used to periodically update the master file.*

Chapter Review

stage 1 LEARNING — MEMORIZATION

"I can recognize and recall information."

Self-Test Questions

1. According to the data storage hierarchy, databases are composed of _____, _____, _____, _____, and _____.
2. An individual piece of data within a record is called a _____.
3. _____ is the right of people not to reveal information about themselves.
4. The four types of database access are _____, _____, _____, and _____.
5. The buying and selling of products and services through computer networks is called _____.
6. _____ files contain software instructions; _____ files contain data.
7. _____ storage means that the computer can go directly to the information you want.

Multiple-Choice Questions

1. Which of the following is not a type of database?
   a. distributed
   b. combination
   c. public
   d. shared
   e. individual

2. Which of the following database models stores data in any form, including graphics, audio, and video?
   a. hierarchical
   b. network
   c. object-oriented
   d. relational
   e. offline

True/False Questions

T F 1. The use of key fields makes it easier to locate a record in a database.

T F 2. A transaction file contains permanent records that are periodically updated.

T F 3. A database is an organized collection of integrated files.

T F 4. A shared database is stored on different computers in different locations connected by a client/server network.

T F 5. A directory (folder) is a storage place for files in one of your drives.

stage 2 LEARNING — COMPREHENSION

"I can recall information in my own terms and explain them to a friend."

Short-Answer Questions

1. What is the difference between sequential and direct-access storage?.
2. What is the difference between master files and transaction files?
3. What is data mining?
4. What is the difference between offline and online storage?
5. What is an ASCII file?
7. What is a distributed database?

Concept Mapping

On a separate sheet of paper, draw a concept map, or visual diagram, linking concepts. Show how the following terms are related.

- Access
- B2B system
- character
- database
- database access
- database management system
- data files
- data storage hierarchy
- direct access storage
- distributed database
- e-commerce
- file
- filename
- hierarchical database model
- individual database
- key field
- master file
- morphing
- network database model
- object-oriented database model
- Oracle
- privacy
- program files
- public databank
- query language
- record
- relational database model
- sequential storage
- shared database
- transaction file

stage 3 LEARNING: APPLYING, ANALYZING, SYNTHESIZING, EVALUATING

"I can apply what I've learned, relate these ideas to other concepts, build on other knowledge, and use all these thinking skills to form a judgment."

Knowledge in Action

Interview someone who works with or manages an organization's database. What types of records make up the database? Which departments use it? What database structure is used? What are the types and sizes of storage devices? Are servers used? Was the database software custom written?

Web Exercises

1. Visit these online stock trading websites and compare their services:
   - www.etrade.com
   - www.datek.com
   - www.ameritrade.com
   - www.schwab.com
   - www.fidelity.com
   - www.daytrade.com
   - www.stocks.com
   - www.daytradingstocks.com
   - www.nettradedirect.com

2. If you encounter a website that has an extensive amount of text and you would like to locate your search string inside the text without having to read it all, press CTRL-F, and a "find" search box will appear.
   If you need to bring up an exact replica of a web browser (which includes all of the pages viewed previously), press CTRL-N, and a new browser will open. This is helpful when you need multiple windows open but don't want to lose your current location.

3. Visit the following websites to read about the petabyte:
   - http://content.techweb.com/wire/story/TWB20010226S0026
   - www.jamesshuggins.com/h/tek1/how_big.htm
   - www.sdsc.edu/GatherScatter/GSfall95/1petabyte.html
   - http://siliconvalley.internet.com/news/article/0,2198,3531_534901,00.html

4. Visit www.cisco.com to learn more about the company that uses its worldwide intranet to connect its distributive databases.

5. Visit these sites about data mining and the search for meaningfulness in large quantities of data.
   - www.almaden.ibm.com/cs/quest
   - www.dmbenchmarking.com
   - www.dmg.org
   - www.siam.org/meetings/sdm01
   - www.spss.com/datamine
   - www.idagroup.com

   What do you think the next technology will be for handling and analyzing massive amounts of data?

6. How much information about you is out there? Run various search strings about yourself to see just how private your life is.

7. Call up your tech support phone number for your computer's manufacturer. The first thing they will ask you for is your serial number. This is the key field required to bring up all of your information in their customer database. With that serial number, they should be able to bring up information and troubleshooting issues for each and every part of your PC. They will know everything that came with your PC down to the exact version of the system restore disk.

8. An extensive database featuring photographs and detailed information of each inmate of the Florida Department of Corrections can be found at the following website:
   www.dc.state.fl.us/inmateinfo/inmateinfomenu.asp
   Run several searches to see how this database works. Is it efficient? How would you improve it?

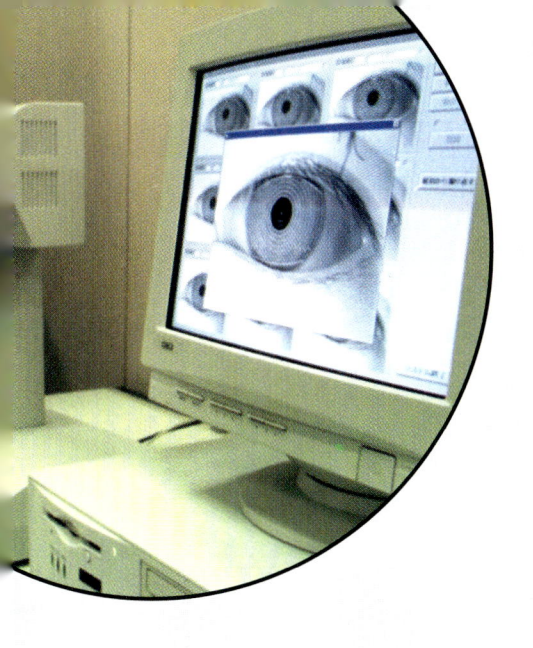

Chapter 8

Society & the Digital Age

Challenges & Promises

Key Questions
You should be able to answer the following questions.

8.1 **The Digital Environment: Is There a Grand Design?** What are the NII, the new Internet, the Telecommunications Act, the 1997 White House plan, and ICANN?

8.2 **Security Issues: Threats to Computers & Communications Systems** What are some characteristics of the key security issues for information technology?

8.3 **Security: Safeguarding Computers & Communications** What are the characteristics of the four components of security?

8.4 **Quality-of-Life Issues: The Environment, Mental Health, & the Workplace** How does information technology create environmental, mental-health, and workplace problems?

8.5 **Economic Issues: Employment & the Haves/Have-Nots** How may technology affect the unemployment rate and the gap between rich and poor?

8.6 **Artifical Intelligence** What are the main areas of artificial intelligence?

8.7 **The Promised Benefits of the Digital Age** What are some benefits of the digital age?

If the Internet is on its way to becoming the dominant mode of information exchange, then it is no longer a luxury but, like the telephone, a necessity."

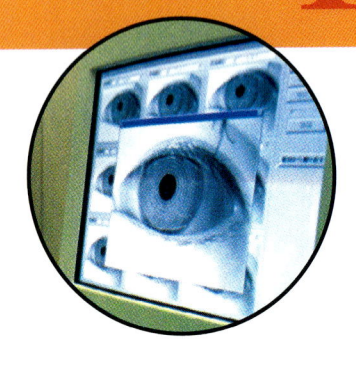

And it follows, in this analyst's opinion, that "anyone without it is in danger of being shut out."[1]

This "digital divide" between those with and without access to information technology is actually narrowing as the Information Age continues to expand productivity and wealth.[2] Still, addressing that divide is one of the most important challenges of our time. Among U.S. households earning more than $75,000 a year, 82% have Internet access, compared with 38% of those in households earning less than $30,000.[3] Internationally, the digital divide is between rich countries and poor countries. In the United States, 10% of the economy is devoted to the purchase of hardware and software, as compared with, say, Bangladesh, where it is one-tenth of 1%.[4] Of the world's approximately 1 billion web pages, more than 80% are in English.[5] With only 5% of the world's population, the U.S. has 50% of the world's Internet-linked home computers.[6]

The digital divide is only one of the many challenges confronting us as infotech sweeps the world. Elsewhere in the book we have considered ergonomics (Chapter 5) and privacy (Chapter 7). In this chapter, we consider some other major issues:

- Is there a grand design for the digital environment?
- Security issues—accidents, hazards, crime, viruses—and security safeguards
- Quality-of-life issues—environment, mental health, the workplace
- Economic issues—employment and the haves/have-nots

8.1 The Digital Environment: Is There a Grand Design?

KEY QUESTIONS
What are the NII, the new Internet, the Telecommunications Act, the 1997 White House plan, and ICANN?

The former buzz word *information superhighway* has lost its luster in favor of other coinages such as the *digital environment*. The presumed goal of this worldwide system of computers and telecommunications is to give us lightning-fast (high-bandwidth) voice and data exchange, multimedia, interactivity, and near-universal, low-cost access—and to do so reliably and securely. Whether you're a Russian astronaut aloft in a spacecraft, a Bedouin tribesman in the desert with a PDA/cell phone, or a Canadian work-at-home mother with her office in a spare bedroom, you'll be able, it is hoped, to connect with nearly anything or anybody anywhere. You'll be able to participate in telephony, teleconferencing, telecommuting, teleshopping, telemedicine, tele-education, televoting, and even telepsychotherapy (already available), to name a few possibilities.

What shape will this digital environment take? Some government officials hope it will follow a somewhat orderly model, such as that envisioned in the National Information Infrastructure (of which new versions of the Internet are a part). Others hope it will evolve out of competition intended by the passage of the 1996 Telecommunications Act. Still others hope that a White House document, *A Framework for Global Electronic Commerce*, offers a realistic policy. Finally, a nonprofit organization, ICANN, is concerned with Internet addresses. What can these attempts to create an all-encompassing design do?

The National Information Infrastructure

As portrayed by government officials, the *National Information Infrastructure (NII)* is a kind of grand vision for existing networks and technologies, as well as technologies yet to be deployed. Services would be delivered by telecommunications companies, cable-television companies, and the Internet for a range of applications—education, health care, information access, electronic commerce, and entertainment.

Who would put the pieces of the NII together? The current national policy is to let private industry do it, with the government trying to ensure fair competition among the carriers—phone, cable, and satellite companies—and compatibility among various technological systems. In addition, NII envisions open access to people of all income levels.

The New Internet: VBNS, Internet2, & NGI

Lately less is being said about NII and more about new Internet networks: *VBNS, Internet2,* and the *Next Generation Internet.*

Does this mean three new networks will be built? Actually, all three names refer to the same network. This high-speed Internet is designed to relieve the congested electronic highway of the older Internet. Here's what the three efforts represent:

- VBNS: **Linking supercomputers and other banks of computers across the nation, *VBNS (Very-High-Speed Backbone Network Service)* is the main U.S. government component to upgrade the "backbone," or primary hubs of data transmission.** Speeds are 1000 times conventional Internet speeds.

    Financed by the National Science Foundation and managed by the telecommunications giant MCI, VBNS is somewhat exclusive. It is being used by just 101 top universities and other research institutions for network-intensive applications, such as transmitting high-quality video for distance education.[7] (Internet2, by contrast, will eventually touch a great many more members.) VBNS has been underway since 1996, and most of the present members are also members of Internet2.

- Internet2: ***Internet2* is a cooperative university/business program to enable high-end users to quickly and reliably move huge amounts of data, using VBNS as the official backbone.** Whereas VBNS provides data transfer at 1000 times commercial Web speeds, Internet2 operates at only 100 times those speeds.

    In effect, Internet2 adds "toll lanes" to the older Internet to speed things up. The purpose is to advance videoconferencing, research, and academic collaboration—to enable a kind of "virtual university." Presently more than 150 universities in partnership with companies such as Cisco Systems and IBM are participants.

- Next Generation Internet: **The *Next Generation Internet (NGI)* is the U.S. government's program to parallel the university/business–sponsored effort of Internet2. It is designed to provide money to six government agencies to help tie the campus high-performance backbones into the broader federal infrastructure.** The intent is that NGI will connect at least 100 sites, including universities, federal national laboratories, and other research organizations. Speeds are 100 times those of the older Internet, and 10 sites are connected at speeds that are 1000 times as fast.

All these networks are modeled after the original Internet, except that they will use high-speed fiber-optic circuits and more sophisticated software. NGI and Internet2 should be available to the public by 2003.

Call your office
A traveling salesman contacts his office via cell phone and laptop computer.

The 1996 Telecommunications Act

After years of legislative attempts to overhaul the 1934 Communications Act, in February 1996 President Bill Clinton signed into law the *Telecommunications Act of 1996*, undoing 60 years of federal and state communications regulations. The act is designed to let phone, cable, and TV businesses compete and combine more freely. The purpose of the law was to cultivate greater competition between local and long-distance telephone companies, as well as between the telephone and cable industries. Under this legislation, different carriers can offer the same services—for example, cable companies can offer telephone services, phone companies can offer cable.

Is the law successful? "The only point on which all parties agree," says Laurence Tribe, Harvard professor of constitutional law, "is that the law isn't working as intended, and that American consumers are still waiting for free and healthy competition in communications services."[8]

The 1997 White House Plan for Internet Commerce

In 1997, the Clinton administration unveiled a document, authored by a White House group, that is significant because it endorsed a governmental hands-off approach to the Internet—or, as the report more grandly calls it, "the Global Information Infrastructure."[9] Behind the title *A Framework for Global Electronic Commerce* was a plan whose gist is this: Government should stay out of the way of Internet commerce.

"Where government is needed," it states, "its aim should be to support and enforce a predictable, minimalist, consistent, and legal environment for commerce."[10] Otherwise, the plan states that private companies, not government, should take the lead in promoting the Internet as an electronic marketplace, in adopting self-regulation, and in devising ratings systems to help parents guide their children away from objectionable online content.

ICANN: The Internet Corporation for Assigned Names & Numbers

Acting on the belief that the Internet moves too quickly to be regulated by government, the White House decided to let Internet users govern themselves. In June 1998, it proposed the creation of a series of nonprofit corporations to manage such complex issues as fraud prevention, privacy, and intellectual property protection. The first such group, **ICANN (Internet Corporation for Assigned Names & Numbers) was established to regulate Internet domain names,** those addresses ending with .com, .org, .net, and so on, that identify a website. ICANN got off to a rocky start, but if it succeeds, according to one report, "it could evolve into the preeminent regulatory body on the Internet, with power extending far beyond the arena of domain names."[11]

So far, ICANN's stewardship has been a bit rocky. Critics have charged that the domain name system is unraveling under increasing strain.[12] For example, since domain names have to be unique, they are inevitably linked to trademark disputes when two or more companies or people want the same name. There are also issues of privacy: Anyone using a search tool such as that available at www.whonami.com can look up the name and address of whoever is registered as the owner of a domain name, which opens the door to spammers and hackers. Resolution of such problems becomes more important if an idea called *Enum* is ever realized. *Enum* is a fancy system for translating phone numbers into Internet addresses, so that the Internet and the global phone system will converge, and everyone will have just a single identification.[13]

CONCEPT CHECK

What is the National Information Infrastructure?

Distinguish among VBNS, Internet2, and the Next Generation Internet.

Describe the 1997 White house plan for internet commerce and its offshoot, ICANN.

8.2 Security Issues: Threats to Computers & Communications Systems

KEY QUESTION
What are some characteristics of the key security issues for information technology?

Security issues go right to the heart of the workability of computer and communications systems. Here we discuss several threats to computers and communications systems.

Errors & Accidents

In general, errors and accidents in computer systems may be classified as human errors, procedural errors, software errors, electromechanical problems, and "dirty data" problems.

- **Human errors:** Quite often, when experts speak of the "unintended effects of technology," what they are referring to are the unexpected things people do with it. Among the ways in which people can complicate the workings of a system are the following:[14]

    (1) Humans often are not good at assessing their own information needs. For example, many users will acquire a computer and communications system that either is not sophisticated enough or is far more complex than they need.

    (2) Human emotions affect performance. For example, one frustrating experience with a computer is enough to make some people abandon the whole system. But throwing your computer out the window isn't going to get you any closer to learning how to use it better.

    (3) Humans act on their perceptions, which in modern information environments are often too slow to keep up with the equipment. Decisions influenced by information overload, for example, may be just as faulty as those based on too little information.

- **Procedural errors:** Some spectacular computer failures have occurred because someone didn't follow procedures. In 1999, the $125 million Mars Climate Orbiter was fed data expressed in pounds, the English unit of force, instead of newtons, the metric unit (about 22% of a pound). As a result, the spacecraft flew too close to the surface of Mars and broke apart.[15]

- **Software errors:** We are forever hearing about "software glitches" or "software bugs." A software bug is an error in a program that causes it not to work properly. In 2001, the nonprofit American Medical College Application Service launched a new web-based application service that was supposed to make medical-school applications for 115 medical schools easier and more efficient than the old paper version. Instead, it was plagued by seemingly endless software bugs. One applicant, Yale University student Amit Sachdeva, found himself spending night after night logging on again and again. His routine: stay awake until 2 a.m. when Internet traffic slowed down, plod through a few pages until an error message appeared, then watch as the system froze and crashed. Time for completion: an estimated 24–36 hours, instead of the 5–8 hours it was supposed to take. "It was incredibly frustrating," said Sachdeva.[16]

- **Electromechanical problems:** Mechanical systems, such as printers, and electrical systems, such as circuit boards, don't always work. They may be faultily constructed, get dirty or overheated, wear out, or become damaged in some other way. Power failures (brownouts and blackouts) can shut a system down. Power surges can also burn out equipment.

- **"Dirty data" problems:** When keyboarding a research paper, you undoubtedly make a few typing errors (which, hopefully, you clean up). So do all the data-entry people around the world who feed a continual stream of raw data into computer systems. A lot of problems are caused by this kind of "dirty data." *Dirty data* is incomplete, outdated, or otherwise inaccurate data.

Natural & Other Hazards

Some disasters do not merely lead to temporary system downtime; they can wreck the entire system. Examples are natural hazards, and civil strife and terrorism.

- **Natural hazards:** Whatever is harmful to property (and people) is harmful to computers and communications systems. This certainly includes natural disasters: fires, floods, earthquakes, tornadoes, hurricanes, blizzards, and the like. If they inflict damage over a wide area, as have ice storms in eastern Canada or hurricanes in Florida, natural hazards can disable all the electronic systems we take for granted. Without power and communications connections, automated teller machines, credit-card verifiers, and bank computers are useless.

- **Civil strife and terrorism:** Wars and insurrections seem to take place mainly in other parts of the world. Yet we are not immune to civil unrest, such as the riot in Los Angeles in 2000 after the hometown Lakers won the National Basketball Association championship. Nor are we immune to acts of terrorism, as has happened several times in the last few years: the 1993 bombing of New York's World Trade Center, the 1995 Timothy McVeigh bombing of the Federal Building in Oklahoma City, and the 2001 hijacked plane crashes into the World Trade Center and the Pentagon. The Pentagon alone has 650,000 terminals and workstations, 100 WANs, and 10,000 LANs, although the September 11, 2001, terrorist attack damaged only a few of them.

Terrorism
The kind of terrorist destruction that occurred in 2001 with the Pentagon can be a major threat to computer systems as well as people.

Crimes against Computers & Communications

A *computer crime* can be of two types. (1) It can be an illegal act perpetrated against computers or telecommunications. Or (2) it can be the use of computers or telecommunications to accomplish an illegal act.

Crimes against information technology include theft—of hardware, of software, of computer time, of cable or telephone services, or of information. Other illegal acts are crimes of malice and destruction. Some examples are as follows:

Ethics

- **Theft of hardware:** Hardware theft can range from shoplifting an accessory in a computer store to removing a laptop or cellular phone from someone's car. Professional criminals may steal shipments of microprocessor chips off a loading dock or even pry cash machines out of shopping-center walls.

- **Theft of software and music:** Generally, software theft involves illegal copying of programs, rather than physically taking someone's floppy disks. Software makers secretly prowl electronic bulletin boards in search of purloined products, then try to get a court order to shut down the bulletin boards. They also look for organizations that "softlift"—companies, colleges, or other institutions that buy one copy of a program and make copies for many computers.

  Many such so-called pirates are reported by co-workers or fellow students to the "software police," the Software Publishers Association. The SPA has a toll-free number (800-388-7478) for reporting illegal copying of software. In the 1990s, two New England college students were indicted for allegedly using the Internet to encourage the exchange of copyrighted software.[17]

  Another type of software theft is copying or counterfeiting of well-known software programs. These pirates often operate in China, Taiwan, Mexico, Russia, and various parts of Asia and Latin America. In some countries, most of the U.S. microcomputer software in use is thought to be illegally copied.

- **Theft of time and services:** The theft of computer time is more common than you might think. Probably the biggest instance is people using their employer's computer time to play games, do online shopping or stock trading, or dip into web pornography. Some people even operate sideline businesses.

  For years "phone phreaks" have bedeviled the telephone companies. For example, they have found ways to get into company voice-mail systems, then use an extension to make long-distance calls at the company's expense. They have also found ways to tap into cellular phone networks and dial for free.

- **Theft of information:** "Information thieves" have infiltrated the files of the Social Security Administration, stolen confidential personal records, and sold the information. On college campuses, thieves have snooped on or stolen private information such as grades. Thieves have also broken into computers of the major credit bureaus and stolen credit information. They have then used the information to charge purchases or have resold it to other people.

- **Crimes of malice and destruction:** Sometimes criminals are more interested in abusing or vandalizing computers and telecommunications systems than in profiting from them. For example, a student at a Wisconsin campus deliberately and repeatedly shut down a university computer system, destroying final projects for dozens of students. A judge sentenced him to a year's probation, and he left the campus.

CLICK-ALONG 8-1
Update: security issues

Crimes Using Computers & Communications

Just as a car can be used to perpetrate or assist in a crime, so can information technology. For example, Craig Pribila, 18, a student at the University of Nevada, Reno, faced a possible sentence of a year in jail after being convicted of charges of using his computer to counterfeit $20 bills and fake driver's licenses.[18]

In addition, investment fraud has come to cyberspace. Many people now use online services to manage their stock portfolios through brokerages hooked into the services. Scam artists have followed, offering nonexistent investment deals and phony solicitations, and manipulating stock prices.

Worms & Viruses

Worms and viruses (collectively called *malware*) are forms of high-tech maliciousness whose destructive effects are called their *payload*. There are about

Survival Tip

Keep Antivirus Software Updated

The antivirus software that comes with your computer won't protect you forever. To guard against new worms and viruses, not just last year's, visit the antivirus software maker's website from time to time.

57,000 known worms and viruses, not to mention others that could afflict you in the future. Researchers say they typically discover between 500 and 800 new ones every month.[19]

A *worm* **is a program that copies itself repeatedly into a computer's memory or onto a disk drive until no memory or disk space is left.** Sometimes it will copy itself so often it will cause a computer to crash. One example that turned up in 2001 was the Nimda worm, which attacked Microsoft Windows NT or 2000 machines; it reportedly crawled through more than a million computers in the United States, Europe, and Asia, clogging Internet traffic and resulting in some computer shutdowns.[20]

A *virus* **is a "deviant" program that copies itself into other programs, floppies, and hard disks, causing undesirable effects, such as destroying or corrupting data.** (See ● *Panel 8.1.*) The famous e-mail Love Bug (its subject line was I LOVE YOU), which originated in the Philippines in May 2000 and did perhaps as much as $10 billion in damage worldwide, was both a worm and a virus.

Worms and viruses are passed in two ways:

- **By diskette:** The first way is via an infected diskette, perhaps obtained from a friend or a repair person.
- **By network:** The second way is via a network, as from e-mail or an electronic bulletin board. This is why, when taking advantage of all the freebie games and other software available online, you should use virus-scanning software to check downloaded files.

● PANEL 8.1
Types of worms and viruses

- **Boot-sector virus:** The boot sector is that part of the system software containing most of the instructions for booting, or powering up, the system. The boot sector virus replaces these boot instructions with some of its own. Once the system is turned on, the virus is loaded into main memory before the operating system. From there it is in a position to infect other files. Any diskette that is used in the drive of the computer then becomes infected. When that diskette is moved to another computer, the contagion continues. Examples of boot-sector viruses: AntCMOS, AntiEXE, Form.A, NYB (New York Boot), Ripper, Stoned.Empire.Monkey.

- **File virus:** File viruses attach themselves to executable files—those that actually begin a program. (In DOS these files have the extensions .com and .exe.) When the program is run, the virus starts working, trying to get into main memory and infecting other files.

- **Multipartite virus:** A hybrid of the file and boot-sector types, the multipartite virus infects both files and boot sectors, which makes it better at spreading and more difficult to detect. Examples of multipartite viruses are Junkie and Parity Boot.

    A type of multipartite virus is the *polymorphic virus,* which can mutate and change form just as human viruses can. Such viruses are especially troublesome because they can change their profile, making existing antiviral technology ineffective.

    A particularly sneaky multipartite virus is the *stealth virus,* which can temporarily remove itself from memory to elude capture. An example of a multipartite, polymorphic stealth virus is One Half.

- **Macro virus:** Macro viruses take advantage of a procedure in which miniature programs, known as macros, are embedded inside common application programs (such as Microsoft Word or Excel). Macros allow users to do multiple steps in response to a user-defined keystroke or command. A macro virus attaches itself to documents created by an application that supports macros. Fortunately, the latest versions of Word and Excel come with built-in macro virus protection.

- **Logic bomb:** Logic bombs, or simply bombs, differ from other viruses or worms in that they are set to go off when certain conditions are met. A disgruntled programmer for a defense contractor created a bomb in a program that was supposed to go off two months after he left. Designed to erase an inventory tracking system, the bomb was discovered only by chance.

- **Trojan horse:** The Trojan horse (named after the horse in the ancient Greek story) covertly places illegal, destructive instructions in the middle of a legitimate program, such as a computer game. Once you run the program, the Trojan horse goes to work, doing its damage (such as formatting your hard drive) while you are blissfully unaware.

The virus usually attaches itself to your hard disk. It might then display annoying messages ("Your PC is stoned—legalize marijuana") or cause Ping-Pong balls to bounce around your screen and knock away text. More seriously, it might add garbage to your files, then erase or destroy your system software. It may evade your detection and spread its havoc elsewhere, since an infected hard disk will infect every floppy disk used by the system.

If you look in the utility section of any software store, you'll see a variety of virus-fighting programs. **_Antivirus software_ scans a computer's hard disk, floppy disks, and main memory to detect viruses and, sometimes, to destroy them.** Such virus watchdogs operate in two ways. First, they scan disk drives for "signatures," characteristic strings of 1s and 0s in the virus that uniquely identify it. Second, they look for suspicious virus-like behavior, such as attempts to erase or change areas on your disks. Examples of antivirus programs are Symantec's Norton AntiVirus, McAfee VirusScan, Panda Antivirus Platinum, and Computer Associates' E Trust EZ Antivirus for Windows, and Virex for Macs.

Computer Criminals

Ethics

What kind of people are perpetrating most of the information-technology crime? Over 80% may be employees; the rest are outside users, hackers and crackers, and professional criminals.

- **Employees:** Says Michigan State University criminal justice professor David Carter, who surveyed companies about computer crime, "Seventy-five to 80% of everything happens from inside."[21] Most common frauds, Carter found, involved credit cards, telecommunications, employees' personal use of computers, unauthorized access to confidential files, and unlawful copying of copyrighted or licensed software.

    Workers may use information technology for personal profit, or steal hardware or information to sell. They may also use it to seek revenge for real or imagined wrongs, such as being passed over for promotion. Sometimes they may use the technology simply to demonstrate to themselves they have power over people.

- **Outside users:** Suppliers and clients may also gain access to a company's information technology and use it to commit crimes. This becomes more likely as electronic connections such as intranets and extranets become more commonplace.

- **Hackers and crackers:** *Hacker* is so overused it has come to be applied to anyone who breaks into a computer system. Some people think it means almost any computer lover. In reality, there is a difference between hackers and crackers, although the term *cracker* has never caught on with the general public.

    Hackers **are people who gain unauthorized access to computer or telecommunications systems, often just for the challenge of it.** Some hackers even believe they are performing a service by exposing security flaws. Whatever the motivation, network system administrators view any kind of unauthorized access as a threat, and they usually try to pursue offenders vigorously. The most flagrant cases of hacking are met with federal prosecution. Former computer science student Ikenna Iffih, who studied at Northeastern University in Boston, could have been sentenced to 20 years in prison for hacking against private and government targets, but under a plea bargain was allowed to serve no more than six months. As a Nigerian national, he also faced deportation.[22]

    Crackers **are people who illegally break into computers for malicious purposes—to obtain information for financial gain, shut down hardware, pirate software, or alter or destroy data.** Sometimes you

Fake-buck buster
This electronic device can quickly (0.7 second) detect whether a U.S. bill is counterfeit or not.

hear of "white hat" hackers (who aren't malicious) versus "black-hat" hackers (who are). The Federal Bureau of Investigation estimates that cybercrime costs Americans more than $20 billion a year—and that more than 60% of such computer crime goes unreported. Says one article:

> On the low end, Web vandals, mostly teenagers derided by older [crackers] as "script kiddies," have attacked and shut down hundreds of busy Web sites, including those of the NASDAQ stock exchange, ABC, the White House, the Senate, and the FBI itself. On the more sophisticated end, a single program triggered by a New Jersey hacker—[the] Melissa virus—caused an estimated $80 million in damage to computer users.[23]

- **Professional criminals:** Members of organized crime rings don't just steal information technology. They also use it the way that legal businesses do—as a business tool, but for illegal purposes. For instance, databases can be used to keep track of illegal gambling debts and stolen goods. Not surprisingly, the old-fashioned illegal booking operation has gone high-tech, with bookies using computers and fax machines in place of betting slips and paper tally sheets.

As information-technology crime has become more sophisticated, so have the people charged with preventing it and disciplining its outlaws. Campus administrators are no longer being quite as easy on offenders and are turning them over to police. Industry organizations such as the Software Publishers Association are going after software pirates large and small. (Commercial software piracy is now a felony, punishable by up to five years in prison and fines of up to $250,000 for anyone convicted of stealing at least 10 copies of a program, or more than $2500 worth of software.) Police departments as far apart as Medford, Massachusetts, and San Jose, California, now have officers patrolling a "cyber beat." They regularly cruise online bulletin boards and chat rooms looking for pirated software, stolen trade secrets, child molesters, and child pornography.

Survival Tip

FBI's Top 20 Security Tips

The FBI's Top 20 computer security tips appear on the nonprofit Sans Institute website (www.sans.org). A free automated Internet scanner is also available for home users by e-mailing info@cisecurity.org with the subject "Top Twenty Scanners."

CONCEPT CHECK

Explain some of the errors, accidents, and hazards that can affect computers.

What are the principal crimes against computers?

What are some crimes using computers?

Describe some types of computer criminals.

8.3 Security: Safeguarding Computers & Communications

KEY QUESTION
What are the characteristics of the four components of security?

The ongoing dilemma of the Digital Age is balancing convenience against security. <u>Security</u> **is a system of safeguards for protecting information technology against disasters, systems failure, and unauthorized access that can result in damage or loss.** We consider four components of security.

Identification & Access

Are you who you say you are? The computer wants to know.

There are three ways a computer system can verify that you have legitimate right of access. Some security systems use a mix of these techniques. The systems try to authenticate your identity by determining (1) what you have, (2) what you know, or (3) who you are.

Survival Tip

Deal with Secure Websites

Be wary of what personal information (e.g., social security or credit card numbers) you give to an unfamiliar website. Best to deal with *secure* websites, those beginning with *https://* (not *http://*) or showing a closed-padlock icon.

- **What you have—cards, keys, signatures, badges:** Credit cards, debit cards, and cash-machine cards all have magnetic strips or built-in computer chips that identify you to the machine. Many require you to display your signature, which may be compared with any future signature you write. Computer rooms are always kept locked, requiring a key. Many people also keep a lock on their personal computers. A computer room may also be guarded by security officers, who may need to see an authorized signature or a badge with your photograph before letting you in.

  Of course, credit cards, keys, and badges can be lost or stolen. Signatures can be forged. Badges can be counterfeited.

- **What you know—PINs, passwords, and digital signatures:** To gain access to your bank account through an automated teller machine (ATM), you key in your PIN. A **PIN (personal identification number) is the security number known only to you that is required to access the system.** Telephone credit cards also use a PIN. If you carry either an ATM or a phone card, never carry the PIN written down elsewhere in your wallet (even disguised).

  A *password* **is a special word, code, or symbol required to access a computer system.** Passwords are one of the weakest security links, says AT&T security expert Steven Bellovin. Passwords (and PINs, too) can be guessed, forgotten, or stolen. To foil a stranger's guesses, Bellovin recommends never choosing a real word or variations of your name, birthdate, or those of your friends or family. Instead you should mix letters, numbers, and punctuation marks in an oddball sequence of no fewer than eight characters.[24]

- **Who you are—physical traits:** Some forms of identification can't be easily faked—such as your physical traits. Biometrics tries to use these in security devices. **Biometrics is the science of measuring individual body characteristics.**

  For example, before University of Georgia students can use the all-you-can-eat plan at the campus cafeteria, they must have their hands read. As one writer describes the system, "a camera automatically compares the shape of a student's hand with an image of the same hand pulled from the magnetic strip of an ID card. If the patterns match, the cafeteria turnstile automatically clicks open. If not, the would-be moocher eats elsewhere."[25]

  Besides handprints, other biological characteristics read by biometric devices are fingerprints (computerized "finger imaging"), voices, the blood vessels in the back of the eyeball (retinal scan), the lips, and even the entire face.

Fingerprint check
This system can be mounted on a door or on a safe. Entry is granted only after the individual's fingerprint is verified.

Some computer security systems have a "call-back" provision. In a *call-back system*, the user calls the computer system, punches in the password, and hangs up. The computer then calls back a certain preauthorized number. This measure will block anyone who has somehow got hold of a password but is calling from an unauthorized telephone.

Encryption

PGP is a computer program written for encrypting computer messages—putting them into secret code. **Encryption is the altering of data so it is not usable unless the changes are undone.** *PGP* (for *Pretty Good Privacy*) is so good that it is practically unbreakable; even government experts can't crack it. Another encryption system is *DES* (for *Data Encryption Standard*), adopted as a federal standard in 1976.

Encryption is clearly useful for some organizations, especially those concerned with trade secrets, military matters, and other sensitive data. A very

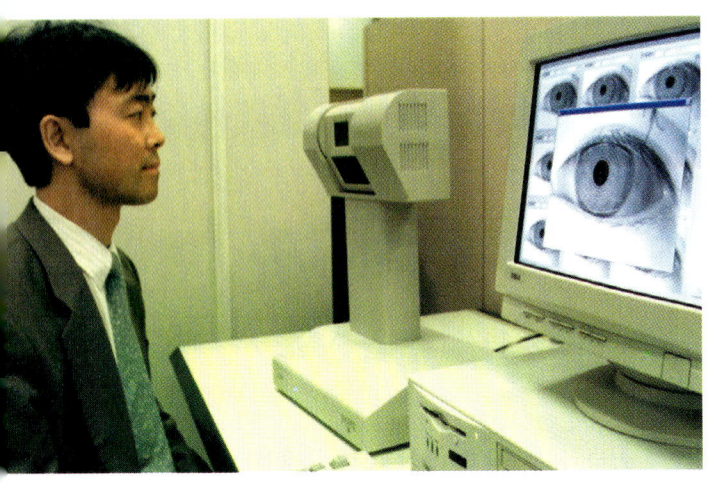

Iris scan
This system verifies the identity of computer users by iris scans.

sophisticated form of encryption is used in most personal computers and is available with every late-model web browser to provide for secure communications over the Internet. In fact, encryption is what has given people confidence to do online shopping or stock trading.

However, from the standpoint of society, encryption is a two-edged sword. For instance, the 2001 attack on the World Trade Center and Pentagon raised the possibility that the terrorists might have communicated with each other using unbreakable encryption programs. (There is no evidence they did.) Should the government be allowed to read the coded e-mail of overseas terrorists, drug dealers, and other enemies? What about the e-mail of all American citizens?

Protection of Software & Data

Organizations go to tremendous lengths to protect their programs and data. As might be expected, this includes educating employees about making backup disks, protecting against viruses, and so on. Other security procedures include the following:

- **Control of access:** Access to online files is restricted to those who have a legitimate right to access—because they need them to do their jobs. Many organizations have a system of transaction logs for recording all accesses or attempted accesses to data.
- **Audit controls:** Many networks have *audit controls* for tracking which programs and servers were used, which files opened, and so on. This creates an *audit trail*, a record of how a transaction was handled from input through processing and output.
- **People controls:** Because people are the greatest threat to a computer system, security precautions begin with the screening of job applicants. Résumés are checked to see if people did what they said they did. Another control is to separate employee functions, so people are not allowed to wander freely into areas not essential to their jobs. Manual and automated controls—input controls, processing controls, and output controls—are used to check if data is handled accurately and completely during the processing cycle. Printouts, printer ribbons, and other waste that may contain passwords and trade secrets to outsiders is disposed of through shredders or locked trash barrels.

Disaster-Recovery Plans

A *disaster-recovery plan* **is a method of restoring information processing operations that have been halted by destruction or accident.** "Among the countless lessons that computer users have absorbed in the hours, days, and weeks after the [1993 New York] World Trade Center bombing," wrote one reporter, "the most enduring may be the need to have a disaster-recovery plan. The second most enduring lesson may be this: Even a well-practiced plan will quickly reveal its flaws."[26] The 2001 attack on the World Trade Center reinforced these lessons in a spectacular way as companies had to scramble for new office space.

Mainframe computer systems are operated in separate departments by professionals, who tend to have disaster plans. Whereas mainframes are usually backed up, many personal computers, and even entire local area networks, are not, with potentially disastrous consequences. It has been reported that, on average, a company loses as much as 3% of its gross sales within eight days of a sustained computer outage. In addition, the average company struck by a computer outage lasting more than 10 days never fully recovers.[27]

A disaster-recovery plan is more than a big fire drill. It includes a list of all business functions and the hardware, software, data, and people to support those functions, as well as arrangements for alternate locations. The disaster-recovery plan includes ways for backing up and storing programs and data in another location, ways of alerting necessary personnel, and training for those personnel.

CONCEPT CHECK

How do computer systems try to authenticate your identity?

Describe encryption.

What security procedures are used to protect software and data?

What is a disaster-recovery plan?

8.4 Quality-of-Life Issues: The Environment, Mental Health, & the Workplace

KEY QUESTION
How does information technology create environmental, mental-health, and workplace problems?

The worrisome effects of technology on intellectual property rights and truth in art and journalism, on censorship, on health matters and ergonomics, and on privacy were explained earlier in this book. Here are some other quality-of-life issues related to information technology.

Environmental Problems

Environmental problems involving information technology take different forms—manufacturing and disposal by-products, electricity demand, and environmental blight.

- **Manufacturing by-products:** Many communities are eager to have computer and chip manufacturers locate there because they perceive them to be "clean" industries. But there have been lawsuits charging that the semiconductor industry has knowingly exposed workers to a variety of hazardous toxins, some of which were linked to miscarriages, and there is speculation that others may be linked to cancer and birth defects.[28]

- **Disposal byproducts:** What to do with the hundreds of millions of obsolete or broken PCs, monitors, printers, cellphones, TVs, and other electronic gadgetry? Much of it winds up in landfills. Most of it contains large amounts of lead and other toxins that can leach into groundwater or produce dioxins and other cancer-causing agents when burned.[29] The problem is worsening as new computer models are introduced on faster cycles—by 2004, the average life of a computer is expected to be only two years—and the National Safety Council estimates that there will be nearly 500 million obsolete computers in the United States by 2007.[30]

  Don't always assume you can get a school or local charity to take your old personal computer; some will, but many are tired of being stuck with junk. Only 11% of old PCs are recycled.[31] Fortunately, the Electronic Industries Alliance Consumer Education Initiative offers a website *(www.eiae.org)* that helps consumers locate donation programs and recycling companies.

- **Electricity demand:** Most Americans take electric power for granted, says one account, but know little about it—which is why they have difficulty conserving it.[32] Nevertheless, the digital economy is putting a severe strain on the nation's (indeed the world's) electric utilities. For example, whereas a typical home will use 1 watt of power per square foot, a so-called server farm, which houses rows and rows of computers in climate-controlled rooms to serve as data centers, will use 75–100 watts per square foot.[33] Most of the extra electricity that the United States requires is being used in computer chips and the optical and wireless connections that unite them. "Micropower" solutions such as solar or wind cells, admirable though they may be, are probably not sufficient to accommodate the 3–4% annual increases in power needs that may be required. Moreover, the copper wires that make up most of the distribution grid aren't sufficient to carry the additional supply that will have to be generated. What appears to be needed are heavy investments in a national grid.[34]

- **Environmental blight:** Call it "techno-blight." This is the visual pollution represented by the forest of wireless towers, roof antennas, satellite dishes, and all the utility poles topped with transformers and strung with electric, phone, cable-TV, and other wires leading off in all directions. As the nation's electrical grid becomes upgraded, so, most likely, will be the increase in obtrusive, ugly technology in our physical environment. Environmentalists worry about its impact on vegetation and wildlife, such as millions of birds and bats that collide with cellular towers. Residents worry about the effect on views and property values.

Mental-Health Problems

Some of the mental-health problems linked to information technology are the following:

- **Isolation?** A few years ago, a Stanford University survey found that, as people spent more time online, they had less time for real-life relationships with family and friends. "As Internet use becomes more widespread," predicted a study coauthor, "it will have an increasingly isolating effect on society."[35] The inclination of heavy Internet users to experience isolation, as well as loneliness and depression, was also supported by a 1998 study by Robert Kraut. Three years later, however, the Carnegie Mellon University professor reported the opposite— that the study's subjects experienced fewer feelings of loneliness and isolation. And he found that, for a new group, the more they used computers, the more integrated they were with others, the better their sense of well-being, and the more positive their moods.[36] Indeed, the Internet has led to a great many web communities, from cancer survivors to Arab immigrants.

- **Gambling:** Gambling is already widespread in North America, but information technology makes it almost unavoidable. Although gambling by wire is illegal in the United States, host computers for Internet casinos and sports books have been established in Caribbean tax havens. Satellites, decoders, and remote-control devices allow TV viewers to do racetrack wagering from home. In these circumstances, law enforcement is extremely difficult.

- **Stress:** In one survey of 2802 American PC users, three-quarters of the respondents (ranging in age from children to retirees) said personal computers had increased their job satisfaction and were a key to success and learning. However, many found PCs stressful: In particular, 59% admitted getting angry at their PCs within the previous year, and

How People Shirk at Work: Recreational Web Surfing	
29.1%	General news
22.5%	Investment
9.7%	Pornography
8.2%	Travel
6.6%	Entertainment
6.1%	Sports
3.1%	Shopping
14.7%	Other

41% said they thought computers had reduced job opportunities rather than increased them.[37] Another survey found that 83% of corporate network administrators reported "abusive and violent behavior" by employees toward computers—including smashing monitors, throwing mice, and kicking system units.[38]

Workplace Problems: Impediments to Productivity

First the mainframe computer, then the desktop stand-alone PC, and now the networked computer were all brought into the workplace for one reason only: to improve productivity. How is it working out? Let's consider three aspects: misuse of technology, fussing with computers, and information overload.

- **Misuse of technology:** "For all their power," says an economics writer, "computers may be costing U.S. companies tens of billions of dollars a year in downtime, maintenance and training costs, useless game playing, and information overload."[39]

  Employees may look busy, as they stare into their computer screens with brows crinkled. But sometimes they're just hard at work playing Quake. Or browsing online malls (forcing corporate mail rooms to cope with a deluge of privately ordered parcels). Or looking at their investments or pornography sites.[40] Indeed, one study found that recreational web surfing accounts for nearly *one-third* of office workers' time online.[41]

- **Fussing with computers:** Another reason for so much wasted time is all the fussing that employees do with hardware, software, and online connections. One study in the early 1990s estimated microcomputer users wasted 5 billion hours a year waiting for programs to run, checking computer output for accuracy, helping coworkers use their applications, organizing cluttered disk storage, and calling for technical support.[42] And that was *before* most people had to get involved with making online connections work.

  Comments technology writer Dan Gillmor, "We would never buy a TV that forced us to reboot the set once a month, let alone once a week or every other day."[43] But until recently, this was the current evolutionary stage of the computer age. Windows XP promises to reduce the rebooting frustrations.

- **Information overload:** The new technology is definitely a two-edged sword. Cellular phones, pagers, fax machines, and modems may untether employees from the office, but these employees tend to work longer hours under more severe deadline pressure than do their tethered counterparts who stay at the office, according to one study.[44]

  To avoid information overload, some people install so-called *Bozo filters,* software for screening out trivial e-mail messages and assigning priorities to the remaining files. But the real change may come as people realize that they need not always be tied to the technological world, that solitude is a scarce resource, and that seeking serenity means streamlining the clutter and reaching for simpler things.

8.5 Economic Issues: Employment & the Haves/Have-Nots

KEY QUESTION
How may technology affect the unemployment rate and the gap between rich and poor?

In recent times, a number of critics have provided a counterpoint to the hype and overselling of information technology to which we have long been exposed. Some critics find that the benefits of information technology are balanced by a real downside. Other critics make the alarming case that technological progress is actually no progress at all—indeed, it is a curse. The

BOOKMARK IT!

PRACTICAL ACTION BOX
When the Internet Isn't Productive: Online Addiction & Other Time Wasters

There is a handful of activities that can drain hours of time, putting studying—and therefore college—in serious jeopardy. They include excessive television watching, partying, and working too many hours while going to school. They also include misuse of the computer.

The Great Campus Goof-Off Machine?

"I have friends who have spent whole weekends doing nothing but playing Quake or Warcraft or other interactive computer games," reports Swarthmore College sophomore Nate Stulman, in an article headed "The Great Campus Goof-Off Machine." He goes on: "And many others I know have amassed overwhelming collections of music on their computers. It's the searching and finding they seem to enjoy: some of them have more music files on their computers than they could play in months." [a]

In Stulman's opinion, having a computer in the dorm is more of a distraction than a learning tool for students. "Other than computer science or mathematics majors, few students need more than a word processing program and access to e-mail in their rooms."

Most educators wouldn't banish computers completely from student living quarters. Nevertheless, it's important to be aware that your PC can become a gigantic time sink, if you let it. Reports Rutgers communication professor Robert Kubey: "About 5% to 10% of students, typically males and more frequently first- and second-year students, report staying up late at night using chat lines and e-mail and then feeling tired the next day in class or missing class altogether." [b]

Internet Addiction/Dependency

"A student e-mails friends, browses the World Wide Web, blows off homework, botches exams, flunks out of school." [c] This is a description of the downward spiral of the "Net addict," often a college student—because schools give students no-cost/low-cost linkage to the Internet—but it can be anyone. Some become addicted (although until recently some professionals felt "addiction" was too strong a word) to chat groups, some to online pornography, some simply to the escape from real life. [d]

Stella Yu, 21, a college student from Carson, California, was rising at 5 A.M. to get a few hours online before school, logging on to the Internet between classes and during her part-time job, and then going home to web surf until 1 A.M. Her grades dropped and her father was irate over her phone bills. "I always make promises I'm going to quit; that I'll just do it for research," she said. "But I don't. I use it for research for 10 minutes, then I spend two hours chatting." [e]

College students are unusually vulnerable to Internet addiction, which is defined as "a psychological dependence on the Internet, regardless of type of activity once 'logged on,'" according to psychologist Jonathan Kandell. [f] The American Psychological Association, which officially recognized "Pathological Internet Use" as a disorder in 1997, defines *Internet addict* as anyone who spends an average of 38 hours a week online. [g] (The average Interneter spends 5½ hours on the activity. [h]) More recently, psychologist Keith J. Anderson of Rensselaer Polytechnic Institute found that Internet-dependent students, who make up at least 10% of college students, spent an average of 229 minutes a day online for nonacademic reasons, compared with 73 minutes a day for other students. As many as 6% spend an average of more than 400 minutes a day—almost seven hours—using the Internet. [i]

What are the consequences of Internet addiction disorder? A study of the freshman dropout rate at Alfred University in New York found that nearly half the students who quit the preceding semester had been engaging in marathon, late-night sessions on the Internet. [j] The University of California, Berkeley, found some students linked to excessive computer use neglected their course work. [k] A survey by Viktor Brenner of State University of New York at Buffalo found that some Internet addicts had "gotten into hot water" with their school for Internet-related activities. [l] "Grades decline, mostly because attendance declines," says psychologist Anderson about Internet-dependent students. "Sleep patterns go down. And they become socially isolated."

The box on page 370 lists questions that may yield insights as to whether you or someone you know is Internet-dependent. [m]

Online Gambling

A particularly risky kind of Internet dependence is online gambling. David, a senior at the University of Florida, Gainesville, owed $1500 on his credit cards as a result of his online gambling habit, made possible by easy access to offshore casinos in cyberspace. He is not alone. A survey of 400 students at Southern Methodist University found that 5% said they gambled frequently via the Internet. [n]

Other students do the kind of de facto gambling known as day-trading—buying and selling stocks on the Internet. This can be risky, too. "It's a lot like going to a casino," says one finance professor. "You can make or lose money within a few seconds, and there are going to be some people who are addicted to it." [o]

two biggest charges (which are related) are, first, that information technology is killing jobs and, second, that it is widening the gap between the rich and the poor.

Technology, the Job Killer?

Certainly, ATMs do replace bank tellers, E-Z pass electronic systems do replace turnpike-toll takers, and Internet travel agents do lure customers away from small travel agencies. The contribution of technological advances to social progress is not purely positive.

But is it true, as technology critic Jeremy Rifkin says, that intelligent machines are replacing humans in countless tasks, "forcing millions of blue-collar and white-collar workers into temporary, contingent, and part-time employment and, worse, unemployment"?[45]

This is too large a question to be fully considered in this book. We can say for sure that the U.S. economy is undergoing powerful structural changes, brought on not only by the widespread diffusion of technology but also by greater competition, increased global trade, the shift from manufacturing to service employment, the weakening of labor unions, more flexible labor markets, more rapid immigration, partial deregulation, and other factors.[46]

A counterargument is that jobs don't disappear, they just change. According to some observers, the jobs that do disappear represent drudgery. "If your job has been replaced by a computer," says Stewart Brand, "that may have been a job that was not worthy of a human."[47]

Gap between Rich & Poor

"In the long run," says M.I.T. economist Paul Krugman, "improvements in technology are good for almost everyone. . . . Unfortunately, what is true in the long run need not be true over shorter periods."[48] We are now, he believes, living through one of those difficult periods in which technology doesn't produce widely shared economic gains but instead widens the gap between those who have the right skills and those who don't.

A 2001 report by the General Accounting Office, the investigative arm of Congress, suggests that education and income remain decisive in determining who goes online and in what way. Internet users are more likely to be white, to be well-educated, and to have higher-than-average household incomes. People who use broadband (about 12% of users), which is more expensive than dial-up but also helps make the Internet more practical, are apt to be in higher income brackets.[49]

Education—especially college—makes a great difference. Every year of formal schooling after high school adds 5–15% to annual earnings later in life.[50] Being well educated is only part of it, however; it's essential to be technologically literate. Employees with technology skills "earn roughly 10–15% higher pay," according to the chief economist for the U.S. Labor Department.[51]

CLICK-ALONG 8-2
Update: quality-of-life & economic issues

CONCEPT CHECK

What are some potential environmental consequences of information technology?

Discuss four types of mental-health problems linked to information technology.

Describe some workplace problems associated with computers.

What are some key economic issues related to information technology?

8.6 Artificial Intelligence

KEY QUESTION
What are the main areas of artificial intelligence?

You're having trouble with your new software program. You call the customer "help desk" at the software maker. Do you get a busy signal or get put on hold to listen to music (or, worse, advertising) for several minutes? Technical support lines are often swamped, and waiting is commonplace. Or, to deal with your software difficulty, do you find yourself dealing with . . . other software?

The odds are good that you will. For instance, a software technology has been patented called *automated virtual representatives* (*vReps*), which offers computer-generated images (animation or photos of real models) that answer customer questions in real time, using natural language.[52] Programs that can walk you through a problem and help solve it are called *expert systems*. As the name suggests, these are systems imbued with knowledge by a human expert. Expert systems are one of the most useful applications of artificial intelligence.

<u>Artificial intelligence (AI)</u> **is a group of related technologies used for developing machines to emulate human qualities, such as learning, reasoning, communicating, seeing, and hearing.** Today the main areas of AI are *robotics*, *expert systems*, and *natural language processing*.

We will consider these and also an area known as *artificial life*.

Robotics

More than 45 years ago, in the film *Forbidden Planet*, Robby the Robot could sew, distill bourbon, and speak 187 languages. We haven't caught up with science-fiction movies, but maybe we'll get there yet.

<u>Robotics</u> **is the development and study of machines that can perform work normally done by people.** The machines themselves are called *robots*. *(See* ● *Panel 8.2.)* Basically, a <u>robot</u> **is an automatic device that performs functions ordinarily executed by human beings or that operates with what appears to be almost human intelligence.** ScrubMate—a robot equipped with computerized controls, ultrasonic "eyes," sensors, batteries, three different cleaning and scrubbing tools, and a self-squeezing mop—can clean bathrooms.

● **PANEL 8.2**
Robots
Left: NASA remote-controlled research robot. *Right:* The sheriff's office bomb squad's robot inspects a suitcase left beside a planter in downtown Jacksonville, Florida.

Survival Tip

Trying Out Simulation: Driving a Train

Want to know how it feels to be the engineer on a train? Microsoft's Train Simulator ($50) lets you drive passenger and freight trains over routes in the United States, England, Austria, and Japan.

Robots are also used for more exotic purposes such as fighting oil-well fires, doing nuclear inspections and cleanups, and checking for mines and booby traps. An eight-legged, satellite-linked robot called Dante II was used to explore the inside of Mount Spurr, an active Alaskan volcano, sometimes without human guidance.

Expert Systems

An _expert system_ **is an interactive computer program used in solving problems that would otherwise require the assistance of a human expert.** Such programs simulate the reasoning process of experts in certain well-defined areas. That is, professionals called *knowledge engineers* interview the expert or experts and determine the rules and knowledge that must go into the system. For example, MYCIN helps diagnose infectious diseases. PROSPECTOR assesses geological data to locate mineral deposits. DENDRAL identifies chemical compounds. Home-Safe-Home evaluates the residential environment of an elderly person. Business Insight helps businesses find the best strategies for marketing a product. Jnana offers legal advice in narrow, highly regulated areas, such as environmental or securities law. REBES (Residential Burglary Expert System) helps detectives investigate crime scenes.

Natural Language Processing

Natural languages are ordinary human languages, such as English. (A second definition, discussed in the Appendix, is that they are fifth-generation programming languages.) _Natural language processing_ **is the study of ways for computers to recognize and understand human language,** whether in spoken or written form.

Think how challenging it is to make a computer translate English into another language. In one instance, the English sentence "The spirit is willing, but the flesh is weak" came out in Russian as "The wine is agreeable, but the meat is spoiled." The problem with human language is that it is often ambiguous; different listeners may arrive at different interpretations.

Most existing language systems run on large computers, although scaled-down versions are now available for microcomputers. A product called Intellect uses a limited English vocabulary to help users orally query databases on both mainframes and microcomputers. LUNAR, developed to help analyze moon rocks, answers questions about geology on the basis of an extensive database. Verbex, used by the U.S. Postal Service, lets mail sorters read aloud an incomplete address and will reply with the correct ZIP code.

Artificial Life, the Turing Test, & AI Ethics

Ethics

What is life, and how can we replicate it out of silicon chips, networks, and software? We are dealing now not with artificial intelligence but with artificial life. _Artificial life_, **or A-life, is a field of study concerned with "creatures"—computer instructions, or pure information—that are created, replicate, evolve, and die as if they were living organisms.** Thus, A-life software (such as LIFE) tries to simulate the responses of a human being.

Of course, "silicon life" does not have two principal attributes associated with true living things—it is not water- and carbon-based. Yet in other respects such creatures mimic life: If they cannot learn or adapt, then they perish.

How can we know when we have reached the point that computers have achieved human intelligence? How will you know, say, whether you're talking to a human being on the phone or to a computer? Clearly, with the strides made in the fields of artificial intelligence and artificial life, this question is no longer just academic.

Is this fish for real?
This sea bream is about 1½ feet long, weighs 5½ pounds, and can swim up to 38 minutes—before recharging. The robot fish, created by Mitsubishi, looks and swims exactly like the real thing.

Ethics

CLICK-ALONG 8-3
More on AI

Interestingly, Alan Turing, an English mathematician and computer pioneer, addressed this very question in 1950. Turing predicted that by the end of the century computers would be able to mimic human thinking and to conduct conversations indistinguishable from a person's. Out of these observations came the Turing test, which is intended to determine whether a computer possesses "intelligence" or "self-awareness."

In the *Turing test*, **a human judge converses by means of a computer terminal with two entities hidden in another location—one a person typing on a keyboard, the other a software program. Following the conversation, the judge must decide which entity is human. In this test, intelligence—the ability to think—is demonstrated by the computer's success in fooling the judge.**

Judith Anne Gunther participated as one of eight judges in the third annual Loebner Prize Competition, which is based on Turing's ideas.[53] (There have been other competitions since.) The "conversations"—each limited to 15 minutes—are restricted to predetermined topics, such as baseball, because even today's best programs have neither the databases nor the syntactical ability to handle an unlimited number of subjects.

Gunther found that she wasn't fooled by any of the computer programs. The winning program, for example, relied as much on deflection and wit as it did on responding logically and conversationally. (For example, to a judge trying to discuss a federally funded program, the computer said: "You want logic? I'll give you logic: shut up, shut up, shut up, shut up, shut up, now go away! How's that for logic?") However, Gunther *was* fooled by one of the five humans, a real person discussing abortion. "He was so uncommunicative," wrote Gunther, "that I pegged him for a computer."

Behind everything to do with artificial intelligence and artificial life—just as it underlies everything we do—is the whole matter of *ethics*. In his book *Ethics in Modeling*, William A. Wallace, professor of decision sciences at Rensselaer Polytechnic Institute, points out that computer software, including expert systems, is often subtly shaped by the ethical judgments and assumptions of the people who create it.[54] In one instance, he notes, a bank had to modify its loan-evaluation software on discovering that the software rejected certain applications because it unduly emphasized old age as a negative factor. Another expert system, used by health maintenance organizations (HMOs), tells doctors when they should opt for expensive medical procedures, such as magnetic resonance imaging tests. HMOs like such systems because they help control expenses, but critics are concerned that doctors will have to base decisions not on the best medicine but simply on "satisfactory" medicine combined with cost constraints.[55]

Clearly, there is no such thing as completely "value-free" technology. Human beings build it, use it, and have to live with the results.

CONCEPT CHECK

Distinguish among the main areas of AI.

What is artificial life?

8.7 The Promised Benefits of the Digital Age

KEY QUESTION
What are some benefits of the digital age?

We have described the benefits of information technology throughout this book. But there's probably no better way to conclude this subject then to discuss some of the promises the future offers.

Information & Education

Getting the right kind of information is a major challenge. Can everything in the Library of Congress be made available online to citizens and companies? What about government records, patents, contracts, and other legal documents? Or geographical maps photographed from satellites?

As one solution, computer scientists have been developing so-called intelligent agents to find information on computer networks and filter it. An _intelligent agent_ **program performs work tasks—such as roaming networks and compiling data—on your behalf.** A software agent is a kind of electronic assistant that will filter messages, scan news services, and perform similar secretarial chores. An agent will also travel over communications lines to computer databases, collecting files to add to a personalized database.

With their potential for finding and processing information, computers are pervasive on campus. More than a third of college courses require the use of e-mail, and many have their own dedicated Web pages.[56] College students spend an average of 5.6 hours a week on the Internet.[57] And as computer prices drop, more and more students have their own PCs. (For the rest, colleges often make public microcomputers available.)

Most students have been exposed to computers since the lower grades. The percentage of elementary schools in the United States with Internet access in 2000 was 97%, according to the National Center for Education Statistics. For secondary schools, 100% had Internet access.[58]

One revolution in education—before, during, and after the college years—is the advent of distance learning, or "cyberclasses," along with the explosion of Internet resources. The home-schooling movement, for example, has come of age, thanks to Internet resources.[59] Many colleges—both individually and in associations such as that of the Western Governors University (backed by 17 states and Guam) and the Community College Distance Learning Network—are offering a variety of Internet and/or video-based online courses.[60] Corporations are also offering training classes via the Internet or corporate intranet.[61]

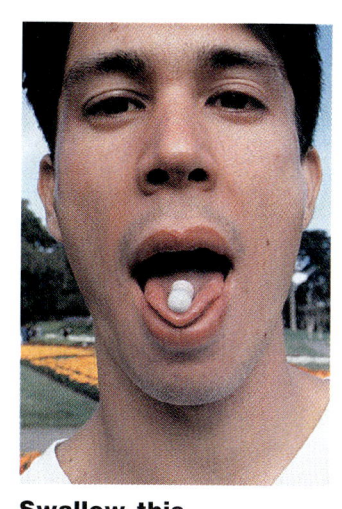

Health

For some time, physicians in rural areas lacking local access to radiologists have used "teleradiology" to exchange digital images such as X-rays via telephone-linked networks with expert physicians in metropolitan areas. Now *telemedicine*—medical care delivered via telecommunications—is moving to an exciting new level, as the use of digital cameras and sound, in effect, moves patients to doctors rather than the reverse.

Commerce & Money

Businesses clearly see the Internet as a way to enhance productivity and competitiveness. However, as we have already observed, the changes go well beyond this.

The thrust of the original Industrial Revolution was separation—to break work up into its component parts so as to permit mass production. The effect of computer networks in the Digital Revolution, however, is unification—to erase boundaries between company departments, suppliers, and customers.[62] Indeed, the parts of a company can now as easily be scattered around the globe as down the hall from one another. Thus, designs for a new product

---

Survival Tip

Libraries Join Forces to Answer Public Queries

The Collaborative Reference Service is a question-and-answer system developed by the Library of Congress to draw on libraries around the world (eventually up to 20 languages). Check out *www.loc.gov/rr/digiref*.

Swallow this

MIT student Bradley Geilfuss prepares to swallow a radio transmitter that will send data about his metabolism via wireless modem to fellow students during his participation in a San Francisco marathon.

can be tested and exchanged with factories in remote locations. With information flowing faster, goods can be sent to market faster and inventories reduced. Says an officer of the Internet Society, "Increasingly you have people in a wide variety of professions collaborating in diverse ways in other places. The whole notion of 'the organization' becomes a blurry boundary around a set of people and information systems and enterprises."[63]

Some areas of business that are undergoing rapid change are sales and marketing, retailing, banking, stock trading, and manufacturing.

> Survival Tip
>
> **Financial Portals**
>
> Online financial portals offer free information. Some of the most important:
>
> AOL Personal Finance
>
> MoneyCentral.msn.com
>
> Yahoo! Finance
>
> Quicken.com
>
> Fidelity.com
>
> Etrade.com
>
> Schwab.com

- **Sales and marketing:** Caradon Everest equips its sales staff with laptops containing software that "configures" customized windows and calculates the prices on the spot, a process that was once handled by the company's technical people and took a week. "The company can also load on product images for multimedia presentations and training programs," says one account. "Using a digital camera, pictures of the customer's house can be loaded into the computer, which can superimpose the company's windows and print out a color preview."[64]

  Other companies are also taking advantage of information technology in marketing. An Atlanta company called The Mattress Firm, for instance, uses geographical information systems (GIS) software called MapLinx that plots drop-off points for a driver's most efficient delivery of mattresses to customers. Moreover, MapLinx depicts sales by neighborhood, showing owner Darin Lewin where his mattresses are selling well and thus where he should spend money to market them. "We can pinpoint our customer," says Lewin.[65]

- **Banking and economy:** The world of cybercash has come to banking—not only smart cards, but Internet banking, electronic deposits, electronic bill paying, online stock and bond trading, and online insurance shopping.

  Some banks are backing an electronic-payment system that will allow Internet users to buy tiny online goods and services with "micropayments" of as little as 25 cents, from participating merchants. For instance, publishers could charge buyers a quarter to buy an article or listen to a song online, a small transaction that until now has not been practical. "It allows you to buy things by the sip rather than the gulp," says futurist Paul Saffo.[66]

- **Stock trading:** Computer technology is unquestionably changing the nature of stock trading. Only a few years ago, hardly anyone had the occupation of "day trader," an investor who relies on quick market fluctuations to turn a profit. Now, points out technology observer Denise Caruso, "anyone with a computer, a connection to the global network, and the requisite ironclad stomach for risk has the information, tools, and access to transaction systems required to play the stock market, a game that was once the purview of an elite few."[67] (Caruso is right that it takes a strong stomach to be a day trader; many have lost their shirts.)

- **Manufacturing:** Computers have been used in manufacturing for some time, most famously, perhaps, in those pictures you see of welding robots on car-assembly lines. But computers, and the Internet in particular, are players in other ways. For example, in Benton Harbor, Michigan, Whirlpool, the world's largest maker of dishwashers and other appliances, deploys the Internet on three fronts—to deal with customers, manage employees, and buy components from other companies. The last of these—streamlining the way metals, plastics, electronics, and resins come into manufacturing plants—is particularly important for Whirlpool. This business-to-business automation saves space and funds, since the company doesn't have to pay for a part until it arrives (an example of a "just-in-time" system).[68]

CLICK-ALONG 8-4
More e-business

Roy Sato, senior animator for a digital animation feature, shows the digital face of Dr. Aki Ross, the film's hero.

Survival Tip

Online Government Help

You can gain access to government agencies through the following websites:

www.firstgov.gov

www.govspot.com

www.hicitizen.com

www.info.gov

Entertainment

Information technology is being used for all kinds of entertainment, ranging from video games to tele-gambling. It is also being used in the arts, from painting to photography. Let's consider just two examples, music and film.

Not only do websites offer guitar chords and sheet music, but promotional sites feature signed and unsigned artists, both garage bands and established professional musicians.[69] There are legal MP3 sites on the Internet but also many illegal ones, which operate in violation of copyright laws protecting ownership of music. Many of these will probably cease to exist as the recording industry develops new Web technologies that frustrate the pirating of music and instead offer fee-based services.[70]

Now that blockbuster movies routinely meld live action and animation, computer artists are in big demand. *Star Wars: Episode I,* for instance, had fully 1965 digital shots out of about 2200 shots. Even when film was used, it was scanned into computers to be tweaked with animated effects, lighting, and the like. Entire beings were created on computers by artists working on designs developed by producer George Lucas and his chief artist.[71] Computer techniques have even been used to develop digitally created actors—called "synthespians." The late John Wayne was recruited for a Coors beer ad.[72] With the soaring demand for computer-generated imagery—not only for movies but also for TV ads and video games—colleges and trade schools have expanded their digital-animation training programs.

Government & Electronic Democracy

"In general, the Internet is being defined not by its civic or political content," says one writer, "but by merchandising and entertainment, much like television."[73] Even so, a Rutgers University study suggests that the Internet has great potential for civic betterment because it is free of government intrusion, is fast and cheap for users (once connected), and facilitates communication among citizens better than mass media such as radio and TV.[74]

If we can shop online, could we not vote online for political candidates? This question took on heightened importance following the vote count in Florida for the 2000 presidential election, when punch-card ballots were found to be unreliable.[75] Beyond Florida, flaws in voting technology left perhaps as many as 6 million votes uncounted in that election, according to a study by researchers at Caltech and M.I.T.[76] The study found that optical scanners are the most reliable method of counting votes. (A poor, mostly minority district in Alabama had the lowest rate of disqualified ballots—0.3%, better than any affluent district—because it used optical-scan systems.[77]) Touch-screen voting, however, remains largely unproven. Internet voting is vulnerable because of the potential for fraud and hackers.[78] There is also, surprisingly, the matter of expense: a 2000 experiment by the Pentagon to let overseas soldiers vote by the Internet netted just 84 ballots—at a cost of nearly $74,000 per voter.[79] True Internet voting, therefore, may still be a decade away. Still, ten states have approved testing of online voting. In 1999, voters in the small city of Piedmont, California, were the first in the state to vote by "touch-screen" technology at their polling sites; the results were tallied in 29 minutes instead of the usual 3 hours. In 2000, 86,000 citizens in Arizona also voted in that state's Democratic presidential primary via the Internet.[80]

CONCEPT CHECK

What are some uses of information technology in education and health?

Discuss some uses of computers in commerce and money.

How are computers being used in music and movies?

Describe some ways computers are being used in the civic realm.

A new way to vote: After you insert your voter ID card, choose your candidates by touching their names on a display screen.

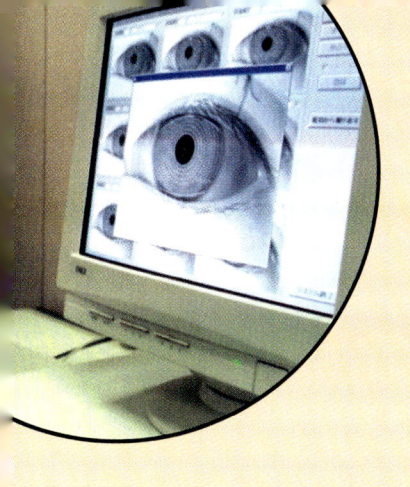

Summary

antivirus software (p. 279, KQ 8.2) Program that scans a computer's hard disk, floppy disks, and main memory to detect viruses and, sometimes, to destroy them. *Why it's important:* Computer users must find out what kind of antivirus software to install in their systems—and how to keep it up to date—for protection against damage or shutdown.

artificial intelligence (AI) (p. 288, KQ 8.6) Group of related technologies used for developing machines to emulate human-like qualities, such as learning, reasoning, communicating, seeing, and hearing. *Why it's important:* Some of the main areas of AI are robotics, natural language processing, expert systems, and artificial life.

artificial life (p. 289, KQ 8.6) Also called *A-life;* field of study concerned with "creatures"—computer instructions, or pure information—that are created, replicate, evolve, and die as if they were living organisms. *Why it's important:* A-life software (such as LIFE) tries to simulate the responses of a human being.

biometrics (p. 281, KQ 8.3) Science of measuring individual body characteristics. *Why it's important:* Biometrics is used in some computer security systems, to restrict user access.

computer crime (p. 276, KQ 8.2) Crime of two types: (1) an illegal act perpetrated against computers or telecommunications; (2) the use of computers or telecommunications to accomplish an illegal act. *Why it's important:* Crimes against information technology include theft—of hardware, of software, of computer time, of cable or telephone services, or of information. Other illegal acts are crimes of malice and destruction.

crackers (p. 279, KQ 8.2) People who illegally break into computers for malicious purposes. *Why it's important:* Crackers attempt to break into computers to obtain information for financial gain, shut down hardware, pirate software, or alter or destroy data.

disaster-recovery plan (p. 282, KQ 8.3) Method of restoring information processing operations that have been halted by destruction or accident. *Why it's important:* Such a plan is important if an organization desires to resume computer operations quickly.

encryption (p. 281, KQ 8.3) Altering data so it is not usable unless the changes are undone. *Why it's important:* Encryption is clearly useful for some organizations, especially those concerned with trade secrets, military matters, and other sensitive data. Some maintain that encryption will determine the future of e-commerce, because transactions cannot flourish over the Internet unless they are secure.

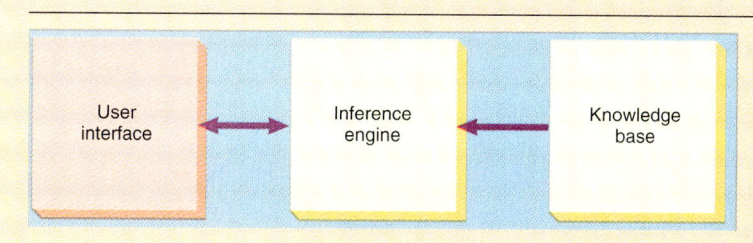

expert system (p. 289, KQ 8.6) Interactive computer program that helps solve problems that would otherwise require the assistance of a human expert. Fundamental to an expert system is a knowledge base constructed by experts that can "learn" by adding new knowledge. *Why it's important:* Expert systems are designed to be users' assistants, not replacements—to help them more easily perform their jobs.

hackers (p. 279, KQ 8.2) People who gain unauthorized access to computer or telecommunications systems, often just for the challenge of it. *Why it's important:* Hackers create problems not only for the institutions that are victims of break-ins but also for ordinary users of the systems.

ICANN (Internet Corporation for Assigned Names & Numbers) (p. 274, KQ 8.1) Organization established to regulate Internet domain names, those addresses ending with .com, .org, .net, and so on, that identify a website. *Why it's important:* ICANN could evolve into the preeminent regulatory body on the Internet, with power extending far beyond the arena of domain names.

intelligent agent (p. 291, KQ 8.7) Program that performs work tasks—such as roaming networks and compiling data—on the user's behalf. *Why it's important: A software agent is a kind of electronic assistant that will filter messages, scan news services, and perform similar secretarial chores. An agent will also travel over communications lines to computer databases, collecting files to add to a personalized database.*

Internet2 (p. 273, KQ 8.1) Cooperative university/business program established to enable high-end users to quickly and reliably move huge amounts of data, using VBNS as the official backbone. *Why it's important: Whereas VBNS provides data transfer at 1000 times commercial Web speeds, Internet2 operates at only 100 times those speeds. In effect, Internet2 adds "toll lanes" to the older Internet to speed things up. The purpose is to advance video-conferencing, research, and academic collaboration—to enable a kind of "virtual university."*

natural language processing (p. 289, KQ 8.6) Study of ways for computers to recognize and understand human language, whether in spoken or written form. *Why it's important: Natural languages make it easier to work with computers.*

Next Generation Internet (NGI) (p. 273, KQ 8.1) U.S. government's program to parallel the university/business–sponsored effort of Internet2. *Why it's important: NGI is designed to provide money to six government agencies to help tie the campus high-performance backbones into the broader federal infrastructure. The intent is that NGI will connect at least 100 sites, including universities, federal national laboratories, and other research organizations. Speeds are 100 times those of the older Internet, and 10 sites are connected at speeds that are 1000 times as fast.*

password (p. 281, KQ 8.3) Special word, code, or symbol required to access a computer system. *Why it's important: Passwords are one of the weakest security links; they can be guessed, forgotten, or stolen.*

PIN (personal identification number) (p. 281, KQ 8.3) Security number known only to the user; it is required to access a system. *Why it's important: PINS are required to access many computer systems, as well as automated teller machines.*

robot (p. 288, KQ 8.6) Automatic device that performs functions ordinarily executed by human beings or that operates with what appears to be almost human intelligence. *Why it's important: Robots are performing more and more functions in business and the professions.*

robotics (p. 288, KQ 8.6) Development and study of machines that can perform work normally done by people. *Why it's important: See robot.*

security (p. 280, KQ 8.3) System of safeguards for protecting information technology against disasters, systems failure, and unauthorized access that can result in damage or loss. Four components of security are identification and access, encryption, protection of software and data, and disaster-recovery plans. *Why it's important: With proper security, organizations and individuals can minimize information technology losses from disasters, system failures, and unauthorized access.*

Turing test (p. 290, KQ 8.6) A test for determining whether a computer possesses "intelligence" or "self-awareness." In the Turing test, a human judge converses by means of a computer terminal with two entities hidden in another location. *Why it's important: Some experts believe that once a computer has passed the Turing test, it will be judged to have achieved a level of human intelligence.*

VBNS (Very-High-Speed Backbone Network Service) (p. 273, KQ 8.1) Main U.S. government component to upgrade the backbone, or primary hubs of data transmission, of the Internet. *Why it's important: Linking supercomputers and other banks of computers across the United States, the VBNS is 1000 times as fast as the conventional Internet. Financed by the National Science Foundation and managed by the telecommunications giant MCI, VBNS is somewhat exclusive. It is being used by just 101 top universities and other research institutions for network-intensive applications, such as transmitting high-quality video for distance education.*

virus (p. 278, KQ 8.2) Deviant program that copies itself into other programs, floppies, and hard disks, causing undesirable effects. *Why it's important: Viruses can cause users to lose data and/or files or can shut down entire computer systems.*

worm (p. 278, KQ 8.2) Program that copies itself repeatedly into a computer's memory or onto a disk drive until no memory or disk space is left. *Why it's important: Worms can shut down computers.*

Chapter Review

stage 1 LEARNING MEMORIZATION
"I can recognize and recall information."

Self-Test Questions

1. The purpose of _____ is to scan a computer's disk devices and memory to detect viruses and, sometimes, to destroy them.
2. So that information processing operations can be restored after destruction or accident, organizations should adopt a _____.
3. _____ is the altering of data so that it is not usable unless the changes are undone.
4. _____ is incomplete, outdated, or otherwise inaccurate data.
5. An error in a program that causes it not to work properly is called a _____.

Multiple-Choice Questions

1. Which of the following are crimes against computers and communications?
    a. natural hazards
    b. software theft
    c. information theft
    d. software bugs
    e. procedural errors

2. Which of the following are methods or means of safeguarding computer systems?
    a. signatures
    b. keys
    c. physical traits of users
    d. worms
    e. VBNS

True/False Questions

T F 1. Viruses cannot be passed from computer to computer through a network.
T F 2. One of the weakest links in a security system is biometrics.
T F 3. The VBNS is faster than the conventional Internet.
T F 4. The Turing test determines whether a computer possesses intelligence.
T F 5. An intelligent agent performs network work tasks on the user's behalf.

stage 2 LEARNING COMPREHENSION
"I can recall information in my own terms and explain them to a friend."

Short-Answer Questions

1. What is the difference between a hacker and a cracker?
2. What does a worm do?
3. Briefly describe and differentiate VBNS, Internet2, and Next Generation Internet.
4. What was the 1996 U.S. Telecommunications Act intended to do?
5. What does ICANN stand for, and what does it do?
6. Name five threats to computers and communications systems.
7. The definition of computer crime distinguishes between two types. What are they?

Concept Mapping

On a separate sheet of paper, draw a concept map, or visual diagram, linking concepts. Show how the following terms are related.

antivirus software	disaster-recovery plan	Internet2	security
biometrics	encryption	NGI	VBNS
computer crime	hacker	password	virus
cracker	ICANN	PIN	worm

stage 3 LEARNING: APPLYING, ANALYZING, SYNTHESIZING, EVALUATING

"I can apply what I've learned, relate these ideas to other concepts, build on other knowledge, and use all these thinking skills to form a judgment."

Knowledge in Action

1. What, in your opinion, are the most significant disadvantages of using computers? What do you think can be done about these problems?

2. What's your opinion about the issue of free speech on an electronic network? Research some recent legal decisions in various countries, as well as some articles on the topic. Should the contents of messages be censored? If so, under what conditions?

3. Research the problems of stress and isolation experienced by computer users in the United States, Japan, and one other country. Write a brief report on your findings.

Web Exercises

1. For more information about Internet2, visit

   www.internet2.edu

   Identify the three primary goals of Internet2. Also read the Frequently Asked Questions section.

2. How secure is your browser? First find out how much information is already known about you by visiting

   www.cryptography.ws/anonymity_test.htm

   Next discover what you can do to protect yourself. What is 128-bit encryption? Is your browser equipped? Visit the following websites to learn more about browser security:

   www.consumersenergy.com/welcome.htm?./products/index.asp?ASID=106

   http://working4u.scudder.com/SU1/test_7/chkbrwsr.htm

   http://iasweb.com/articles/webshopping.html

   http://retire.comerica.com/warn.htm

3. The CIH (Chernobyl) virus affected many Windows 95 users. This virus struck intitially on April 26, 1999, and can still strike if your system is still infected. If your PC has Windows 95 installed, or you have had files sent from other users of Windows 95, it may be infected with the virus, and you may not know you have it. Visit this website for an in-depth report about what this virus does, its origins, and how to fix or protect against it:

   www.symantec.com/avcenter/venc/data/cih.html

4. Here's a way to semi-encrypt an e-mail message to a friend. Type out your message; then go to the Edit menu and choose Select All. Next go to the Format menu, select Font; then choose a font such as Wingdings that doesn't use letters. When your friend receives your message, he or she need only change the font back to an understandable one (such as Times New Roman) in order to read it. This certainly isn't high-level encryption, but it's a fun activity to try.

5. Internet addiction can be a serious problem. Some games such as Everquest calculate how many days of your life you have spent playing online. These days are calculated as 24-hour days. Most humans do not stay awake for 24 hours. Visit these following bulletin boards to read about various players and how they can't shake their Everquest habit:

   www.sharkygames.com/forum_frag/2a/5.shtml

   www.cdmag.com/Home/home.html?article=/articles/028/064/everquest_column.html

   www.victoriapoint.com/internetaddiction/board/posts/26.html

   Then visit the following website that discusses general Internet addiction:

   www.victoriapoint.com/internetaddiction

6. If you're spending too much time indoors using a PC or watching TV, you might want to consider going outside. Visit the following websites to learn about indoor air versus outdoor air.

   www.epa.gov/children/air.htm

   www.who.int/inf-fs/en/fact201.html

   Also, to read more about Internet addiction that causes isolation from the real world, visit

   www.sciencenews.org/20000226/fob8.asp

   The following two websites discuss computer health and safety:

   www.ics.uci.edu/~chair/comphealth2.html

   www.stanford.edu/dept/EHS/work/ergo/keys.html

Appendix

Systems & Programming
Development, Programming, & Languages

Key Questions
You should be able to answer the following questions.

- **A.1** **Systems Development: The Six Phases of Systems Analysis & Design** What are the six phases of the systems development life cycle?
- **A.2** **Programming: A Five-Step Procedure** What is programming, and what are the five steps in accomplishing it?
- **A.3** **Programming Languages** What is a programming language, and how is a high-level language converted to a low-level language?
- **A.4** **Object-Oriented Programming & Visual Programming** How do OOP and visual programming work?
- **A.5** **Internet Programming: HTML, XML, VRML, Java, & ActiveX** What are the features of HTML, XML, VRML, Java, and ActiveX?

Organizations can make mistakes, of course, and big organizations can make *really big* mistakes.

California's state Department of Motor Vehicles' databases needed to be modernized, and in 1988 Tandem Computers said it could do the job. "The fact that the DMV's database system, designed around an old IBM-based platform, and Tandem's new system were as different as night and day seemed insignificant at the time to the experts involved," said one writer investigating the project later.[1] The massive driver's license database, containing the driving records of more than 30 million people, first had to be "scrubbed" of all information that couldn't be translated into the language used by Tandem computers. One such scrub yielded 600,000 errors. Then the DMV had to translate all its IBM programs into the Tandem language. "Worse, DMV really didn't know how its current IBM applications worked anymore," said the writer, "because they'd been custom-made decades before by long-departed programmers and rewritten many times since." Eventually the project became a staggering $44 million loss to California's taxpayers.

This example shows how important planning is, especially when an organization is trying to launch a new kind of system. How do you avoid such mistakes? By employing systems analysis and design.

A.1 Systems Development: The Six Phases of Systems Analysis & Design

KEY QUESTION
What are the six phases of the systems development life cycle?

You may not have to wrestle with problems on the scale of motor-vehicle departments. That's a job for computer professionals. You're mainly interested in using computers and communications to increase your own productivity. Why, then, do you need to know anything about systems analysis and design?

In many careers, you may find your department or your job the focus of a study by a systems analyst. Knowing how the procedure works will help you better explain how your job works or what goals your department is supposed to achieve. In progressive companies, management is always interested in suggestions for improving productivity. This is the method for expressing your ideas.

The Purpose of a System

A *system* **is defined as a collection of related components that interact to perform a task in order to accomplish a goal.** A system may not work very well, but it is nevertheless a system. The point of systems analysis and design is to ascertain how a system works and then take steps to make it better.

An organization's computer-based information system consists of hardware, software, people, procedures, and data, as well as communications setups. These work together to provide people with information.

Getting the Project Going: How It Starts, Who's Involved

To get a project rolling, all it takes is a single individual who believes that something badly needs changing. An employee may influence a supervisor. A customer or supplier may get the attention of someone in higher management. Top management on its own may decide to take a look at a system that looks inefficient. A steering committee may be formed to decide which of many possible projects should be worked on.

Participants in the project are of three types:

- **Users:** The system under discussion should *always* be developed in consultation with users, whether floor sweepers, research scientists, or customers. Indeed, inadequate user involvement in analysis and design can be a major cause of a system's failing for lack of acceptance.
- **Management:** Managers within the organization should also be consulted about the system.
- **Technical staff:** Members of the company's information systems (IS) department, consisting of systems analysts and programmers, need to be involved. For one thing, they may well have to carry out and execute the project. Even if they don't, they may have to work with outside IS people contracted to do the job.

Complex projects will require one or several systems analysts. A <u>**systems analyst**</u> **is an information specialist who performs systems analysis, design, and implementation.** His or her job is to study the information and communications needs of an organization and determine what changes are required to deliver better information to people who need it. "Better" information means information that is summarized in the acronym "CART"—complete, accurate, relevant, and timely. The systems analyst achieves this goal through the problem-solving method of systems analysis and design.

The Six Phases of Systems Analysis & Design

<u>**Systems analysis and design**</u> **is a six-phase problem-solving procedure for examining an information system and improving it.** The six phases (actually the phases overlap, so the number varies according to practitioner) make up what is called the systems development life cycle. The <u>**systems development life cycle (SDLC)**</u> **is the step-by-step process that many organizations follow during systems analysis and design.**

Whether applied to a Fortune 500 company or a three-person engineering business, systems analysis and design consists of six phases. *(See* • *Panel A.1.)*

● **PANEL A.1**
Systems development life cycle
An SDLC typically includes six phases.

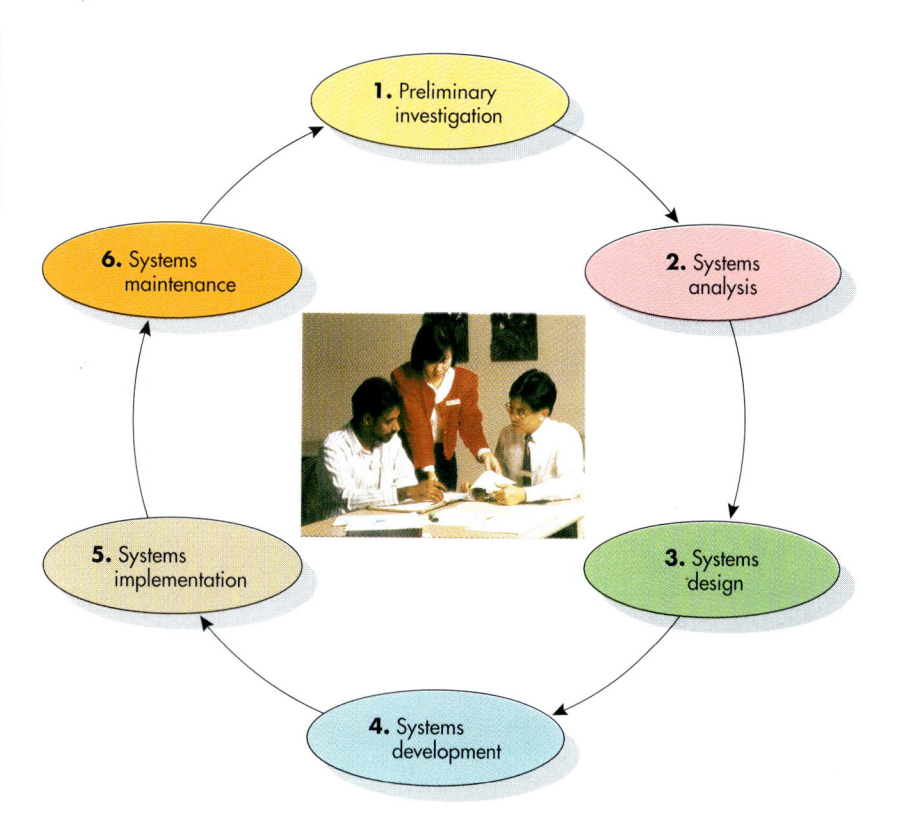

1. Preliminary investigation
2. Systems analysis
3. Systems design
4. Systems development
5. Systems implementation
6. Systems maintenance

301

Phases often overlap, and a new one may start before the old one is finished. After the first four phases, management must decide whether to proceed to the next phase. *User input and review is a critical part of each phase.*

The First Phase: Conduct a Preliminary Investigation

The objective of Phase 1, <u>preliminary investigation</u>, is to conduct a preliminary analysis, propose alternative solutions, describe costs and benefits, and submit a preliminary plan with recommendations.

- **Conduct the preliminary analysis:** In this step, you need to find out what the organization's objectives are and the nature and scope of the problem under study. Even if a problem pertains only to a small segment of the organization, you cannot study it in isolation. You need to find out what the objectives of the organization itself are. Then you need to see how the problem being studied fits in with them.

- **Propose alternative solutions:** In delving into the organization's objectives and the specific problem, you may have already discovered some solutions. Other possible solutions can come from interviewing people inside the organization, clients or customers affected by it, suppliers, and consultants. You can also study what competitors are doing. With this data, you then have three choices. You can leave the system as is, improve it, or develop a new system.

- **Describe the costs and benefits:** Whichever of the three alternatives is chosen, it will have costs and benefits. In this step, you need to indicate what these are. Costs may depend on benefits, which may offer savings. There are many possible kinds of benefits. A process may be speeded up, streamlined through elimination of unnecessary steps, or combined with other processes. Input errors or redundant output may be reduced. Systems and subsystems may be better integrated. Users may be happier with the system. Customers or suppliers may interact better with the system. Security may be improved. Costs may be cut.

- **Submit a preliminary plan:** Now you need to wrap up all your findings in a written report. The readers of this report will be the executives who decide in which direction to proceed—make no changes, change a little, or change a lot—and how much money to allow the project. You should describe the potential solutions, costs, and benefits and indicate your recommendations.

The Second Phase: Do an Analysis of the System

The objective of Phase 2, <u>systems analysis</u>, is to gather data, analyze the data, and write a report. In this second phase, you will follow the course selected by management on the basis of your Phase 1 feasibility report. We are assuming that they have ordered you to perform Phase 2—to do a careful analysis or study of the existing system in order to understand how the new system you proposed would differ. This analysis will also consider how people's positions and tasks will change if the new system is put into effect.

- **Gather data:** In gathering data, you will review written documents, interview employees and managers, develop questionnaires, and observe people and processes at work.

- **Analyze the data:** The next step is to come to grips with the data you have gathered and analyze it. Many analytical tools, or modeling tools, are available. *Modeling tools* enable a systems analyst to present graphic, or pictorial, representations of a system. Some of these tools involve creating flowcharts and diagrams on paper. Examples of modeling tools are *grid charts, decision tables, data flow diagrams, system flowcharts,* and *connectivity diagrams.*

- **Write a report:** Once you have completed the analysis, you need to document this phase, by writing a report to management. This report should have three parts. First, it should explain how the existing system works. Second, it should explain the problems with the existing system. Finally, it should describe the requirements for the new system and make recommendations on what to do next.

At this point, not a lot of money will have been spent on the systems analysis and design project. If the costs of going forward seem prohibitive, this is a good time for the managers reading the report to call a halt. Otherwise, you will be asked to move to Phase 3.

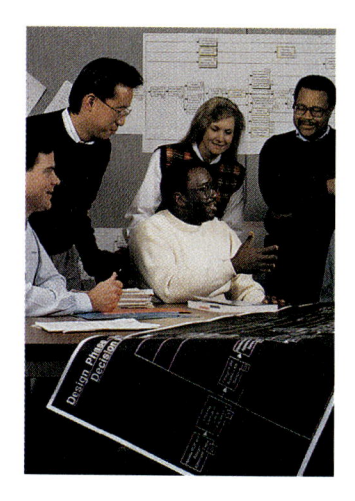

The Third Phase: Design the System

The objective of Phase 3, <u>systems design</u>, is to do a preliminary design and then a detailed design, and write a report. In this third phase of the SDLC, you will essentially create a "rough draft" and then a "detail draft" of the proposed information system.

- **Do a preliminary design:** A *preliminary design* describes the general functional capabilities of a proposed information system. It reviews the system requirements and then considers major components of the system. Usually several alternative systems (called *candidates*) are considered, and the costs and the benefits of each are evaluated.

  Some tools that may be used in the design are *CASE tools* and *project management software*.

  CASE (computer-aided software engineering) tools are programs that automate various activities of the SDLC in several phases. This technology is intended to speed up the process of developing systems and to improve the quality of the resulting systems. CASE tools, also known as *automated design tools*, may be used at other stages of the SDLC as well. Examples of such programs are Excelerator, Iconix, System Architect, and Powerbuilder.

  CASE tools may also be used to do prototyping. In *prototyping*, workstations, CASE tools, and other software applications are used to build working models of system components, so that they can be quickly tested and evaluated. Thus, a *prototype* is a limited working system developed to test out design concepts. A prototype, which may be constructed in just a few days, allows users to find out immediately how a change in the system might benefit them.

  Project management software consists of programs used to plan, schedule, and control the people, costs, and resources required to complete a project on time.

- **Do a detail design:** A *detail design* describes how a proposed information system will deliver the general capabilities described in the preliminary design. The detail design usually considers the following parts of the system in this order: output requirements, input requirements, storage requirements, processing requirements, and system controls and backup.

- **Write a report:** All the work of the preliminary and detail designs will end up in a large, detailed report.

The Fourth Phase: Develop the System

In Phase 4, <u>systems development</u>, the systems analyst or others in the organization develop or acquire the software, acquire the hardware, and then test the system. Depending on the size of the project, this is the phase that will probably involve the organization in spending substantial sums of money. It could also involve spending a lot of time. However, at the end you should have a workable system.

- **Develop or acquire the software:** During the design stage, the systems analyst may have had to address what is called the "make-or-buy" decision, but that decision certainly cannot be avoided now. In the *make-or-buy decision*, you decide whether you have to create a program—have it custom-written—or buy it, meaning simply purchase an existing software package. Sometimes programmers decide they can buy an existing program and modify it.

    If you decide to create a new program, then the question is whether to use the organization's own staff programmers or hire outside contract programmers (outsource it). Whichever way you go, the task could take many months.

    Programming is an entire subject unto itself, and we address it in the next major section.

- **Acquire hardware:** Once the software has been chosen, the hardware to run it must be acquired or upgraded. It's possible your new system will not require obtaining any new hardware. It's also possible that the new hardware will cost millions of dollars and involve many items: microcomputers, mainframes, monitors, modems, and many other devices. The organization may find it's better to lease rather than to buy some equipment, especially since, as we mentioned (Moore's law), chip capability has traditionally doubled every 18 months.

- **Test the system:** With the software and hardware acquired, you can now start testing the system. Testing is usually done in two stages: unit testing, then system testing.

    In *unit testing*, individual parts of the program are tested, using test (made-up, or sample) data. If the program is written as a collaborative effort by multiple programmers, each part of the program is tested separately.

    In *system testing*, the parts are linked together, and test data is used to see if the parts work together. At this point, actual organization data may be used to test the system. Tests with erroneous data and massive amounts of data check whether the system can be made to fail ("crash").

    At the end of this long process, the organization will have a workable information system, one ready for the implementation phase.

The Fifth Phase: Implement the System

Whether the new information system involves a few handheld computers, an elaborate telecommunications network, or expensive mainframes, the fifth phase will involve some close coordination in order to make the system not just workable but successful. **Phase 5, _systems implementation_, consists of converting the hardware, software, and files to the new system and training the users.**

- **Convert to the new system:** *Conversion*, the process of converting from an old information system to a new one, involves converting hardware, software, and files. *Hardware conversion* may be as simple as taking away an old PC and plunking a new one down in its place. Or it may involve acquiring new buildings and putting in elaborate wiring, climate-control, and security systems.

    Software conversion means making sure the applications that worked on the old equipment can be made to work on the new.

    File conversion, or *data conversion*, means converting the old files to new ones without loss of accuracy. For example, can the paper contents from the manila folders in the personnel department be input to the system with a scanner? Or do they have to be keyed in manually, with the consequent risk of introducing errors?

There are four strategies for handling conversion: *direct, parallel, phased,* and *pilot.*

Direct implementation means the user simply stops using the old system and starts using the new one. The risk of this method should be evident: What if the new system doesn't work? If the old system has truly been discontinued, there is nothing to fall back on.

Parallel implementation means that the old and new systems are operated side by side until the new system has shown it is reliable, at which time the old system is discontinued. Obviously there are benefits to this cautious approach. If the new system fails, the organization can switch back to the old one. The difficulty with this method is the expense of paying for the equipment and people to keep two systems going at the same time.

Phased implementation means that parts of the new system are phased in separately—either at different times (parallel) or all at once in groups (direct).

Pilot implementation means that the entire system is tried out but only by some users. Once the reliability has been proved, the system is implemented with the rest of the intended users. The pilot approach still has its risks, since all of the users of a particular group are taken off the old system. However, the risks are confined to a small part of the organization.

In general, the phased and pilot approaches are the most favored methods. Phased is best for large organizations in which people are performing different jobs. Pilot is best for organizations in which all people are performing the same task (such as order takers at a direct-mail house).

- **Train the users:** Various tools are available to familiarize users with the new system. They run from documentation (instruction manuals) to videotapes to live classes to one-on-one, side-by-side teacher-student training. Sometimes the organization's own staffers do the training; sometimes it is contracted out.

The Sixth Phase: Maintain the System

Phase 6, <u>systems maintenance</u>, involves adjustment and improvement of the system by conducting system audits and periodic evaluations and by making changes based on new conditions. After conversion and the user training, the system won't just run itself. There is a sixth—and never-ending—phase in which the information system must be monitored to ensure that it is successful. Maintenance includes not only keeping the machinery running but also in updating and upgrading the system to keep pace with new products, services, customers, government regulations, and other requirements.

A.2 Programming: A Five-Step Procedure

KEY QUESTIONS
What is programming, and what are the five steps in accomplishing it?

To see how programming works, we must understand what a program is. A **<u>program</u> is a list of instructions that the computer must follow in order to process data into information.** The instructions consist of *statements* used in a programming language, such as BASIC. Examples are programs that do word processing, desktop publishing, or payroll processing.

The decision whether to buy or create a program forms part of Phase 4 in the systems development life cycle. *(See ● Panel A.2, next page.)* Once the decision is made to develop a new system, the programmer goes to work.

A program, we said, is a list of instructions for the computer. **<u>Programming</u>, also called software engineering, is a multistep process for creating that list of instructions.**

The five steps are as follows.

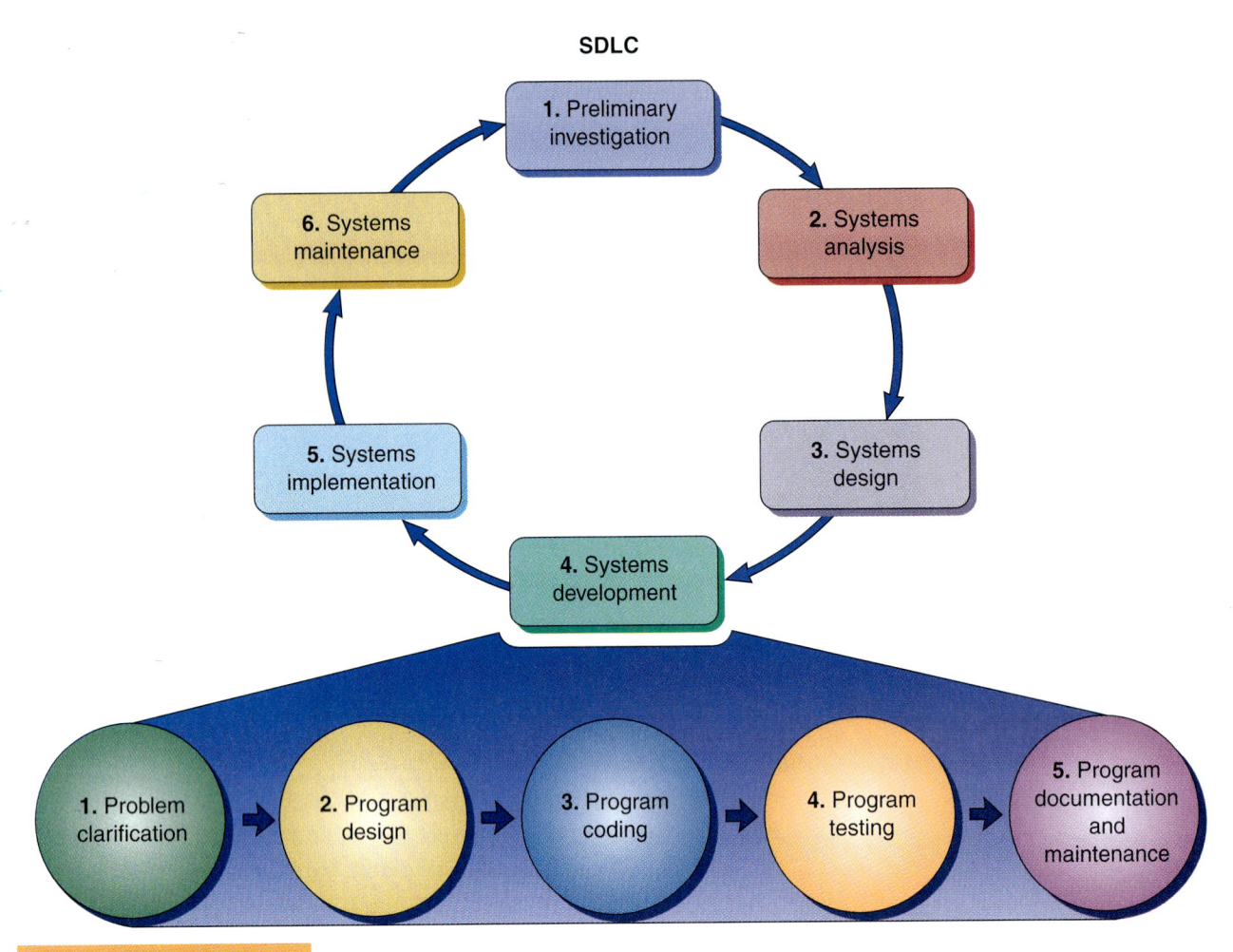

SDLC

```
        1. Preliminary
         investigation
       ↗              ↘
6. Systems          2. Systems
maintenance          analysis
    ↑                    ↓
5. Systems           3. Systems
implementation         design
       ↖              ↙
        4. Systems
        development
```

1. Problem clarification → 2. Program design → 3. Program coding → 4. Program testing → 5. Program documentation and maintenance

● **PANEL A.2**
Where programming fits in the systems development life cycle
The fourth phase of the six-phase systems development life cycle includes a five-step procedure of its own. These five steps constitute the problem-solving process called *programming*.

1. **Clarify the problem**—include needed output, input, processing requirements.
2. **Design a solution**—use modeling tools to chart the program.
3. **Code the program**—use a programming language's syntax, or rules, to write the program.
4. **Test the program**—get rid of any logic errors, or "bugs," in the program ("debug" it).
5. **Document and maintain the program**—include written instructions for users, explanation of the program, and operating instructions.

Coding—sitting at the keyboard and typing words into a computer—is what many people imagine programming to be. As we see, however, it is only one of the five steps. Coding consists of translating the logic requirements into a programming language—the letters, numbers, and symbols that make up the program.

A.3 Programming Languages

KEY QUESTIONS
What is a programming language, and how is a high-level language converted to a low-level language?

A *programming language* is a set of rules and symbols that tells the computer what operations to do. Examples of well-known programming languages are BASIC, COBOL, and C. (See ● Panel A.3.) Not all languages are appropriate for all uses. Some, for example, have strengths in mathematical and statistical processing. Others are more appropriate for database management. Thus, in choosing a language, the programmer must consider what purpose the program is to serve and what languages are already being used in the organization or field.

> ● **PANEL A.3**
> **Some common programming languages**
> These are but a few of many languages.
>
> The most important high-level languages that you may come across are the following, ranging from oldest to newest: *FORTRAN, COBOL, BASIC* (and *Visual Basic*), *Pascal, C* (and *C++*), and *Ada*.
>
> - **FORTRAN—the language of mathematics and the first high-level language:** Developed in 1954 by IBM, *FORTRAN* (for *FORmula TRANslator*) was the first procedural language and is still the most widely used language for mathematical, scientific, and engineering problems. It is also useful for complex business applications, such as forecasting and modeling. However, because it cannot handle a large volume of input/output operations or file processing, it is not used for more typical business problems. The newest version of FORTRAN is FORTRAN 95.
>
> - **COBOL—the language of business:** Formally adopted in 1960, *COBOL* (for *COmmon Business Oriented Language*) is the most frequently used business programming language for large computers. Its most significant attribute is that it is extremely readable. For example, a COBOL line might read: MULTIPLY HOURLY-RATE BY HOURS-WORKED GIVING GROSS PAY
>
>   Writing a COBOL program resembles writing an outline for a research paper. The program is divided into four divisions, which in turn are divided into sections, which are divided into paragraphs, which are divided into sections, which are divided into statements.
>
> - **BASIC—the easy language:** BASIC was developed by John Kemeny and Thomas Kurtch in 1965 for use in training their students at Dartmouth College. By the late 1960s, it was widely used in academic settings on all kinds of computers, from mainframes to PCs. Now its use has extended to business.
>
>   *BASIC* (for *Beginner's All-purpose Symbolic Instruction Code*) has been the most popular microcomputer language and is considered the easiest programming language to learn. The interpreter form is popular with first-time and casual users because it is interactive, meaning that user and computer can communicate with each other during the writing and running of the program.
>
>   Today there is no one version of BASIC, but one of the current evolutions is Visual Basic, which is the most popular visual programming language.
>
> - **PASCAL—the simple language:** Created in 1970 and named after the 17th-century French mathematician Blaise Pascal, *Pascal* is an alternative to BASIC as a language for teaching purposes and is relatively easy to learn. A difference from BASIC is that Pascal uses structured programming. It also has extensive capabilities for graphics programming.
>
> - **C and C++—for portability and applications software development:** "C" is this language's entire name, and it does not "stand" for anything. Developed at Bell Laboratories in the early 1970s, *C* is a general-purpose language that works well for microcomputers and is portable among many computers. It is widely used for writing operating systems, utilities, spreadsheet programs, database programs, and some scientific uses. C is also the programming language used most commonly in commercial software development, including games, robotics, and graphics.
>
>   Developed in the early 1980s, *C++* ("C plus plus")—the plus signs stand for "more than C"—combines the traditional C with object-oriented programming (OOP) capability, a technique we discuss later in the chapter. C++ is used for developing applications software.
>
> - **Ada—for weapons and commercial uses:** *Ada* is an extremely powerful structured programming language designed by the U.S. Department of Defense to ensure portability of programs from one application to another. Ada was named for Countess Ada Lovelace, considered the world's "first programmer." Based on Pascal, Ada was originally intended to be a standard language for weapons systems. However, it has been used successfully in commercial applications.
>
>   The advantage of Ada is that it is a structured language, with a modular design; thus, pieces of a large program can be written and tested separately. Moreover, because it has features that permit the compiler to check it for errors before the program is run, programmers are more apt to write error-free programs.

In writing a program, you have to follow the correct *syntax,* the rules of the programming language. Programming languages have their own grammar just as human languages do. But computers are probably a lot less forgiving if you use these rules incorrectly. Even a typographical error can constitute faulty syntax.

Programming languages are also called *high-level languages.* For the computer to be able to understand them, they must be translated into the low-level language called machine language. **Machine language is the basic language of the computer, representing data as 1s and 0s.** Machine-language programs vary from computer to computer; that is, they are *machine-dependent.*

PANEL A.4
Low-level and high-level languages

Machine language

```
11110010 01110011 1101 001000010000 0111 000000101011
11110010 01110011 1101 001000011000 0111 000000101111
11111100 01010010 1101 001000010010 1101 001000011101
11110000 01000101 1101 001000010011 0000 000000111110
11110011 01000011 0111 000001010000 1101 001000010100
10010110 11110000 0111 000001010100
```

⬇

COBOL

```
MULTIPLY HOURS-WORKED BY PAY-RATE GIVING GROSS-PAY ROUNDED
```

A high-level language allows users to write in a familiar notation, rather than numbers or abbreviations. (See ● *Panel A.4.*) Most high-level languages are not machine-dependent—that is, they can be used on more than one kind of computer. Examples are FORTRAN, COBOL, BASIC, Pascal, and C. For a high-level language to work on a computer, it needs a *language translator* to translate it into machine language. Depending on the language, either of two types of translators may be used—a *compiler* or an *interpreter*.

- Compiler—execute later: A <u>compiler</u> is a language translator that converts the entire program of a high-level language into machine language BEFORE the computer executes the program. The programming instructions of a high-level language are called the *source code*. The compiler translates it into machine language, called *object code*. The object code can then be saved and executed later (as many times as desired), rather than run right away. (See ● *Panel A.5.*)

  Examples of high-level languages using compilers are COBOL, FORTRAN, Pascal, and C.

- Interpreter—execute immediately: An <u>interpreter</u> is a language translator that converts each high-level language statement into machine language and executes it IMMEDIATELY, statement by statement. In contrast to the compiler, no object code is saved. Therefore, interpreted code generally runs more slowly than compiled code. However, code can be tested line by line.

  An example of a high-language language using an interpreter is BASIC.

A.4 Object-Oriented Programming & Visual Programming

KEY QUESTION
How do OOP and visual programming work?

Two developments have made programming a bit easier—*object-oriented programming* and *visual programming*.

Object-Oriented Programming

Imagine you're writing a program in BASIC or a similar language, creating your coded instructions one at a time. As you work on some segment of the program (such as how to compute overtime pay), you may think, "I'll bet some other programmer has already written something like this. Wish I had it. It would save a lot of time."

Fortunately, a kind of recycling technique now exists. This is object-oriented programming, as used in C++, for example. Let us explain this approach in four steps:

● **PANEL A.5**
Compiler

Source Code
(high-level language)

```
IF COUNT = 10
   GOTO DONE

ELSE
   GOTO AGAIN

ENDIF
```

Language translator program →

```
10010101001010001010100
10101010010101001001010
10100101010001010010010
```

Object Code
(machine languge)

Conventional Programs

Object-Oriented Programs

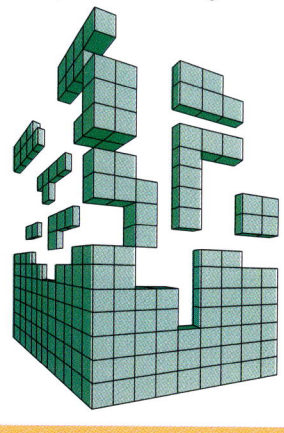

● **PANEL A.6**
Conventional versus object-oriented programs

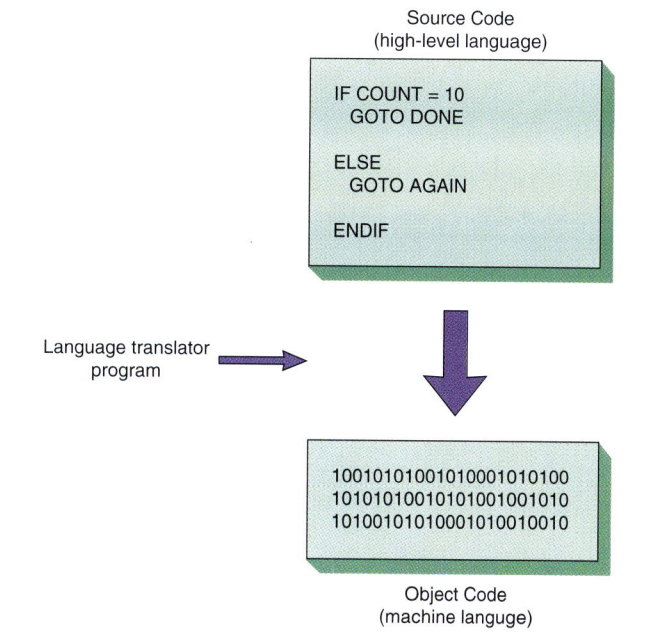

1. **What OOP is:** In *object-oriented programming* (OOP, pronounced "oop"), **data and instructions for processing that data are combined into a self-sufficient "object" that can be used in other programs.** *(See ● Panel A.6.)* Objects provide powerful tools for programmers.

2. **What an "object" is:** An *object* is a block of preassembled programming code in a self-contained module. The module contains both (1) a chunk of data and (2) processing instructions that may be performed on that data.

3. **When an object's data is processed—sending the "message":** Once the object becomes part of a program, the processing instructions are activated when a "message" is sent. A *message* is an alert sent to the object when the program needs to perform an operation involving that object.

4. **How the object's data is processed—the "methods":** The message need only identify the operation. The processing instructions that are part of the object contain information on how the operation is actually to be performed. These instructions are called the *methods*.

Once you've written a block of program code (that computes overtime pay, for example), it can be reused in any number of programs. Thus, with OOP, in contrast to traditional programming, you don't have to reinvent the wheel each time.

Object-oriented programming takes longer to learn than traditional programming because it involves internalizing a whole new way of thinking. The beauty of OOP, however, is that it speeds up development time and lowers costs, because an object can be used repeatedly in different applications and by different programmers.

Visual Programming

Visual programming **is a method of creating programs in which the programmer makes connections between objects by drawing, pointing, and clicking on diagrams and icons and by interacting with flowcharts.** Thus, the programmer can create programs by clicking on icons that represent common programming routines. Visual programming enables users to think more about the problem solving than about handling the programming language. There is no need to learn syntax or write code.

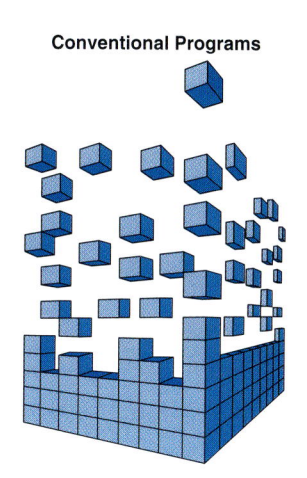

The most popular visual programming language is Visual Basic. Developed by Microsoft Corp. in the early 1990s, *Visual Basic* offers a visual environment for program construction, allowing you to build various components using buttons, scroll bars, and menus.

A.5 Internet Programming: HTML, XML, VRML, Java, & ActiveX

KEY QUESTION
What are the features of HTML, XML, VRML, Java, and ActiveX?

Many of the thousands of Internet data and information sites around the world are text-based only. However, the World Wide Web permits virtually unlimited use of graphics, animation, video, and sound.

One way to build such multimedia sites on the Web is to use some fairly recently developed programming languages and standards: HTML, XML, VRML, and Java.

HTML—for Creating 2-D Web Documents & Links

HTML (Hypertext Markup Language, discussed in Chapter 2) lets people create onscreen documents for the Internet that can easily be linked by words and pictures to other documents. Not considered a "real" programming language, HTML is a type of code that embeds simple commands within standard ASCII text documents to provide an integrated, two-dimensional display of text and graphics. In other words, a document created in any word processor and stored in ASCII format can become a Web page with the addition of a few HTML commands.

One of the main features of HTML is the ability to insert hypertext links into a document. Hypertext links enable you to display another Web document simply by clicking on a link area—usually underlined or highlighted—on your current screen. One document may contain links to many other related documents. The related documents may be on the same server as the first document, or they may be on a computer halfway around the world. A link may be a word, a group of words, or a picture.

XML—for Making the Web Work Better

The chief characteristics of HTML are its simplicity and its ease in combining plain text and pictures. But, in the words of journalist Michael Krantz, "HTML simply lacks the software muscle to handle the business world's endless and complex transactions."[2]

Enter XML. Whereas HTML makes it easy for humans to read Web sites, *XML (extensible markup language)* makes it easy for machines to read Web sites. At present, when you use your browser to find a Web site, search engines generate too many options, so that it's difficult to turn up the specific site you want—say, one with a recipe for a low-calorie chicken dish for 12. Says Krantz, "XML makes Web sites smart enough to tell other machines whether they're looking at a recipe, an airline ticket, or a pair of easy-fit blue jeans with a 34-inch waist." To do so, XML lets Web site developers put "tags" on their Web pages that describe information in, for example, a food recipe as "ingredients," "calories," "cooking time," and "number of portions." Thus, your browser can read those tags and search much more effectively for that low-calorie poultry recipe for 12.

VRML—for Creating 3-D Web Pages

VRML rhymes with "thermal." *VRML (Virtual Reality Modeling Language)* is a type of programming language used to create three-dimensional Web pages. Even though VRML's designers wanted to let nonprogrammers create their own virtual spaces quickly and painlessly, it's not as simple to describe a three-dimensional scene as it is to describe a page in HTML. However, many existing modeling and CAD tools now offer VRML support, and new VRML-centered software tools are arriving.

Java—for Creating Interactive Web Pages

Available from Sun Microsystems and derived from C++, Java is a major departure from the HTML coding that makes up most Web pages. Sitting atop markup languages such as HTML and XML, *Java* is an object-oriented programming language that allows programmers to build applications that can run on any operating system. With Java, big applications programs can be broken into mini-applications, or "applets," that can be downloaded off the Internet and run on any computer. Moreover, Java enables a Web page to deliver applets that, when downloaded, can make Web pages interactive.

ActiveX—also for Creating Interactive Web Pages

ActiveX was developed by Microsoft as an alternative to Java for creating interactivity on Web pages. Indeed, Java and ActiveX are the two major contenders in the Web-applet war for transforming the Web into a complete interactive environment.

ActiveX is a set of controls, or reusable components, that enables programs or content of almost any type to be embedded within a Web page. Whereas Java requires you to download an applet each time you visit a Web site, with ActiveX the component is downloaded only once, then stored on your hard drive for later and repeated use.

Thus, the chief characteristic of ActiveX is that it features *reusable* components—small modules of software code that perform specific tasks (such as a spelling checker), which may be plugged seamlessly into other applications. With ActiveX you can obtain from your hard disk any file that is suitable for the Web—such as a Java applet, animation, or pop-up menu—and insert it directly into an HTML document.

Programmers can create ActiveX controls or components in a variety of programming languages, including C, C++, Visual Basic, and Java. Thousands of ready-made ActiveX components are now commercially available from numerous software development companies.

BOOKMARK IT!

PRACTICAL ACTION BOX
More Tips for Creating Your Own Website

Now that you're on the Internet, you can promote your business or yourself by creating your own website.

To do so, you once needed to learn the basics of HTML. No more. Although knowledge of HTML's page-layout codes still helps, now you can point-and-click to create web pages with easy-to-use web-authoring tools. Moreover, many locations on the Internet will post your web pages for free or at low cost.

Web-Authoring Tools

All the following let you create web pages, using icons and menus:[a]

- **Internet service providers, portals, and other free sources:** Internet service providers such as America Online, Earthlink, AT&T WorldNet Service, and Juno Web offer simple page-creation programs that let you type in text, specify background colors, and submit photographs. Often this service is free when you sign up with the ISP.

    Portals such as Yahoo!, Lycos, Go Network, and About.com also allow you to create your website at no charge, if you're willing to register (giving personal information) and allow the portal to run advertising on your site.

    With these programs from ISPs and portals, you can create a web page in a couple of hours. However, the appearance of your page will be limited in the number of colors, borders, fonts, and page templates available. Another alternative is Homestead.com, a free web publishing service that offers many specific templates.

- **Browsers:** Both Netscape Navigator and Microsoft Explorer offer web-authoring tools. For example, Netscape provides Composer, which allows you to design web pages by choosing items from menus; the program automatically inserts HTML tags for each component you select.

- **Commercial application software:** Most commercial application programs allow you to save documents in HTML format. With the word processing program Microsoft Word, you can use the Help feature to find out how to translate your document into a web page, although not all elements will look the same.

- **Web-authoring software:** Several web-authoring software packages available for $50—$150 can serve not only professional webmasters but also beginners. In this category are Microsoft FrontPage, Claris Home Page, Adobe Page Mill, Corel Web Designer, IBM's TopPage, Ixla's WebEasy, and Macromedia Dreamweaver.

Creating a Good Web Page

The design templates and options offered allow you to do a lot of things. But your objective is to get people to *use* your website. Some suggestions:

- **Decide what you want to say:** The first thing to do is decide on the purpose of your website. Entertain your friends? Post news for your extended family? Attract customers to your small business? You can get ideas by looking at other people's sites, but ultimately you need to have a clear idea of the goal of your site and know what you want to say.[b]

- **Be concise:** "For the next several years, the vast majority of users will access the Internet through slow modems," writes Jakob Nielsen, author of *Designing Web Usability: The Practice of Simplicity.* "All [web] pages must download quickly, or users may not only become reluctant to follow the links, but they may also have trouble navigating the site and finding areas they've previously visited. People get lost more often on slow sites than on fast ones and are more likely to leave them and never come back."[c] The implications: People want to get their information quickly, so you have to write concise text that is easy to scan, using highlighted keywords, subheads, and bulleted lists. Give facts, not fluff. Be aware that most users will spend only a few seconds at the site.

- **Build links:** The software allows you to simply highlight a block of text, then type in the address of any link you want. Links, of course, are what will help people find your site, using search engines or connections from other web pages.

- **Make your site follow good design principles:** Avoid going hog wild with crazy type fonts. Don't put in sound files that play automatically. Use black type for extended sections of text. Don't have bloated graphics, such as files exceeding 30 kilobytes, which take too long to load. (Incidentally, if you don't have a scanner, you can scan photos onto a floppy disk at a copy shop such as Kinko's, then upload them to the site later.)

Publishing Your Web Page

Once you've created your website, you'll need to "publish" it—upload it to a web server for it to be viewed on the Internet. You can get upload instructions from your ISP. Some ISPs will give you free space on their servers.

Note: Don't include any information on your site that you wouldn't put on a sign on your front door. That includes the names or location of family members shown in any photographs, especially those of children.

Notes

Chapter 1

1. Kevin Maney, "In the Future, You'll Pluck Your Info from Thin Air," *USA Today*, July 20, 2001, pp. 1B, 2B.
2. Kevin Maney, "Net's Next Phase Will Weave Through Your Life," *USA Today*, March 2, 2001, pp. 1B, 2B.
3. Don Clark, "Managing the Mountain," *Wall Street Journal*, June 21, 1999, p. R4; "Does America Have ADD?" *U.S. News & World Report*, March 26, 2001, p. 14; and Jon Swartz, "Email Overloads Taxes Workers and Companies," *USA Today*, June 26, 2001, p. 1A.
4. Dave Wilson, "Some Are Losing It, Bit by Bit," *Los Angeles Times*, July 17, 2001, pp. A1, A8.
5. Katie Hafner, "Teenage Overload, or Digital Dexterity?" *New York Times*, April 12, 2001, pp. D1, D5.
6. Fred Abatemarco, "Hand Weapons of the Modern Age," *Popular Science*, September 1999, pp. 9–10.
7. Vivian S. Toy, "Teen-Agers and Cell Phones: It's All Talk, and All the Time," *New York Times*, August 2, 1999, pp. A1, A17.
8. B. Chaney, "Stupid Things About Smart Phone," *Fortune*, Summer 2001, p. 26.
9. Michael Specter, "Your Mail Has Vanished," *New Yorker*, December 6, 1999, pp. 96–103.
10. Data from International Data Corp., in "Like It or Not, You've Got Mail," *Business Week*, October 4, 1999, pp. 178–184.
11. "Like It or Not, You've Got Mail," 1999.
12. Robert Rossney, "E-Mail's Best Asset–Time to Think," *San Francisco Chronicle*, October 5, 1995, p. E7.
13. Adam Gopnik, "The Return of the Word," *New Yorker*, December 6, 1999, pp. 49–50.
14. Peter H. Lewis, "The Good, the Bad, and the Truly Ugly Faces of Electronic Mail," *New York Times*, September 6, 1994, p. B7.
15. Gopnik, 1999.
16. David A. Whittle, *Cyberspace: The Human Dimension* (New York: W. H. Freeman, 1997).
17. David A. Whittle, quoted in "Living Online," *The Futurist*, July–August 1997, p. 54.
18. Edward Iwata, "The Net at 30," *USA Today*, December 14, 1999, pp. 1A, 2A.
19. Kevin Maney, "The Net Effect: Evolution or Revolution?" *USA Today*, August 9, 1999, pp. 1B, 2B.
20. Center for Communication Policy (www.ccp.ucla.edu), reported in David Plotnikoff, "Study Asks How Wired World Has Changed the Way We Live," *San Jose Mercury News*, June 13, 1999, pp. 1F, 7F.
21. December 10–13, 1999, survey by The Strategis Group, reported in Dru Sefton, "The Big Online Picture: Daily Web Surfing Now the Norm," *USA Today*, March 22, 2000, p. 3D.
22. Teenage Research Unlimited, reported in Eryn Brown, "The Future of Net Shopping? Your Teens," *Fortune*, April 12, 1999, p. 152.
23. Kevin Murphy, Gartner Group, quoted in Timothy J. Mullaney, "Death to E-Words Everywhere," *Business Week*, November 15, 1999, p. 10.
24. Adapted from Dan Gillmor, "Electronic Appliances We Have—We Need Embedded Values, Too," *San Jose Mercury News*, October 3, 1999, pp. 1E, 3E.
25. Andy Reinhardt, Steven V. Brull, Peter Burrows, and Catherine Yang, "The Soul of a New Refrigerator," *Business Week*, January 17, 2000, p. 42; Cox News Service, "Internet Can Be Found in Tiny Places," *San Francisco Chronicle*, October 11, 1999; and Steven Butler, "Smart Toilets and Wired Refrigerators," *Newsweek*, June 7, 1999, p. 48.
26. David Einstein, "Custom Computers," *San Francisco Chronicle*, April 15, 1999, pp. B1, B3.
27. Einstein, 1999.
25. Tammy Joyner, "Turning Old Jobs into Hot Ones," *San Francisco Chronicle*, January 17, 2000, p. B3.
28. Laurence Hooper, "No Compromises," *Wall Street Journal*, November 16, 1992, p. R8.
29. Tom Forester and Perry Morrison, *Computer Ethics: Cautionary Tales and Ethical Dilemmas in Computing* (Cambridge, MA: The MIT Press, 1990), pp. 1–2.
30. Norman Solomon, "The Media's Role in the Commercialization of Cyberspace," *San Francisco Chronicle*, January 27, 2000, p. A25.
31. Jane Costello, "Airlines Find Lost Cellphones Are a Real Hang-Up," *Wall Street Journal*, August 7, 2001, p. B7.
32. Sandra Blakeslee, "Car Calls May Leave Brain Short-Handed," *New York Times Magazine*, April 6, 2001, pp. 72–74.
33. E. Z. Zechmeister and S. E. Nyberg, *Human Memory: An Introduction to Research and Theory* (Pacific Grove, CA: Brooks/Cole, 1982).
34. F. P. Robinson, *Effective Study*, 4th ed. (New York: Harper & Row, 1970).
35. B. K. Broumage and R. E. Mayer, "Quantitative and Qualitative Effects of Repetition on Learning from Technical Text," *Journal of Educational Psychology*, 1982, 78, 271–278.
36. R. J. Palkovitz and R. K. Lore, "Note Taking and Note Review: Why Students Fail Questions Based on Lecture Material," *Teaching of Psychology*, 1980, 7:159–161.
37. J. Langan and J. Nadell, *Doing Well in College: A Concise Guide to Reading, Writing, and Study Skills* (New York: McGraw-Hill, 1980), pp. 93–100.

Bookmark It! Box

a. Study by Impulse Research for *Iconoclast* newsletter, reported in D. Plotnikoff, "E-Mail: A Critical Medium Has Reached Critical Mass," *San Jose Mercury News*, April 4, 1999, pp. 1F, 2F.
b. K. Clark, "At Least the Coffee and Pens Are Still Free," *U.S. News & World Report*, June 7, 1999, p. 70.
c. S. Shostak, "You Call This Progress?" *Newsweek*, January 18, 1999, p. 16.
d. A. Markels, "Don't Manage by E-Mail," *San Francisco Examiner*, August 11, 1996, p. B-5, reprinted from *Wall Street Journal*; "Don't Overuse Your E-Mail," *CPA Client Bulletin*, March 1999, p. 3; B. Fryer, "E-Mail: Backbone of the Info Age or Smoking Gun?" *Your Company*, July/August 1999, pp. 73–76; S. Armour, "Boss: It's in the E-Mail," *USA Today*, August 10, 1999, p. 3B; L. Guernsey, "Attachments #@%&#@ Are Full #+@&*¢# of Surprises," *New York Times*, July 22, 1999, p. D11.

Chapter 2

1. Graham T. T. Molitor, "Five Forces Transforming Communications," *The Futurist*, September-October 2001, pp. 32–37.
2. Forrester Research, cited in Molitor, 2001
3. Charles Smith, "Ready or Not, Here Comes . . . Ubicomp," *San Francisco Chronicle*, August 11, 2001, pp. 1, 4.
4. Molitor, 2001, p. 33.
5. Some of this discussion was adapted from Kate Murphy, "Cruising the Net—in Hyperdrive," *Business Week*, January 24, 2000, pp. 170–172.
6. William J. Holstein and Fred Vogelstein, "You've Got a Deal!" *U.S. News & World Report*, January 24, 2000, pp. 34–40.
7. Alec Klein, "A Revolution in a Twinkling," *Washington Post*, July 26, 2001, p. E1.
8. Michelle Slatalla, "The Office Meeting That Never Ends," *New York Times*, September 23, 1999, pp. D1, D8.
9. Jeanne Hinds, quoted in Slatalla , 1999.
10. September 2000 IDC study, reported by Sally McGrane, "A Little E-Mail (Or a Lot of It) Eases the Workday," *New York Times*, March 8, 2001, p. D8.
11. Editors of *PC Computing*, "End E-Mail Insanity Forever," *PC Computing*, November 1999, pp. 170–198; Jennifer Powell, "E-Mail: Common Sense Plays a Big Role," *Smart Computing*, March 2000, p. 47; and Kenneth M. Morris, *User's Guide to the Information Age* (New York: Lightbulb Press, 1999), pp. 111, 153.
12. David Lazarus, "Fan Spam Is Hard to Shake," *San Francisco Chronicle*, February 7, 2000, pp. C1, C2.
13. Julian Haight, quoted in Lazarus, 2000, p. C1.
14. Tom Spring, "First Look at Netscape 6.1," *PC World*, July 13, 2001.
15. Weise, "Web Changes Direction to People Skills," 2000; and Saul Hansell, "Obsessively Independent, Yahoo Is the Web's Switzerland," *New York Times*, August 23, 1999, pp. C1, C10.
16. Janet Kornblum, "Portals Suffer as Internet Surfers Get More Savvy," *USA Today*, July 2, 1999, p. 1B; and Dan Gillmor, "Small Portals Prove that Size Matters," *San Jose Mercury News*, December 6, 1998, pp. 1E, 3E..
17. Elizabeth Weise, "Successful Net Search Starts with a Need," *USA Today*, January 24, 2000, p. 3D; Timothy Hanrahan, "The Best Way to . . . Search Online," *Wall Street Journal*, December 6, 1999, p. R25, and Weise, "Web Changes Direction to People Skills," 2000.
18. Timothy Hanrahan, "The Best Way to Search Online," *Wall Street Journal*, December 6, 1999, p. R25; Reva Basch, "Cutting Through the Clutter," *Smart Computing*, July 1999, pp. 88–91; Matt Lake and Dylan Tweney, "Find It on the Web," *PC World*, June 1999, pp. 168–182; and Matt Lake, "Desperately Seeking Susan OR Suzie NOT Sushi," *New York Times*, September 3, 1998, p. D1.
19. Lake, 1998.
20. Lake and Tweney, 1999.
21. Laurie Bryan, "Wired for Shopping," *San Francisco Chronicle*, February 9, 2000, zone 7, pp. 1, 4.
22. Morris, 1999, p. 100.
23. Megan Doscher, "The Best Way to Find Love," *Wall Street Journal*, December 6, 1999, p. R34.
24. Sharon Cleary, "The Best Way to Find an Old Friend," *Wall Street Journal*, December 6, 1999, pp. R26, R45.
25. Wendy M. Grossman, "Language Is a Virus," *PC Computing*, March 2000, p. 62.
26. William M. Bulkeley, "The Best Way to Go to School," *Wall Street Journal*, December 6, 1999, pp. R18, R22.
27. Bulkeley, 1999; and Faith Bremner, "On-line College Classes Get High Marks Among Students," *USA Today*, November 16, 1998, p. 16E.
28. Lorrie Grant, "Internet Has Become Integral Part of Everyday Life," *USA Today*, April 21, 1999, p. 6B.
29. Haya El Nasser, "Main Street Enters Mainstream," *USA Today*, November 16, 1999, p. 3A.
30. Denise Caruso, "On-Line Day Traders Are Starting to Have an Impact on a Few Big-Cap Internet Stocks in What Some Call a 'Feeding Frenzy,'" *New York Times*, December 14, 1998, C3.

31. Gary McWilliams, "The Best Way to Find a Job," *Wall Street Journal*, December 6, 1999 pp. R16, R22.
32. Del Jones, "E-Purchasing Saves Businesses Billions," *USA Today*, February 7, 2000, pp. 1B, 2B.
33. Ellen Laird, "Internet Plagiarism: We All Pay the Price," *Chronicle of Higher Education*, June 13, 2001, p. B5.
34. Bruce Leland, quoted in Peter Applebome, "On the Internet, Term Papers Are Hot Items," *New York Times*, June 8, 1997, sec. 1, pp. 1, 20.
35. Eugene Dwyer, "Virtual Term Papers" [letter], *New York Times*, June 10, 1997, p. A20.
36. John Yankey, "Cheating, Digital Plagiarism Rise as Tech Hits Classroom," *Reno Gazette-Journal*, July 30, 2001, p. 23E.
37. William L. Rukeyser, "How to Track Down Collegiate Cyber-Cheaters" [letter], *New York Times*, June 14, 1997, sec. 4, p. 14.
38. David Rothenburg, "How the Web Destroys the Quality of Students' Research Papers," *Chronicle of Higher Education*, August 15, 1997.

Bookmark It! Box

a. Karen Jacobs, "The Best Way to Pick a Provider," *Wall Street Journal*, December 6, 1999, pp. R8, R10; Tracy Baker, "Free Access to the Internet," *Smart Computing*, June 1999, pp. 44–46; Tina Kelley, "Choosing an ISP: Convenience, Cost, and Service," *San Jose Mercury News*, May 30, 1999, pp. 1F, 4F; reprinted from the *New York Times*; and "Going Online: Best Ways to Get Started," *Consumer Reports*, July 1998, p. 64.

Chapter 3

1. Alan Robbins, "Why There's Egg on Your Interface," *New York Times*, December 1, 1996, sec. 3, p. 12.
2. Joshua Quittner, "Aqua: The Movie," *Time*, January 31, 2000, p. 82.
3. Hamm, Burrows, and Reinhardt, 2000; and Lawrence M. Fisher, "Sun Plans to Start Shipping Operating System Next Month," *New York Times*, January 27, 2000, p. C12.
4. Irving Wladawsky-Berger, quoted in Deborah Solomon, "Could Linux Outdo Windows?" *USA Today*, March 9, 2000, pp. 1B, 2B.
5. Stanley Holmes, "Companies Form Group to Promote Linux," *Los Angeles Times*, March 10, 2000, pp. C2, C4; Sam Jaffe, "Will Linux Investors Be Left Out in the Cold?" *Business Week*, March 2000, pp. 178–180; David Kirkpatrick, "Dell to Wintel: Your Hegemony Is Over," *Fortune*, February 21, 2000, pp. 50, 52; Lawrence M. Fisher, "Looking for a New Life in Linux," *New York Times*, February 14, 2000, p. C4; Lawrence M. Fisher, "A View that Needs No Windows," *New York Times*, February 6, 2000, sec. 3, p. 2; and Bloomberg News, "Linux-Related Shares Soar on News of Revised Version," *Los Angeles Times*, February 3, 2000, p. C7.
6. http://www.itworld.com/Comp/2384/NWW010312005048151
7. David Rynecki, "Is Palm's IPO Really the One to Catch?" *Fortune*, February 7, 2000, pp. 213–214; Roy Furchgott, "Using the Net to Soup Up Your Palm Top Computer," *Business Week*, November 8, 1999, pp. 165–166; Walter S. Mossberg, "A Pilot Rival Organizes Your Life, Then Morphs into Something Else," *Wall Street Journal*, September 16, 1999, p. B1; and Kevin Maney, "Palm Has the Whole World in Its Hand," *USA Today*, September 14, 1999, pp. 1B, 2B.
8. Margaret Trejo, quoted in Richard Atcheson, "A Woman for Lear's," *Lear's*, November 1993, p. 87.
9. John Ennis, quoted in Peter Plagens and Ray Sawhill, "Throw Out the Brushes," *Newsweek*, September 1, 1997, pp. 76–77.
10. Anita Hamilton, "Scheduling Snafu," *Time*, May 10, 1999, p. 96.
11. Meredith McCarty Whalen, Clare, Gillan, Amy Mizoras, "ASPs: Delivering Applications as a Service," an IDC White Paper, 2000, www.itpapers.com
12. J. William Gurley, "The New Market for 'Rentalware,'" *Fortune*, May 10, 1999, p. 142.
13. Gary Bloom, quoted in Lawrence M. Fisher, "Software Evolving into a Service Rented Off the Net," *New York Times*, December 20, 1999, p. C37.

Bookmark It! Box

a. "XP and OS X," *Consumer Reports*, October 2001, p. 59; David Pogue, "Windows XP: Microsoft's New Look for Fall, in Size XXL," *New York Times*, September 6, 2001, pp. D1, D9; J. D. Biersdorfer, "Dear User: This Bootleg Copy Will Self-Destruct in 30 Days," *Family PC*, September 2001, pp. 48–49; Lawrence J. Magid, "Yet Another Step to Start Windows XP," *Los Angeles Times*, July 19, 2001, p. T8; Walter S. Mossberg, "Microsoft Cracks Down on Sharing Windows among Home Users," *Wall Street Journal*, July 5, 2001, p. B1; Michael J. Miller, "Windows XP: Worth the Wait?" and "The Incredible Expanding OS," *PC Magazine*, May 8, 2001, p. 7; J. P. Vellotti, "XP," *PC Magazine*, May 8, 2001, pp. 120–130; Oliver Kaven, "Mac OS," *PC Magazine*, May 8, 2001, p. 126; and Edward C. Baig, "New Windows Puts Emphasis on Ease," *Reno Gazette-Journal*, April 30, 2001, p. 3E, reprinted from *USA Today*.

Chapter 4

1. Michael S. Malone, "The Tiniest Transformer," *San Jose Mercury News*, September 10, 1995, pp. 1D, 2D; excerpted from *The Microprocessor: A Biography* (New York: Telos/Springer Verlag, 1995).
2. Malone, 1995.
3. Laurence Hooper, "No Compromises," *Wall Street Journal*, November 16, 1992, p. R8.
4. Data from PC Data Inc., cited by Gary McWilliams, "Reversing Course, Home-PC Prices Head Higher," *Wall Street Journal*, January 13, 2000, pp. B1, B4; and John Simons, "Cheap Computers Bridge Digital Divide," *Wall Street Journal*, January 27, 2000, p. A22.
5. Henry Norr, "The NC—with a Twist," *San Francisco Chronicle*, February 28, 2000, pp. B1, B2.
6. Henry Norr, "Why Thin Computing Is In," *San Francisco Chronicle*, February 28, 2000, p. B2.
7. Walter S. Mossberg, "Mossberg's Mailbox," *Wall Street Journal*, April 13, 2000, p. B9.
8. Edward Baig, "Be Happy, Film Freaks," *Business Week*, May 26, 1997, pp. 172–173.
9. Keith L. Alexander, "DVD Sales Energize Home Video Market," *USA Today*, February 28, 2000, pp. C1, C6.

Bookmark It! Box

a. Walter S. Mossberg, "How to Buy a Laptop: Some Basic Guidelines in a Dizzying Market," *Wall Street Journal*, October 21, 1999, p. B1.
b. Stephen H. Wildstrom, "How to Shop for a Laptop," *Business Week*, April 3, 2000, p. 25.
c. "The New Laptops," *Consumer Reports*, May 2000, pp. 12–16; Walter S. Mossberg, "Buying Your Next PC? Get the Most Memory, Not the Fastest Chip," *Wall Street Journal*, April 6, 2000, p. B1; and Bill Howard, "Notebook PCs," *PC Magazine*, August 1999, pp. 154–155.

Chapter 5

1. Lorrie Grant, "Let Your Fingers Do Shopping . . . in Store," *USA Today*, July 28, 1999, p. 3B.
2. Salina Khan, "Kiosks Offer Maps, Hotel, Dining Info," *USA Today*, April 4, 2000, p. 5B.
3. Lorrie, Grant, "Kiosks Let You Bypass Check-In Lines," *USA Today*, p. 7E, April 10, 2001.
4. "Coming to an ATM Near You: Movie Previews," *Reno Gazette-Journal*, May 2, 2000, pp. 1E, 6E; Marc Gunther, "Take Your $20, and a Coupon," *Fortune*, April 3, 2000, p. 48; and "Cash Crop," *New York Times Magazine*, August 15, 1999, p. 23.

Bookmark It! Box

a. Marty Jerome, "Boot Up or Die," *PC Computing*, April 1998, pp. 172–86.

Chapter 6

1. "What Does 'Digital' Mean in Regard to Electronics?" *Popular Science*, August 1997, pp. 91–94.
2. Simon Romero, "Weavers Go Dot-Com, and Elders Move In," *New York Times*, March 28, 2000, pp. A1, A4.
3. U.S. Department of Labor, cited in Stephanie Armour, "Telecommuting Gets Stuck in the Slow Lane," *USA Today*, June 25, 2001, pp. 1A, 2A.
4. Armour, June 25, 2001; Stephanie Armour, "More Bosses Keep Tabs on Telecommuters," *USA Today*, July 24, 2001, p. 1B; Jim Hopkins, "How Solo Workers Keep from Feeling Deserted," *USA Today*, May 9, 2001, p. 9B; Bonnie Harris, "Companies Turning Cool to Telecommuting Trend," *Los Angeles Times*, December 28, 2000, pp. A1, A6; and Kemba J. Dunham, "Telecommuters' Lament," *Wall Street Journal*, October 31, 2000, pp. B1, B18.
5. David Kline, quoted in W. James Au, "The Lonely Long-Distance Worker," *PC Computing*, February 2000, pp. 42–43.
6. David Leonhardt, "Telecommuting to Pick Up as Workers Iron Out Kinks," *New York Times*, December 20, 1999, p. C6.
7. Maryanne Murray Buechner, "Superconnected," *Time*, March 22, 1999.
8. Mike Romano, "Brave New Home," *U.S. News & World Report*, April 5, 1999, pp. 60–62.
9. Dori Jones Yang, "A Boob Tube with Brains," *U.S. News & World Report*, March 13, 2000, pp. 42–43.
10. Hank Hogan, "HDTV Is Here, but Can You Afford It?" *High-Tech Careers*, June/July 2000, pp. 17–20.
11. Neil Hickey, "The Digital Newsroom: Ready or Not," *Columbia Journalism Review*, February 2000, p. 56.
12. David Perlman, "Satellite Network Captures Volcano Drama in Hawaii," *San Francisco Chronicle*, February 17, 1997, p. A4.
13. Stephen P. Rizzo, "Why Is Broadband So Narrow?" *Forbes*, September 10, 2001, pp. 50–52; Randall E. Stross, "America's Bad Call," *U.S. News & World Report*, September 4, 2000, p. 47; Edward Harris, "As 3G Waits in Wings, Need for 2G Gear May Rise," *Wall Street Journal*, August 16, 2001, p. B3; Stephen Komarow, "Oh the Places Your Cellphone Can Go," *USA Today*, June 26, 2001, p. 14E; Stephen H. Wildstrom, "Tearing Down the Wireless Babel," *BusinessWeek*, June 18, 2001, p. 28; Andy Reinhardt, "Wireless Web Woes," *BusinessWeek e.biz*, June 4, 2001, pp. EB22–EB27; Eric Knorr, "Mobile Web vs. Reality," *Technology Review*, June 2001, pp. 56–61; Marty Jerome, "3G Cometh Not So Fast," *Smart Business*, June 2001, p. 64; Mary E. Behr and Angela Graven, "Choose Your Weapon," *PC Magazine*, April 24, 2001, pp. 111–135; Andy Reinhardt, "All That Money on 3G—and for What?" *BusinessWeek*, March 26, 2001, p. 60; Kevin Maney, "Bombarded by Wireless Technology," *USA Today*, March 21, 2001, p. 6B.
14. Stephen H. Wildstrom, "Why Bluetooth Has a Black Eye," *BusinessWeek*, July 2, 2001, p. 24; Edward C. Baig, "The Era of L:iving Wirelessly," *USA Today*, June 26, 2001, pp. 1E,

2E; "Wireless Tangle," *Popular Science,* May 2001, p. 40; Jared Sandberg, "Raft of New Wireless Technologies Could Lead to Airwave Gridlock," *Wall Street Journal,* January 1, 2001, pp. B1, B10; Stephen H. Wildstrom, "Teething Pains for Bluetooth," *BusinessWeek,* February 12, 2001, p. 20.
15. Yahoo!, cited in Del Jones, "Cyber-porn Poses Workplace Threat," *USA Today,* November 27, 1995, p. B1.
16. Lawrence J. Magid, "Be Wary, Stay Safe in the On-line World," *San Jose Mercury News,* May 15, 1994, p. 1F.
17. Peter H. Lewis, "Limiting a Medium without Boundaries," *New York Times,* January 15, 1996, pp. C1, C4.
18. John M. Broder, "Making America Safe for Electronic Commerce," *New York Times,* June 22, 1997, sec. 4, p. 4; Margaret Mannix and Susan Gregory Thomas, "Exposed Online," *U.S. News & World Report,* June 23, 1997, pp. 59–61; and Noah Matthews, "Shareware," *San Jose Mercury News,* October 12, 1997, p. 4F.
19. Survey by Equifax and Louis Harris & Associates, cited in Bruce Horovitz, "80% Fear Loss of Privacy to Computers," *USA Today,* October 31, 1995, p. 1A.

Chapter 7
1. John Chambers, quoted in Del Jones and Beth Belton, "Cisco Chief: Virtual Close to Hit Big," *USA Today,* October 12, 1999, p. 3B.
2. James A. Larson, *Database Directories* (Upper Saddle River, NJ: Prentice Hall PTR, 1995)
3. Michael J. Hernandez, *Database Design for Mere Mortals,* copyright 1997 by Michael J. Hernandez, Addison-Wesley, Reading, MA: p. 8.
4. Hernandez, p. 11.
5. Hernandez, p. 12
6. See J. Quittner, "Tim Berners-Lee," *Time,* March 29, 1999, pp. 193–194.
7. Forrester Research Inc., reported in "E-Commerce: It's Clicking" [editorial], *Business Week,* January 11, 1999, p. 154.
8. Sarah E. Hutchinson and Stacey C. Sawyer, *Computers, Communications, and Information: A User's Introduction,* rev. ed. (Burr Ridge, IL: Irwin/McGraw-Hill, 1998), pp. E1.1–E1.3.
9. Jeff Bezos, quoted in K. Southwick, interview, October 1996, *www.upside.com.*
10. D. Levy, "On-line Gamble Pays Off with Rocketing Success," *USA Today,* December 24, 1998, pp. 1B, 2B.
11. Richard Williamson, "Happy Holidays," *ZDNet,* October 3, 2001.
12. Jonathan Berry, John Verity, Kathleen Kerwin, and Gail DeGeorge, "Database Marketing," *BusinessWeek,* September 5, 1994, pp. 56–62.
13. Cheryl D. Krivda, "Data-Mining Dynamite," *Byte,* October 1995, pp. 97–103.
14. Krivda, 1995.
15. Edmund X. DeJesus, "Data Mining," *Byte,* October 1995, p. 81.
16. Forester Research, cited in Denise Caruso, "Taking Stock of the Differences Between the Consumer Internet Market and Its Business-to-Business Cousin," *New York Times,* February 28, 2000, p. C5; and Gartner Group, cited in William J. Holstein, "Rewiring the 'Old Economy,'" *U.S. News & World Report,* April 10, 2000, pp. 38–40.
17. Don Tapscot, "Virtual Webs Will Revolutionize Business," *Wall Street Journal,* April 24, 2000, p. A38.
18. Carolyn Said, "Online Middlemen," *San Francisco Chronicle,* April 10, 2000, pp. C1, C3; Holstein, 2000; Claudia H. Deutsch, "Another Economy on the Supply Side," *New York Times,* April 8, 2000, pp. B1, B4; and Kelly Zito, "Online Exchange for Shops," *San Francisco Chronicle,* March 9, 2000, pp. B1, B4.
19. Tapscot, 2000.
20. William Safire, "Art vs. Artifice," *New York Times,* January 3, 1994, p. A11.
21. Cover, *Newsweek,* June 27, 1994; and cover, *Time,* June 27, 1994.
22. Jonathan Alter, "When Photographs Lie," *Newsweek,* July 30, 1990, pp. 44–45.
23. Fred Ritchin, quoted in Alter, 1990.
24. Robert Zemeckis, cited in Laurence Hooper, "Digital Hollywood: How Computers Are Remaking Movie Making," *Rolling Stone,* August 11, 1994, pp. 55–58, 75.
25. Woody Hochswender, "When Seeing Cannot Be Believing," *New York Times,* June 23, 1992, pp. B1, B3.
26. Bruce Horowitz, "Believe Your Eyes? Ads Bend Reality," *USA Today,* April 24, 2000, pp. 1B, 2B.
27. Penny Williams, "Database Dangers," *Quill.* July/August 1994, pp. 37–38.
28. Lynn Davis, quoted in Williams, 1994.

Bookmark It! Box
a. Kathryn Rambo, quoted in Ramon G. McLeod, "New Thieves Prey on Your Very Name," *San Francisco Chronicle,* April 7, 1997, pp. A1, A6.
b. Rambo, quoted in T. Trent Gegax, "Stick 'Em Up! Not Anymore. Now It's Crime by Keyboard," *Newsweek,* July 21, 1997, p. 14.
c. McLeod, 1997.

Chapter 8
1. Katie Hafner, "We're Not All Connected, Yet," *New York Times,* January 27, 2000, pp. D1, D9.
2. Government Accounting Office study, reported in "Government Study Finds Digital Divide Narrowing in United States," *San Francisco Chronicle,* February 23, 2001, p. B3.
3. Pew Internet and American Life Project, reported in "Number of New Internet Users Is Growing," *New York Times,* February 18, 2001, p. C1; Anick Jesdanum, "56 Percent of U.S. Adults Use Internet, Study says," *San Francisco Chronicle,* February 19, 2001, pp. B1, B3; and Janet Kornblum, "Web Users Look Like America, Only Richer," *USA Today,* February 19, 2001, p. 3D.
4. Michael Dertouzos, "The Net Revolution Spawns a 'Fast Caste,'" *Los Angeles Times,* January 20, 2000, p. A15.
5. Survey by Inktomi Corp. and NEC Research Institute, reported in Ashley Dunn, "It's a Very Wide Web: 1 Billion Pages' Worth," *Los Angeles Times,* January 20, 2000, p. C7.
6. Alan Murray, "Trying to Make World Safe for E-Commerce," *Wall Street Journal,* November 29, 1999, p. A1.
7. Scott Carlson, "High-Speed Network Will Serve Universities for 3 More Years," *Chronicle of Higher Education,* April 21, 2000, p. A49.
8. Laurence H. Tribe, "The FCC vs. the Constitution," *Wall Street Journal,* September 5, 1997, p. A8.
9. John M. Broder, "Let It Be," *New York Times,* June 30, 1997, pp. C1, C9.
10. *A Framework for Global Electronic Commerce,* quoted in Steven Levy, "Bill and Al Get It Right," *Newsweek,* July 7, 1997, p. 80.
11. Mike France, "What's in a Name.com? Plenty," *Business Week,* September 6, 1999, pp. 86–90.
12. Randall E. Stross, "Dot Unamelt," *U.S. News & World Report,* July 2, 2001, p. 41; Gary Chapman, "What's in a Web Domain Name? For a System Under Strain, It Spells Trouble," *Los Angeles Times,* June 28, 2001, p. B3.
13. Thomas E. Weber, "How a Phone Number May One Day Become Your Internet Address," *Wall Street Journal,* January 8, 2001, p. B1.
14. We are grateful to Prof. John Durham for contributing these ideas.
15. John Allen Paulos, "Smart Machines, Foolish People," *Wall Street Journal,* October 5, 1999, p. A26.
16. Katherine S. Mangan, "Online Medical-School Application Becomes a Nightmare for Students," *Chronicle of Higher Education,* July 13, 2001, p. A33; and Sally McGrane, "Glitches Stymie Medical School Applicants," *New York Times,* July 5, 2001, p. D3.
17. Janet Rae-Dupree and Richard J. Newman, "A Twisted Kind of Love," *U.S. News & World Report,* May 15, 2000, p. 24; Brad Stone, Mark Hosenball, and Stefan Theil, "Bitten by Love," *Newsweek,* May 15, 2000, pp. 42–43; and Lev Grossman et al., "Attack of the Love Bug," *Time,* May 15, 2000, pp. 49–56.
18. Edward Iwata, "Mutating Computer Virus Hits," *USA Today,* May 19, 2000, p. 1A.
19. Henry Norr, "Vigilance Against Viruses," *San Francisco Chronicle,* September 24, 2001, pp. E1,E2.
20. J. D. Biersdorfer, "Ridding the Computer of Malicious Worm," *New York Times,* August 2, 2001, p. D4; and Jon Swartz, "Nimda Called Most Serious Internet Attack on Business," *USA Today,* September 26, 2001, p. 5B.
21. Steven Bellovin, cited in Jane Bird, "More Than a Nuisance," *The Times* (London), April 22, 1994, p. 31.
22. Eugene Carlson, "Some Forms of Identification Can't Be Handily Faked," *Wall Street Journal,* September 14, 1993, p. B2.
23. Justin Matlkick, "Security of Online Markets Could Well Be at Stake," *San Francisco Chronicle,* September 16, 1997, p. A21.
24. John Holusha, "The Painful Lessons of Disruption," *New York Times,* March 17, 1993, pp. C1, C5.
25. The Enterprise Technology Center, cited in "Disaster Avoidance and Recovery Is Growing Business Priority," special advertising supplement in *LAN Magazine,* November 1992, p. SS3.
26. John Holusha, "The Painful Lessons of Disruption," *New York Times,* March 17, 1993, pp. C1, C5.
27. The Enterprise Technology Center, cited in "Disaster Avoidance and Recovery Is Growing Business Priority," special advertising supplement in *LAN Magazine,* November 1992, p. SS3.
28. David Lazarus, "Toxic Technology," *San Francisco Chronicle,* December 3, 2000, pp. B1, B4, B8.
29. Associated Press, "Disposal Is a Problem as LCDs Displace Tube Monitors," *Wall Street Journal,* August 13, 2001, p. B6; Henry Norr, "Drowning in E-Waste," *San Francisco Chronicle,* May 27, 2001, pp. E1, E5; Lawrence Magid, "Don't Be So Quick to Toss Out Your PC," *San Francisco Chronicle,* October 29, 2000, pp. D-5, D-7; John Yaukey, "Upgrading the Junk Computer Contagion," *Reno Gazette-Journal,* August 28, 2000, p. 4E: and William J. Holstein, "Take My Personal Computer—Please!" *U.S. News & World Report,* June 5, 2000, p. 51.
30. National Safety Council, reported in Norm Alster and William Echikson, "Are Old PCs Poisoning Us?" *BusinessWeek,* June 12, 2000, pp. 78, 80.
31. Environmental Helth Center, National Safety Council, reported in John R. Quain, "Upgrades Create Digital Landfill," *Popular Science,* February 2001, p. 38.
32. David Ferrell, "Electricity a Mystery to Many Consumers," *Los Angeles Times,* December 28, 2000, pp. A1, A11.
33. Todd Wallack, "Troubles Crop Up for Server Farms," *San Francisco Chronicle,* June 4, 2001, pp. B1, B2; Jon Swartz and Michelle Kessler, "Power Woes Could Leave Web Surfers in the Dark," *USA Today,* May 9, 2001, p. 1B; Neal Templin, "Power-Hungry Web 'Server Farms' Find Cooler Reception in California," *Wall Street Journal,* February 28, 2001, pp. B1, B4; and Barnaby J. Feder, "Digital Economy's Demand for Steady Power Strains

Utilities," *New York Times*, July 3, 2000, pp. C1, C4.
34. Queena Sook Kim, "Nation's Aging Grid May Cramp Power-Delivery Growth," *Wall Street Journal*, August 14, 2001, p. B2; Roger Anderson, "Wattage Where It's Needed," *New York Times*, June 6, 2001, p. A31; Charles Wardell, "Blackout," *Popular Science*, May 2001, pp. 63–67; and Peter Huber and Mark Mills, "Got a Computer?" More Power to You," *Wall Street Journal*, September 7, 2000, p. A26.
35. Lutz Erbring, coauthor of Stanford University survey of 4113 people about Internet impact on daily activities, quoted in Joellen Perry, "Only the Cyberlonely," *U.S. News & World Report*, February 28, 2000, p. 62.
36. Study by Robert Kraut, *Journal of Social Issues*, reported in Deborah Mendenhall, "Web Doesn't Promote Isolation, Study Says," *San Francisco Chronicle*, August 22, 2001, p. C3, reprinted from *Pittsburgh Post-Gazette*; and Lisa Guernsey, "Cyberspace Isn't So Lonely After All," *New York Times*, July 26, 2001, pp. D1, D5. Also see study by Jeffrey Cole, UCLA Center for Communication Policy, reported in Greg Miller and Ashley Dunn, "Net Does Not Exact a Toll on Social Life, New Study Finds," *Los Angeles Times*, October 26, 2000, pp. C1, C8.
37. Survey by Microsoft Corp., reported in Don Clark and Kyle Pope, "Poll Finds Americans Like Using PCs, but May Find Them to Be Stressful," *Wall Street Journal*, April 10, 1995, p. B3.
38. Survey by Concord Communications, reported in Matt Richtel, "Rage Against the Machine: PCs Take Brunt of Office Anger," *New York Times*, March 11, 1999, p. D3.
39. Jonathan Marshall, "Some Say High-Tech Boom Is Actually a Bust," *San Francisco Chronicle*, July 10, 1995, pp. A1, A4.
40. Eleena de Lisser, "One-Click Commerce: What People Do Now to Goof Off at Work," *Wall Street Journal*, September 24, 1999, pp. A1, A8.
41. Surfwatch Checknet, cited in Keith Naughton, Joan Raymond, Ken Shulman, and Diane Struzzi, "CyberSlacking," *Newsweek*, November 29, 1999, pp. 62–65.
42. STB Accounting Systems 1992 survey, reported in Del Jones, "On-line Surfing Costs Firms Time and Money," *USA Today*, December 8, 1995, pp. 1A, 2A.
43. Dan Gillmor, "Online Reliability Will Carry a Price," *San Jose Mercury News*, July 18, 1999, pp. 1E, 7E.
44. Daniel Yankelovich Group report, cited in Barbara Presley Noble, "Electronic Liberation or Entrapment," *New York Times*, June 15, 1994, p. C4.
45. Jeremy Rifkin, "Technology's Curse: Fewer Jobs, Fewer Buyers," *San Francisco Examiner*, December 3, 1995, p. C-19.
46. Michael J. Mandel, "Economic Anxiety," *Business Week*, March 11, 1996, pp. 50–56; Bob Herbert, "A Job Myth Downsized," *New York Times*, March 8, 1996, p. A19; and Robert Kuttner, "The Myth of a Natural Jobless Rate," *Business Week*, October 20, 1997, p. 26.
47. Stewart Brand, in "Boon or Bane for Jobs?" *The Futurist*, January-February 1997, pp. 13–14.
48. Paul Krugman, "Long-Term Riches, Short-Term Pain," *New York Times*, September 25, 1994, sec. 3, p. 9.
49. Government Accounting Office study, 2001.
50. Beth Belton, "Degree-based Earnings Gap Grows Quickly," *USA Today*, February 16, 1996, p. 1B.
51. Alan Kruger, quoted in Lyn-Nell Hancock, Pat Wingert, Patricia King, Debra Rosenberg, and Alison Samuels, "The Haves and the Have-Nots," *Newsweek*, February 27, 1995, pp. 50–52.
52. Sabra Chartrand, "Software to Provide 'Personal' Attention to Online Customers with Service Untouched by a Human," *New York Times*, August 20, 2001, p. C8.
53. Judith Anne Gunther, "An Encounter with AI," *Popular Science*, June 1994, pp. 90–93.
54. William A. Wallace, *Ethics in Modeling* (New York: Elsevier Science, Inc., 1994).
55. Laura Johannes, "Meet the Doctor: A Computer that Knows a Few Things," *Wall Street Journal*, December 18, 1995, p. B1.
56. Campus Computing Project 1997 survey, reported in Lisa Guernsey, "E-Mail Is Now Used in a Third of College Courses, Survey Finds," *Chronicle of Higher Education*, October 17, 1997, p. A30; and Edward C. Baig, "A Little High Tech Goes a Long Way," *Business Week*, November 10, 1997, p. E10.
57. Student Monitor LLC, cited in Danielle Sessa, "For College Students, Web Offers a Lesson in Discounts," *Wall Street Journal*, January 21, 1999, p. B7.
58. National Center for Education Statistics, in "Internet Access Booms in Schools," *USA Today*, July 17, 2001, p. 1C.
59. Nanette Asimov, "Home-Schoolers Plug into the Internet for Resources," *San Francisco Chronicle*, January 29, 1999, pp. A1, A15.
60. Mary Beth Marklein, "Distance Learning Takes a Gigantic Leap Forward," *USA Today*, June 4, 1998, pp. 1D, 2D; and Godie Blumenstyk, "Leading Community Colleges Go National with New Distance-Learning Network," *Chronicle of Higher Education*, July 10, 1998, pp. A16–A17.
61. Rebecca Quick, "Software Seeks to Breathe Life into Corporate Training Classes," *Wall Street Journal*, August 6, 1998, p. B8.
62. Myron Magnet, "Who's Winning the Information Revolution," *Fortune*, November 30, 1992, pp. 110–117.
63. Tony Rutkowski, quoted in Patricia Schnaidt, "The Electronic Superhighway," *LAN Magazine*, October 1993, pp. 6–8.
64. Hal Lancaster, "Technology Raises Bar for Sales Job; Know Your Dress Code," *Wall Street Journal*, January 21, 1997, p. B1.
65. Ingrid Wickelgren, "Treasure Maps for the Masses," *Business Week/Enterprise*, 1996, pp. ENT22–ENT24.
66. Paul Saffo, quoted in Jared Sandberg, "CyberCash Lowers Barriers to Small Transactions at Internet Storefronts," *Wall Street Journal*, September 30, 1996, p. B6.
67. Rebecca Buckman and Aaron Lucchetti, "Electronic Networks Threaten Trading Desks on Street," *Wall Street Journal*, December 23, 1998, pp. C1, C15.
68. Gene Bylinsky, "The E-Factory Catches On," *Fortune*, July 23, 2001, pp. 200[B]–200[H].
69. Lee Gomes, "Free Tunes for Everyone!" *Wall Street Journal*, June 15, 1999, pp. B1, B4.
70. Deidre Pike, "Reno Musicians: It's Business, Even Artists Need to Pay Bills," *Reno Gazette-Journal*, May 29, 2000, pp. 1E, 3E.
71. Bruce Haring, "Digitally Created Actors: Death Becomes Them," *USA Today*, June 24, 1998, p. 8D.
72. David Ansen and Ray Sawhill, "The New Jump Cut," *Newsweek*, September 2, 1996, pp. 64–66.
73. *The State of "Electronically Enhanced Democracy": A Survey of the Internet* (New Brunswick, NJ: Rutgers University, Douglass Campus, Walt Whitman Center, Department of Political Science, 1998).
74. "Minus a Daily Newspaper, City Turns Online," *San Francisco Chronicle*, September 8, 1998, p. A22; reprinted from *New York Times*.
75. California Institute of Technology and Massachusetts Institute of Technology, study reported in Florence Olsen, "MIT and Caltech Researchers Propose Shifts in Voting Technology," *Chronicle of Higher Education*, August 3, 2001, p. A37; Katharine Q. Seelye, "Study Says 2000 Election Missed Millions of Votes," July 17, 2001, p. A17; Richard Winton, "Balloting Study Calls for Updating Equipment," *Los Angeles Times*, July 17, 2001, p. A6; and Florence Olsen, "Computer Scientists and PoliticalScientists Seek to Create a Fiasco-Free Election Day," *Chronicle of Higher Education*, April 20, 2001, pp. A51–A53.
76. Study by House Government Reform Committee, U.S. Congress, reported in Laura Parker, "Technology Can Reduce Voting Flaws, Study Says," *USA Today*, July 10, 2001, p. 9A.
77. Edward Tenner, "The Perils of High-Tech Voting," *New York Times*, February 5, 2001, p. A27; and John Carey, "Is There Any Help for the 'Hanging Chad'?" *BusinessWeek*, November 27, 2000, pp. 54–56.
78. Associated Prress, "Pentagon's E-Voting Test Blasted Over Cost Per Ballot," *San Francisco Chronicle*, August 11, 2001, p. A4.
79. Frank Eltman, "Government of the People, Via the Net," *San Francisco Chronicle*, May 31, 2000, p. D3.
80. Michael Cornfield, quoted in Jon Swartz, "Electronic Engineering," *San Francisco Chronicle*, March 27, 1999, pp. D1, D3.

Bookmark It! Box
a. Nate Stulman, "The Great Campus Goof-Off Machine," *New York Times*, March 15, 1999, p. A25.
b. Robert Kubey, "Internet Generation Isn't Just Wasting Time" (letters), *New York Times*, March 21, 1999, sec. 4, p. 14.
c. Marco R. della Cava, "Are Heavy Users Hooked or Just Online Fanatics?" *USA Today*, January 16, 1996, pp. 1A, 2A.
d. Kenneth Hamilton and Claudia Kalb, "They Log On, but They Can't Log Off," *Newsweek*, December 18, 1995, pp. 60–61; Kenneth Howe, "Diary of an AOL Addict," *San Francisco Chronicle*, April 5, 1995, pp. D1, D3.
e. Stella Yu, quoted in Hamilton and Kalb, 1995.
f. Jonathan Kandell, quoted in J. R. Young, "Students Are Unusually Vulnerable to Internet Addiction, Article Says," *Chronicle of Higher Education*, February 6, 1998, p. A25.
g. American Psychological Association, reported in R. Leibrock, "AOLaholic: Tales of an Online Addict," *Reno News & Review*, October 22, 1997, pp. 21, 24.
h. Hamilton and Kalb, 1995.
i. Keith J. Anderson, reported in Leo Reisberg, "10% of Students May Spend Too Much Time Online," *Chronicle of Higher Education*, June 16, 2000, p. A43.
j. R. Sanchez, "Colleges Seek Ways to Reach Internet-Addicted Students," *San Francisco Chronicle*, May 23, 1996, p. A16; reprinted from the *Washington Post*.
k. Sanchez, 1996.
l. P. Belluck, "The Symptoms of Internet Addiction," *New York Times*, December 1, 1996, sec. 4, p. 5.
m. Questionnaire adapted from chart, "Characteristics of 'Internet Dependent' Students," from Keith J. Anderson, Rensselaer Polytechnic Institute, in Reisberg, 2000.
n. Ben Gose, "A Dangerous Bet on Campus," *Chronicle of Higher Education*, April 7, 2000, pp. A49–A51.
o. Jeremy Siegel, Wharton School, University of Pennsylvania, quoted in David Segal, "The Minefield of Internet Trading," *San Jose Mercury News*, September 6, 1998, pp. 1F, 6F; reprinted from the *Washington Post*.

Index

Boldface numbers indicate pages on which key terms are defined.

Access rights
  identification systems and, 280–281
  procedures for controlling, 282
Active-matrix display, 170, **195**, **202**
ActiveX controls, 66, 311
Ada programming language, 307
Addicts, Internet, 286
Address book feature, 45
Addresses
  e-mail, 42–45
  website, 53–54
AGP (accelerated graphics port) bus, **159**, **171**
Amazon.com, 255–256
Amplitude, 214
Amplitude modulation, 216
Analog cellular phones, **227**, **238**
Analog data, 214
Analog signals, **213**–214, **238**
  converting into digital signals, 216
  modem conversion of, 214–216
Analog television, 220
Analytical graphics, **115**, **131**
Analyzing systems. *See* Systems analysis
Anderson, Keith J., 286
Animation, 66, **75**, 293
Anonymous FTP sites, 70
Antivirus software, **279**, **295**
Applets, **66**, **75**
Appliances, information, 220
Application service providers (ASPs), **127**, 128, 130, **131**
Application software, **18**, **24**, 84, 101–130
  business categories for, 101–103
  computer-aided design programs, **125**–126
  database software, **116**–119
  desktop publishing programs, **123**–124
  drawing programs, **124**
  e-mail programs, 42
  file types and, 104
  financial software, **121**–122
  painting programs, **124**–125
  presentation graphics software, **119**–121
  project management software, **125**
  review questions on, 139–140
  specialty software, **119**–125
  spreadsheet programs, **113**–116
  tutorials and documentation for, 103
  types of, 104–105
  video/audio editing software, 125
  web authoring software, 125, 312
  word processing software, **105**–112
  See also System software
Arithmetic operations, 153
Arithmetic/logic unit (ALU), **153**, **171**
ARPANET, 34
Artificial intelligence (AI), **288**–290, **295**
  ethics of, 290
  expert systems, **289**
  natural language processing, **289**
  robotics, **288**–289
  Turing test and, **290**
Artificial life (A-life), **289**, **295**
ASCII coding scheme, **147**, 148, **171**
ASCII files, 248
ASPs (application service providers), **127**, 128, 130, **131**
ATMs (automated teller machines), 182, 185, 264
Attachments, e-mail, 45, 46
Auctions, online, 73
Audio
  input, 190–191
  output, 198
  software for editing, 125
  streaming, **67**
Audio board, 190–191
Audio CDs, 165
Audio files, 248
Audio-input devices, **190**–191, **202**
Audit controls, 282
Audit trail, 282
Automated design tools, 303
Automated teller machines (ATMs), 182, 185, 264
Automated virtual representatives (vReps), 288
B2B commerce. *See* Business-to-business (B2B) commerce
Backbone, 39, **232**, **238**
Backups, 162–163
Band, **221**, **238**
Bandwidth, **32**, **75**, **221**
Banking, online, 73, 292
Bar charts, 115
Bar codes, **189**, **202**
Bar-code readers, **189**, **202**
BASIC programming language, 307
Batteries, notebook computer, 170
Bays, **148**, **171**
Berners-Lee, Tim, 255

Bezos, Jeffrey, 255–256
Binary coding schemes, 147–148
Binary system, **146**–147, **171**, 213
Bioinformaticist, 19
Biometrics, **192**, **202**, **281**, **295**
Bitmap, 188
Bits (binary digits), 33, **147**, **171**
Blocking software, 236
Bloom, Benjamin, 29
Bluetooth wireless digital standard, **229**, **238**
Bookmarks, 58, 59
Boolean operators, 65
Booting, **85**, **131**
Boot-sector virus, 278
Bozo filters, 285
Bps (bits per second), **33**, **75**
Brand, Stewart, 287
Brenner, Viktor, 286
Bricklin, Daniel, 113
Bridges, **234**, **238**
Broadband connections, **32**, 38, **75**, **221**, **238**
Broadband wireless digital services, **228**, **238**
Broadcast radio, **223–224**, **238**
Browser software. *See* Web browsers
Bugs, software, 275–276
Buses, **154**, 159, **171**
Business, online, 72–73, 291–292
  See also E-commerce
Business webs (b-webs), 257
Business Week, 4
Business-to-business (B2B) commerce, 73, 75, **257**–258, **265**
Buying considerations
  for notebook computers, 170
  for personal computers, 145–146
  for printers, 199
Bytes, 15, **147**, **171**

C programming language, 307
C++ programming language, 307
Cable modem, 35, **37**, **75**
Cache, 63, **156**, **171**
CAD (computer-aided design) programs, **125**–126, **131**
CAD/CAM (computer-aided design/computer-aided manufacturing) programs, **126**, **131**
CADD (computer-aided design and drafting) programs, 126
Call-back system, 281
Carrier waves, 214
Caruso, Denise, 73, 292
Case (system cabinet), **13**, **24**, 148
CASE (computer-aided software engineering) tools, **303**

CD drive, **15**, **24**
CDMA wireless digital standard, 228
CD-R disks, **167**, **171**
CD-ROM disks, **165**–166, **171**
CD-ROM drive, 166
CD-RW disks, **167**, **172**
Cells
  communication, 227
  spreadsheet, **113**, **131**
Cellular telephones, 3
  analog, **227**
  digital, 227–228
Censorship issues, 236–237
Central processing unit (CPU), **11**, **24**, **153**–154, **172**
  data processing by, 153–154
  internal management of, 85–86
Chambers, John, 251
Channels, 68
Character-recognition devices, 189–190
Characters, **246**, **265**
Chats, 71
Cheating, 74
Checkbook management, 122
Chips, **13**, **24**, **143**, **172**
  CISC, **151**
  CMOS, **155**
  flash memory, **155**
  making of, 144
  memory, **13**
  microprocessor, **145**, 150–151
  RAM, **154**–155
  RISC, **151**
  ROM, **155**
CISC (complex instruction set computing) chips, **151**, **172**
Civil strife, 276
Classrooms, online, 72, 291
Client/server LANs, 233
Clients, **9**, **24**, 127
  thin vs. fat, 127–129, 145
Clinton, Bill, 274
Clip art, 111
Clipboard, 108
Closed architecture, 158
Closing windows, 93, 94
CMOS (complementary metal-oxide semiconductor) chips, **155**, **172**
Coaxial cable, **223**, **239**
COBOL programming language, 307
Codec techniques, 230
Collaborative computing, 218
Collaborative Reference Service, 291
Columns
  spreadsheet, 113
  text, 111
Command-driven interface, 88
Comments, inserting, 112

317

Commerce, electronic. *See*
  E-commerce
Commercial software, 102
Communications, 11, 12
  censorship and, 236–237
  ethical issues in, 236–237
  practical uses of, 217
  privacy issues and, 237
Communications channels,
  221–230, **239**
    compression methods and,
      229–230
    radio frequency spectrum
      and, **221**, 222
    wired vs. wireless, 223–230
Communications hardware, 17
Communications satellites, 38,
  75, **224**–225, **239**
Communications technology, 3,
  24, 211–234
    bandwidth and, **222**
    cellular telephones, 227–228
    compression methods and,
      229–230
    contemporary examples of,
      3–6
    cyberethics and, 236–237
    fusion with computer technology, 214–215
    Global Positioning System,
      225–226
    home networks, 219–220
    information appliances, 220
    networks, **231**–236
    overview of developments in,
      20–21
    pagers, **226**
    radio frequency spectrum
      and, **221**, 222
    review questions on,
      243–244
    satellites and, 224–225
    smart television and, 220
    telecommuting, **219**
    timeline of progress in,
      212–215
    videoconferencing, **217**–218
    virtual office, **219**
    wired vs. wireless, 223–230
    workgroup computing, **218**
    See also Information technology
Compcierge, 19
Compilers, **308**
Comprehension, 29
Compression, **229**–230, **239**
  lossless vs. lossy, 230
Computer crime, **276**–277,
  279–280, **295**
Computer Ethics (Forester and
  Morrison), 22
Computer technology, 3
  contemporary examples of,
    3–6
  fusion with communications
    technology, 214–215
  overview of developments in,
    20–21
  timeline of progress in,
    212–215
  See also Information technology
Computer-aided design (CAD)
  programs, **125**–126, **131**
Computer-created search sites,
  63, 64

Computers, **3**, **24**
  basic operations of, 10–11
  crashing of, 163–164
  crimes related to, 276–277,
    279–280
  custom-built, 10, 18
  durability of, 200
  environmental issues related
    to, 283–284
  guarding against theft, 200
  hardware of, **10**, 11–17
  health issues related to,
    200–201
  network, **127**–129
  purpose of, 10
  recycling, 18, 283
  software of, **10**, 17–18
  types of, 7–10
Connectivity, **20**, **24**, 217
Content templates, 121
Context-sensitive help, 94
Control unit, **153**, **172**
Convergence, **21**, **24**
Conversion, system, 304–305
Cookies, **237**, **239**
Copy command, 108
Copyright, **102**, **131**
CPU. *See* Central processing
  unit
Crackers, **279**–280, **295**
Creating documents, 107–108
Credit bureaus, 263
Credit cards, 264
Crime, computer, 276–277,
  279–280
CRT (cathode-ray tube),
  194, **202**
CSnet, 34
Cursor, **107**, **131**
Custom-built PCs, 10, 18
Cut command, 108
Cybercash, 292
Cyberethics, 236–237
Cyberspace, **4**–5, **25**
*Cyberspace: The Human
  Dimension* (Whittle), 4
Cycles, 151

Daisy chains, 157
Data, **10**, **25**
  accuracy and completeness
    of, 260–261
  conversion of, 304
  digital vs. analog, 213–214
  hierarchical organization of,
    246–247
  importing and exporting, 104
  manipulation of, 258–260
  permanent storage of, 11,
    14–15
  protecting, 200, 282
Data access area, 161
Data files, **248**, 249, **265**
Data mining (DM), **256**–257, **265**
Data storage hierarchy, **246**–247,
  265
Data transmission, 33–34
Database files, **104**, **132**
Database management system
  (DBMS), **250**, **265**
Database server, 233
Database software, **116**–119, **132**
  benefits of, 116
  features of, 116–118
  illustrated overview of, 117

personal information
  managers, 118–119
Databases, **116**, **132**, **246**–270,
  265
  accessing data on, 249
  accuracy of information on,
    260–261
  business-to-business systems
    and, **257**–258
  data mining and, **256**–257
  data storage hierarchy and,
    246–247
  e-commerce and, **255**–256
  ethical issues related to,
    258–261
  file types and, 247–248
  hierarchical, **252**, 253
  identity theft and, 263
  key field in, **247**
  management systems for,
    250
  network, **252**, 253
  object-oriented, **254**
  privacy issues and, 261
  relational, **116**, **254**, 255
  review questions on,
    269–270
  storing data from, 249–250
  types of, 250–251
Davis, Lynn, 261
Dedicated fax machines, **190**,
  202
Dedicated ports, 158
Default settings, **111**–112, **132**
Deleting text, 108
Democracy, online, 293–294
DES (Data Encryption Standard),
  281
Design templates, 121
*Designing Web Usability: The
  Practice of Simplicity*
  (Nielsen), 312
Desktop, **88**, **132**
Desktop PCs, **8**, **25**
Desktop publishing (DTP),
  123–124, **132**
  Web page editors and, 217
Detail design, 303
Developing information systems. *See* Systems development
Device drivers, 84, **87**, **132**
Diagnostic routines, 85
Dialog boxes, 96
Dial-up connection, 34, 35, 36
Digital cameras, **191**, **202**
Digital convergence, 212
Digital data, 213
Digital divide, 272
Digital environment, 272–275
Digital signals, **213**, **239**
  converting analog signals
    into, 216
  modem conversion of,
    214–216
Digital television (DTV), 212,
  220, **239**
Digital wireless services,
  227–228, **239**
Digital-data compression, **229**
Digitized media
  audio, 190–191
  photographs, 191
  video, 191
Digitizer, **187**, **203**
Digitizing tablet, **187**, **203**

DIMM (dual inline memory
  module), 155
Direct access storage, **249**, **266**
Direct implementation, 305
Directories, Web, **62**, **75**
Dirty data problems, 276
Disaster-recovery plans,
  282–283, **295**
Diskettes, 161–163
  formatting, **86**, 162
  maintenance tips, 200
Disks
  floppy, **161**–163
  hard, **163**–165
  HiFD, **163**
  optical, **165**–167
  SuperDisks, **163**
  Zip, **162**–163
Display screens, **193**–195, **203**
  health issues related to, 201
  notebook computer, 170
  touch screens, 186, **186**
  videographic standards for,
    195, 196
Distance learning, **72**, **75**, 291
Distributed databases, **251**, **266**
Document files, **104**, **132**
  creating, 107–108
  editing, 108, 110
  formatting, **110**–112
  illustrated overview of, 109
  inserting comments into, 112
  printing, 112
  saving, **112**
  toggling between, 108
  tracking changes in, 112
  Web, 112
Documentation, **103**, **132**
Domain, **42**–43, **75**
DOS (Disk Operating System),
  95, **132**
Dot pitch (dp), **194**, **203**
Dot-matrix printers, **196**, **203**
Downlinking, 224
Downloading data, **34**, **75**
Dpi (dots per inch), 188, **196**
Draft quality output, 196
DRAM chips, 154
Drawing programs, **124**, **132**
Drive bays, 148
Drive gate, 161
Drop-down menu, 91
DSL (digital subscriber line), 35,
  37, **75**
Dumb terminal, **184**–185, **203**
DVD drive, **15**, **25**
DVD-R disks, **167**, **172**
DVD-ROM disks, **167**, **172**
Dwyer, Eugene, 74

EBCDIC coding scheme, **147**,
  148, **172**
E-commerce (electronic commerce), 51, **72**–73, **76**, **255**, **266**
  accountants for, 19
  changes created by, 292
  databases and, 255–256
  developments in, 72–73
  government and, 274
Economic issues, 285, 287
Editing
  documents, 108, 110
  video clips, 125
Education
  distance learning and, **72**, 291

information technology and, 291
EIDE controllers, 165
Einstein, David, 10, 18
Electrical power issues, 150, 284
Electromagnetic spectrum, 221, 222
Electromechanical problems, 276
Electronic democracy, 293–294
Electronic imaging, **189**, **203**
Ellison, Larry, 127
E-mail (electronic mail), 3–**4**, **25**, 41–50
    addresses, 42–45
    attachments, 45, 46
    filters for sorting, 49
    instant messaging and, **45**, 47
    junk, 49–50
    mailing lists, 47
    netiquette, **48**
    privacy issues, 50, 237
    remote access to, 45
    replying to, 43, 44
    sending and receiving, 44
    software and services, 42
    tips on using, 5
Embedded computers, 145
Emoticons, **48**, **76**
Encryption, **281**–**282**, **295**
Enhanced paging, 226–227
ENIAC computer, 19, 152
Entertainment, 51, 72, 293
Enum system, 275
Environmental problems, 283–284
ERP (enterprise resource planning) software, **129**, **132**
Errors, computer, 275–276
Ethics, **22**, **25**
    artificial intelligence and, 290
    censorship issues and, 236–237
    information accuracy and, 260–261
    media manipulation and, 258–260
    plagiarism and, 74
    privacy and, 237, 261
Ethics in Modeling (Wallace), 290
Executable files, 248
Expansion, **150**, **158**, **172**
Expansion bus, 159
Expansion cards, **158**–**160**, **172**
Expansion slots, **13**, **25**, **158**, 159, **172**
Expert systems, 288, **289**, **295**
Exporting data, **104**, **132**
Extensible markup language (XML), 310
Extension names, 247
Extranets, **235**, **239**

FAQs (Frequently Asked Questions), **48**, **76**
Fat client, 127–128, 129
Favorites, 58, 59
Fax machines, **190**, **203**
Fax modems, **190**, **203**
Federal Bureau of Investigation (FBI), 280
Federal Communications Commission (FCC), 220

Federal privacy laws, 262
Fiber-optic cable, **223**, **239**
Fields, **246**, **266**
File allocation table (FAT), 249
File server, **233**, **240**
File Transfer Protocol (FTP), 69–**70**, **76**
File virus, 278
Filenames, **247**, **266**
Files, **104**, **133**, **246**, **266**
    converting, 304
    data, **248**, 249
    importing and exporting, 104
    managing, 86
    program, **248**
    types of, 104, 247–248
Film making, 293
Filters, e-mail, 49
Finance, online, 73, 292
Financial software, **121**–**122**, **133**
Find command, 108
Firewalls, **235**–**236**, **240**
Flaming, **48**, **76**
Flash memory cards, **169**, **173**
Flash memory chips, **155**, **173**
Flat-panel displays, **194**, **204**
Floppy disks, **161**–**163**, **173**
    formatting, 86, 162
    maintenance tips, 200
Floppy-disk cartridges, **161**, **173**
Floppy-disk drive, **15**, **25**
Flops, **152**, **173**
Fly-out menu, **91**, **133**
Folders, 88
Fonts, **110**, **133**
Footers, 111
Forester, Tom, 22
Formatting
    disks, **86**, **133**, 162
    documents, **110**–**112**
Formulas, **113**, **133**
FORTRAN programming language, 307
Frame-grabber video cards, 191
Frames, **60**, **76**
Framework for Global Electronic Commerce, A, 272, 274
Freeware, **103**, **133**
Frequency
    modem technology and, 214
    radio signals and, 221, 222
Frequency modulation, 216
Frontside bus, 159
FTP (File Transfer Protocol), 69–**70**, **76**
Function keys, **106**, **133**
Functions, 115

Gambling, online, 284, 286
Gateways, **234**, **240**
Geographical information system (GIS), 292
Geostationary earth orbit (GEO), 225
Gibson, William, 4
Gigabits per second (Gbps), **34**, **76**
Gigabyte (G, GB), 15, **147**, **173**
Gigaflops, 152
Gigahertz (GHz), 13, 151, **152**, **173**
Gillmor, Dan, 285
Global Positioning System (GPS), **225**–226, **240**

Government
    federal privacy laws and, 262
    information technology and, 293–294
    Internet regulation and, 274
Grammar checker, **110**, 111, **133**
Grant, Lorrie, 72
Graphical user interface (GUI), **88**–**94**, **133**
Graphics, analytical, **115**
Graphics cards, **159**, **173**, 195
Groupware, **105**, **133**, 218
GSM wireless digital standard, 228
Gunther, Judith Anne, 290

Hackers, **279**–**280**, **295**
Hamilton, Anita, 127
Handheld computers, 9
    operating systems for, 98–99
    See also Notebook computers
Hard disks, **163**–**165**, **173**
    crashing of, 163–164
    nonremovable, **164**–165
    removable, **165**
Hardcopy output, **193**, **196**–**198**, **204**
Hard-disk controller, **165**, **173**
Hard-disk drive, **15**, **25**
Hard goods, 73
Hardware, **10**, **11**–**17**, **25**, 141–170
    buying, 145–146
    cache, **156**
    communications, 17
    converting, 304
    CPU, **153**–**154**
    expansion cards, **159**–160
    health issues related to, 200–201
    input, 11–**12**, **183**, 184–192
    memory, 13–**14**, 154–155
    microchips, 143–145
    miniaturization of, 143, 145
    motherboard, 150
    obtaining for systems, 304
    output, 15–**16**, **183**, 193–199
    ports, **156**–158
    power supply, **150**
    processing, 13–14, 150–154
    protecting, 200
    review questions on, 179–180, 209–210
    secondary-storage, 14–15, 161–169
    system unit, 146–160
    theft of, 200, 276
    transistors, **142**–143
Head crash, **163**–164, **173**
Headers, 111
Health issues
    computer use and, 200–201
    data mining and, 257
    medical technology and, 291
    online information about, 72
Help command, **94**, **133**
Hierarchical databases, **252**, 253, **266**
HiFD Disks, **163**, **174**
High-definition television (HDTV), 212, **220**, **240**
High-level programming language, 307–308
History list, 58, 59

Home networks, 219–220
Home page, **52**, **76**
    personalizing, 55
    web portal, 61–62
HomeRF wireless digital standard, 229
Host computer, **232**, **240**
HTML (Hypertext Markup Language), **51**–**52**, **76**, 310
HTTP (HyperText Transfer Protocol), **53**, **76**
Human errors, 275
Human-biology input devices, 192
Human-organized search sites, 62–63, 64
Hybrid search sites, 63, 64
Hyperlinks, 51, 54, 56–57, 60
Hypertext, **51**, **76**
Hypertext index, 62
Hypertext Markup Language (HTML), **51**–**52**, **76**, 310
HyperText Transfer Protocol (HTTP), **53**, **76**

ICANN (Internet Corporation for Assigned Names & Numbers), **274**, **295**
Icons, **88**, **133**
iDEN wireless digital standard, 228
Identification systems, 280–281
Identity theft, 263–264
Illustration software, 124–125
Image files, 248
Image-editing software, 125
Imaging systems, **188**, **204**
Impact printers, **196**, **204**
Implementing systems. *See* Systems implementation
Importing data, **104**, **134**
Income gap, 287
Individual databases, **250**–**251**, **266**
Information, **10**, **25**
    accuracy and completeness of, 260–261
    manipulation of, 258–260
    privacy issues and, 261
    theft of, 277
    See also Data
Information appliances, 220
Information overload, 2, 22, 285
Information superhighway, 272
Information technology, **2**, **25**
    business and, 291–292
    contemporary examples of, 3–6
    crime related to, 276–277, 279–280
    economic issues and, 285, 287
    education and, 291
    entertainment industry and, 293
    government and, 293–294
    medicine and, 291
    overview of developments in, 20–21
    quality-of-life issues and, 283–285
    security issues and, 275–283
    See also Communications technology; Computer technology

Infrared ports, **158**, **174**
Infrared wireless transmission, **223**, **240**
Initializing disks, 86, 162
Ink-jet printers, **197**, **204**
Input, 10, 12, **25**
Input hardware, 11–12, **183**, 184–192, **204**
 audio-input devices, 190–191
 digital cameras, **191**
 digitizing tablets, **187**
 health issues related to, 200–201
 human-biology input devices, 192
 keyboards, **184–185**
 pen-based computer systems, **186**, 187
 pointing devices, **185**–187
 radio-frequency identification devices, **192**
 review questions on, 209–210
 scanning devices, 188–190
 sensors, **192**
 source data-entry devices, **188**–192
 touch screens, **186**
 types of, 184
 video-input cards, 191
 voice-recognition systems, **191**–192
 Webcams, 191
Inserting text, 108
Installation process, 17
Instant messaging (IM), **45**, 47, **76**
Integrated circuits, 20, **143**, **174**
Intelligence, artificial. *See* Artificial intelligence
Intelligent agents, **291**, **296**
Intelligent smart cards, 168
Intelligent terminal, **185**, **204**
Intel-type chips, **151**, **174**
Interactive TV, 220
Interactivity, **20**, **25**
 of web pages, 60
Interfaces, 84
 command-driven, 88
 graphical user, 88–94
 menu-driven, 88
Internet, 5–6, **25**, 31–82
 addiction to, 286
 bandwidth and, 32
 brief history of, 34–35
 business conducted on, 72–73, 291–292
 censorship issues, 236–237
 discussion groups, 47
 domain abbreviations on, 43
 e-mail and, 41–50
 FTP sites on, **69**–70
 influence of, 6
 ISPs and, 39–41
 netiquette, **48**
 new versions of, 273
 newsgroups on, **70**–71
 physical connections to, **33**–38
 plagiarism issues, 74
 privacy issues, 237
 programming languages for, 310–311
 pursuing personal interests via, 71–72
 real-time chat on, **71**

 regulation of, 274–275
 review questions about, 80–82
 Telnet feature, **70**
 trends and statistics, 6, 32
 See also World Wide Web
Internet2, **273**, **296**
Internet phones, 3
Internet Relay Chat (IRC), 71
Internet Service Providers (ISPs), **39**–41, **76**
 comparison of, 41
 tips on choosing, 40
 Web authoring tools, 312
Internet telephony, **68**, **77**
Internet terminal, **185**, **204**
Interpreters, **308**
Intranets, **235**, **240**
Investment software, 122
Investments, 73, 292
ISA (industry standard architecture) bus, **159**, **174**
ISDN (Integrated Services Digital Network), 35, **36**, **77**
Isolation, 284
ISPs. *See* Internet Service Providers

Java programming language, **66**, **77**, 311
JavaScript, 66
Job hunting, 73
JPEG compression standard, **230**, **240**
Junk e-mail, 49–50
Justification, 111

Kandell, Jonathan, 286
Kemeny, John, 307
Kernel, 85
Key field, **118**, **134**, **247**, 249, **266**
Keyboards, **12**, **26**, **184**–185, **204**
 injuries from using, 201
 layout and features, 106–107
 notebook computer, 170
 specialty, 184–185
 traditional, 184
Keywords, **62**, **77**
Kilobits per second (Kbps), **34**, **77**
Kilobyte (K, KB), 15, **147**, **174**
Kiosks, 182
Kline, David, 219
Knowledge engineers, 289
Krugman, Paul, 287
Kurtch, Thomas, 307

Language translators, 84, **308**
LANs. *See* Local area networks
Laptop computers. *See* Notebook computers
Laser printers, **196**–197, **204**
Laws, privacy, 262
Learning
 critical-thinking skills for, 29
 distance, **72**, 291
Legacy systems, 95
Lewis, Peter, 4, 237
Library of Congress, 291
Licenses, software, 102
Life
 analog basis of, 213–214
 artificial, 289

Light pen, **186**–187, **205**
Line graphs, 115
Line-of-sight communication, 223
Line-of-sight systems, 192
Links. *See* Hyperlinks
Linux, **98**, **134**
Liquid crystal display (LCD), **194**, **205**
List-serves, **47**, **77**
Local area networks (LANs), **8**, **26**, 219, **232**, 233–234, **240**
 client/server, 233
 components of, 234, 235
 operating systems for, 96
 peer-to-peer, **233**
Logic bomb, 278
Logical operations, 153
Log-on procedures, **40**, **77**
Long-distance wireless communications, 225–228
Lossless compression, 230
Lossy compression, 230
Low-earth orbit (LEO), 225
Lucas, George, 293

Machine cycle, **153**, **174**
Machine language, **307**
Macintosh operating system (Mac OS), **95**–96, **134**
 OS X version of, 95, 96, 101
Macro virus, 278
Macros, 106–**107**, **134**
Magid, Lawrence, 236
Magnetic tape, **168**, **174**
Magnetic-ink character recognition (MICR), **189**, **205**
Mail, electronic. *See* E-mail
Mail servers, 41, 233
Mailing lists, 47
MailStation device, 41
Main memory, 154–155
Mainframes, **8**, **26**
Maintenance, system, 305
Make-or-buy decision, 304
Malone, Michael, 142
Malware, 277
Management
 database, 250
 file, 86
 task, 86
Managers, 301
Maney, Kevin, 2, 6
Manufacturing systems, 292
Margins, 111
Marketing
 data mining used in, 256
 online, 292
Mark-recognition devices, 189
Master file, **249**, **267**
Matchmaking services, 71
Maximizing windows, 93, 94
Media manipulation, 258–260
Medical technology, 291
 See also Health issues
Medium-earth orbit (MEO), 225
Megabits per second (Mbps), **34**, **77**
Megabyte (M, MB), 15, **147**, **174**
Megaflops, 152
Megahertz (MHz), 13, 151, **151**, **174**
Memorization, 22–23, 29
Memory
 flash, 155

 main, 154–155
 read-only, 155
 virtual, 156
Memory bus, 159
Memory chips, **13**, **26**, 154–155
Memory hardware, 13–14
Memory modules, 155
Mental-health problems, 284
Menu, **91**, **134**
Menu-driven interface, 88
Message, 309
Metasearch sites, 63, 64
Methods, 309
Metropolitan Area Exchanges (MAEs), 39
Metropolitan area networks (MANs), **232**, **241**
Microchips. *See* Chips
Microcomputers, **8**, **26**
 networking, 8
 types of, 8–9
 See also Personal computers
Microcontrollers, **9**, **26**, 145
Microprocessors, 142, **145**, 150–152, **174**
Microsoft Internet Explorer, 53
Microsoft Windows. *See* Windows operating systems
Microwave radio, **224**, **241**
MIDI board, 191
Midsize computers, 8
Miniaturization, 20
Minicomputers, 8
Minimizing windows, 93–94
MIPS, **152**, **174**
Mobility, 145
Modeling tools, 302
Modem cards, 160
Modems, **17**, **26**, **214**, **241**
 cable, 35, **37**
 dial-up, 34, 35, 36
 signal conversion by, 214–216
Molitor, Graham, 32
Money
 online use of, 73
 software for managing, 121–122
Monitors, **16**, **26**, 194–195, 196
 See also Display screens
Moore, Gordon, 151
Moore's law, 150–151
Morphing, **258**, **267**
Morrison, Perry, 22
Mossberg, Walter, 158, 170
Motherboard, **13**, 14, **26**, 150
Motorola-type chips, **151**, **174**
Mouse, **12**, **26**, 185
 functions performed by, 87–88, 89
 variant forms of, 185–186
 See also Pointing devices
Mouse pointer, 185
Movie industry, 293
Moving windows, 94
MPEG compression standard, **230**, **241**
Multifunction printers, **198**, **205**
Multimedia, **5**, 21, **26**
 World Wide Web and, 51, 65–67
Multimedia computer, 145
Multipartite virus, 278
Multitasking, 2, 23, **86**, **134**

Music
    online access to, 293
    theft of, 277
My Computer icon, 92

National Information Infrastructure (NII), 273
National Science Foundation (NSF), 34–35, 273
Natural hazards, 276
Natural language processing, **289**, **296**
Near-letter-quality (NLQ) output, 196
Netiquette, **48**, **77**
Netscape Navigator, 53
NetWare, **96**, **135**
Network Access Points (NAPs), 39
Network computers, **127**–**129**, **135**, 185
Network databases, **252**, 253, **267**
Network interface cards, **160**, **175**, 234
Network operating system (NOS), 96–98, 234
Networks, 4, 26, **231**–**236**, **241**
    benefits of, 231
    client/server, 9
    components of, 234, 235
    extranets, **235**
    firewalls for, **235**–**236**
    home, 219–220
    intranets, **235**
    local area (LANs), 219, **232**, 233–234
    metropolitan area (MANs), **232**
    types of, 232
    wide area (WANs), **232**
Neuromancer (Gibson), 4
Newbies, 48
Newsgroups, **70**–**71**, **77**
Newsreader program, **71**, **77**
Next Generation Internet (NGI), **273**, **296**
Nielsen, Jakob, 312
Node, **232**, **241**
Nonimpact printers, **196**–**198**, **205**
Nonremovable hard disks, **164**–165, **175**
Nonvolatile memory, 169
Notebook computers, **9**, **27**
    durability of, 200
    guarding against theft, 200
    tips on buying, 170
    See also Handheld computers
Nua Internet Surveys, 6

Object, 309
Object code, 308
Object-oriented databases, **254**, **267**
Object-oriented programming (OOP), **308**–**309**
Office suite, **104**, **135**
Offline storage, **249**, **267**
One-way communications, 225, 226
Online, **3**, **27**
Online auctions, 73

Online classes, 72
Online game player, 185
Online services, 39
    See also Internet Service Providers
Online software, 127
Online storage, 169, **250**, **267**
Open architecture, 158
Open-source software, **98**, **135**
Operating systems (OS), 84, **85**–86, 87–**101**, **135**
    DOS, **95**
    functions of, 85–86
    handheld computer, 98–99
    Linux, **98**
    Macintosh, **95**–96, 101
    NetWare, **96**
    network, 96–98
    Palm OS, **99**
    Pocket PC, **99**
    Solaris, 97–98
    Unix, **97**
    upgrading, 100–101
    user interface for, **87**–**94**
    Windows, 96–97, 99, 100
Operators, Boolean, 65
Optical cards, **169**, **175**
Optical character recognition (OCR), **190**, **205**
Optical disks, **165**–167, **175**
Optical mark recognition (OMR), **189**, **205**
Option Red supercomputer, 152
Outline feature, 108
Output, **11**, 12, **27**
Output hardware, 15–16, **183**, 193–**199**, **205**
    display screens, **193**–195
    health issues related to, 200–201
    printers, 196–198, 199
    review questions on, 209–210
    sound-output devices, 198
    types of, 193
    video, **198**
    voice-output devices, 198

Page description language (PDL), **196**–197, **205**
Pagers, **226**, **241**
Painting programs, **124**–125, **135**
Palm OS, **99**, **135**
Palmtops, 9
Parallel implementation, 305
Parallel ports, **156**–157, **175**
Parity bit, **175**
PASCAL programming language, 307
Passive-matrix display, 170, **195**, **205**
Passwords, 40, 264, **281**, **296**
Payload, 277
PC cameras, 218
PC cards, **160**, **175**
PC/TV terminal, 185
PCI (peripheral component interconnect) bus, **159**, **175**
PCs. *See* Personal computers
PDAs. *See* Personal digital assistants
Peer-to-peer LANs, **233**, **241**
Pen-based computer systems, **186**, **205**
People controls, 282

Peripheral devices, **16**, **27**
Personal computers (PCs), 8
    buying, 145–146, 170
    crashing of, 163–164
    custom-built, 10, 18
    durability of, 200
    environmental issues related to, 283–284
    guarding against theft, 200
    health issues related to, 200–201
    recycling, 18, 283
    types of, 8–9
    See also Computers; Notebook computers
Personal digital assistants (PDAs), 9, 27, 98, 185
Personal-finance managers, **121**–122, **135**
Personal information managers (PIMs), **118**–119, 125, **135**, 251
Personalization, 21
Personalized TV, 220
Pervasive computing, 2
Petabyte (P, PB), **147**, **175**
PGP (Pretty Good Privacy), 281
Phased implementation, 305
Photographs, manipulation of, 259–260
Photolithography, 144
Physical connections, **33**–**38**, **77**
    broadband, 32, 38
    cable modems, 37
    dial-up modems, 34, 35, 36
    high speed phone lines, 36–37
    wireless systems, 38
Pie charts, 115
Pilot implementation, 305
PIN (personal identification number), **281**, **296**
Pirated software, **103**, **135**
Pixels, **194**, 196, **205**
Plagiarism, 74
Platform, **85**, 95, **135**
Plug and Play, **158**, **175**
Plug-ins, **65**–66, **77**
Pocket PC operating system, **99**, **134**
Point of presence (POP), 39
Pointer, **87**, **135**
Pointing devices, **185**–187, **205**
    mouse and its variants, 185–186
    notebook computer, 170
    pen input devices, 186–187
    touch screens, 186
Pointing stick, **186**, **206**
Point-of-sale (POS) terminal, 185
Polymorphic virus, 278
Pop-up menu, **92**, **135**
Portability, 21, 145
Portable computers. *See* Handheld computers; Notebook computers
Ports, **156**–**158**, **175**
Power supply, **150**, **175**
Preliminary design, 303
Preliminary investigation, **302**
Presentation graphics software, **119**–121, **136**
    templates, 121
    views, 120, 121
Previewing documents, 112
Primary storage, **11**, 13, **27**, 154–155

Print Scrn key, 184
Print server, 233
Printers, **16**, **27**, **196**–**198**, **206**
    buying considerations, 199
    dot-matrix, **196**
    ink-jet, **197**
    laser, **196**–197
    multifunction, **198**
    thermal, **198**
Printing documents, 112
Privacy, **237**, **241**, **261**, **267**
    databases and, 261
    e-mail messages and, 50, 237
    federal laws on, 262
    Web cookies and, 237
Procedural errors, 275
Processing, **11**, 12, **27**
    hardware components for, 13–14, 150–154
    speed of, 20, 151–152, 156
Processor chips, 13, 145, 150–151
Productivity software, **104**, **136**
Program files, **248**, **267**
Programming, 305
    Internet, 310–311
    object-oriented, 308–**309**
    steps in process of, 306
    visual, **309**–310
Programming languages, **306**–**308**
    common types of, 307
    for Web pages, 310–311
Programs, 10, 84, **305**
    creating, 305–306
    utility, 84, 87, **137**
    See also Software
Project management software, **125**, **136**, 303
PROM (programmable read-only memory), 155
Proprietary software, 102
Protocols, 53, 77
Prototype, 303
Prototyping, 303
Public databanks, **251**, **267**
Public-domain software, **102**–103, **136**
Publishing websites, 312
Pull-down menu, **91**, **136**
Pull-up menu, 91, **92**, **136**
Push technology, **68**, **77**

Quality-of-life issues, 283–285
    environmental problems, 283–284
    mental-health problems, 284
    workplace problems, 285
Queries, database, 118
Quittner, Joshua, 255

Radio buttons, **60**, **78**
Radio frequency spectrum, **221**, **222**, **242**
Radio transmission, 223–224
Radio-frequency identification technology (RF-ID), **192**, **206**
RAM (random access memory), 13, 154–155
    See also Memory
RAM chips, **154**–155, **175**
Raster images, 124
RDRAM chips, 154
Read/write head, **161**, **176**

Reading data, **155**, **176**
Read-only memory (ROM), **155**
  CD-ROM disks, 165–166
RealAudio, 67
Real-time chat (RTC), **71**, **78**
Recalculation, **115**, **136**
Records, **246**, **267**
Recycling computers, 18, 283
Refresh rate, **194**, **206**
Registers, **154**, **176**
Relational databases, **116**, **136**, **254**, 255, **267**
Relationships, online, 71–72
Releases, software, 102
Removable hard disks, **165**, **176**
Removable-pack hard-disk system, **176**
Rentalware, **103**, **136**
Replace command, 108
Research, Web-based, 74
Resolution, 188, **194**, **206**
Retailing, online, 72–73
Review questions
  on application software, 139–140
  on communications technology, 243–244
  on databases, 269–270
  on hardware, 179–180, 209–210
  on the Internet, 80–82
  on security, 297–298
  on system software, 139–140
Ribbon cable, 15
Rifkin, Jeremy, 287
RISC (reduced instruction set computing) chips, **151**, **176**
Robbins, Alan, 84
Robotics, **288**–**289**, **296**
Robots, **288**–**289**, **296**
Rollover feature, **88**, **136**
ROM (read-only memory), **155**, **176**
ROM chips, 155
Root record, 252
Rothenberg, David, 74
Routers, **234**, **242**
Row headings, 113
Rukeyser, William, 74

Saffo, Paul, 292
Safire, William, 259
Sales, online, 292
Satellites, communications, 38, **75**, **224**–**225**
Saving documents, **112**, **136**
Scanners, **188**–**189**, **206**
Scanning devices, 188–190
  bar-code readers, **189**
  character-recognition devices, 189–190
  fax machines and modems, 190
  imaging systems, **188**–189
  mark-recognition devices, 189
Science, data mining in, 257
Screens. *See* Display screens
Scroll arrows, **60**, **78**
Scrolling, **60**, **78**, 107–**108**, **136**
SCSI controllers, 165
SCSI ports, **157**–158, **176**
SDLC. *See* Systems development life cycle
SDRAM chips, 154

Search command, 108
Search engines, 62–63, **78**
  general guide to, 64
Secondary storage, **11**, 14–15, **27**, 160–169
  direct access storage, **249**
  offline, **249**
  online, 169, **250**
  sequential storage, **249**
Secondary storage hardware, 14–15, **161**–**169**, **176**
  flash memory cards, **169**
  floppy disks, **161**–163
  hard disks, **163**–165
  magnetic tape, **168**
  optical cards, **169**
  optical disks, **165**–167
  smart cards, **168**
Second-generation (2G) technology, 227–228
Sectors, **161**, **176**
Security, 200, 275–283, **280**, **296**
  access rights and, 280–281
  civil strife and, 276
  computer crime and, **276**–**277**, 279–280
  disaster-recovery plans and, **282**–283
  encryption and, 281–282
  errors/accidents and, 275–276
  identification systems and, 280–281
  natural hazards and, 276
  procedures for, 282
  review questions on, 297–298
  terrorism and, 276
  viruses and, 278–279
  worms and, 278
Semiconductor, **143**, **176**
Sensors, **192**, **206**
Sequential storage, **249**, **268**
Serial ports, **156**, **176**
Servers, **9**, **27**, 127
Service programs, 87
Services, theft of, 277
Set-top box, 185
Shared databases, **251**, **268**
Shareware, **103**, **136**
Shopping, online, 72–73
Short-range wireless communications, 229
Silicon, 20, **143**, **177**
SIMM (single inline memory module), 155
Site, **52**, **78**
Slide shows, 120
Smart cards, **168**, **177**
Smart television, 220
Soft goods, 73
Softcopy output, **193**–195, **206**
Software, **10**, **17**–**18**, **27**, 84
  antivirus, **279**
  blocking, 236
  commercial, 102
  converting, 304
  ERP, **129**
  errors in, 275–276
  freeware, **103**
  guarding against damage to, 200
  make-or-buy decision for, 304
  online, 127
  open-source, **98**

pirated, **103**
productivity, **104**
protecting, 282
public-domain, **102**–103
rentalware, **103**
review questions on, 139–140
shareware, **103**
theft of, 277
tutorials and documentation, 103
user interface for, 87–94
versions and releases, 102
See also Application software; System software
Software engineering. *See* Programming
Software license, **102**, **137**
Software platform, **85**, 95
Software Publishers Association (SPA), 277
Solaris, 97–98
Solid state technology, **143**, 155, **177**
Sorting
  database records, 118
  e-mail messages, 49
Sound cards, **16**, **27**, **159**, **177**
Sound manipulation, 259
Sound-output devices, **198**, **206**
Source code, 308
Source data-entry devices, **188**–**192**, **206**
  audio-input devices, **190**–191
  digital cameras, **191**
  human-biology input devices, 192
  radio-frequency identification devices, **192**
  scanning devices, 188–190
  sensors, **192**
  video-input cards, 191
  voice-recognition systems, **191**–192
  Webcams, 191
Source program files, 248
Spacebar, 106
Spacing, text, 110
Spam, **49**–**50**, **78**
Speakers, **16**, **27**
Special effects, 293
Special-purpose keys, **106**, **137**
Specialty keyboards, 184–185
Specialty software, 119–125
Speech-recognition systems, 191–192
Speed
  data transmission, 33–34
  processing, 20, 151–152, 156
Spelling checker, **108**, 110, **137**
Spreadsheets, **113**–116, **137**
  creating charts from, 115
  features of, 113–115
  illustrated overview of, 114
SRAM chips, 154
Standard-definition television (SDTV), **220**, **242**
Start menu, 92
Start page, 55
Statements, 305
Stealth virus, 278
Stock trading, 292
Storage, 11, 12
  direct access, **249**
  offline, **249**
  online, 169, **250**
  primary, **11**, 13, 154–155

secondary, **11**, 14–15, 160–169
sequential, **249**
volatile, **154**
Streaming audio, **67**, **78**
Streaming video, 66–67, **78**
Stress, 284
Students
  distance learning for, 72
  plagiarism issues for, 74
  websites for, 82
Stulman, Nate, 286
Supercomputers, **7**, **28**
SuperDisks, **163**, **177**
Supervisor, **85**–**86**, **137**
Surge protector, **177**
SVGA (super video graphics array), **195**, **206**
SXGA (super extended graphics array), **195**, **207**
Syntax, 307
System, **300**
System board. *See* Motherboard
System bus, 159
System cabinet, **13**, 148
System clock, **151**, **177**
System software, 17–**18**, **28**, 84–101
  components of, 84–85
  device drivers, 84, **87**
  language translators, 84
  operating systems, 84, **85**–86, 87–101
  review questions on, 139–140
  utility programs, 84, **87**
  See also Application software
System testing, 304
System unit, **13**, 146–160
  illustrated, 149
Systems analysis, **302**–303
Systems analysis and design, **301**
  participants in, 301
  purpose of, 300
  six phases of, 301–302
  See also Systems development life cycle
Systems analyst, **301**
Systems design, **303**
Systems development, **303**–304, 306
Systems development life cycle (SDLC), **301**–305
  analyzing the system, 302–303
  designing the system, 303
  developing/acquiring the system, 303–304, 306
  implementing the system, 304–305
  maintaining the system, 305
  participants in, 301
  phases in, 301–302
  preliminary investigation phase, 302
Systems implementation, **304**–305
Systems maintenance, **305**

T1 line, 35, **37**, **78**
Tape cartridges, **168**, **177**
Task management, 86
Taskbar, **93**, **137**
Tax programs, 122
Taxonomy of Educational Objectives (Bloom), 29

TDMA wireless digital standard, 228
Technical staff, 301
Teenage Research Unlimited (TRU), 6
Telecommunications Act (1996), 237, 274
Telecommunications technology. *See* Communications technology
Telecommuting, **219**, **242**
Teleconferencing. *See* Videoconferencing
Telemedicine, 291
Telephones
    cellular, 3, 227–228
    Internet, 3
Telephony, Internet, **68**
Television
    contemporary technologies for, 220
    manipulation of content on, 260
Telework, 219
Telnet, **70**, **78**
Templates
    document, **110**, **137**
    presentation graphics, 121
    worksheet, 115
Terabyte (T, TB), 15, **147**, **177**
Teraflops, 152
Term papers, 74
Terminals, **8**, **28**, 184–185
Terrorism, 276
Testing information systems, 304
TFT (thin-film transistor) display, 195
Theft
    hardware, 200, 276
    identity, 263–264
    information, 277
    software, 277
    time and services, 277
Thermal printers, **198**, **207**
Thesaurus, **110**, **137**
Thin client, 128, 129, 145
Thinking skills, 29
3G wireless technology, 228
Time, theft of, 277
Toolbar, **93**, **137**
Torvalds, Linus, 98
Touch screens, **186**, **207**, 293–294
Touchpad, **186**, **207**
Tower PCs, **8**, **28**, 148
Trackball, **185**, **207**
Tracking document changes, 112
Tracks, **161**, **177**
Traditional computer keyboards, 184
Train simulator program, 289
Transaction file, **249**, **268**
Transceiver, 224
Transistors, 20, **142**–143, 150–151, **177**
Transmitting data, 33–34
Tribe, Laurence, 274
Trojan horse, 278
Turing, Alan, 290
Turing test, **290**, **296**
Tutorial, **103**, **137**
Twisted-pair wire, **223**, **242**
2G wireless technology, 227–228

Two-way communications, 225, 226–227
Two-way paging, 226–227

Ultra ATA controllers, 165
Undo command, 108
Unicode, **148**, **177**
Unit testing, 304
Universal broadband, 38
Unix, **97**, **137**
Upgrading, **150**, 158, **177**
Uplinking, 224
Uploading data, **34**, **78**
UPS (uninterruptible power supply), **178**
URL (Universal Resource Locator), **53**–54, **78**
USB ports, **158**, **177**
USEnet, **71**, **78**
User ID, 42
User interface, **87**–**94**, **137**
User name, 40
Users
    participation in systems development by, 301
    training on new systems, 305
Utility programs, 84, **87**, **137**
UXGA (ultra extended graphics array), **195**, **207**

Vacuum tubes, 142
Values, spreadsheet, **113**, **138**
VBNS (Very-High-Speed Backbone Network Service), **273**, **296**
V-chip technology, 237
Vector images, 124
Versions, software, 102
Video, **198**, **207**
    analog to digital, 191
    manipulation of, 260
    software for editing, 125
    streaming, 66–67
Video cards, **16**, **28**, 191, 195
Video display terminals (VDTs), 184
    See also Display screens
Video files, 248
Video-capture cards, 191
Videoconferencing, **198**, **207**, 217–218, **242**
Videophones, 218
Viewable image size (vis), 194
Views, presentation graphics, 120, 121
Virtual memory, **156**, **178**
Virtual office, **219**, **242**
Virtual Reality Modeling Language (VRML), 310–311
Virtual set designer, 19
Viruses, **278**–279, **296**
    as attached files, 45
    protection against, 279
Visual Basic programming language, 307
Visual programming, **309**–310
Voice-output devices, **198**, **207**
Voice-recognition systems, **191**–192, **208**
Volatile storage, **154**, **178**
Voltage regulator, **178**
Voting technology, 293–294
VRML (Virtual Reality Modeling Language), 310–311

Wafers, 144
Wallace, William A., 290
Wavetable synthesis, 159
Web browsers, **53**, **79**
    cache folder for, 63
    features of, 54–55, 58–60
    navigating the Web with, 54–60
    plug-ins for, **65**–66
    rating systems in, 236–237
    web-authoring tools and, 217, 312
Web files, 248
Web pages, **52**, **79**
    designing, 69, 312
    hyperlinks on, 54, 56–57, 60
    interactivity of, 60
    multimedia effects on, 66–67
    programming languages for, 310–311
    publishing, 312
    software for authoring, 125
    tools for creating, 217, 312
    See also Websites
Web portals, **61**–62, **79**
Web radio, 67
Web servers, 53–54, 233
Web terminal, 185
Web-authoring tools, 217, 312
Webcams, 191, 218
Webcasting, **68**, **79**
Websites, **52**, **79**
    bookmarking, 58, 59
    creating and publishing, 312
    especially for students, 82
    history of visiting, 58, 59
    tips for finding, 63, 65
    See also Web pages; World Wide Web
WebTV, 212
What-if analysis, **115**, **138**
Whittle, David, 4
Wide area networks (WANs), **232**, **242**
WiFi wireless digital standard, **229**, **242**
Wildstrom, Stephen, 217
Williams, Penny, 260
Window (computer display), **45**, **79**, **93**, **138**
Windows operating systems, 96
    Pocket PC, **99**, **134**
    Windows 95/98, **96**, **134**
    Windows 2000, **97**, **134**
    Windows CE, **99**, **134**
    Windows Millennium Edition, **96**, **134**
    Windows NT, **97**, **134**
    Windows XP, **97**, 100, **134**
Wired communications channels, 223
Wireless Application Protocol (WAP), 229
Wireless communications, 38, 223–230
    long-distance, 225–228
    short-range, 229
Wireless pocket PC, 185
Wizards, **110**, **138**
Word processing software, **105**–112, **138**
    creating documents, 107–108
    editing documents, 108, 110
    formatting documents, 110–112
    illustrated overview of, 109

keyboard layout and functions, 106–107
printing documents, 112
saving documents, 112
tracking document changes, 112
Web documents and, 112
web-authoring tools and, 217, 312
Word size, **153**, **178**
Word wrap, 108
Workbook, 113
Workgroup computing, **218**, **242**
Workplace problems, 285
Worksheet files, **104**, 113, **138**
Workstations, **8**, **28**
World Wide Web (WWW), **5**–6, **28**, 51–69
    addiction to, 286
    addresses used on, 53–54
    browser software, **53**, 54–55
    business conducted on, 72–73, 291–292
    censorship issues, 236–237
    distinctive features of, 51–52
    hyperlinks on, 51, 54, 56–57, 60
    interactivity of, 60
    multimedia on, 51, 65–67
    navigating, 54–60
    plagiarism issues, 74
    portal sites on, 61–62
    privacy issues, 237
    programming languages for, 310–311
    pursuing personal interests via, 71–72
    push technology and, **68**
    review questions about, 80–82
    search engines on, 62–63
    telephone calls via, 68
    terminology associated with, 52–54
    tips for searching, 63, 65
    See also Internet; Websites
Worms, **278**, **296**
Write-protect notch, **161**, **178**
Writing data, **155**, **178**

XGA (extended graphics array), **195**, **208**
XML (extensible markup language), 310

Zemeckis, Robert, 260
Zip disks, **162**–163, **178**
Zip-disk drive, **15**, **28**

Photo Credits

PHOTOS

Page 1 Don Mason/The Stock Market; **2** Don Mason/The Stock Market; **3** AP/Wide World Photos; **4** Inge Yspeert/Corbis; **5** Don Mason/The Stock Market; **7** AP/Wide World Photos; **8** (top) Tom Tracy/Photophile; **8** (middle right) Susan Friedman; **8** (bottom middle) Bill Rogers; **8** (bottom right) Bill Rogers; **9** (top left) AP/Wide World Photos; **9** (top right) AP/Wide World Photos; **9** (bottom left) Courtesy of Intel; **18** (top) AP/Wide World Photos; **18** (bottom) Brian Williams; **19** (left) Courtesy of Unisys Archives; **19** (right) Mark Richards/PhotoEdit; **20** (top) Don Mason/The Stock Market; **20** (bottom) Eric Draper/AP/Wide World; **21** AP/Wide World Photos; **24** (top) Don Mason/The Stock Market; **24** (bottom) Courtesy of Intel; **25** Don Mason/The Stock Market; **26** (far right) Don Mason/The Stock Market; **26** (middle left) Tom Tracy/Photophile; **26** (bottom middle right) Courtesy of Intel; **27** (top left) Bill Rogers; **27** (far left) Don Mason/The Stock Market; **27** (middle left) AP/Wide World Photos; **28** (top left) AP/Wide World Photos; **28** (far right) Don Mason/The Stock Market; **28** (middle left) Bill Rogers; **41** AP/Wide World Photos; **58** Richard Nowitz/Corbis; **64** (**Panel 2.10**) Table adapted from Elizabeth Weise, "Successful Net Search Starts with Need," *USA Today*, January 24, 2000, p3D. Copyright 2000, USA Today. Reprinted with permission; **75** (bottom) Richard Nowitz/Corbis; **83** Lori Adamson-Peck; **83** Lori Adamson-Peck; **98** (middle) AP/Wide World Photos; **99** (bottom) Courtesy of Handspring; **99** Lori Adamson-Peck; **100** (top) Lori Adamson-Peck; **113** Mark Richards/PhotoEdit; **126** © Ed Kashi/Corbis; **131** Lori Adamson-Peck; **132** (far left) Lori Adamson-Peck; **133** Lori Adamson-Peck; **134** Lori Adamson-Peck; **141** Courtesy of Intel; **142** Courtesy of Intel; **143** Courtesy of IBM Archives; **144** (top right) J. Kyle Keener; **144** (left) Courtesy of Intel; **144** (bottom left) Courtesy of Intel; **150** (top) Courtesy of Intel; **151** (bottom) Courtesy of Intel; **152** Courtesy of Intel; **155** (top) Will & Deni McIntyre/Photo Researchers Inc.; **155** (bottom) Tom Pantages; **157** (top & bottom) Brian Williams; **159** Brian Williams; **160** John S. Reid; **162** (top) Brian Williams; **163** (bottom) John S. Reid; **163** Courtesy of IBM Corp.; **165** (top) Courtesy of IBM Corp.; **166** Brian Williams; **169** (bottom) AP/Wide World Photos; **169** Courtesy of IBM Corp.; **170** (top) Courtesy of Intel; **170** (bottom) Courtesy of Intel; **171** (top) Courtesy of Intel; **171** (bottom) Brian Williams; **172** (top left) Courtesy of Intel; (far left) Courtesy of Intel; **172** (bottom) Brian Williams; **172** (top) John S. Reid; **173** (middle) Brian Williams; **173** (bottom) Courtesy of IBM Corp; **173** (far right) Courtesy of Intel; **173** (far left) Courtesy of Intel; **174** (top) Courtesy of Intel; **178** (bottom) John S. Reid; **181** Charles Gupton/Stock Boston; **182** (top) Charles Gupton/Stock Boston; **182** (bottom left) David Young-Wolff/PhotoEdit; **182** (bottom right) Wernher Krutein/Photovault; **183** (top & bottom) Brian Williams; **184** (top) Wernher Krutein/Photovault; **186** (top) © Susan Friedman; **186** (middle) Courtesy of IBM Corp.; **186** (bottom) Brian Williams; **186** (bottom) Courtesy of AT&T Global Info/Solution; **187** (top left) Courtesy of Compaq; **187** (top right) Courtesy of Aqcess Technologies, Inc. Irvine, CA; **187** (bottom left) Courtesy of FTG Data Systems; **188** Melissa Farlow; **189** Charles Gupton/Stock Boston; **190** Courtesy of IBM Corp.; **190** (middle) Stephen Weistead/LWA/Stock Market; **190** (top) Photodisc; **190** Copyright 2001, USA Today. Reprinted with permission; **191** Courtesy of IBM Corp.; **191** Courtesy of IBM Corp.; **192** Tom Burdete/USGS; **195** (left) AP/Wide World Photos; **195** (right) Courtesy of ViewSonic Corp.; **198** Don Mason/The Stock Market; **200** Charles Gupton/Stock Boston; **202** (top) Charles Gupton/Stock Boston; **202** (middle) Charles Gupton/Stock Boston; **202** (bottom right) Courtesy of CalComp Ultraslate; **202** (top) Courtesy of Intel and Konica; **203** (middle) Courtesy of Calcomp; **203** (bottom) Wernher Krutein/Photovault Ultraslate; **204** (top) Courtesy of ViewSonic Corp.; **204** (top left) Courtesy of Kinesis; **205** (top) Courtesy of Aqcess Technologies, Irivine CA; **206** (bottom) Courtesy of IBM Corp.; **207** Brian Williams; **207** AP/Wide World Photos; **208** (bottom) Courtesy of IBM Corp.; **211** Brian Williams; **212** (top) Brian Williams; **217** Brian Williams; **218** Hulton Getty/Stone; **219** AP/Wide World Photos; **222** Artwork from Mike Snider, "Wireless World Has Fewer Wide-Open Spaces," *USA Today*, July 31, 2001, p. 3D. Copyright 2001, USA Today. Reprinted with permission; **223** (top) Courtesy of AT&T; **223** (middle top) Courtesy of AT&T; **223** (middle bottom) Courtesy of AT&T; **224** (bottom) Brian Williams; **224** (top) Brian Williams; **224** (bottom) AP/Wide World Photos; **225** (bottom left) AFP/Corbis; **228** © Moshe Shai/Corbis; **234** (top) Rudi Meisal Visum; **237** Ian Shaw/Stone; **238** Brian Williams; **239** (left) Courtesy of AT&T; **239** (right) Brian Williams; **242** (left) Brian Williams; **242** (bottom left) Mark Richards/PhotoEdit; **256** Brian Williams; **258** Elastic Reality Inc.; **259** Paul Higdon/The New York Times Pictures; **260** (left) Courtesy of Princeton Video Inc.; **260** (right) AP/Wide World Photos; **262** Susan Friedman; **267** Elastic Reality Inc.; **268** Susan Friedman; **271** Yoshida-Fujifotos/The Image Works; **272** Yoshida-Fujifotos/The Image Works; **274** (top) PhotoEdit; **276** (top) AP/Wide World Photos; **280** (left) Fujifotos/The Image Works; **281** Hironri Miyata/Fujifotos/The Image Works; **282** Yoshida-Fujifotos/The Image Works; **286** Yoshida-Fujifotos/The Image Works; **288** (middle center) Mark Richards/PhotoEdit; **288** (bottom right) Spencer Grant/Stock Boston; **290** Fujifotos/The Image Works; **291** (bottom) Denise Rocco/U.C.Berkeley; **293** AP/Wide World Photos; **294** MarissaRoth/The New York Times Pictures; **295** (top left) Yoshida-Fujifotos/The Image Works; **295** (middle left) Hironri Miyata/Fujifotos/The Image Works; **296** Mark Richards/PhotoEdit; **301** Photodisc; **302** Photodisc; **302** Photodisc; **304** Photodisc; **304** Photodisc; **305** Photodisc.